W9-CEN-249

modern earth science

Ramsey

Phillips

Watenpaugh

ABOUT THE COVER

The cover of your MODERN EARTH SCIENCE text is part of a color composite photo from Landsat-1 satellite. This photo of rock units in Arizona was taken from an altitude of 914 kilometers. The California Institute of Technology's Jet Propulsion Laboratory has applied special techniques to enhance and distinguish different rock types. The area covered contains a section of the Coconino Plateau (green and white area, bottom left corner of cover) and part of the Grand Canyon (upper white stretch). The green and white areas in the Coconino Plateau indicate limestone and the yellow areas indicate reddish sandstone and conglomerates exposed on the plateau or in the canyon. The red tone characterizes vegetation.

modern
earth
science

William L. Ramsey

Clifford R. Phillips

Frank M. Watenpaugh

Holt, Rinehart and Winston, Publishers
New York   London   Toronto   Sydney

THE AUTHORS

WILLIAM L. RAMSEY
Head of the Science Department,
Helix High School, La Mesa, California

CLIFFORD R. PHILLIPS
Science Department Chairman,
Mount Miguel High School,
Spring Valley, California

FRANK M. WATENPAUGH
Science Department, Helix High School,
La Mesa, California

Photo Credits:

Book cover and frontispiece: NASA; *Unit openers:* Unit 1: NASA: Unit 2: Elyse Lewin/Imagebank; Unit 3: Manuel Rodriquez; Unit 4: Hank Morgan/Rainbow; Unit 5: Ward's Natural Science Est.; Unit 6: Dan McCoy/Rainbow; Unit 7: David Hiser/ Imagebank. *Chapter openers:* Chap. 1: California Institute of Technology; Chap. 2: J. Ward/Imagebank; Chap. 4: Harvey Lloyd/Imagebank; Chap. 5: NASA; Chap. 6 and 8: HRW Photos by Russell Dian; Chap. 7: Manfred Kage/Peter Arnold; Chap. 9: Vance Henry/Taurus Photos; Chap. 10: George Hall/Woodfin Camp and Assoc.; Chap. 11: HRW Photo by Richard Weiss; Chap. 12 and 18: Donald Young/Sierra Club; Chap. 13: Dan McCoy/Rainbow; Chap. 14: Jay Lurie; Chap. 15: Granger Collection; Chap. 16 and 23: Southern Stock Photos; Chap. 17: Richard Steedman/Imagebank; Chap. 19: Icelandic Tourist Office; Chap. 20: Hank Morgan/Rainbow; Chap. 21: Jacques Jangoux/Peter Arnold; Chap. 22: R.K. Pilsbury/Bruce Coleman; Chap. 24: The Bettmann Archive. *Special Features:* p. 112 (all) Georg Gerster/Photo Researchers; p. 113 (all) NASA; p. 228 (t) National Park Service, (m) Soil Conservation Service; p. 229 (t) H. Reinhard/Bruce Coleman, (m,r) Hank Morgan/Rainbow, (m,l) John Running, (b,l) HRW Photo by Russell Dian; p. 312 (t) Robert Asher, (b,l) Hosking/Uniphoto; p. 313 (t) John Running, (b,l) Soil Conservation Service, (b,r) Robert Asher; p. 372 (t,l) U.S. Geological Survey, (b,l) NASA; p. 373 (t) Shostal Assoc., (b,l) Dan McCoy/Rainbow, (b,r) Bob Evans/Peter Arnold; p. 416 (b) A.H. Kroel and E.S. Barghoorn, *Science* p. 396, Oct. 1977; p. 417 (t) Tass/Sovfoto, (m) Walter H. Hodge/Peter Arnold, (b) Yoram Kahana/Peter Arnold; p. 500 (t) Robert Frerck/Dimension, (b,l) Moos-Hake, Greenberg/Peter Arnold; p. 501 (t) Robert Asher, (b) L.L.T. Rhodes/Taurus Photos; p. 520 (t) Yoram Kahana/Peter Arnold; p. 521 (t) HRW Photo by Russell Dian, (b,l) Soil Conservation Service, (b,r) Paul Fusco/Magnum.

Other credits appear with the photographs.

The earth is a fragile planet. It might seem as if very little could happen to disturb this giant sphere as it continuously orbits the sun. And the solid body of the earth will probably continue to orbit the sun as it has for over four billion years! But the special conditions that exist on the earth and make up what we call the earth's "environment" are a different matter. The atmosphere, the sea, streams and lakes, soil, and even the rock of the crust are parts of the total environment that hang in a delicate state of balance. The earth's environment is a result of certain processes that have a history as old as the planet itself. Such processes as those involving movements of air and water are related to one another in a variety of complex ways. The ways in which they work to create their share of our familiar environment is often a result of delicate balances between them. During the past few years, there have been clear signs that the earth's human population has seriously disturbed some important balances in the earth's environment. The study of earth science can help to provide you with an understanding of the nature of the earth's environment and how people have affected it.

The position of the earth in the universe is first established in Unit 1. Unit 2 explores the general characteristics of the earth as a planet. Unit 3 deals with the ways by which the earth's surface is sculptured into its many landforms. Unit 4 describes the characteristics of the oceans as one of the most vital of all the earth's surface features. Unit 5 is devoted to the earth's history, describing the development of North America in a more detailed view of the entire story. Unit 6 is concerned with the atmosphere, emphasizing the nature of weather and climate. The final unit includes just one chapter. This chapter emphasizes the fact that many of the earth's resources are not renewable. Following each of the six units is a special section that introduces many interesting discoveries, phenomena, and careers in earth science.

At the beginning of each chapter, objectives have been added to guide the student in the basic concepts of that chapter.

To assist students in the mastery of technical terms associated with the text discussion, these terms are printed in *italics* where they occur and are clearly defined. In order to help evaluate behavioral skills, each chapter includes a number of marginal notes to the student which are intended to extend the learning process beyond the realm of the textbook. At the end of each chapter the vocabulary is reviewed in a section that students may use as a self-quiz to test their understanding. A complete *Glossary* also appears at the back of the book. In addition to the *Vocabulary Review* at the end of each chapter, there are two groups of *Questions*. Those in *Group A* are based directly on the text and may be used as a self-quiz by the student. The questions in *Group B* are more difficult and often require interpretation of the text material.

The *Appendix* includes a map of the United States denoting national parks and geological phenomena of special interest to earth science students. Also included in the *Appendix* is a complete key to the identification of minerals.

In addition to the text, there is a workbook, *Activities and Investigations for MODERN EARTH SCIENCE*. The activities give further insight into the basic concepts of the chapter. The laboratory experiments give direct experience with the spirit of scientific inquiry.

In regard to the sections in *MODERN EARTH SCIENCE* dealing with geologic evolution and the age of the earth, the authors have used scientific data to present this material as theory rather than fact. The information as presented allows for the widest possible interpretation. Every attempt has been made to present this material in a non-dogmatic approach.

ACKNOWLEDGEMENTS
Special thanks are due to the consultants who have made suggestions and criticisms on the MODERN EARTH SCIENCE text.

Dr. Robert W. Ridky, Science Teaching Center, University of Maryland, College Park, Maryland. (Unit 1)

Sister de Montfort Babb, I.H.M., Maria Regina High School, Uniondale, New York. (Units 2, 3, 5)

Dr. Robert E. Boyer, Geology Department, University of Texas, Austin, Texas. (Unit 4)

Dr. William R. Pampe, Department of Geology, LaMar University, Beaumont, Texas. (Unit 6 and 7)

contents

Unit 1
THE EARTH IN THE UNIVERSE

1 ☐ design of the universe   **2**
Observing the Universe. A Closer Look at the Universe. Universal Laws and the Solar System.

2 ☐ the sun   **21**
Structure of the Sun. The Energy of the Sun. Other Stars in Our Galaxy. Life History of a Star. Origins of the Universe.

3 ☐ the earth as a member of the solar system   **47**
Origin of the Solar System. Member of the Solar System. The Earth and Its Motions.

4 ☐ the moon   **75**
Past and Present on the Moon. Motions of the Moon. Tides. The Calendar.

5 ☐ probing the secrets of space   **93**
Getting into Space. Interplanetary Travel. In Orbit around the Earth.

Pioneers in a New Frontier   **112–113**

Unit 2
THE PLANET EARTH

6 ☐ models of the planet earth   **116**
A Closer Look at our Planet. Mapping the Earth's Surface. Topographic Maps.

7 ☐ earth chemistry   **138**
Atoms. Molecules.

8 ☐ materials of the earth's crust   **153**
Minerals. Rocks.

9 □ the restless earth   **182**
　　　Plate Tectonics: A Theory. Volcanism.

10 □ movements of the earth's crust   **207**
　　　Changing the Shape of the Crust. Earthquakes.
　　　Mountain Building.

Our Dynamic Planet   **228–229**

Unit 3
FORCES THAT SCULPTURE THE EARTH

11 □ weathering and erosion   **232**
　　　Weathering. Erosion. Erosion Shapes the Land.

12 □ water and rock   **253**
　　　The Water Cycle. Run-Off of Water over the Land
　　　Surface. Water beneath the Earth's Surface.

13 □ ice and rock   **275**
　　　Origin and Movement of Glaciers. The Work of
　　　Glaciers. Lakes. The Ice Ages.

14 □ shorelines   **296**
　　　The Attack on the Shore. How a Shoreline Develops.

Time and Change   **312–313**

Unit 4
THE EARTH'S ENVELOPE OF WATER

15 □ the oceans   **316**
　　　Characteristics of the Sea. The Shape of the Crust
　　　beneath the Sea. Sea Floor Sediments.

16 □ sea water   **335**
　　　Physical Properties of Sea Water. Chemical Properties
　　　of Sea Water. The Sea as a Source of Wealth.

17 ☐ motions of the sea   **352**
General Circulation in the Sea. Waves.

Earth, the Water Planet   **372–373**

Unit 5
THE RECORD OF EARTH HISTORY

18 ☐ the rock record   **376**
The Pattern of the Past. Measuring Geologic Time.
The Fossil Record.

19 ☐ earth history   **396**
The Geologic Calendar. Division of Geologic Time.
Modern North America.

Digging Up the Past   **416–417**

Unit 6
THE EARTH'S ATMOSPHERE

20 ☐ air and its movements   **420**
The Air. The Atmosphere and the Sun. The Winds.

21 ☐ water in the atmosphere   **439**
Atmospheric Moisture. Condensation of Water Vapor.
Precipitation of Moisture.

22 ☐ weather   **457**
Air Masses. Weather Fronts. Weather Instruments.
The Weather Map. Local Weather.

23 ☐ elements of climate   **479**
Controls of Climate. Climatic Regions. North
American Climates.

Weather Woes   **500–501**

Unit 7
PEOPLE AND THEIR PLANET

24 □ earth's resources and energy needs   **504**
The Crust as a Source of Materials. Energy from the
Earth's Crust. The Need for Alternate Energy Sources.

Polluting Our Planet   **520–521**

Appendix: A. How to Use This Book   **525**
**   B. Geology in the National Parks**   **529**
**   C. Identification of Minerals**   **534**
**   D. Temperature and Measurement**
**   Conversion Tables**   **549**

Glossary   **551**
Index   **566**

modern earth science

1
design of the universe

objectives

- ☐ Describe star patterns, called constellations, in the night sky.

- ☐ List and describe three basic types of galaxies.

- ☐ Describe the electromagnetic spectrum and the use of telescopes.

- ☐ Explain the changing patterns of movement of the celestial bodies.

- ☐ Describe the ancient Greek model of the solar system.

- ☐ Explain Kepler's Law of Planetary Motion.

- ☐ Explain Newton's Law of Gravitation.

- ☐ List two ways that a person's weight would differ depending on their location on Earth.

Since the first successful orbiting of the moon, it has become possible for humans to actually see Earth as a planet. From their spacecraft the astronauts were able to view the entire earth from thousands of miles above its surface. Not too many years ago we could only use our imagination to picture our home planet. But now it is fairly common to see pictures of the whole earth as a planet, taken from far out in space. And there are no great surprises in the way it appears. The earth appears to be surrounded by a dim blue light. White clouds which appear in patches or continuous sheets, partly cover the surface. Continents and oceans can be seen through gaps in the clouds. But from deep in space, no mountains, rivers or evidence of humans can be seen. But one important fact becomes very plain to everyone when they see the brilliantly colored earth suspended in the blackness of space. The earth, like her sister planets, is clearly seen to be another isolated planet moving around an ordinary star which we call our sun.

It is easy for the inhabitants of the earth to think of their planet as almost a complete environment in itself. Yet humans have always been curious about the environment of space which lies beyond the earth. One way to begin a study of the earth is to first become familiar with the plan or design of the universe, of which the earth is just one small part.

OBSERVING THE UNIVERSE

The night sky. What could you discover about the universe by using nothing but your eyes and mind? You probably

2

have already made some important observations just by looking at the stars at night. When you gaze into the sky on a clear moonless night, it is possible to observe with the unaided eye close to 3000 individual stars. Many of these visible stars seem to be arranged in permanent patterns or groups among the scattering of the stars. Men have recorded this observation from the earliest written history. Groups of stars were given names suggested by their arrangement after animals, gods, and legendary heroes. Even today the star patterns, called *constellations*, are useful in locating and identifying individual stars. They are still known by their ancient names. These include *Ursa Major*, the great bear; *Cygnus* (sig-nus), the swan; *Pegasus*, the winged horse; and *Perseus* (per-soos), the slayer of monsters. Individual constellations are now easily identified by using imaginary boundary lines to outline the region of the sky they occupy. These boundaries divide the sky just as land boundaries divide the surface of the earth into separate countries. Thus, the constellations can be used as guides in making sky maps for astronomy. See Figure 1–1.

In addition to studying the arrangement of stars we can also see that stars differ in brightness. Many of the stars seen with the unaided eye are named by their degree of brightness and location in a constellation. The letters of the Greek alphabet are used to rank stars with the brightest usually labeled *alpha*, the second brightest *beta*, and so on. Thus, Beta (β) Pegasae is the second brightest star in the Pegasus constellation. With a few exceptions, this system of naming individual stars is quite common. Many of the brightest stars in the sky also have names not connected with the constellation in which they are found. The five brightest stars of the northern sky, in order of decreasing brightness, are called *Sirius*, *Vega* (vee-ga), *Capella*, *Arcturus* (ahrk-*tyoo*-rus) and *Rigel* (rhi-jil).

If we observe carefully the stars and their constellation, we will also see that each night they appear to move in a general counter-clockwise direction. It is the turning of the earth from west to east on its axis that makes the constellations appear to move across the sky. The entire pattern of stars is carried across the sky as if painted on a moving roof over the entire earth. The stars seem as if they were on a transparent globe surrounding the earth. If the axis of the earth were extended out into space, it would also form the axis of this imaginary globe on which the stars appear to be located. See Figure 1–2.

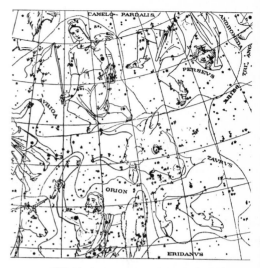

FIG. 1–1. A portion of a star map showing the relative locations of a number of winter constellations in the Northern Hemisphere. (Museum of Natural History)

FIG. 1–2. As seen from the earth, stars appear to move as if they were fixed on an imaginary transparent globe around the earth.

From the northern half of the earth, all stars appear to rotate from east to west around a point in the sky called the *celestial northpole.* "Celestial" refers to any object or position in the sky. A star located directly on the celestial north pole would not appear to move. The North Star, called *Polaris,* is very close to the celestial north pole. Its nightly movement is very slight. All stars close to the celestial north pole appear to move in circular paths around the North Star. Those stars which are farther away from the celestial pole can be seen by an observer on earth for only a part of their total path. Figure 1–3 shows the apparent movement of stars as streaks on the film made during a time exposure.

As the earth moves in its orbit around the sun, the portion of the nighttime sky visible from earth changes in appearance from one part of the year to another. See Figure 1–4.

Stars and Galaxies. The arrangement of stars in constellations such as the Big Dipper is easily recognized. A careful observer, however, can discover still another much larger pattern. The ancient Greeks first observed a misty, cloud-like streak extending across the sky like part of a great circle. Because of its cloudy appearance they called this region in the sky the *Milky Way.*

By using field glasses or binoculars the modern day stargazer can see that the Milky Way is actually made up of millions of distant stars that give it its cloudy appearance. As an observer views the stars in either direction away from the Milky Way, the number of stars he can see in a given area decreases. Thus, the dim light in the region of the Milky Way is not spread evenly throughout the sky.

The observations have led to the conclusion that most of the visible stars lie along a flat circular disc. As a result, when a person looks out through our universe in a direction along the plane of this flat disc, he sees the many stars of the Milky Way. But when he looks out at right angles to the plane of the Milky Way, the sky appears much darker. There the stars are more thinly distributed in space.

Today we know that the system of the Milky Way contains about 100 billion stars. We also know it has the form of a huge disc somewhat bulged out on both top and bottom. This entire system is called *Our Galaxy* (gal-ak-see). It is so large that astronomers usually express its size in light-years. A light-year is the distance that light travels

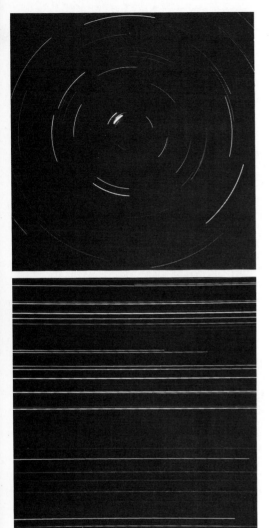

FIG. 1–3. Star streaks made by pointing a camera at the night sky and leaving the shutter open. The picture at the top was made by pointing a camera at the celestial north pole. At what location on the earth's surface must the other picture have been made? (Yerkes Observatory)

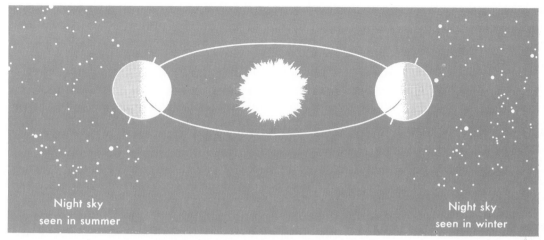

Night sky
seen in summer

Night sky
seen in winter

in one year at a speed of 300,000 km/sec (186,000 mi/sec). If you think of the fact that light can travel completely around the earth in one-tenth of a second, you can get an idea of how far light travels in one year. Astronomers have estimated the size of Our Galaxy to be 100,000 light-years in diameter and 10,000 light-years thick. Within this system the sun and its solar system have been accurately measured to occupy a position about 30,000 light-years from the center. See Figure 1–5.

Within Our Galaxy the stars are moving. Each star follows a path that carries it slowly around the center of the galaxy. The sun, for example, makes one complete

FIG. 1–4. Because the earth revolves around the sun, the appearance of the night sky changes with the seasons.

Observe
On a clear moonless night locate the Milky Way in the sky. Describe its appearance.

FIG. 1–5. The probable appearance of Our Galaxy is shown from a top view and side view below. The arrow and small circle indicate the location of our solar system.

Sun's position

Spiral arms

Disk

FIG. 1–6. The spiral galaxy in the constellation of Andromeda. (Mt. Wilson and Palomar Observatory)

FIG. 1–7. A spiral galaxy in the constellation of Ursa Major, the Big Dipper. (Mt. Wilson and Palomar Observatory)

FIG. 1–8. An elliptical galaxy. Some of the individual stars can be distinguished around its edge. (Yerkes Observatory)

orbit around the galaxy every 240 million years. Stars closer to the center of the galaxy move faster. As we look at these stars, they seem to be either catching up with the sun or moving ahead. Stars farther from the center of the galaxy appear to be moving back toward the sun or falling behind. All the other stars that we see in Our Galaxy seem to be moving in a random way even though their movement is parallel to the sun. Some stars also move along in clusters and whirl around each other within the cluster.

Hazy patches of light often appear in telescopic photographs of the night sky. Many of these have been observed to be other galaxies. These other star systems are separated from Our Galaxy by thousands or millions of light-years. Each one seems to contain billions of stars, just as our own galaxy. Millions of these other galaxies have been observed. Only sixteen are within three million light-years of Our Galaxy.

One of the closest, the Andromeda (an-*drom*-uh-da) galaxy, is about two million light-years away. On a good telescopic photograph, we can see that the Andromeda galaxy has a spiral shape with curved arms reaching outward from a dense central nucleus of stars. See Figure 1–6. Astronomers believe that Our Galaxy would appear similar to Andromeda if it could be observed from another part of the universe.

All of the galaxies we can observe can be divided according to their shape into three basic types. Among the nearest galaxies, the most common are those which consist of a central round body surrounded by a flat disc. See Figure 1–7. These are called *spiral galaxies* because this disc is usually composed of arms which spiral around the center. Our Galaxy and the Andromeda galaxy are of this type. Some spiral galaxies have one bar passing through the center body with the spiral arms reaching out from the bar. These are called *barred, spiral galaxies*. Study of these galaxies has shown that they rotate on an axis that passes through the central body.

A second type of galaxy is nearly round in shape. These are called *elliptical galaxies* because they usually appear pumpkin-shaped. See Figure 1-8. The angle at which they are seen by the observer has a great effect on their apparent shape. Some appear perfectly round. They are brightest near the center and fade out toward the edges. Most galaxies seem to be either the spiral or elliptical type.

A third class of galaxies has no regular shape. These are called *irregular galaxies* and appear as hazy clouds with an uneven outline. See Figure 1–9. This group is generally rare, and makes up about 3 percent of all observed galaxies.

All of the billions upon billions of stars in the universe seem to be organized into galaxies. The galaxies themselves appear to form in scattered clusters. Our Galaxy belongs to the Local Group, a cluster which contains seventeen other galaxies. Beyond the Local Group are several thousand more clusters, a number of which contain more than a thousand galaxies. Some of these clusters are at least three billion light-years away. Very bright, single galaxies which may be as far as ten billion light-years away have been observed. There appear to be galaxies which exist at distances in space that only the most powerful telescopes are able to reach. Thus the earth appears to be surrounded by a universe without any known boundary. As far as astronomers can tell, all of the known universe is populated by billions of giant star groups.

FIG. 1–9. An irregular galaxy. (Mt. Wilson and Palomar Observatory)

A CLOSER LOOK AT THE UNIVERSE

Windows in the sky. Light is the key to our detailed knowledge of the universe. Light received from objects far out in space, has helped scientists discover what lies within the part of the universe that surrounds the earth. Almost all the celestial bodies either radiate or reflect light. But the term "light" is used here to mean more than the light which can be seen with the eye. It includes all energy which travels through space in the form of waves. This energy is called *electromagnetic radiation.* It includes both visible and invisible light waves. Electromagnetic radiation, such as X-rays, ultraviolet, infrared, and radio waves, is invisible to the eye. What we see is radiation in a certain limited range. The various kinds of electromagnetic radiations are usually described according to their *wavelength.* A wavelength is the distance from one wave crest to the next as is shown in Figure 1–9A. For example, visible light is made up of radiations with wavelengths from about 0.0000004 meter or 4×10^{-7}m. (about 1/70,000 inch) to about 0.0000007 meter or 7×10^{-7}m. The complete range of wavelengths extending from 3×10^{7} meters to 3×10^{-17} meters, makes up the *electromagnetic spectrum.* See Figure 1–10. All wavelengths travel at the speed of light.

Investigate
Find out what "airglow", a twilight phenomena, has to do with Atmospheric pollution.

FIG. 1–9A. The closer the crests of a wave the shorter the wavelength.

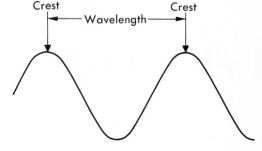

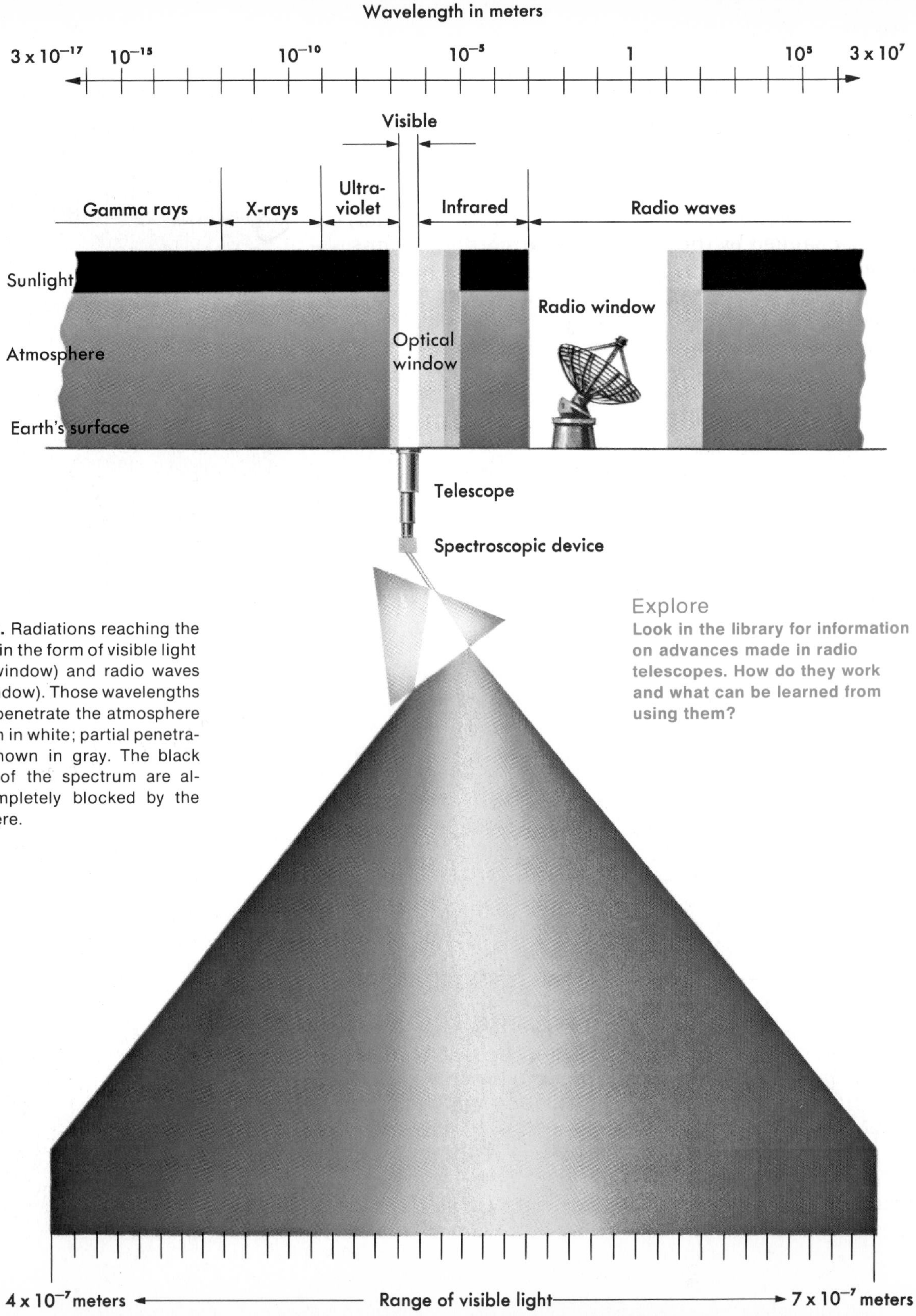

Wavelength in meters

3×10^{-17} 10^{-15} 10^{-10} 10^{-5} 1 10^{5} 3×10^{7}

Visible

Gamma rays | X-rays | Ultra-violet | Infrared | Radio waves

Sunlight

Atmosphere

Earth's surface

Optical window

Radio window

Telescope

Spectroscopic device

Explore
Look in the library for information on advances made in radio telescopes. How do they work and what can be learned from using them?

FIG. 1–10. Radiations reaching the earth are in the form of visible light (optical window) and radio waves (radio window). Those wavelengths that can penetrate the atmosphere are shown in white; partial penetration is shown in gray. The black portions of the spectrum are almost completely blocked by the atmosphere.

4×10^{-7} meters ← Range of visible light → 7×10^{-7} meters

Most stars radiate a wide variety of wavelengths. The sun, for example, produces almost a complete spectrum of radiations from X-rays to radio waves. However, the earth's atmosphere blocks out most solar radiations with the exception of visible light and some radio waves. Other types of radiation are either blocked out completely or greatly changed by the gases in the atmosphere. Thus observers from earth are allowed only two "windows" for their view of the universe. One of them is the optical telescope which makes use of visible light. The other, is the radio telescope which uses some of the shorter radio waves, as shown in Figure 1–10.

Optical telescopes use either lenses or mirrors to concentrate the incoming visible light. The most familiar type of telescope is the *refractor*. Figure 1–11. In the refractor telescope, light passes through the lenses and is bent or refracted to project an image of the object. This image can then be viewed directly or photographed.

The other kind of telescope is the *reflector*. Figure 1–12. This telescope has a curved mirror that reflects light to form an image of the object viewed. The largest telescopes are reflectors because mirrors can be supported from behind easily. Large lenses tend to sag out of shape from their own weight. Both types of telescopes can be trained on a particular point in the sky and can track their targets in spite of the movement caused by the earth's rotation. This is an important fact to remember since astronomical telescopes are used mostly as cameras. A photograph made over a period of time as the telescope slowly follows its target will record many objects too dim to be seen otherwise.

Modern radio telescopes are very sensitive radio wave receivers with large antennas. Some of these radio telescope antennas are dish-shaped and can be aimed at a particular spot in the sky. Other antennas are firmly attached to the ground and are aimed at the sky by the earth's rotation. See Figure 1–13. Compared to optical telescopes, a radio telescope gives a less clear picture of the location of objects being viewed. One reason for this is that radio waves are so long, compared to visible light, that nearby radio sources cannot be separated unless the antenna is very large.

A radio telescope has clear advantages in not being affected by clouds or city lights. It is also able to detect objects at distances far beyond the range of the best optical telescopes. However, radio telescopes are not without their

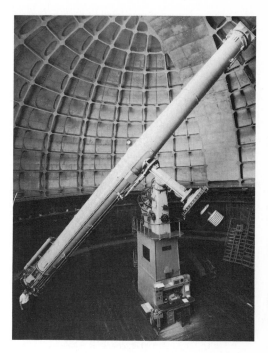

FIG. 1–11. A large refracting telescope. (Lick Observatory)

FIG. 1–12. The 200-inch Hale reflecting telescope of the Palomar Observatory. (Mt. Wilson and Palomar Observatory)

FIG. 1–13. Above, a moveable radio telescope antenna at Green Bank, West Virginia. Below, a fixed antenna at Arecibo, Puerto Rico. Signals are received by the dish-shaped hole in the ground and reflected to the part suspended above. This type of antenna is aimed by the earth's rotation. (top: NSF Photo) (bot: Official US Army Photo)

FIG. 1–14. Signals collected by the antenna of a radio telescope are usually amplified and fed into a computer to eliminate interference. A recorder produces a graph of the wanted signals.

troubles. Man-made radio interference is one of the hardest problems. To block out unwanted interference, the signals received are usually fed first into powerful amplifiers. Then they may be passed into a computer to sort out the wanted signals from the unwanted interference. See Figure 1–14.

UNIVERSAL LAWS AND THE SOLAR SYSTEM

Cycles of change. The only direct way you can learn about the universe surrounding the earth is to look into the sky. You know that part of the universe exists because you can see the sun, moon, and stars. But the most important observation you can make is that there is an ever changing pattern of movements by these celestial bodies.

The sun follows a daily path across the sky, but not always rising and setting at the same place. It rises and sets at a more northerly point each day until summer arrives. Then it moves in a more southerly path each day until winter begins. Thus the sun marks the yearly passage of the seasons. See Figure 1–15.

Each day the moon also rises and sets. Although it always follows the same path across the sky, it rises later each day. The moon also presents an ever changing appearance as the half facing the earth becomes fully lighted.

Even the seemingly changeless pattern of stars in the night sky slowly turns. The entire roof of stars circles around a certain position in the north sky. The pattern of

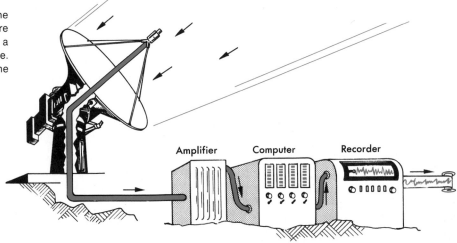

Amplifier   Computer   Recorder

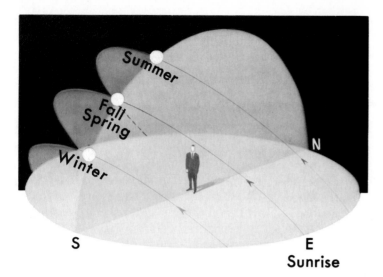

FIG. 1–15. The path of the sun in summer as it reaches its most northern position, carries it higher in the sky than in winter. During which season will the noon-day sun cast its longest shadow?

stars in the night sky also changes with the seasons as new stars come into view and others disappear.

Of all our observations of the sky, one of the strangest is the movement of a few bright objects. They appear to be stars, but follow their own paths against the fixed background of the actual stars. These objects were named planets, which means "wanderer." Some of the planets follow odd, looping paths which appear to carry them forward, then back among the stars. Other planets never move high in the sky but rise and set close to the horizon.

The ever changing pattern of movements by the sun, moon, stars, and planets repeats in a series of endless cycles when observed for a long time. These movements are the visible evidence of the universe beyond the earth.

Sky cycles and scientific models. The cycles of change we observe in the sky have always interested curious men. The superstitions which made gods out of the celestial bodies have never satisfied the best thinkers. These men have tried to picture a universe that would explain the changing appearance of the sky. It may be true that the beginnings of modern science go back to the time men first tried to think logically about the universe.

When faced with a problem which demands a logical explanation, men often invent models to help their thinking. The word "model," used in this way, means an idea that gives us a clearer picture of some part of nature. For example, a model which someone thought of many hundreds of years ago pictured the earth at the center

Measure

How far north or south does the sunset-point move on the horizon in one week? You might want to set up a Polaroid camera to shoot the sun each day from the same position, to obtain a permanent record.

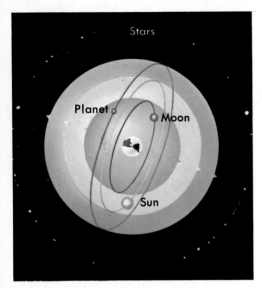

FIG. 1–16. An ancient Greek model of the universe. Separate spheres surrounded the earth and carried the sun, moon, planets and stars.

Mikolai Kopernik

FIG. 1–17. Nicolaus Copernicus (1473–1543), was the first to offer evidence that the earth moves around the sun. (The Granger Collection)

of the universe. This model explained the motions of the sun, moon, stars, and planets by having them circle around the earth.

Almost everyone agreed with the ancient Greek idea of an earth-centered universe as the astronomer Ptolemy (127–151 A.D.) described it. He had worked out a complicated system in which the earth was surrounded by a number of turning transparent globes, one inside the other. On their separate spheres, the sun, moon, planets, and stars were carried around the earth. Figure 1–16. This was a clever plan to explain the easily seen motions of the stars and their companions (the planets). But it did have some trouble explaining the back and forth movement of the planets. Ptolemy explained this motion by a system in which the planets moved in small circles. Their centers, in turn, revolved around the earth. Ptolemy's whole plan meant that the earth had to remain at rest while the other transparent spheres carrying the rest of the universe slowly turned at different speeds.

The ancient view of the universe described by Ptolemy was very complicated. But it did correctly describe the observed cycles of the sky. Thus for more than fourteen centuries, the idea that the earth lay at the center of things was not seriously questioned. However, in the sixteenth century, a Polish churchman and mathematician named Nicolaus Copernicus (koh-*per*-ni-kus) challenged the idea. He stated his belief that the sun, not the earth, was the center around which all the planets revolved. This was such a revolutionary idea that Copernicus hesitated to publish his work until the year of his death in 1543. It took many years of bitter argument before people could accept the thought that the earth was not the center of the universe but only one of a family of planets controlled by the sun.

Basic plan of the solar system. The sun and a large number of objects orbiting around it make up the *solar system.* Members of the solar system may be as small as a grain of sand or more than ten times the size of the earth. All members have the common characteristic of following a regular path around the sun. The exact way in which the members of the solar system move around the sun is described by three basic rules. These rules are usually called Kepler's Laws of Planetary Motion. Johannes Kepler, the sixteenth-century astronomer and mathematician, first discovered them from observations available to him. *Sci-*

FIG. 1–18. An ellipse is an oval shape that can be constructed from a plane section of a cone.

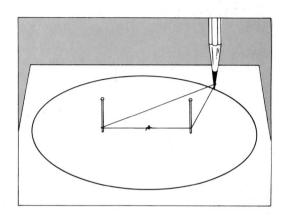

FIG. 1–18A. This figure represents the general shape of the path followed by all members of the solar system.

entific laws are rules that correctly describe some part of the natural world. When used to describe the motion of the planets, Kepler's laws can be stated as follows:

1. The path each planet follows around the sun is an ellipse. An ellipse looks like a circle but has the same shape as a cone if it were sliced through. See Figure 1–18. If an ellipse is drawn as in Figure 1–18A the positions of each of the two pins are called the focus points. The sun is always at one of the focus points of a planet's elliptical orbit. As a result, with the sun at one focus, the planet's distance from the sun constantly changes.

When the earth is closest to the sun, the distance between the earth and sun is 147,000,000 km (91,300,000 mi). The earth at its nearest approach to the sun is said to be at *perihelion* (pehr-i-hee-lee-on). At *aphelion* (a-fee-lee-on) the earth is at its farthest point from the sun. The distance is 152,000,000 km (94,500,000 mi). The earth's average distance from the sun is 150,000,000 km (92,900,000 mi). Actually the earth's orbit is very nearly a perfect circle, as shown in Figure 1–19.

2. The second law of planetary motion concerns the speed with which a planet moves in its orbit. Each planet moves in its orbit in such a way that an imaginary line joining it to the sun sweeps over equal areas in equal periods of time. See Figure 1–20. A planet moves fastest

Explain
Look up information on Bode's Law and explain how it was used to predict the presence of a planet between Mars and Jupiter.

FIG. 1–19. Though the earth's orbit is an ellipse, it is nearly a perfect circle as shown in this drawing which is made to scale.

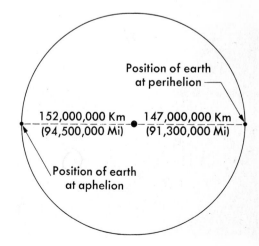

Position of earth at perihelion

152,000,000 Km (94,500,000 Mi)   147,000,000 Km (91,300,000 Mi)

Position of earth at aphelion

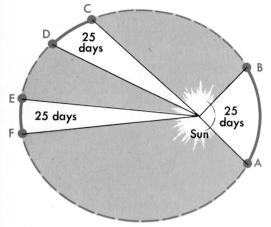

FIG. 1-20. A line joining a planet and the sun will sweep over equal areas in equal periods of time. The light areas in the diagram are equal because the time taken to cover the parts of the orbit A to B, C to D and E to F are the same. The elliptical shape of the orbit is exaggerated in this drawing.

FIG. 1-21. Sir Isaac Newton (1642–1727), laid the foundation for much of the modern scientific view of the universe. (The Granger Collection)

when it is nearest the sun and slowest when farthest away.

3. Kepler's third law relates the distance between a planet and the sun to the time taken to complete one revolution of the planet's orbit. The relationship can be stated as: The square of a planet's period of revolution is proportional to the cube of its average distance from the sun. This means that a planet having an average distance from the sun four times that of the earth would have a period of revolution of eight years. That is, the period would be equal to the square root of the cube of its average distance from the sun. The average distance cubed of the planet in question would be 64 or ($4^3 = 4 \times 4 \times 4$). Its period of revolution in years would be the square root of 64. This gives eight years as the period ($\sqrt{64} = 8$).

The discovery of Kepler's laws more than three hundred years ago represented a great advance in our knowledge of the planetary system. However, these rules of planetary motion were worked out entirely from observations of the changing position of the planets. Kepler was not able to explain why these regular and predictable motions of planets took place. Thus, according to Kepler's rules, the model which pictured the earth and other planets moving around the sun was not complete. There was a need to search for something in nature that could explain the very definite rules that described the motion of the planets.

Gravity. In many ways the solar system is like a giant machine. It has a large number of parts which move smoothly through space year after year. But unlike the ordinary machines we know, the solar system does not seem to need a source of energy. What then provides the energy needed to move the earth and other planets around the sun in endless cycles? This question went unanswered until the brilliant English scientist Sir Isaac Newton provided a clear understanding of the rules that describe almost all motion, including that of the planets.

Newton demonstrated that any moving body will change its motion only if some outside force acts upon it. Normally we expect moving objects to stop eventually due to friction. When a moving object comes in contact with another surface, friction acts to take away its energy. However, a planet meets practically no friction in space. There a body can move at almost a constant speed without ever needing a push or pull to keep it going. Newton showed that only motion in a straight line is possible without some outside force present. Since the members of the solar system move in curving paths around the sun, some outside force must exist to cause this behavior. Newton identified and described this force as *gravity*. His great achievement was the discovery of a way to describe gravity and thereby provide the means for predicting its effect.

Newton's Law of Gravitation is based on his interpretation that a force exists which pulls every particle of matter toward every other particle. But the size of this attracting force is the result of two factors. The gravitation force becomes greater if the amount of matter (mass) in the two bodies gets bigger. The other factor which determines the size of the gravity force is the distance between the two objects being attracted toward each other. Gravity force becomes larger if the objects move closer together. In other words, the bigger the objects and the closer they come to each other, the greater will be the gravity force pulling them together. See Figure 1–22.

A mathematical statement describing Newton's Law of Gravitation follows:

Gravitational force (F) is proportional to

$$\frac{Mass_1(M)_1 \times Mass_2(M)_2}{Distance^2(d^2)} \text{ or } F = G\frac{M_1 \times M_2}{d^2}$$

where (G) is a universal constant. A constant is a quantity having a fixed value.

The Law of Gravitation explains why the planets revolve around the sun in an unbroken pattern. If the sun were not present, the planets would move in a straight line. But the sun is there, and it exerts forces on the planets. In turn, the planets exert forces on the sun. But the sun has so much more mass than the planets that it is affected only slightly by them. The planets however, are moved quite a bit by the sun's force. This force deflects the movement of a planet from its straight-line path. and sends the planet into its orbit. This resulting orbiting is due to the sum of the forces acting on the planet. The movement of the planet toward the sun is due to its gravitational pull.

The earth's gravity. Just as the sun controls the movement of planets, the earth has a strong gravitational influence upon objects close to it. The earth's gravitational force is generally thought of as the force responsible for an object's weight. When we weigh an object it is the force of the earth's gravitational pull that is being measured.

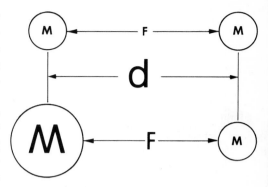

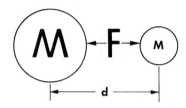

FIG. 1–22. The gravitational force which will attract two bodies toward each other (F) is directly proportional to the product of their masses ($M_1 \times M_2$) and inversely proportional to the square of the distance (d^2) between their centers. In the diagram, the size of the letter represents a relative quantity.

Discover
Look up some references in the Physics area and find out how Lord Cavendish used a laboratory balance to determine the mass of the earth.

activity

Use the following method to measure the density of some common substances.

Materials: 50 ml graduated cylinder, lead sinker, iron bolt, brass screw, aluminum nail, small rock, metric balance, water

1. Weigh the empty graduated cylinder. Record as mass of cylinder.
2. Fill the graduated cylinder to the 40 ml mark and weigh. Record as mass of cylinder plus water.
3. Subtract the two masses. Record as the mass of water.
4. Density is the mass divided by volume. (Density = $\frac{mass}{volume}$) Calculate the density of water. The volume was 40 ml.
5. Weigh the lead sinker. Record as mass of sinker.
6. Carefully drop the sinker in the cylinder of water.
7. The water will rise to a new level. The difference in the original level (40 ml) and the new level is equal to the volume of the sinker. Record this volume.
8. Calculate the density of lead.
9. Repeat steps 5–8 for the other objects.

FIG. 1–23. The density of some planets compared to the earth's, which is given as one. Jupiter has about one quarter the density of Earth. However, the large mass of Jupiter makes its surface gravity 2.6 times greater than Earth's gravity.

For objects on the earth's surface, any change in weight is generally the result of the amount of matter or mass the object contains. If your weight changes, you can assume that it is your mass that has changed and not the earth's gravitational attraction. However, there are actually some conditions on earth which could cause your weight to change without any change in your mass.

The earth's rotation on its axis affects the force of gravity felt. To illustrate this, picture yourself standing at a point on the earth's equator. The speed of the earth's rotation on its axis is carrying you around at about 1609 km/hr (1000 mi/hr). Like the planets orbiting the sun, you would move in a straight line and would fly off into space if it were not for gravity pulling you toward the earth's center.

Part of the earth's gravitational force on an object at the equator is used only to keep it moving in a curved path. The remaining gravitational force is a measure of the weight of the object. Now suppose that you are standing at one of the earth's poles. Here your speed due to the earth's rotation is zero. At either pole, the force of gravity is not needed to keep you moving in a curved path. Thus the force of gravity responsible for your weight would then be slightly greater at the poles than at the equator. Actually, your weight would be only about 0.3 percent more at the poles than at the equator. However, this difference does seem to have an effect on the shape of the earth.

If you could measure your weight very accurately, you would find that the attraction due to the earth's gravitational force is likely to be very slightly greater at ground level than at the top of a tall building. This increase in weight results from your shorter distance to the earth's center. But even when the earth's rotation and changes in altitude are taken into account, it is still found that the earth's surface gravity is not the same everywhere. It is believed that these gravity differences are caused by the unevenness of the earth's shape or by the nature of the underlying rocks. Some types of rocks beneath the surface may have a high density. A substance with a high

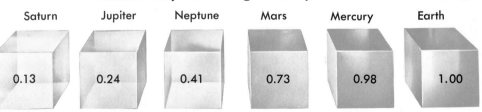

Saturn	Jupiter	Neptune	Mars	Mercury	Earth
0.13	0.24	0.41	0.73	0.98	1.00

density contains a large mass for a certain volume. The metal lead has a high density because a piece of lead has greater mass than an equal volume of almost any other substance. Since the density of rocks beneath the surface is reflected in gravity measurements, the changes in the earth's surface gravity give us some clues about the rocks beneath the surface. Thus it follows that as the density or mass of a certain volume of the underlying rocks decreases, the surface gravity decreases. See Figure 1–23.

Looking at our earth. Our knowledge of the universe beyond the earth gives us a better understanding of the planet that is our home. We can imagine a bright sphere rolling through space controlled by the forces that operate through the whole universe. However, this picture must also remind us that we are actually passengers on this planet spaceship. While we may be able to send space vehicles to explore our near neighbors, the earth is still our home. And we cannot change any of the material laws that govern the planet and allow it to support its many kinds of life. This means that we must be careful not to alter the environment in which we live. Only the knowledge we gather by carefully studying the earth can provide the way to keep it a balanced spaceship, fit for its living passengers.

VOCABULARY REVIEW

Match the word or words in the column on the right with the correct phrase in the column on the left. *Do not write in this book.*

1. A permanent pattern of stars in the sky.
2. A star located very close to the celestial pole.
3. Our Galaxy and a cluster of seventeen other galaxies.
4. A unit used in measuring very large distances.
5. Second brightest star in the Pegasus constellation.
6. Distance from one wave crest to the next wave crest.
7. Energy which travels through space as waves.
8. Rules which have been found to correctly describe some part of the natural world.
9. The shape of the path a planet makes as it goes around the sun.
10. Closest approach of a planet to the sun.
11. A measure of the earth's gravitational pull
12. An idea which gives an accurate picture of some part of nature.

a. weight
b. scientific laws
c. light year
d. model
e. wavelength
f. ellipse
g. aphelion
h. constellation
i. Beta Pegasae
j. electromagnetic radiation
k. perihelion
l. Local Group
m. Polaris
n. gravity

QUESTIONS

Group A

Select the best term to complete the following statements. *Do not write in this book.*

1. Something which cannot be seen when viewing the entire earth from space is (a) a light blue film surrounding the sphere (b) white clouds partly covering the surface (c) mountains and rivers (d) continents and oceans.

2. Which one of the following is not a constellation? (a) Ursa Major (b) Cygnus (c) Pegasus (d) Canopus.

3. From the northern half of the earth, all stars appear to rotate about the celestial north pole in a direction of (a) east to west (b) east to north (c) west to east (d) west to south.

4. Which one of the following is *not* a characteristic of our Galaxy? It (a) is composed of about 100 billion stars (b) contains a large amount of matter between the stars (c) is about 100,000 light years across and about one-fifth as thick (d) has our sun located near the center of the Galaxy.

5. The galaxy of which our sun is a member is thought to be similar in shape to the galaxy named (a) Andromeda (b) Polaris (c) Cygnus (d) Perseus.

6. Our Galaxy belongs to a cluster of galaxies called the Local Group which consists of our galaxy and (a) three others (b) eleven others (c) seventeen others (d) twenty-one others.

7. The type of galaxy to which our own galaxy belongs is called a(n) (a) spiral galaxy (b) barred galaxy (c) elliptical galaxy (d) irregular galaxy.

8. The rarest type of galaxy is the (a) spiral (b) barred (c) elliptical (d) irregular.

9. The most distant galaxies from earth that have been observed are about (a) two million light years away (b) one billion light years away (c) five billion light years away (d) ten billion light years away.

10. The electromagnetic radiation given off by our sun contains (a) visible light only (b) mostly X-rays (c) mostly ultraviolet (d) a mixture of radiation from X-rays to radio waves.

11. The two windows to the universe available to us from earth are (a) X-rays and radio waves (b) ultraviolet and visible light (c) visible light and radio waves (d) infrared and visible light.

12. The largest optical telescopes collect light with (a) mirrors (b) lenses (c) antennas (d) amplifiers.

13. To record many things for later study most optical telescopes use (a) large lenses (b) large mirrors (c) cameras (d) light filters.

14. One advantage of optical telescopes over radio telescopes is (a) they are not affected by clouds (b) they may detect objects at greater distances (c) they can pinpoint the source more closely (d) they are not affected by city lights.

15. The component which does not belong in a radio telescope is (a) an antenna (b) a spectrograph (c) a computer (d) an amplifier.

16. Which term has no relation to the others? (a) Ptolemy (b) Copernicus (c) transparent spheres (d) stationary earth.

17. The idea *not* a part of the ancient Greek model of the universe is that (a) the sun, moon, and stars were located on transparent spheres (b) the planets moved in

small circles on their spheres (c) the earth rotated on its axis (d) the earth was the center of the universe.

18. The Ptolemaic model of the universe was not challenged for more than a thousand years because (a) it explained all the observed celestial cycles accurately (b) there were no great scientists living at the time (c) it was the only explanation available (d) the model was too complicated to understand.

19. Copernicus' theory differed from the ancient Greek model of the universe in that he (a) placed the sun, planets, and stars at varying distances from the stationary earth (b) placed the moon at the center of the universe (c) suggested that the earth moves in a similar manner to the other planets (d) suggested that the sun moved in a circle.

20. The idea which is not a part of Kepler's Laws is that (a) the planets move in elliptical orbits (b) a planet's speed varies as it orbits the sun (c) the planets rotate on their axes at various rates (d) the squares of the periods of revolution are proportional to the cubes of their average distances from the sun.

21. Kepler's Laws accurately described the motions of planets but were not complete in that they (a) did not explain why the planets move (b) were contrary to the accepted model of the solar system (c) required the earth to move in a manner different from the other planets (d) did not describe the motion of all members of the solar system about the sun.

22. According to Sir Isaac Newton, any moving body will (a) slow down and come to rest (b) be acted upon by friction (c) change its motion only if some outside influence acts on it (d) eventually lose energy.

23. According to Newton, a moving body will move in a curved path only if (a) it is a planet (b) a force acts continuously on it (c) it is slowing down (d) it is speeding up.

24. Gravity, a force studied by Newton, explains (a) why planets move through space without noticeable friction (b) why the sun is the center of the solar system (c) why planets move in orbits about the sun (d) why planets rotate on their axes.

25. The force of gravity between any two objects depends upon their masses and (a) their size (b) their speed (c) the material of which they are made (d) the distance between them.

26. The law of gravitation might be expressed in formula as

(a) $F = \dfrac{M_1 M_2}{d^2}$ (b) $M_1 = \dfrac{G M_2}{d_2}$ (c) $G = \dfrac{d^2}{M_1 M_2}$

(d) $F = G\,\dfrac{M_1 M_2}{d^2}$

27. The orbit a planet follows around the sun is due mostly to the balance between gravitational pull of the sun and (a) the pull of the other planets (b) the tendency for a planet to move in a straight line due to its original motion (c) Kepler's third law (d) its tendency to move in a circle.

28. The main reason you would weigh less at the earth's equator than at the north pole is that (a) you are closer to the center of the earth (b) the warmer air at the equator is rising (c) the air at the equator is always moist (d) due to the earth's increased speed of rotation near the equator, some of the force of gravity is used to keep you in a curved path.

29. Clues about the nature of rock layers under the earth's surface are given by changes in earth's gravity due to (a) altitude (b) density differences (c) the moon (d) earth's rotation.

30. In order to keep the earth as an environment useful for life, we must (a) change the material laws to better fit humans (b) go back to the material laws which existed several centuries ago (c) limit competing forms of life (d) learn the rules that provided the present environment.

✓ = do only ones checked

Group B

1. Name four constellations and tell what they are supposed to look like.

2. What is the relationship of the Milky Way to Our Galaxy?

3. How does the motion of the planets differ from that of the stars?

4. Give an example of a scientific model and explain how models are used by scientists.

5. Why was the Ptolemaic model of the universe so widely accepted for more than fourteen centuries, even when the Copernican model provided a truer picture?

6. State briefly Kepler's Laws of Planetary Motion.

7. How is the solar system like a giant machine? How does it differ from human-made machines?

8. How did Sir Isaac Newton add to our understanding of the motion of the planets about the sun?

9. What factors other than mass and distance from the earth's center may change the weight of an object?

10. Why doesn't the surface of the earth receive all of the radiations given off by the sun?

11. Actually, astronomical telescopes are used mostly as cameras. Explain.

12. Give two advantages and two disadvantages that radio telescopes have over optical telescopes.

13. Star patterns in the night sky called constellations were recognized and named by people living in ancient times. What value do they have today?

14. Describe the three basic types of galaxies.

15. Why are Kepler's Laws considered scientific laws while Copernicus' description of the solar system is termed a scientific model?

16. Why do we say that the earth is surrounded by a universe without any known boundary?

17. Using the mathematical statement that describes Newton's Law of Gravitation, determine a numerical answer for what happens to the force of gravity when (a) the mass of M_1 is doubled (b) the mass of both M_1 and M_2 is doubled (c) the distance between M_1 and M_2 is doubled.

18. List at least three serious problems that exist as a result of our lack of knowledge of the natural laws controlling the earth's environment.

2
the sun

objectives

- ☐ Describe the role of each part of the sun.
- ☐ Compare and contrast sunspots, prominences, and solar flares.
- ☐ Describe two ways solar energy can be used in homes.
- ☐ Explain why Einstein's equation led to a better understanding of the source of the sun's energy.
- ☐ Identify the reaction that produces the sun's energy.
- ☐ Describe several ways that light is used to measure the distance to stars and the characteristics of stars.
- ☐ Classify star populations according to temperature and brightness.
- ☐ Describe the life history of a star.
- ☐ List three theories of the origin of the universe.

The sun is an example of a very common type of star. To us, however, the sun is very different from the billions of other stars that exist in the universe. Those stars are so far away that they have no significant influence on us. The sun, however, is 150 million kilometers (93 million miles) from the earth and sends a constant stream of radiant energy to the earth. This radiant energy from the sun is the most important outside influence upon the earth.

The relative closeness of the sun gives us the opportunity to study a star at fairly close range. It is from the study of the sun that men have gained much detailed knowledge of stars in general. This chapter deals with a general picture of the sun as a representative of the family of stars. Based on this information, it is possible to describe the general life cycle of any star.

STRUCTURE OF THE SUN

The sun as seen from the earth. To an observer on earth the sun is a blinding, brilliant source of light. So intense is this light that looking directly at it without some protection for the eyes can cause permanent damage. But the visible light from the sun is only a small part of its total energy output. All of the energy given off by the sun is in the form of electromagnetic radiations. We can usually identify these radiations individually by their wavelengths, and together they make up part of the electromagnetic spectrum. (See page 8.)

Of the total light energy produced on the surface of the sun, the most intense falls in the visible light range of the spectrum. See Figure 2–1. However, the largest portion of the sun's electromagnetic spectrum is made up of wave-lengths above and below those of visible light. Since we cannot see these radiations, they are less obvious. In fact, most of the shortest wave and longest wave radiations from the sun never reach the earth's surface. Also, as you have already learned, the atmosphere absorbs almost all radiation except for visible light and some radio waves. Clouds also reflect a large amount of solar energy back into space. Only about half the total solar energy directed toward the earth actually reaches the surface of the earth.

Parts of the sun. The most important single characteristic of the sun is its very high temperature. We can determine the sun's temperature because there is a relationship between the temperature of a body giving off radiant energy and the wavelengths it produces. As the temperature of a body increases, the wavelengths it radiates grow shorter. When we apply the above rule, it shows us that the temperature of the sun reaches thousands of degrees near its surface. Its deep interior cannot be observed directly, but there is evidence that temperatures near the center are in millions of degrees.

The sun's extremely high temperatures indicate that it is made up entirely of gases. No solids or liquids can exist even at the lowest temperatures found in the sun. However, the intensely hot ball of gases that makes up the body of the sun is not the same temperature in every area. Since the sun's source of energy is in its interior, there is a cooling of the gases toward the surface. This tends to divide the sun into several layers.

The largest part of the sun is taken up by a central *core*. Around the core, there is a deep zone of radiation. See

FIG. 2–1. This graph shows that solar energy produces the highest intensity of radiation in the visible wavelength portion of the electro-magnetic spectrum.

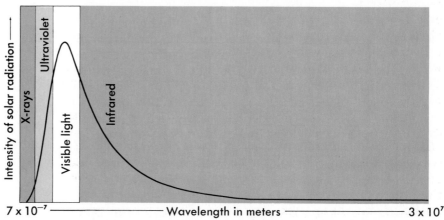

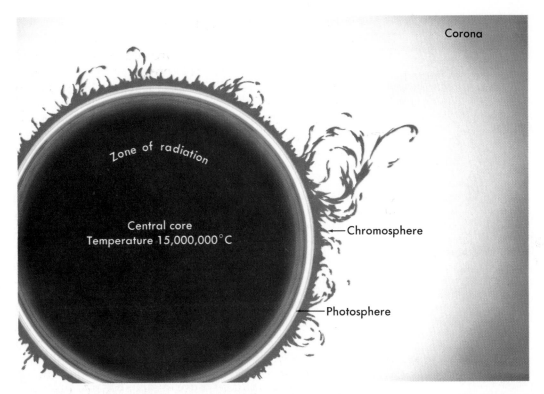

Figure 2–2. Reactions which produce the sun's energy take place within the core. The temperature in this region is estimated to be about 15 million degrees centigrade. Energy produced in the core bombards the atoms in the surrounding zone of radiation. These atoms then produce short wavelength radiations (X-ray and ultraviolet.) These radiations pass outward and heat a cooler gas layer near the surface. In this layer many of the short wavelength radiations are absorbed and the energy is given off again as a visible light. Since this layer is the source of the sun's visible light, it is called the *photosphere* (light sphere). It is the brilliant surface of the photosphere that you see when you look at the sun with your eyes properly protected. Temperature at the surface of the photosphere is about 6000°C (about 10,800°F).

Above the photosphere the solar gases are thinner and give off much less radiant energy. A gas layer about 16,000 km (10,000 mi) thick lies just above the surface of the photosphere. It produces a relatively weak red light and is called the *chromosphere*. Some distance beyond the chromosphere is a very thin cloud of gas which produces a faint white light. This is called the *corona* (crown)

FIG. 2–2. The above is a drawing of the sun showing the temperature in the central core. The interior is dark because the invisible radiations from the inside do not produce visible light until they are near the surface.

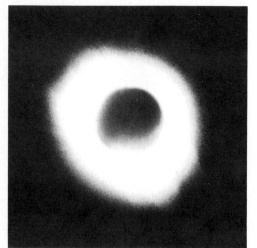

FIG. 2-3. The photograph above shows the sun during a total eclipse when the corona becomes visible (Los Alamos Scientific Laboratory)

FIG. 2-4. Below, left, is a photograph of the sun showing many large sunspots. (Yerkes Observaory) Below, right, is a sunspot photograph taken through a telescope carried by a high altitude balloon. (Mt. Wilson and Palomar Observatory)

of the sun. The corona reaches outward a vast distance from the sun until it thins out completely and disappears into deep space. Both the chromosphere and corona are normally invisible to the eye because their weak light is lost in the sun's total brilliance. It is only during the brief minutes of a total eclipse of the sun that we can view the chromosphere and corona without the use of special instruments. See Figure 2-3.

Disturbances on the sun. The gases of the sun are never still. Energy pouring out from the core sets the atoms and similar particles of the gases into wild motion. But near the surface of the sun we can see great clouds of hot gases boiling up from below. Since the hotter gases give off more light, the surface of the photosphere has an uneven appearance. Large dark areas of cooler gases often appear on the sun's surface. These are called *sunspots*. See Figure 2-4. They only appear to be dark because their gases are about 1500°C cooler than the surrounding ones. These cooler gases are not as bright and therefore appear dark against the brilliant background.

Sunspots seem to be connected with *magnetic disturbances* on the sun. They may be seen in groups but are more often seen in pairs where each sunspot probably acts as the pole of a magnet. Each visible sunspot is believed to mark a region of the photosphere where the magnetic effect blocks off the very hot gases. The cooler region of a sunspot gives it this darker appearance. Sunspots range in size from about 800 km (500 mi) to more than 80,000 km (50,000 mi) in diameter. Many sunspots are

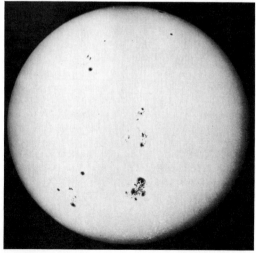

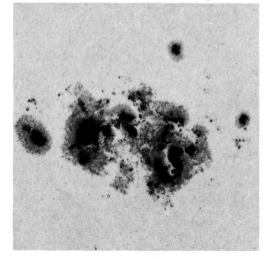

much larger than the earth's diameter and may last for months. Small sunspots however, usually last only a few days.

For reasons not understood, sunspots generally follow an eleven year cycle. Approximately every eleven years the number of sunspots reaches a maximum. This is only a very general rule. No one has yet explained what causes the eleven year cycle. The peaks in sunspot activity occur in a range from as little as seven years to as many as seventeen years apart.

Like storms on earth, sunspots may move slowly on the surface of the sun. Their individual motions are almost unnoticed when compared to the solar rotation, which carries them across the face of the sun. Since sunspots remain in relatively the same location on the sun's surface, observation of their daily movement will show how fast the sun rotates on its axis. See Figure 2-5.

Observations have shown that the sun turns on its axis once about every twenty-seven days. Since the sun is a ball of gas and not a solid, the time required for its rotation is not the same at all points. Places closest to the sun's equator have the shortest period of rotation. A point on the sun's equator takes only 25.33 days to make one rotation. A point halfway between the equator and the poles takes 28 days, while points near the poles take about 33 days.

Disturbances other than sunspots can also be seen taking place on the sun. Great arches of hot gas may be observed streaming between sunspots, as shown in Figure 2-6. These streams of gas rising above the sun's surface are called *prominences*. They represent gases flowing in response to the magnetic disturbances which accompany the sunspots. Frequently the disturbances which fling hot gas up from the sun's surface are much more violent than the relatively quiet looping of gases that form prominences. Most violent are the sudden eruptions called *solar flares*, which spray fountains of very hot gases far out into space.

The most remarkable thing about the solar flares is the direct effect they have on the earth. It has been discovered that these flares produce a strong burst of radiations. They are of three different types: ultraviolet, radio, and visible light. If a solar flare occurs in such a way that its radiations move toward the earth, the various radiations will reach the earth's atmosphere in about 8½ minutes. Powerful ultraviolet radiation usually causes electrical effects in the earth's upper atmosphere. These violent solar

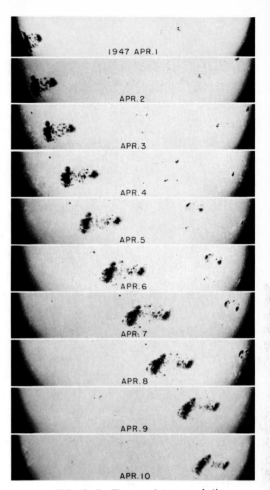

FIG. 2-5. These pictures of the movement of sunspots were taken at the same hour on 10 successive days during the sun's rotation. (Mt. Wilson and Palomar Observatory)

activity

Use Fig. 2-5 to answer the following questions.

1. What fraction of a circle does the largest sunspot make from the sightings on April 1 to April 10?
2. How many days of movement does this fraction represent?
3. At this rate of movement, how many days would it take for the largest sunspot to move completely around the sun?
4. On the basis of the sunspot's motion, does this place the sunspots near the sun's equator, pole or halfway between? Explain.

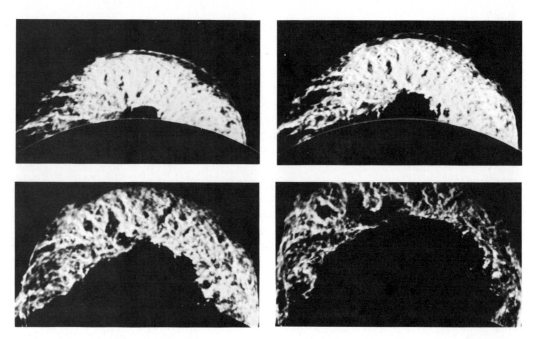

FIG. 2–6. Four successive stages in the development of a large solar prominence are shown above. Because the light from the prominence is faint, a special photo filter is used to block out the light from the sun. (Mt. Wilson and Palomar Observatory)

FIG. 2–7. This photograph of Aurora Borealis, or Northern Lights, was taken in Alaska. (National Bureau of Standards)

eruptions interrupt all forms of shortwave radio communication and announce the beginning of a *magnetic storm* on the earth. The effects of the ultraviolet radiations disappear in about an hour.

Actually the magnetic storm has only begun. In addition to the radiations, solar flares also produce slower moving atomic particles with electrical charges. These particles reach the earth about 13 to 26 hours after the flare takes place. The intensity of the solar flare determines the number and speed of the particles. As they pass through the upper atmosphere the particles may generate powerful electrical currents. Radio communications may be interrupted, and particularly severe magnetic storms may disrupt some telephone service and cause electrical power failure.

Auroras (the northern and southern lights) are the most spectacular results of many magnetic storms. The electrically charged particles from the sun are drawn to the poles by the earth's magnetic field. These particles may collide with the gas molecules in the upper atmosphere. This may cause the gas molecules to give off green, red, and occasionally blue or violet light. These displays of light, called *aurora borealis* in the north and *aurora australis* in the south, take many forms. The most common are sheets or curtains of light that shift and change degrees of brightness. See Figure 2–7. Auroras are most easily seen from locations near the poles. In severe magnetic storms they can be seen over greater areas. Figure 2–8 shows the average frequency per year of auroras in the Northern Hemisphere.

THE ENERGY OF THE SUN

Solar energy. In 15 minutes, the earth receives enough energy from the sun to satisfy the energy needs of the entire world for a year. Our energy problem would be solved if we could capture and use a small fraction of the solar energy received by the earth. However, some problems must be solved before solar energy can become an important part of our energy supply. The huge amount of energy received from the sun is spread over the entire lighted half of the earth's surface. Solar energy must be collected and concentrated to be used for most purposes. For example, to heat a typical house in the central United States would require solar energy collected from at least 14 square meters (300 sq ft). Solar energy is not dependable. Cloudy days reduce the amount of energy available and at night there is none available. Ways are being developed to solve these problems and make solar energy an immediate source of energy. Solar collectors are now being placed on buildings to trap the sun's heat, to help warm the interior, or to heat water. Another approach uses special cells that convert solar energy into electricity.

At the present time, solar energy is used only as an addition to the more common sources of energy. But in the future, the sun may become a main source of energy. Houses could be equipped with solar collectors to supply heat. A heat storage system would supply warmth on cloudy days and at night. Solar cells could be used to supply much of the electrical energy needed. In addition to the direct use of solar energy by homes, large solar "farms" could be built in sunny locations. A solar farm would be a large area filled with solar energy collectors that supply energy for an electrical power plant. It may also be possible to put a large collection of solar cells into a satellite orbit around the earth. This kind of solar energy collector would always be exposed to full sunlight. Its electrical power could be transmitted as microwaves back to the earth.

The solar energy received by the earth is only a small part of the total energy produced by the sun. The sun has

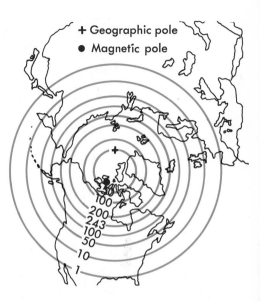

FIG. 2–8. The average frequency of auroras that an observer is likely to see in the Northern Hemisphere is indicated by the number on the circle.

been giving off huge amounts of energy for billions of years. Where does it come from?

The first explanation that is usually considered is that the sun's energy comes from something burning. Fire is the most familiar source of energy. But no ordinary fuel can produce the tremendous energy that pours out of the sun. If the entire sun were made of coal and oxygen, for example, it would burn itself out in less than two years. The secret of the sun's energy supply is found in a principle developed by Albert Einstein. This principle is expressed in the equation $E = mc^2$. E is energy, m is mass representing matter, and c is the speed of light. The equation shows that matter can be changed into energy. It also shows that a small amount of matter becomes a large quantity of energy. The sun produces energy by continuously changing some of its matter into energy. According to Einstein's principle, this conversion could account for the sun's ability to produce large amounts of energy without losing much of its matter. It is believed that this conversion is done under the high temperatures within the sun's interior. At these temperatures, light atoms come together to form heavier ones. It is the atomic reactions that are the source of solar energy.

Nuclear reactions in the sun. Deep within the sun, matter is under a pressure more than one billion times greater than normal air pressure on the earth. Atoms are crowded together so tightly that only their *nucleus* remains intact. But the very high temperature causes these atomic nuclei to move about at high speed. Now and then two of these moving nuclei collide in such a way that they join. The process by which two nuclei join to form a single heavier nucleus is called nuclear *fusion.* The new, heavier nucleus has slightly less mass than the total mass of the two smaller ones from which it was formed. This difference in mass is converted into energy according to the $E = mc^2$ equation. Any process which releases energy as a result of fusion of atomic nuclei under high temperature is a *thermonuclear reaction.* The most powerful nuclear bombs get their energy from thermonuclear reactions. In a nuclear bomb the reactions occur instantaneously. In the sun, the reactions go on continuously.

Only the nuclei of certain light atoms can join in fusion reactions within the sun. The lightest of all atoms is hydrogen. Since it is also the most abundant substance on the sun, there is no doubt that it is the basic "fuel" for the

FIG. 2–9. The diagram below illustrates a three-step nuclear fusion reaction which probably furnishes part of the sun's energy. Notice that the only new substance produced is helium.

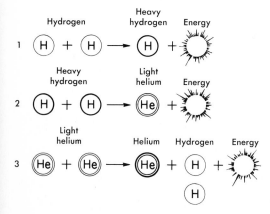

thermonuclear reactions. Research has shown that there are probably two major nuclear reactions that take place within the sun. Most important is the reaction which converts hydrogen directly into helium. This process, shown in Figure 2–9, seems to supply most of the sun's energy. To produce the sun's energy, some 600 million tons of hydrogen must be converted to helium each second.

A second reaction also converts hydrogen into helium but involves other kinds of nuclei as well. This reaction is shown in Figure 2–10. In both reactions only hydrogen is completely used up. Helium is produced, along with the conversion of small amounts of mass into large quantities of energy. It is estimated that the sun used 657 million tons of its hydrogen each second. During the same time 652½ million tons of helium are produced. The difference of 4½ million tons represents the matter changed into the huge amount of energy released by the sun each second.

The earth's share of the energy produced by the sun's nuclear reactions is the most important single influence upon this planet. Without solar energy the earth would be a dead planet. Since our continued existence depends upon the steady flow of energy from the sun, we are deeply interested in questions about the sun's life. How long can the sun continue to produce energy? When did the sun begin? How will it end? Such questions as these about the star called the sun can only come from general knowledge of the entire population of stars throughout the universe.

OTHER STARS IN OUR GALAXY

How far away are the stars? All stars except the sun appear to us on earth as points of light. Yet we must depend upon this small amount of light as the only important source of information about the stars. Starlight is useful in measuring the immense distance between the earth and the stars. One method for measuring stellar (star) distances makes use of *parallax*. Parallax describes what appears to be movement of an object but is actually caused by a change in the position of the observer. For example, more than 6000 stars have been observed seemingly changing their positions among the background of stars while the earth moves in its orbit. Parallax of a star is measured from two points in the earth's orbit six months apart. This time period is necessary so that observations can be made from the two most widely separated points

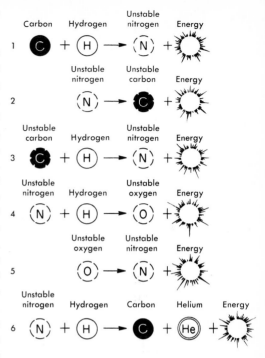

FIG. 2–10. This series of nuclear fusion reactions probably furnishes some of the sun's energy. Note that although atoms of carbon, nitrogen, and oxygen are involved, the final result is the fusion of four atoms of hydrogen to form a single atom of helium. In each case energy is produced because the products of the reaction weigh a little less than the original atoms.

Measure
Find out how to measure short distances by parallax in order to see how it works at astronomical distances.

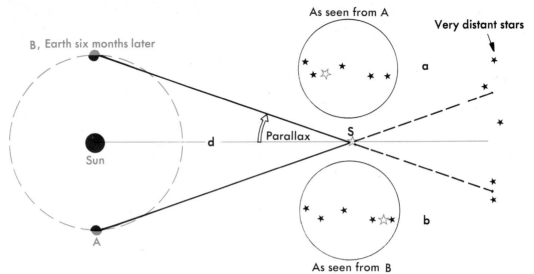

FIG. 2–11A. Measurement of stellar parallax can be used to help calculate the distance to a star. The right triangle formed by the star (S), the sun, and the earth can be used to determine the distance (d). What is the distance along a line joining the sun with either point A or B?

FIG. 2–11B. The farther away a star is, the smaller its angle of parallax. Because the lines of sight to very distant stars are almost parallel, there is no measurable angle of parallax.

in the earth's orbit at which the maximum parallax angle can be measured. See Figure 2–11A.

By using some of the principles of geometry and measuring the parallax angle of a star we can determine its distance from the earth. However, this method is only effective for those stars found within about 300 light-years of the earth. The apparent movements of these relatively closer stars produce parallax angles that can be fairly accurately measured. More distant stars have such a small apparent movement that their parallax angles can barely be measured, even with the largest telescope. See Figure 2–11B.

How bright are the stars? Knowing the distance to a star is necessary to solve the problem of a star's true bright-

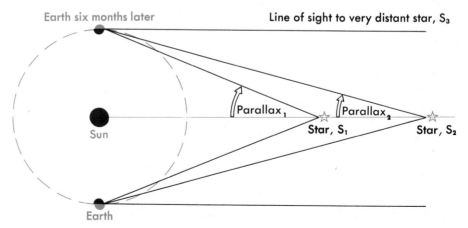

ness. Just as a bright street light looks dimmer when seen from a distance, a bright star is dim if it is very far away. The observed brightness of a star as it appears from the earth is called its *apparent magnitude*. A star that appears dim may actually be a very bright, but distant star. A magnitude scale is often used to describe how bright the star would look if seen at a standard distance of 32.6 light years. This is called *absolute magnitude*. It describes the true brightness of a star. To find the absolute magnitude of a star, its distance from the earth must be determined.

The magnitude scale is a useful system for estimating the brightness of a star seen from earth. In fact, all celestial bodies that reflect or radiate visible light are given a number on the magnitude scale. These numbers represent a relative measurement of light given off. This system first came into use more than 2000 years ago. The Greek astronomer Hipparchus classified about 1000 stars according to their apparent brightness (magnitude). The brightest in appearance he labeled *first magnitude*. The faintest stars visible to the unaided eye he labeled the the sixth magnitude.

Today a very similar classification, based on more precise measurements with the use of the telescope, has been developed. The present magnitude scale has been lengthened to include at one extreme the brightest object in the sky (the sun), and at the other the magnitude of the faintest stars that can be photographed with the 200-inch-telescope on Mt. Palomar. Table 2–1 lists magnitude data for some common celestial objects, including the limits of various instruments. You can see that the lower magnitudes are associated with the brightest object.

Any increase or decrease in brightness by a magnitude of 1, will change the apparent brightness of a star by a factor of 2.5. For example, a fifth-magnitude star appears 2.5 times brighter than a sixth-magnitude star. How many times brighter is a star of the first magnitude compared to a star of the fifth magnitude?

For stars whose distance is figured by parallax measurement, it is a simple matter to determine their absolute magnitudes. More distant stars need other methods. For example, one type of star grows brighter, then dimmer in a regular pattern. These stars are known as *cepheid* (*sef*-i-id) *variables*. The time between the bright and dim phases of the cepheid variable stars is related to their true brightness. By measuring the period of a cepheid we can find its absolute magnitude. Then if we compare this result with

Table 2–1 Scale of Magnitudes

Object	Apparent Magnitude
Sun	− 26.5
Full moon	− 12.5
Venus (at brightest)	− 4
Jupiter; Mars (at brightest)	− 2
Sirius	−1.4
Aldebaran: Altair	1.0
Naked-eye limit	6.5
Binocular limit	10
6-in. telescope limit	13
200-in. (visual) limit	20
200-inch photographic limit	23.5

Increasing Brightness →

Measure
Use a photographic light meter to measure how the apparent brightness of a light bulb diminishes with distance.

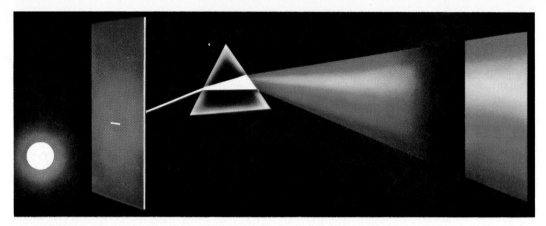

FIG. 2–12A. A continuous spectrum is produced by incandescent solids, liquids, and gases under high pressure. The very dense incandescent gases in the main body of the sun and stars produce continuous spectra. (Adapted from THE UNIVERSE, Time, Inc.)

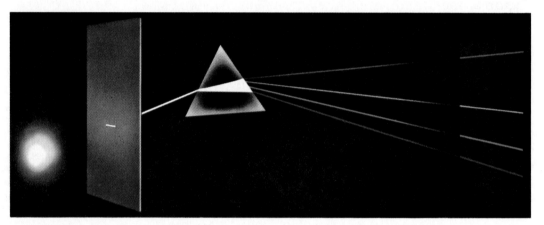

FIG. 2–12B. A bright-line spectrum is produced by incandescent gases of low density. Each chemical substance yields a characteristic pattern of lines that differ from all others.

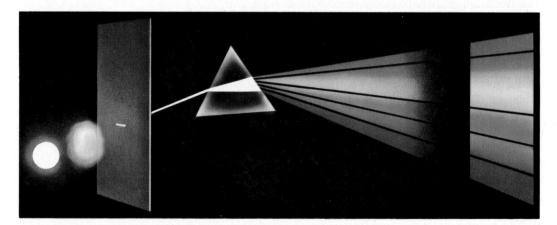

FIG. 2–12C. A dark-line spectrum is produced by a nonluminous (cooler) gas in front of an incandescent source of a continuous spectrum.

the observed visible brightness (apparent magnitude), we can calculate the approximate distance of the star. This method makes it possible to find the distance to any cepheid, such as Polaris (the North Star), 300 light-years away. Distances to other galaxies which contain cepheid variable stars can also be determined in this way.

Analyzing starlight. Using the proper instruments to analyze light from stars, we can get much information other than the star's distance. Next to the telescope, the most valuable instrument for carrying out the analysis of starlight is the *spectrograph*. This is a device for separating light received by telescopes into its different wavelengths. Each wavelength produces a different color. See Figure 2–12A.

Spectrograph analysis of starlight can yield several important kinds of information. For example, the spectrum of a star tells us what chemical elements are to be found within it. Each substance in the star gives off identifying wavelengths of light. See Figure 2–12B. Gaps which appear in the spectrum as dark lines show the composition of cooler gases in front of a light source such as a star. This is shown in Figure 2–12C.

The spectrograph can also be used to indicate the direction a star is moving in relation to the earth. This is made possible by the *Doppler effect*. A common example of the Doppler effect occurs with sound waves. The pitch (number of vibrations/sec) of a locomotive horn rises as the train approaches, then falls as the train passes. The pitch of the sound changes because sound waves coming from the horn are closer together as the train approaches. Then the sound waves are farther apart as the train moves farther away. See Figure 2–13. The Doppler effect is observed when either the observer or the source of a wave is moving, or both.

Light waves also exhibit the Doppler effect. Whether they are spread out or crowded together depends upon if their source is moving toward or away from the earth. When a spectrograph is used to observe a light source, the Doppler effect can be seen as a shift in the lines of the spectrum. As a star approaches the earth, the Doppler effect causes its light to shift toward the blue end of the spectrum. As a star recedes from the earth its light is shifted toward the red end of the spectrum. See Figure 2–14. Astronomers can determine the amount and direction of motion relative to the earth by measuring the shift.

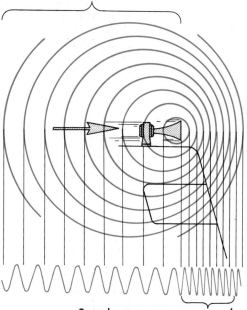

Pitch falls as train passes

Sound waves are compressed as train approaches—pitch rises

FIG. 2–13. The pitch of the horn on an approaching locomotive rises because more sound waves arrive each second than when the locomotive is moving away. This is an example of the Doppler effect.

Experiment
Use a portable tape recorder to record various sounds (a passing fire engine) which illustrate the Doppler effect. Examine how the pitch of a given sound changes with speed.

FIG. 2–14. A Doppler shift is shown in the photographs of spectra of the star Arcturus (bands a and b) taken six months apart. The expanded diagram above the photograph shows the gradual shift of the dark-line spectrum as the position of a star changes. Astronomers use this to measure relative velocity of the earth with respect to a star. (Mt. Wilson and Palomar Observatory)

Star populations. There is a wide range of values in the temperature and absolute magnitude for the observable stars. Stars have been discovered thousands of times brighter than the sun. Others are only a small fraction of the sun's brightness. Temperatures range from about 3000°C to as high as 50,000°C. These are temperatures for the star's outer layers only.

A useful way to summarize these different characteristics of stars is by use of a diagram such as that shown in

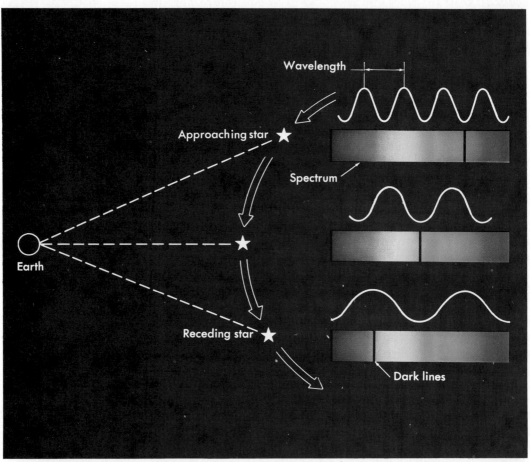

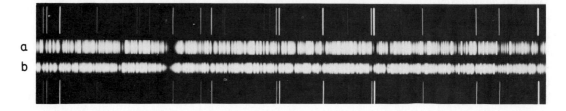

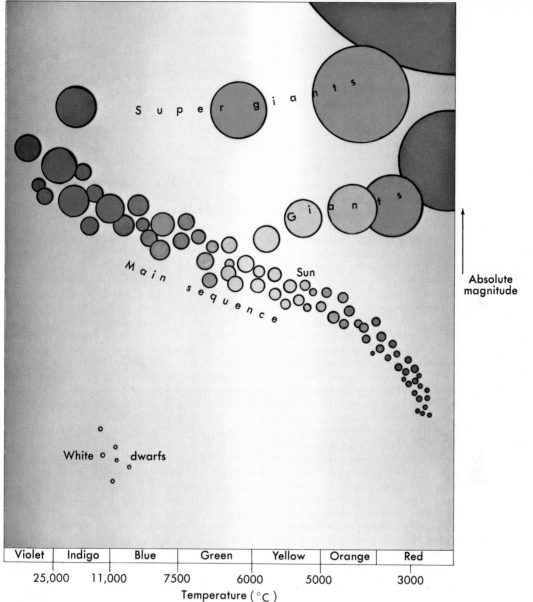

| Violet | Indigo | Blue | Green | Yellow | Orange | Red |

Temperature (°C)

25,000   11,000   7500   6000   5000   3000

Absolute
magnitude

Figure 2–15. When each star is placed on this diagram according to its absolute magnitude (vertical scale) and surface temperature (horizontal scale), it falls into a definite pattern. Most are found in a narrow band running diagonally across the diagram. This is called the *main sequence*. Two groups of stars appear on the upper right of the diagram. These are bright stars but with relatively low temperatures. Their brightness comes from the great size, allowing them to give off large amounts of light. They are called *red giants* because of the reddish color of this light.

FIG. 2–15. This diagram groups stars according to their temperature (color) and absolute magnitude.

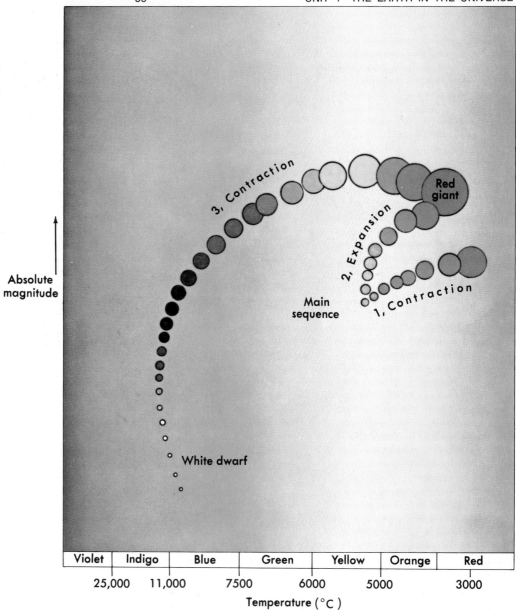

Absolute magnitude

3, Contraction

Red giant

2, Expansion

Main sequence

1, Contraction

White dwarf

Violet	Indigo	Blue	Green	Yellow	Orange	Red

25,000   11,000   7500   6000   5000   3000

Temperature (°C)

FIG. 2–16. Diagram of the three stages in the life cycle of a star.

A third group of stars appears on the lower left of the diagram. These stars are barely visible because of their very small size. They have been named *white dwarfs* because the light they give off has an overall white color.

It is believed that the main groups of stars which appear on the brightness-temperature diagram represent different stages in star development. A star probably begins in its shrinking stage with medium brightness and low temperature. As its nuclear reactions increase, the

NIXON

young star gets hotter. The heat produced by these reactions moves the star to some position along the main sequence. There it remains for most of its life. This is shown in Step 1 in Figure 2–16. In its later stages, heat in the interior increases and the outer layers expand. The main sequence star then becomes a red giant. (Step 2). What happens next is not well understood, but it is thought that in time the interior of the red giant probably begins to cool. The star eventually shrinks to become a white dwarf. (Step 3).

If this picture of star development and classification is correct, most observed stars should be classified as belonging to the main sequence. This is because the main sequence represents the lengthy mature stage of a star's life. Observations support this picture. Most stars do fit on the main sequence. The sun, for example, is a main sequence star apparently in about the middle of its development. Only a small number of stars can be classified as red giants. White dwarfs are also rare but this is believed to be a result of their faint light. They may well be as abundant as main sequence stars.

Many stars seem to develop in an unusual way and cannot fit easily into the main classifications which have just been described. For example, there are *super giant* stars which are even larger than the red giants. The exact position of these overly large stars in the total scheme of star development is not known.

FIG. 2–17. The Horse Head nebula is a dark cloud of cool gas and dust. (Mt. Wilson and Palomar Observatory)

LIFE HISTORY OF A STAR

Birth of a star. In the space between stars there is a very thin scattering of gas and dust. It is so thin that there are fewer particles along a line between the sun and its nearest neighbor star (about four light years away) than the number of molecules in a spoonful of air at the earth's surface. However, the spaces between the stars are so vast that the total amount of gas and dust is nevertheless very large. In the neighborhood of the sun in our Galaxy the space between the stars seems to hold about half as much matter as the stars themselves. Spectrographic analysis of these gases as starlight passes through them, shows that they are made up of about three-quarters hydrogen. The remaining one-quarter is mostly helium with small amounts of heavier elements. In a few places the gases and dust completely cut off the light from behind the

FIG. 2–18. A star cluster. (Mt. Wilson and Palomar Observatory)

FIG. 2–19. The changing positions of the double star, Kruger 60, observed over an interval of 12 years. (Yerkes Observatory)

stars. This gives the appearance of a dark "hole" in the background of stars as shown in Figure 2–17. When there is an accumulation of gases and dust between the stars, they form a cloud called a *nebula* (neb-you-la). Many of these clouds may form the source material for the birth of new stars.

Each particle of gas and dust within a nebula has a very weak gravitational attraction for the other. If some of the particles happen to crowd together, their gravitational attraction increases. This in turn tends to attract more material at an ever increasing speed. Eventually a part of the original cloud will become a rapidly shrinking and increasingly thicker region.

One result of this shrinking action of the cloud is the production of heat. This comes from the release of gravitational energy as the particles draw closer together. As each particle is drawn into the growing body of matter, it collides with the particles already there and contributes more energy. The temperature at the center of the shrinking cloud steadily increases. At a certain temperature the hydrogen atoms present begin to experience the nuclear reactions which release energy. At this point a new star with its own energy supply comes into existence.

The young star will continue to shrink until a state of balance is reached. Heat produced by the nuclear reactions taking place within the star cause its gases to expand. Gravity then forces the star's gases to draw closer together. Finally a condition will be reached in which these two processes are in balance. Then the star will have completed its birth stages and will begin a period during which it quietly produces energy as it carries on nuclear reactions.

Many of the clouds from which stars are formed seem to produce more than a single star. Close groupings of

stars known as *star clusters* appear to be a number of stars formed within the same cloud. See Figure 2–18. Stars are also very commonly found in pairs or groups of three or four. More than half of all the stars that you can see in the night sky are *double stars.* Mutual gravity makes these stars move around each other as shown in Figure 2–19. Double stars were probably formed from the same source.

Some stars do not seem to settle down and shine with a steady light. Changes occurring in the internal nuclear reactions of these stars apparently cause them to grow bright, then dim, and then grow bright again in a regular cycle. The reason for this is not understood. Many of these *variable stars* have been observed in our Galaxy and in distant galaxies.

Death of a star. The life span of a star is set by the amount of hydrogen it has at birth and the rate at which it consumes this hydrogen. When a star uses up about ten percent of its original hydrogen it reaches a critical time. Helium accumulates as nuclear reactions use up this hydrogen (see pages 28–29). The helium which is produced tends to sink toward the star's center because it is heavier than hydrogen. This puts greater pressure on the remaining hydrogen within the interior. Increased pressure on the hydrogen speeds up the release of energy. The star's interior is heated and the outer layers expand. These outer layers of gas cool and take on a reddish color because they are farther from the fierce heat of the interior.

As the star continues to use up hydrogen and accumulate helium, another serious problem eventually arises. After about 40 percent of the original hydrogen supply is used up, enough helium has gathered to create very high pressure in the interior. When the pressure reaches a certain level, the helium itself starts to experience nuclear reactions. A new source of energy has been provided and the star's core becomes even hotter.

However, as the helium is used up, it forms heavier nuclei which accumulate at the star's center. When the star has used up most of its hydrogen and helium, its main supply of nuclear fuel is gone. It then begins to shrink under its own gravity. The final end of the star seems to depend upon its original size. Stars which are about the size of the sun or smaller may have either of two possible fates. Some collapse until they are about the size of the earth and continue to shine with a feeble white light. These are the white dwarfs already mentioned.

FIG. 2–20. Two photographs of Nova Hercules taken two months apart. It was discovered to be a nova when its brightness increased from a magnitude of 15 to a magnitude of 2, and then returned to its normal size shown above. (Lick Observatory)

Other dying stars apparently can collapse more completely until they shrink to a diameter of 20 to 30 kilometers (12 to 18 miles) in diameter. These are called *neutron stars* because their enormous pressures cause all their atoms to become changed into neutrons. Such strange stars may also be called *pulsars* since they frequently give off radio signals in regular pulses. Very large stars, more than twice the mass of the sun, may have a very unusual death. They might collapse so completely that no trace remains except their gravity. This could create a strange region of space sometimes called a "black hole." It is not known whether black holes really exist, since it is not clear how astronomers might observe them.

The time it takes for a particular star to complete its life span depends not only on the amount of hydrogen present but also upon the size of the star. Very large stars create higher pressures at their centers that use up the nuclear fuel rapidly. These very large and heavy stars probably last only one or two million years. Smaller stars use their nuclear fuel more slowly and may continue to produce energy for about ten billion years.

Some stars do not end their existence by quietly fading away. For reasons not well understood, a star may release its energy in a seeming explosion. One kind of variable star known as a *nova* flares up for a few days or even hours, then returns to its normal state. The term nova means "new ones" because it was originally believed that they were not in the sky before one was first seen. However, later observations proved that this "new star" was actually not new at all, but rather an old star that because of some sudden change, abruptly increased in brightness. During these episodes a nova seems to fling matter from its outer layers into space. About one hundred novae have been observed among the stars of our Galaxy. See Figure 2–20.

A typical nova is a star that is observed to suddenly increase in brightness from about 30 times to 100,000 times that of the sun. More spectacular and even more rare is the supernova. The last supernova that occurred in our Galaxy was observed in 1054 A.D. It is recorded in historical descriptions. A cloud of gas and dust can still be seen at the location where this supernova is known to have occurred. See Figure 2–21. A supernova explosion is connected with the last stages in the formation of a neutron star. Such a star has been detected at the center of the gas cloud formed from the supernova explosion of 1054.

FIG. 2–21. The Crab nebula is a great expanding cloud of gas representing the remains of a supernova explosion. (NASA)

ORIGINS OF THE UNIVERSE

Does the universe have a life history? Since stars seem to go through a definite cycle of development, we naturally wonder if the whole universe has some sort of life history. At the present time there is no final answer to this question. An important piece of information which must first be investigated is whether the universe has an outer boundary. Individual stars are not important in studying this question. We need to focus our study on the arrangements of galaxies.

The most powerful optical telescopes have detected millions of galaxies, some of which may be as far as 1200 million light years away. When viewed through a spectrograph, light from all of these galaxies indicates a movement toward the longer wavelength, or the "red shift." This apparent Doppler effect seems to indicate that all other galaxies are moving away from us at very high speeds. The most distant galaxies are moving away at the greatest speed. The fastest speed yet measured is 76,000 mi/sec. The galaxies appear to move as if the entire universe were expanding from some central point.

The idea of an expanding universe has lead to three main theories that attempt to account for the universe as it now exists. See Figure 2–22.

1. The first theory states that all the matter now in the galaxies was once packed together in a single gigantic

body. About 10 billion years ago an explosion we could never imagine sent fragments flying outward to form the still expanding universe as we know it now. This is often called the "Big Bang Theory." According to this theory, the galaxies will continue moving away from each other forever into limitless space.

2. A second theory proposes that new matter is constantly being created everywhere in the universe. Galaxies continuously move out and are lost in endless space. But new galaxies are at the same time being formed out of newly created matter. Thus the total number of galaxies remains the same. This theory is usually called the "Steady State Theory."

3. A third theory refers to a universe which expands after a big bang, then contracts and explodes again. This expanding and contracting is repeated every eighty billion years. Often called the "Pulsating Theory," it is different from the first two theories because it suggests a universe with an outer boundary that moves in and out with a regular beat.

FIG. 2–22. The illustrations below describe the three possible models for an evolving universe. Changes in density are indicated by a shading effect. In which model does the density of the universe constantly decrease?

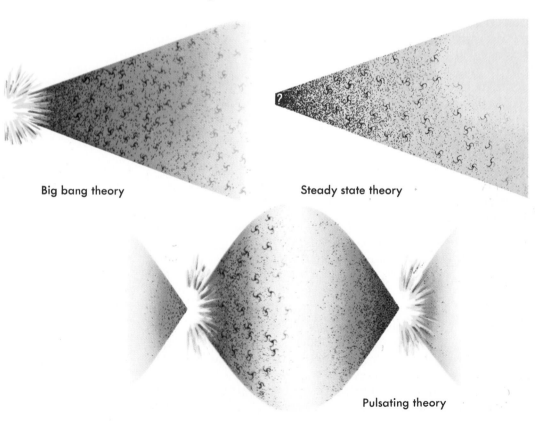

Big bang theory

Steady state theory

Pulsating theory

One of the few ways to test any of these theories is to find the number of most distant galaxies. All of the above theories correctly account for the observations of nearby galaxies, but each theory predicts differences in appearance and arrangement of very distant galaxies. Unfortunately, present-day optical telescopes cannot penetrate far enough into space to confirm these predictions.

Radio telescopes can detect more distant objects than optical telescopes and may eventually provide some answers to the origins of the universe. Already radio telescopes have discovered very strange objects which may be a key to the missing information. These objects are called *quasi* (resembling) *stellar* (star) *radio sources* or *quasars*. One of the most puzzling things about quasars is their ability to produce enormous amounts of energy. They give off more energy than can ever be accounted for by any known process including nuclear reactions. Perhaps these very puzzling bodies will help us finally understand the birth and life history of the universe.

VOCABULARY REVIEW

Match the word or words in the column on the right with the correct phrase in the column on the left. *Do not write in this book.*

1. Form of energy sent out by the sun.
2. Source of sun's visible light.
3. Thin gases making up the outer most portion of the sun.
4. Streams of gases arching from the sun's surface.
5. Sudden eruptions of the sun's surface.
6. Northern lights.
7. Joining together of atomic nuclei.
8. Clouds of gases and dust in space.
9. A variable star which suddenly becomes very bright.
10. Brightness of a star as it appears from earth.
11. A pattern found in plotting magnitude versus temperature of stars.
12. Region in space which produces energy at an unexplainable high rate.
13. An instrument that separates light into individual wavelengths.
14. The spreading out or crowding together of waves due to motion of the source.
15. True brightness of stars.

a. apparent magnitude
b. nebulae
c. quasar
d. magnitude
e. aurora australis
f. radiation
g. absolute magnitude
h. main sequence
i. core
j. chromosphere
k. Doppler effect
l. photosphere
m. prominences
n. aurora borealis
o. solar flares
p. spectrograph
q. star cluster
r. corona
s. fusion
t. nova

QUESTIONS

Group A

Select the best term to complete the following statements. *Do not write in this book.*

1. The energy sent out from the sun is the most intense in the (a) X-ray region (b) ultraviolet region (c) visible region (d) infrared region.

2. Of the total amount of solar energy directed toward the earth, the amount that actually reaches the surface is approximately (a) one third (b) two thirds (c) three fourths (d) one half.

3. Reactions that produce the sun's energy take place within its (a) core (b) zone of radiation (c) photosphere (d) chromosphere.

4. The sun's zone of radiation produces short wavelength radiations such as (a) infrared waves (b) radio waves (c) X rays (d) visible rays.

5. The part of the sun that is made of relatively cool gases which give off a weak red light is the (a) corona (b) chromosphere (c) zone of radiation (d) photosphere.

6. Sunspots appear dark because they (a) show the dark core of the sun (b) are cooler than the surrounding gases (c) are similar to clouds on the earth (d) do not contain gases.

7. Observation of sunspots indicates that the sun rotates on its axis (a) at a rate which is slower at its equator than at its poles (b) at a constant rate of 28 days (c) at a rate which is faster at its equator than at its poles (d) at a constant rate of 25.33 days.

8. Prominences appear as arching streams of gases due to (a) the magnetic fields of sunspots (b) the cooling of the gases as they rise above the sun's surface (c) solar flares (d) the sun's rotation.

9. Great magnetic storms result on earth due to (a) sunspots (b) prominences (c) radio waves from the sun (d) solar flares.

10. Solar energy can be used directly in the home by solar collectors producing heat, or by solar cells producing (a) microwaves (b) radio waves (c) heat (d) electricity.

11. Early ideas concerning fire as the source of the sun's energy were disputed when it was shown that (a) the sun is made of gases which will not burn (b) the sun is too cool to burn (c) the burning process would give off much less energy than the sun actually produces (d) gravity could be changed into energy.

12. The present theory of how the sun's energy is produced is based on Einstein's discovery of the relationship between energy and (a) fire (b) matter (c) gravity (d) light.

13. The reaction taking place in the sun's core involves the continuous fusing of atomic nuclei due to (a) extremely high pressures (b) extremely high pressures and temperatures (c) presence of radioactive materials (d) presence of heavy nuclei.

14. The basic fuel for the thermonuclear reactions going on within the sun is (a) hydrogen (b) helium (c) carbon (d) oxygen.

15. Although the matter existing in the space between stars is spaced far apart in our Galaxy, the total amount found between stars seems to be (a) similar to the number of molecules in a spoonful of air (b) about as thin as the air we breathe (c) about equal to the amount found in the stars themselves (d) about half as much as that in the stars themselves.

16. When clouds of cosmic dust and gases begin to shrink, becoming more dense (a) light is given off (b) heat is generated (c) gravity is decreased (d) the gases become cool.

17. A new star is born when cosmic dust and gases shrink (a) to a solid ball (b) sufficiently to burn (c) sufficiently to cause nuclear reactions (d) to a very large but thick gas ball.

18. A star which is quietly producing radiant energy maintains a balance between (a) mass and energy (b) shrinking due to gravity and shrinking due to cooling effects (c) shrinking due to gravity and expansion due to attraction of more gases (d) shrinking due to gravity and expansion due to energy released by nuclear reactions.

19. Closely grouped stars called star clusters were probably formed (a) at different times (b) from the same cloud of dust and gases (c) from different galaxies (d) under very unusual conditions since few exist.

20. Stars whose light appears to pulsate in a cyclic manner are called (a) nebulae (b) variable stars (c) double stars (d) star clusters.

21. The length of time a star can exist seems to be determined mainly by (a) where it is formed (b) its color (c) the amount of hydrogen it starts with (d) the amount of helium it contains at birth.

22. Two crises mark the start of the death of a star. The first occurs when approximately 10% of the hydrogen is consumed, the second when (a) helium begins to be consumed (b) all the hydrogen has been consumed (c) all the helium is consumed (d) heavy nuclei begin to be consumed.

23. Stars whose life spans reach ten billion years or more would most likely start out as (a) red giants (b) yellow giants (c) white dwarfs (d) nova.

24. A supernova explosion is connected with the last stages in the formation of (a) neutron star (b) red giant (c) blue giant (d) white dwarf.

25. The brightness of a star as it appears at a distance of 192 million million miles is called its (a) magnitude (b) apparent magnitude (c) absolute magnitude (d) maximum magnitude.

26. Since most stars fall in the pattern of stars called the "main sequence", it is felt that this is additional evidence that (a) all stars are born as white dwarfs (b) all stars eventually become red giants (c) stars are born of different sizes (d) the main sequence represents the stages of development of most stars.

27. The group of stars that would fit into the main sequence are the (a) red giants (b) white dwarfs (c) super giants (d) stars like our sun.

28. Neutron stars are produced when a normal star collapses to the size of (a) the earth (b) 20 to 30 kilometers in diameter (c) a red dwarf (d) a white dwarf.

29. The one theory of the origin of the universe which assumes that the universe has an outer boundary is the (a) big bang theory (b) pulsating theory (c) steady state theory (d) quasar theory.

30. In order to determine which of the theories of the origin of the universe is most accurate, scientists would have to determine (a) which galaxies are the oldest (b) the location of the center of the universe (c) the location of all the most distant galaxies of the universe (d) if new stars are being born regularly.

Group B

1. Almost all the sun's radiation reaching us on the surface of the earth is in the form of visible light and radio waves. What other radiations are given off by the sun and why are they not received at the earth's surface?

2. How can it be determined that the sun contains most of the same chemical elements as the earth?

3. Describe the sequence of events by which the sun appears to produce energy. Use the proper names for the parts of the sun involved.

4. How are sunspots related to the magnetic disturbances found on the sun?

5. Explain how a solar farm might be used to supply solar energy.

6. What role is played by solar flares in producing magnetic storms on earth? How might astronauts in flight be affected by such flares?

7. How did Einstein's discovery of the relationship of matter and energy enable scientists to better understand how the sun could continue to give off such huge quantities of energy?

8. Describe the life history of a typical star such as our own sun.

9. What conditions are found in the core of the sun which can cause nuclear reactions to occur?

10. Why do scientists believe stars start out as accumulations of gases containing large amounts of hydrogen?

11. How can we explain the existence of multiple star systems?

12. What similarity can you see between the formation of multiple star systems and the formation of our solar system with its one star and a number of planets?

13. Scientists constantly seek patterns in the data they collect. What pattern is found when the absolute magnitude is compared to the surface temperature for a large number of stars?

14. What relationship is felt to exist between the main sequence grouping of stars and the life history of stars?

15. Suppose as a scientist you are told of the discovery of a galaxy more distant than any others yet discovered. Suppose also that it is found to be moving at a rather slow rate back toward the center of the universe. Considering each theory of the universe separately, how would you explain this phenomenon?

16. What information can be obtained by the use of a spectrograph in astronomy?

17. Until just recently, all super novas were thought to be the result of the complete destruction of a variable star by explosion. What discovery has caused a change in this explanation?

18. Describe the nature of a pulsar.

Mercury
Earth
Venus   Mars   Saturn
Uranus
Jupiter   Neptune
Pluto

3
the earth as a member of the solar system

Moving around the sun are nine planets. They are accompanied by thirty-two moons. In addition to the planets, there are millions of smaller objects ranging in size from a mountain to a single molecule. All of these have at least one thing in common. They all move under the influence of the sun's gravity. The sun and the great number of large and small bodies held by its gravitational influence make up the solar system.

In some ways, the solar system resembles a complicated machine. Each member of the solar system, like each part of a working machine, moves in a definite and regular way. In fact, the planet Earth moves with such precision that it can be used to measure time in the same way that the moving parts of a watch mark off time. The use of the earth as a timekeeper will be discussed later in this chapter.

How did the solar system come into existence? Why should the sun be surrounded by a system of smaller objects? The answers to these questions can only be found through exploration of the earth's neighbors in space. During the past few decades, astronauts have landed on the moon. Space vehicles have landed on some of the planets closest to the earth. Other space probes have flown near the more distant planets. Exploration of the solar system has greatly advanced our knowledge of its origin. Scientists now believe they have a good understanding of how the solar system formed.

ORIGIN OF THE SOLAR SYSTEM

The growth of a theory. Between four and five billion years ago, a cloud of dust and gas formed in one of the

objectives

☐ Describe the nebular theory of the origin of the solar system.

☐ Compare similarities and differences of the inner planets.

☐ Describe the asteroids.

☐ Compare the outer planets with each other; and with the inner planets.

☐ Describe meteors, meteorites, and comets.

☐ Explain the summer and winter solstices and the vernal and autumnal equinoxes.

☐ Compare apparent solar time and mean solar time.

☐ Explain why we use standard time and have the International Date Line.

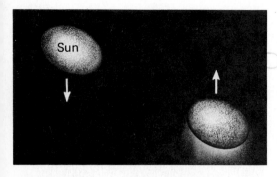

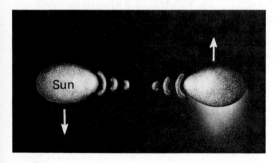

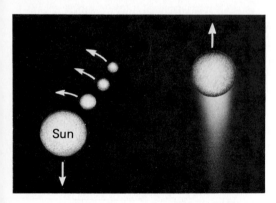

FIG. 3–1. Top, a large body passes close to the sun. Middle, material is pulled out as the gravity of each body affects the other. Bottom, the material drawn out from the sun cools to become the planets.

spiral arms of our galaxy. As the cloud shrank, its center became so dense and hot that a star, the sun, was formed. Smaller bodies also began to grow at various places within the cloud. They formed the planets and their satellites. This is a very brief description of an accepted scientific theory of the origin of the solar system. It is usually called the *nebular* or *dust cloud theory.*

The nebular theory was first suggested by the French astronomer Rene Descartes in 1644. Almost a hundred years later, a second theory was proposed by Georges Buffon, another French scientist. It was called the *near collision theory.* In Buffon's theory, the sun was created before the rest of the solar system. A large body passed close to the sun and pulled out material. This material cooled and became the planets. See Figure 3–1. Buffon proposed that a comet was the passing body whose gravity pulled material from the sun.

For the next two hundred years, scientific thought on the origin of the solar system was based on either the nebular theory or the near collision theory. First one theory, then the other became popular. New ideas and information supported both theories, making each theory better able to account for the solar system.

Until recently, it was impossible to choose between these two opposing theories. Too few facts about the members of the solar system were available to us. The earth was the only part of the solar system that could be studied first-hand. But there was no geological record left from the first few hundreds of millions of years of earth's history. These early theories were based on limited observations of the planets and how they moved.

Exploration of the moon and other planets, along with advances in understanding the way stars are formed has helped settle the question. Almost all scientists now support the nebular theory. They agree that the sun and the planets were formed from a rotating disk of dust and gas. The other theory, that the planets were formed from material torn from the sun, has a flaw. It has been shown that this material would probably spread out in space before planets could grow from it.

The solar system is born. In its modern form, the nebular theory begins with a slowly rotating cloud of gas and dust. The cloud was called a *nebula.* With continued shrinking, the center of the cloud became the young sun. In time, smaller clumps of various sizes developed within the

cloud. See Figure 3-2. Each of these clumps lacked enough internal pressure to start the nuclear reactions necessary to form a star. These smaller centers of condensing matter became the planets. At this time, there is no evidence to show exactly how the nebula first formed. There is also no evidence that shows when the sun began to shine or when the planets grew out of the gas and dust.

When the sun first began to shine, it produced great amounts of energy. This is typical of young stars. Its great heat must have driven the remaining gas and dust back into space. Only the material that had condensed into solid bodies remained. The sun's heat also drove off the lighter materials in the planets closest to it. The more distant planets were not so directly exposed to the new star's energy. They kept their thick shells of gas and light materials.

The end result is the solar system we know today. There are a group of inner planets that are alike in their high densities. A group of much larger outer planets are rich in light materials that give them low densities. The entire system of planets still revolves around the sun in the shape of a flat disc.

The nebular theory provides an explanation for the origin of the solar system that agrees with our observations. At the same time, it proposes no unusual events that would probably not be part of the normal development of stars. If the solar system did come into existence

FIG. 3-2. According to the nebular theory, the solar system began as a cloud of gases. Condensation of this cloud took place in stages until individual planets were produced.

as the nebular theory suggests, it is likely that many other stars could also have developed planetary systems like the sun. However, even the closest stars are so far away that planets moving around them cannot be directly observed.

It may be possible to use indirect evidence to discover other solar systems. Large planets might be detected by their gravitational influence on the movements of the parent star. The presence of planets might also be discovered by changes in the amount of light coming from a star. Planets moving around a star would regularly reduce the amount of light reaching us as they pass through our line of sight to the star. A very small number of stars

FIG. 3–3. In these two diagrams, the planetary orbits are shown in their correct size relationships to each other. The top diagram is of the inner planets, the lower one shows the outer planets.

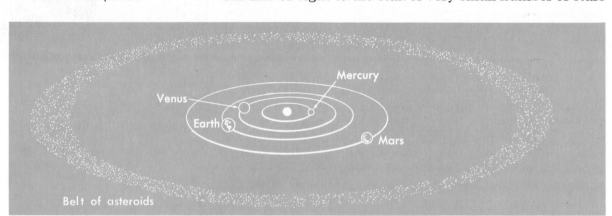

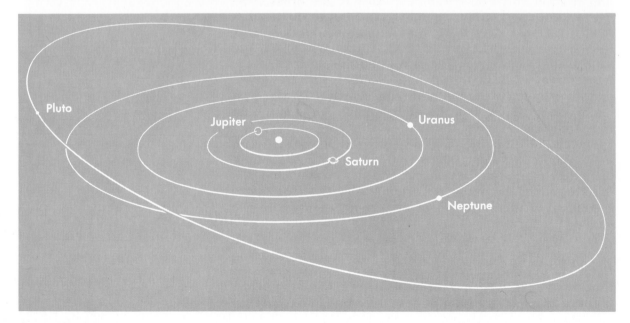

do show evidence of having smaller bodies in orbit around them. These may prove to be systems of planets like the sun's solar system.

If it is common for stars to have planetary systems, there must be billions of planets in the universe. Some might be very similar to Earth.

MEMBER OF THE SOLAR SYSTEM

The inner planets. The four planets closest to the sun—Mercury, Venus, Earth, and Mars—are minor planets in terms of size. This is probably the result of their losing material when it was driven off by the sun's heat in the early development of the solar system. Pluto, the most distant planet from the sun is also small. But Pluto seems to be a special case and will be discussed later.

The smallest of the inner planets is *Mercury*. It is not much larger than the moon. It is also the closest planet to the sun. See Figure 3–3. Mercury is so close to the sun that it is difficult to see even though it is fairly bright. We must look in the direction of the sun to view this planet. We can see it in the night sky only shortly after sunset or shortly before sunrise. Then it is in a position just above the horizon for a brief time. Since the orbit of Mercury lies between the earth and sun, this planet can sometimes be observed by telescope as it crosses the face of the sun. Such a crossing is called a *transit*. See Figure 3–4. Venus is the only other planet known to make transits across the sun.

Through a telescope, a few faint markings on the sur-

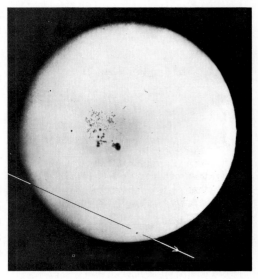

FIG. 3–4. The arrow in the lower part of the photograph indicates the path of Mercury as it crosses in front of the sun. The planet itself appears as a small black dot near the bottom edge of the sun. (Yerkes Observatory)

FIG. 3–5. Photograph of Mercury made during a space probe at a distance of 200,000 km. (NASA)

face of Mercury can be seen. However, during the past few years spacecraft carrying cameras and instruments have flown within a few hundred kilometers of Mercury. For the first time, the surface of Mercury has been seen clearly in the photographs that were sent back to Earth. Instruments have accurately measured conditions on Mercury such as temperature and magnetism. As a result of these close-up observations of Mercury, much of its mystery has been stripped away. It is now as familiar as the moon was before astronauts landed there.

Photographs of the surface of Mercury show a landscape covered with craters and large basins. See Figure 3–5. The surface of this small planet has apparently changed very little during the past four billion years. Unlike the earth, Mercury has no atmosphere or water to wear away its surface features. The craters of Mercury are a record of a time in its early history when it collided with many large objects. Mercury may show evidence of a time when all of the inner planets were bombarded with material left over after the solar system was formed.

Mercury rotates on its axis once every 58.65 days. It takes 88 days to make one trip in its orbit around the sun. This means that the planet turns on its axis exactly three times, while moving around the sun twice. During the long period of daylight, the surface of Mercury is strongly heated by the nearby sun. This heat creates extremely harsh conditions. Temperatures on the sunlit side reach 500°C. On the dark side, temperatures fall to below −200°C.

Mercury shows evidence that its light and heavy materials have been separated. It appears to have a heavy core, probably made of iron. Surrounding this heavy core is a layer of lighter, rocky material. This separation into

FIG. 3–6. Three phases of Venus. (Mt. Wilson and Palomar Observatory)

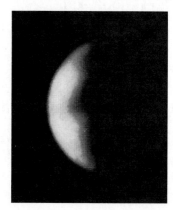

a core and outer layer may have taken place while the planet was molten and in its early stages of development. The presence of a heavy iron core could account for two characteristics of Mercury. It is unusually heavy for its size and it has a magnetic field. Magnetism seems to be a feature found in planets with a metal core.

The next closest planet to the sun, *Venus*, has only recently become less of a mystery. Venus is the earth's closest neighbor planet. However, since Venus is covered with a heavy cloud layer, most of its characteristics are completely concealed. Telescopes do show it to be a bright disc moving through phases like the moon. See Figure 3–6. Both Venus and Mercury pass through these phases because different amounts of their sunlit side become visible as they move between the earth and sun.

Until about twenty years ago, Venus was thought to be the planet that most closely resembled the earth. It was often called the "earth's twin." Venus does have nearly the same size and density as the earth. Although it is closer to the sun, its cloud layer reflects much of the sun's light. As a result it receives about as much solar energy as the earth. However, recent observations have shown that Venus is very different from the earth. Space vehicles have carried instruments close to this cloudy planet. Other spacecraft have landed on its surface. Radar has been used to penetrate the clouds and examine the shape of its surface.

There is still much to learn about Venus. Present knowledge shows that it is not at all like the earth. Its atmosphere is made up of carbon dioxide. The heavy carbon dioxide gas creates a pressure on the surface of Venus. This pressure is 100 times greater than the pressure found on the earth's surface. Pressure such as this is the same as that felt 990 meters below the surface of the ocean. The temperature on the surface of Venus is about 500°c. The planet's heavy atmosphere keeps heat spread evenly over the entire surface. There is little difference in the temperature on the light and dark sides of Venus or between its equator and poles. High temperatures on Venus result from a special property of its carbon dioxide atmosphere. Solar energy can pass through this atmosphere to the planet's surface. Here, most of the heat is absorbed and does not escape back into space. Spacecraft landing on Venus must have a heavy protective cover to shield against the great pressure and heat found there.

Features found on the surface of Venus are concealed

FIG. 3–7. Clouds in the atmosphere of Venus show a swirling pattern when photographed in invisible ultraviolet light. In ordinary light, the clouds appear to be a uniform pale yellow color. (Jet Propulsion Lab)

FIG. 3–8. Radar-signal echoes have produced a map of Venus. The encircled area shows the position of the heavily cratered area on the planet's surface. (Jet Propulsion Lab)

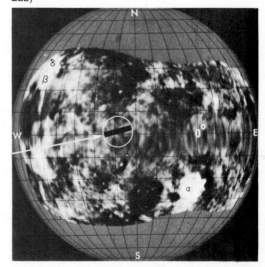

activity

Mercury and the earth move around the sun in their own orbits. See the diagram below. Only part of the earth's orbit is shown.

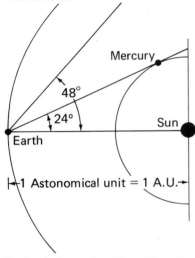

The line to the sun from the earth makes an angle with the line to Mercury from the earth. This angle is never more than 24° and can be found by careful observation and measurement. Be very careful not to look at the sun.

A. Make a drawing like the one above, so that the earth to sun distance is 10 cm. The circle representing the orbit of Mercury must just touch the side of the 24° angle. See the diagram. At one time, it was impossible to measure the distance between the earth and the sun. So the distance was simply called "one astronomical unit" (1 A.U.). In your drawing, 1 A.U. would be represented by 10 cm.

B. Measure the distance between Mercury and the sun in centimeters.

1. What is the length of this line?

2. If one A.U. is ten centimeters, how many A.U. is the Mercury to sun distance?

3. If the earth to sun distance is 150 million kilometers, what is the Mer-

from our direct observation by clouds. Radar, which is not blocked by clouds, does show the presence of craters and volcanoes. See Figure 3–8. The clouds of Venus are almost as much of a mystery as its landscape. They are made up of very small drops, like mist. This layer of mist is about four times thicker than is usual for clouds on the earth. Unlike clouds on Earth, there is little water in the clouds of Venus. The upper layers seem to contain sulfuric acid but the composition of the deep layers is unknown.

Apparently there are swift winds within the atmosphere of Venus. Clouds can be seen to move near the equator at speeds that carry them around the planet in four days. Such rapid movement of clouds is difficult to explain when compared with the slow turning of Venus.

Venus rotates in the opposite direction as the earth. It takes 247 days to complete one turn on its axis. Its orbit around the sun is completed in 224.7 days. This means that the time from when the sun rises in the west and sets in the east is 117 days. A visitor from Earth would probably never notice the difference because of the gloomy atmosphere.

The differences between Venus and Earth seem to result from the carbon dioxide blanket that surrounds Venus. The earth has probably produced as much carbon dioxide as Venus. However, unlike Venus, the carbon dioxide has become locked up in rocks and oceans. As a result there is little carbon dioxide gas in the earth's atmosphere. Venus must have had a different history that produced a dense and cloudy atmosphere making it such a hot, barren planet.

Moving farther from the sun, next among the inner planets is *Earth.* The earth's characteristics are discussed in Unit II.

Mars is the farthest from the sun of the inner planets. And in many ways it is the most interesting. It is about half the size of the earth, with a total mass of one-tenth that of the earth. It also moves in an orbit more than one and half times farther from the sun. As a result it receives much less solar energy. Surface temperatures during the day never rise above 0°C. They drop to around −100°C during the night. Mars is similar to the earth in its period of rotation (24.6 hours) and in the tilt of its axis. This causes Mars to have seasons like those on Earth.

A series of space missions to Mars have made this planet almost as well known as the moon. It seems to be a planet that is still developing. Its surface is covered with a variety of features that give evidence it is far from a dead planet. There are many large volcanoes. One of these, called Olympus Mons, is 16 km high. It covers an area the size of the state of Ohio. See Figure 3–9. There are also many large canyons. One runs for almost 5000 km along the Martian equator. Fields of sand dunes cover some parts of the surface. Sand and dust also partly fill or cover many of the craters that are scattered over the surface of Mars. One of the most surprising discoveries on the Martian surface was finding many features that seem to have been made by running water. Although there is now no liquid water on Mars, there is evidence that great floods were a part of its past history. See Figure 3–10.

At the present time, the main cause of change on the surface of Mars is its atmosphere. The atmosphere is thin and made up almost entirely of carbon dioxide. There is also a small amount of nitrogen and other gases. The pressure at the surface of Mars is about a hundredth of that at the surface of the earth. However, strong winds are caused by daily and seasonal temperature changes. During some parts of the Martian year, winds produce dust storms that can cover the entire planet. Because of the lack of rain, these dust clouds take many weeks to settle out of the atmosphere. Clouds made of ice crystals form in the atmosphere during the cold night and then melt during the day.

Unlike the earth, where water is the main agent chang-

cury to sun distance in kilometers? Venus has the same kind of limit, but the maximum angle is 48° between the earth-sun line and the earth-Venus line.

C. On your drawing of the orbit of Mercury, add the 48° angle and draw the circle to represent the orbit of Venus.

4. What length is the Venus to sun distance in your drawing?

5. How many A.U. is the Venus to sun distance?

6. Using 150 million kilometers as the earth to sun distance, what is the Venus to sun distance?

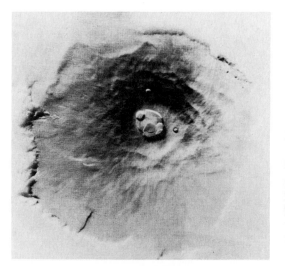

FIG. 3–9. A gaint volcano on the surface of Mars, Olympus Mons, is about 600 km in diameter. The circular area near the center was the opening from which the lava flowed out. Cliffs around the base are 2 km high. (Jet Propulsion Lab)

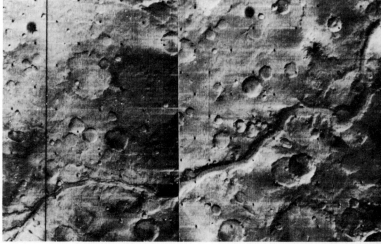

FIG. 3–10. A twisting channel on the surface of Mars that may have been made by flowing water. The channel is more than 1,000 km long. (Jet Proplusion Lab)

FIG. 3–11. An arm from a space vehicle that landed on the surface of Mars exposes some Martian soil. The sponge-like appearance of the rocks is believed to result from bubbles of gas that formed when the rocks were made from hardened lava. (NASA)

ing the surface, the entire surface of Mars shows the effect of wind. Deep layers of material blown by the wind are everywhere. The surface of Mars is now dry, but in the past, water combined with surface rock to produce the red color we see. This has given Mars the name, the "Red Planet." It is the red dust coming from rocks that hangs in the Martian atmosphere, giving it a pink color.

The water that once flooded Mars is now almost entirely ice. The ice caps found at the poles hold a large amount of frozen water. There also seems to be an unknown amount of ice frozen beneath the ground. This is similar to the permanent frost layer found near the earth's polar regions.

FIG. 3–12. Trails left by asteroids during a 2½ hour exposure are marked by arrows. (Lowell Observatory)

It may be that in the past Mars was very much like the earth. It was much warmer and had a heavier atmosphere and abundant liquid water. Mars might also be in the early stages of development that will eventually give it a thicker atmosphere and a warmer climate. Final answers to questions about the past and future of this unique planet must come from further exploration.

The miniature planets. Between the orbit of Mars and the next farthest planet, Jupiter, there is a gap of 560 million km. It seems as if there should be another planet to fill this gap. But instead of a single, large, body, the space is filled with thousands of smaller bodies known as *asteroids.*

Although they are very difficult to find because of their small size, asteroids can usually be identified in photographs. See Figure 3–12. The total number of asteroids that have been discovered is about two thousand. It is estimated that there are at least 50,000 individual asteroid bodies. Only a few of those that have been discovered are very large. Several are more than 160 km across. A few hundred are between 16 and 160 km in diameter. The remainder are less than 16 km in diameter. It is assumed that they are masses of rock, perhaps mixed with metal. Most are quite uneven and probably look like pieces of jagged rock.

The origin of these asteroids is unknown. One theory proposes that the asteroids represent a planet that never grew. According to this theory, something in the early development of the solar system caused the asteroids to remain apart instead of joining to form a large body. Other theories propose that the asteroids are either pieces of two planets that collided, or are fragments of a planet broken apart by gravitational forces. A planet that passed too close to Jupiter could be shattered by Jupiter's gravity.

Most of the asteroids now follow their own orbit somewhere between Mars and Jupiter. However, some asteroids have orbits that carry them outside of this region. These wandering asteroids occasionally collide with the planets or other bodies whose orbits they may cross. Many of the craters seen on the inner planets and their satellites have probably been made by such collisions. The earth shows scars from such past collisions. See Figure 3–13. At the present time, no asteroid is known to have an orbit that could cause it to collide with the earth in the future.

The outer planets. There are five outer planets in the solar system: Jupiter, Saturn, Uranus, Neptune, and Pluto. They are very different from the earth and its neighbors. Some of these distant planets have been visited by unmanned spacecraft. Information gained from these space probes has increased our knowledge of the outer planets.

Jupiter, for example, was known from earth-based observations to be a huge planet. It is more than eleven times the diameter of the earth. Thirteen hundred planets the size of the earth could fit within a sphere the size of Jupiter. However, the mass of Jupiter is only about 300 times that of the earth. This means that Jupiter's density is much less. Even with its low density, Jupiter is so large that its mass makes up nearly three-fourths of the entire mass of the solar system. Jupiter takes nearly twelve

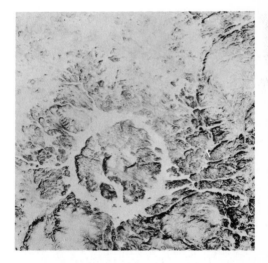

FIG. 3–13. An ancient crater in northeastern Quebec, Canada, photographed from a satellite. This crater was made more than 200 million years ago when the earth collided with a large body. The ring-shaped scar today is filled by a lake whose snow-covered surface outlines the crater. (NASA)

FIG. 3–14 Bands of clouds and the Great Red Spot cover the face of Jupiter as seen from a space probe. The small dark spot is the shadow of one of Jupiter's larger satellites. (NASA)

years to complete one revolution around the sun. However, it rotates so rapidly that it takes less than ten hours to complete one turn on its axis. Rapid rotation makes the region around Jupiter's equator bulge out and it flattens its poles. Jupiter is more than five times farther from the sun than the earth. Thus it receives only 1/27th as much energy on each square meter of surface as the earth. Yet Jupiter gives off radio waves and heat that have energy equal to two and half times the energy it receives from the sun.

Like Venus, Jupiter does not reveal any solid surface. Only the top of a sea of shifting clouds can be seen from Earth. These clouds are spread out in bands of constantly changing patterns and color. See Figure 3–14. Jupiter has 13 satellites. Four of these satellites are about the size of our moon and can be seen with binoculars. The other nine are much smaller and may be asteroids captured by Jupiter's gravity.

Information received from spacecraft exploring Jupiter shows that it is made mostly of hydrogen and helium. It also contains small amounts of methane, ammonia, and water. The clouds around Jupiter form a layer at least 200 km deep. Beneath this layer of clouds, the pressure and temperature become very high. There is no solid surface under the clouds. The clouds gradually change into a heavy, hot liquid. About 2900 km down, the pressure is 100,000 times the pressure at the earth's surface. At this depth, the temperature is about 6600°C. At these high pressures and temperatures, hydrogen and helium are squeezed into a liquid. Most of Jupiter is probably made of this heavy liquid form of hydrogen and helium. Jupiter may have a small solid core, about the size of the earth, containing iron and other heavy materials.

The source of heat that causes Jupiter to pour out more energy than it receives from the sun is not known. It may be heat left from the time when the planet was formed more than four billion years ago. Whatever its source, this heat keeps the interior of Jupiter slowly moving, like a kettle of thick soup. The heat also keeps the outer cloud layer constantly moving and produces violent storms. A particularly large and long-lasting storm may account for Jupiter's *Great Red Spot*. This spot is an oval-shaped region about as wide as the earth and about thirteen times as long. It has been seen on Jupiter for more than three hundred years. Photographs taken from space show it to be a whirling column of colored gases that extends beyond the surrounding clouds. Some scientists believe that it is a giant storm that will one day disappear.

The rest of the outer planets, with the exception of Pluto, seem to resemble Jupiter. *Saturn*, Jupiter's outer neighbor, is the second largest planet. Like Jupiter, it is surrounded by a heavy cloud layer. Its structure seems to resemble that of Jupiter in that it has a thick, dense, liquid layer surrounding a small, solid core. However, Saturn

Compare
Determine the speed of each planet in kilometers per day (use the formula Velocity = 2πr/time). How does the planet's distance from the sun affect its speed?

FIG. 3–15. Saturn shown with its rings in favorable viewing position. Notice the dark, middle band dividing the rings into two parts. (Mt. Wilson and Palomar Observatory.)

does possess an unusual feature—its famous rings. See Figure 3–15. The rings begin about 11,200 km above the clouds and they extend out more than 64,000 km.

At one time, the rings of Saturn were thought to be solid. But it was later found that the inner edges turned faster than the outer parts. Each part of the rings moves around the planet at the same speed as an independent satellite would at the same distance. Therefore, the rings must be composed of a great number of separate bodies all moving in orbits around Saturn. Most of the bodies that make up the rings seem to be no larger than sand grains. The origin of the rings is not known. One possible explanation is the existence of a former satellite that came too close to its parent planet. It was then torn to fragments by the strong gravitational forces.

In addition to its rings, Saturn has ten satellites. The largest of these satellites is between the size of Mercury and Mars. The other nine are not as large.

The planets beyond Saturn. All of the inner planets and two outer planets, Jupiter and Saturn, were known to the ancient Greeks. The three outermost planets—Uranus, Neptune and Pluto—were not discovered until after the invention of the telescope. *Uranus* and *Neptune* are so far away that little is known about them. Although not as large as Jupiter or Saturn, both Uranus and Neptune are giants compared to Earth. Both planets seem to have a rocky, solid core surrounded by icy and liquid material. Uranus has rings like Saturn. Its axis is tilted so much that its rotation gives it the appearance of rolling rather than spinning like a top, as with the other planets.

Beyond Neptune, on the fringes of the solar system, is the small planet *Pluto.* Very little is known about Pluto. It is so far away that it is barely visible, even with the most powerful telescopes. Even the size and mass of Pluto are in doubt. It may be about the same size as the earth. It certainly receives little energy from the sun and therefore must be very cold. Even Pluto's right to be considered as one of the original planets is open to question. Unlike the orbits of the other planets, which all tend to lie in a flat plane, Pluto's orbit has a strange tilt. See Figure 3–3 on page 50. This, along with some other unusual features of Pluto's orbit, has led to the theory that Pluto may once have been a satellite of Neptune. Somehow Pluto might have escaped Neptune's influence and fallen into its own independent orbit around the sun.

Meteors and comets. During any 24 hour period, the earth collides with thousands of millions of small bodies. Most of these objects weigh less than one gram (about 1/30 ounce) but they strike the earth's atmosphere with speeds as high as 50 miles per second. The very high speeds at which these solid objects travel create such intense heat from friction with the air, that they are vaporized. A brilliant light is given off by these vapor trails. This vaporizing body is called a *meteor*. Almost all meteors are completely vaporized before they penetrate deeply into the earth's atmosphere. The visible trail of light left as they vaporize is generally about 100 km (60 mi) above the earth's surface.

Meteors seem to come from two different sources. Most are probably small bits of material like those in the belt of asteroids between Mars and Jupiter. Various gravitational influences may cause them to follow orbits which eventually bring them into contact with the earth. Other meteors appear to be material left behind as comets have passed. Each time the earth crosses an orbit known to have been followed by a comet, swarms of meteors are usually seen. Several of these events occur at regular times each year and are known as "meteor showers." See Figure 3–16. One of the most spectacular meteor showers is the *Perseids* (*per*-see-ids). It is seen each year in mid-August.

A few meteors colliding with the earth each day of the year are large enough to have some fragments survive their plunge through the atmosphere. Solid fragments of meteors which reach the ground are known as *meteorites*. Most are never found because they either fall into the sea or some remote place. Of the thousands which have been examined, most are like stone. Some, however, are composed of iron and nickel. Rarely, a very large meteorite weighing 10,000 tons or more collides with the earth. This type of meteorite produces great craters at the point of impact, such as the one shown in Figure 3–17.

Unlike meteors, *comets* are relatively rare within the region of the solar system. It is believed that a thin shell of comets surrounds the entire solar system. There may be billions of these small bodies made up of such solidified gases as ammonia, carbon dioxide, and ammonia mixed with ordinary ice. Normally they continue moving in great orbits so far from the sun that they give no light. Occasionally one of these objects is disturbed in its orbit. Then it may move toward the sun. When it comes within the orbit

FIG. 3–16. Photograph showing trails of many meteors during the shower of October 1946. The meteor trails are shown radiating away from the earth toward the upper right. Other streaks are star trails. (Griffith Observatory)

FIG. 3–17. Meteor Crater in northern Arizona. (American Meteorite Museum)

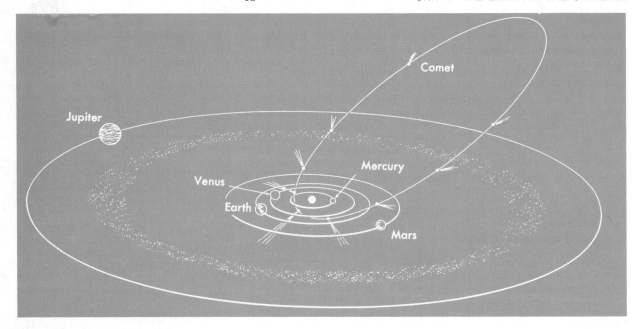

FIG. 3–18. Diagram showing the behavior of a comet as it nears the sun. Notice that the tail of the comet always points away from the sun.

FIG. 3–19. Diagram showing possible orbits for a comet as it moves toward the sun. It is generally agreed that most comet orbits are of either type A or B.

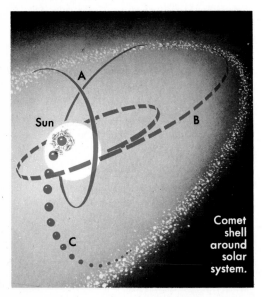

Comet shell around solar system.

of Jupiter, the sun's energy forces the body of the comet to expand and its gases to begin to evaporate. This gas along with solid particles trails out behind the head in a tail. The tail may be millions of kilometers long. The comet's tail is pushed out by the pressure of radiations and particles coming from the sun. Thus, the tail is always directed away from the sun. Energy from the sun also causes the material in the comet's head and tail to give off light. See Figure 3–18.

In its approach to the sun, a comet may follow any one of three paths shown in Figure 3–19. **(A)** It may make one loop around the sun and return to the edges of the solar system. **(B)** It can fall into a new, very elliptical orbit around the sun. **(C)** Or it may come so close to the sun that it disintegrates. Depending upon its particular orbit, a comet may appear only once or it may be seen at regular intervals. Astronomers have observed about 45 known comets that return at regular intervals ranging from 5 to 10 years. A well known comet is Halleys' Comet. It is known to return every 76 years, and is expected to be seen again in 1986. See Figure 3–20.

THE EARTH AND ITS MOTIONS

The effects of the earth's revolution. Even as you read this sentence, you are traveling with the earth in an elliptical orbit around the sun at a speed of 106,000 km/hr (66,000 mi/hr). Each time the earth completes one orbit or revolution around the sun, you have traveled with the earth a distance of 929,000,000 km (590,000,000 mi): The length of time it takes the earth to complete one revolution is called a *year*. The most ancient way of measuring this period of time is based on the passage of the seasons. The seasons occur because the earth revolves with its polar axis at an angle to the plane of its orbit around the sun. The earth's axis is tilted 23½° from a perpendicular to this plane. This tilt is called the inclination of the earth's axis. Other planets also show inclination. See Figure 3–21.

As the earth moves around the sun, the North Pole is tilted first toward the sun then away from it. This is because the tilt of the earth's axis remains in the same direction in space. Study of Figure 3–22 may help you to understand this point.

In June of each year the North Pole is tilted the full 23½° toward the sun. On June 21 or 22, the sun's most vertical ray shines directly on the Tropic of Cancer. See Figure 3–23. The Tropic of Cancer is an imaginary line which marks the most northern point where the sun can ever be seen directly overhead. At any location north of the Tropic of Cancer, which includes the United States, the mid-day sun is always seen in the southern part of the sky.

On June 21 or 22, when the sun's vertical ray is directly over the Tropic of Cancer, the sun stops its apparent northward movement in the sky. This is when the Northern Hemisphere has its longest day of the year. It is called

FIG. 3–20. Halley's Comet. The tail of a comet may extend more than 100 million miles into space. Note that the stars in the background are visible through parts of the comet's tail. (Yerkes Observatory)

TABLE 3–1 CHARACTERISTICS OF THE SUN AND ITS FAMILY OF PLANETS

Average Distance From Sun

	Millions of Km	Millions of Miles	Average Surface Temp. (°C)	Diameter (Km)	(Miles)	Mass (Earth=1)	Revolution	Period of Rotation	Number of Satellites
Sun	—	—	5,600	1,380,000	856,000	333,400	—	25 days	Many
Mercury	57.9	35.9	350	4,880	3,030	.055	88 days	58.6 days	0
Venus	108	67.0	480	12,100	7,520	.815	224.7 days	247 days	0
Earth	150	93.0	22	12,756	7,908.7	1.00	365.26 days	23h 56m 94s	1
Mars	228	141	−23	6,787	4,208	.108	687 days	24h 37m 23s	2
Jupiter	778	482	−150	142,000	88,000	318	11.86 days	9h 50m 30s	13
Saturn	1430	887	−180	120,000	74,400	95.2	29.46 yrs.	10h 14m	10
Uranus	2870	1780	−210	51,800	32,100	14.6	84.01 yrs.	11 hrs.	5
Neptune	4500	2790	−220	49,500	30,700	17.2	164.8 yrs.	16 hrs.	2
Pluto	5900	3660	−230(?)	6,000(?)	3,700(?)	0.1(?)	247.7 yrs.	6 days 9 hrs.	0

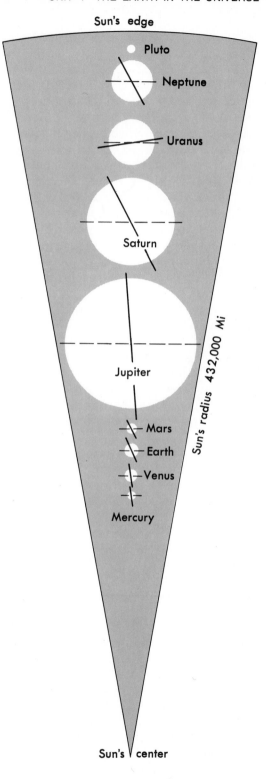

FIG. 3–21. The planets are shown here with their correct relative sizes and the inclination of their axes with relation to the sun.

FIG. 3–22. As the earth moves around the sun it maintains the same orientation in space. This means that the earth's axis is always inclined in the same direction in space throughout its orbit.

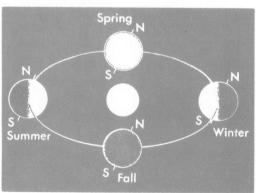

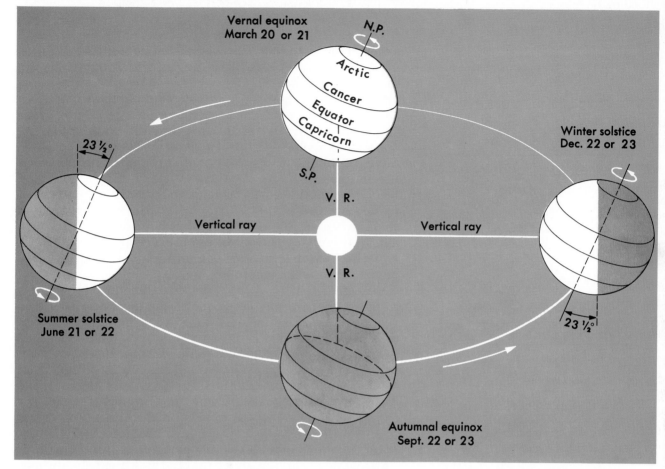

the *summer solstice*. "Solstice" means "sun stands still". It refers to the position of the sun during its apparent maximum movement northward and southward in the course of a year. Regions of the earth's surface within the Arctic Circle have 24 hours of sunlight at the time of the summer solstice as shown in Figure 3-23.

By September the earth is one-fourth farther around the orbit than it was in June. Then the North Pole is no longer tilted toward the sun. See Figure 3-23. By then the Northern Hemisphere is no longer receiving most of the sun's most vertical rays. This area begins to cool off. The sun's vertical rays bring the greatest amount of heat to earth. Since they are concentrated on a smaller surface area they pass through a minimum amount of atmosphere. Fall begins at the time of the *autumnal equinox*, either September 22 or 23. On that date, the sun's vertical ray has moved directly overhead at mid-day for an observer at the equator. See Figure 3-23(A,B,C).

FIG. 3–23. This illustration shows how the earth's revolution causes the sun's vertical ray to appear to move from the Tropic of Cancer to the Tropic of Capricorn. Through how many degrees does the vertical ray appear to travel from June to December?

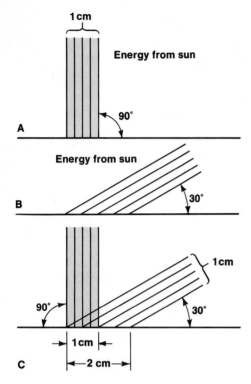

FIG. 3–23 (A,B,C). A given beam of light striking the earth vertically, covers a smaller area than when it strikes at any angle other than 90 degrees.

FIG. 3–24. The circles made by the poles during the precession of the earth's axis are not smooth.

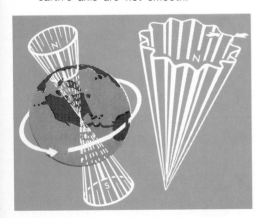

By December the earth has progressed halfway around its orbit. It is then directly opposite its location in June. Now the South Pole is tilted the full 23½° toward the sun and the Southern Hemisphere receives the sun's most vertical rays. When the *winter solstice* occurs, on either December 21 or 22, the vertical ray is directly over the Tropic of Capricorn. During this time the Southern Hemisphere has its longest day. Regions in the Antarctic Circle have 24 hours of daylight and those within the Arctic Circle have 24 hours of darkness.

By March the earth has completed three-quarters of its trip around the sun. The *vernal equinox* occurs near the start of spring, on either March 21 or 22. The sun is again directly overhead to an observer at the equator. In June the summer solstice is again reached and the cycle of the seasons is complete.

Keeping all this in mind, we can accurately define a tropical year as the length of time between two consecutive vernal equinoxes — 365 days, 5 hours, 48 minutes, and 46 seconds.

Actually a tropical year is somewhat short of the time it takes the earth to make one revolution in its orbit. This is a result of the slow wobbling of the earth as it turns on its axis. This motion, called *precession*, is caused mainly by the moon's gravitational effects on the earth. Precession of the earth is similar to the spinning of a top. When a top spins, its upper end moves in a small circle, turning much more slowly than the entire top. In the same way, the axis of the earth moves in a slow circle, taking 26,000 years to complete one turn. See Figure 3–24.

The equinoxes are dependent upon the tilt of the earth's axis. Therefore, the slow wobbling of the axis due to precession affects the date on which the equinoxes occur. This, in turn, changes the length of a tropical year. The effect of precession is to make the tropical year 20 minutes shorter than it would otherwise be.

Because it is difficult to measure a tropical year with great accuracy, astronomers sometimes use the *sidereal* (sy-*deer*-ee-al) *year*. It too is based on one revolution of the earth. However, a sidereal year is measured by reference points among the stars rather than by the position of the sun. Stars are so far away that precession makes little difference in their observed positions.

The effects of the earth's rotation. For people on earth, the most important effect of the earth's rotation on its axis is

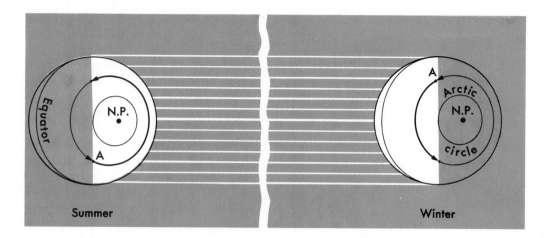

Summer   Winter

the rising and setting of the sun. The period of rotation between any two rising or setting suns is called *"one day."* In one day or about every 24 hours each point on the earth's surface moves through 360°. But at any given time, only one-half of its surface is in daylight, while the other half is in darkness. Since the earth rotates from west to east, the sun appears to rise in the east and travel across the sky, setting in the west. In summer the sun seems to take longer to cross the sky. Day is then longer than night. In winter, the opposite is true. The reason for this can be seen in Figure 3–25.

Two times during the year, day and night are of equal length everywhere on earth. Both times this takes place at an *equinox,* a word meaning "equal night." An equinox occurs when the boundary line that separates the daylight from night-time halves of the earth passes directly through both north and south poles. This is possible only when the earth's axis is not tilted toward or away from the sun. Twice each year this happens, on the two equinoxes already described. See Figure 3–26.

FIG. 3–25. Changes in the length of daylight are shown in this top view of the northern Hemisphere. Locations along line A have their longest daylight period in the summer when the North Pole is tilted toward the sun.

FIG. 3–26. Day and night are equal in length throughout the earth only on the two equinoxes. The view shown here is what you would see from a distant point to the right of the earth on the first day of spring or autumn.

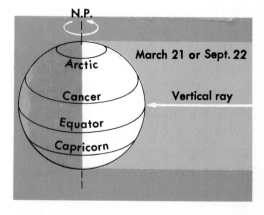

Time systems. The daily apparent movement of the sun across the sky is the basis for our system of timekeeping. The simplest way to use the sun's "motion" to mark the passage of time is with a sundial. Time measured with a sundial is called *apparent solar time* because it is determined entirely by the shadows cast as a result of the sun's apparent position in the sky. It would seem that not much could go wrong with a sundial as long as the sun shines. However, a sundial will sometimes run either too fast or too slow. This happens because the speed at which the sun crosses the sky is not the same throughout the year. During times of the year when the sun is moving fastest, a sundial will register shorter hours as the shadow moves rapidly across its face. At other times the sun moves more slowly and a sundial shows longer hours.

It might seem at first glance that problems with sundials are connected with the shorter and longer days of winter and summer. But this is not true. Since the earth rotates at an almost constant speed, the sun should always appear to cross the sky at the same rate. Differences in the time of rising and setting should only affect the length of day and night. The correct explanation for the sun's apparent changes in speed lies partly in the changing speed of the earth in its orbit around the sun. When the earth is moving fastest in its orbit, it moves farthest ahead during one rotation on its axis. Therefore, it must rotate farther to bring the sun back to a given position in the sky. This makes the sun appear to move slowly across the sky. When the earth is moving more slowly in its orbit, the sun appears to cross the sky more rapidly.

The angle at which we view the sun also makes it appear to change its speed. In summer or winter, near the time of a solstice, the sun is moving almost directly from east to west across the sky. Thus it will appear to cover a given distance in a short time. On the other hand, near an equinox, in spring or fall, the sun appears to move farthest to the north or south at the same time as it crosses the sky. To cover the extra distance, the sun needs more time to complete a given part of its daily trip. The effect of the speed of the earth in its orbit and the angle of movement of the sun combine to make sundials measure off days which differ in length from one part of the year to another.

To avoid irregularities in the length of day, ordinary clocks are set to run on a 24-hour day, which is the average of the days measured by sundials throughout a year. This kind of time is called *mean solar time* or average solar time. Since clocks measure the same amount of time from one noon to the next all year, while sundials do not, clocks usually disagree slightly with sundials. This difference between clock time (mean solar time) and sun time (apparent solar time) is never greater than 16 minutes.

Standard time. The use of mean solar time for clocks means that all days are the same length. But it does not determine the start of a day. Suppose each city and town

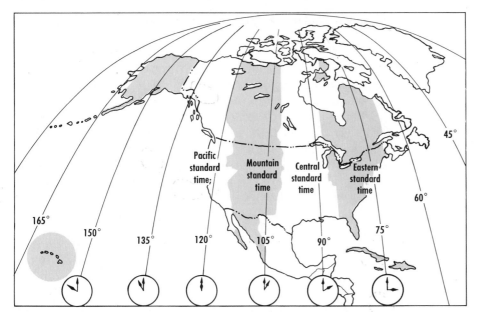

Pacific standard time;

Mountain standard time

Central standard time

Eastern standard time

45°

60°

75°

165°

150°

135°

120°

105°

90°

FIG. 3–27. Standard time zones for North America and Hawaii are shown on this map.

sets its clocks to read 12 noon when the sun is nearly over-head. Then 12 o'clock noon would be a different time for each location as the sun moved west and appeared over-head. This would mean that each locality would have its own time. Actually this situation did exist before 1883. Then the railroads in the United States decided to end the confusion created by differing timetables in various places. They agreed to set up four *standard time zones* across the country. Within each zone, the time was to be that of the most important city within that zone. This was the beginning of the modern system of standard time.

Today the earth's surface is marked off into 24 standard time zones. Standard time in each zone is figured from the mean solar time for the center of that zone. It is one hour earlier than the zone to the east and one hour later than the zone to the west.

In the United States, other than Alaska and Hawaii, there are four standard time zones, as shown in Figure 3–27. Most parts of Alaska and all of Hawaii are in a time zone which is two hours earlier than Pacific Standard. The boundaries between time zones are generally arranged so that all of each state has the same standard time.

During the summer, when the daylight period is longer, many states and cities adopt *Daylight Saving Time*. Clocks are set one hour ahead of standard time in April, then back again one hour in the fall. This allows people to enjoy an extra hour of daylight during the summer evenings. A simple rule for knowing when to add or sub-tract an hour is: *"Spring" ahead* and *"Fall" behind.*

The International Date Line. Suppose you were to take a non-stop trip around the world on a fast aircraft. You take off from San Francisco at sunset and fly west over the Pacific toward the setting sun. Suppose also that the speed of the aircraft is fast enough to keep up with the speed of the earth's rotation. Under these conditions, the sun never appears to set but remains suspended on the hori-zon as you fly toward it. Since you are flying west, you keep turning your watch back one hour as you cross each time zone boundary. Time would then appear to stand still for you. When you finally arrived back in San Francisco, the time and the day would still be the same as they were when you left. If you left on Friday, it would still be Friday for you when you returned, since your watch never advanced to midnight. But for everybody else twenty-four hours would have passed and it would be Saturday.

A little thought will probably show you that the same time problem would exist no matter how long it took to circle the earth. The loss or gain of a day would make up for the hour by hour loss or gain of twenty-four hours caused by crossing all twenty-four time zone boundaries. To prevent such confusion, an imaginary line has been established called the International Date Line. It runs through the Pacific Ocean bending around various land areas and islands to avoid cutting across inhabited re-gions. See Figure 3–28.

To anyone passing over the *International Date Line,* the calendar immediately changes one day. If the line is crossed going from east to west the calendar is set ahead one day. For example, 6 p.m. Thursday becomes 6 p.m. Friday. Going from west to east, a day is repeated when the line is crossed. Friday at 6 p.m. becomes Thursday, 6 p.m. This takes care of setting the calendar ahead one day when the earth is circled going west, and setting it back one day when it is circled going east.

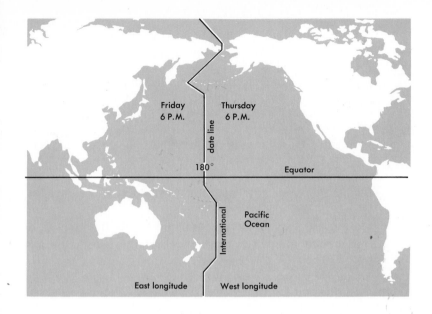

FIG. 3-28. The international date line coincides with or runs close to the 180° meridian, always avoiding land.

VOCABULARY REVIEW

Match the word or words in the column on the right with the correct phrase in the column on the left. *Do not write in this book.*

d **1.** The sun and the great number of bodies held by its gravitational field.

h **2.** Planet most distant from the sun.

k **3.** The planet closest to the sun.

f **4.** A large volcano on Mars.

b **5.** Large number of bodies between Mars and Jupiter.

p **6.** Planet with 13 satellites.

n **7.** Planets with concentric rings.

j **8.** Body whose tail points away from the sun.

a **9.** Where the sun is directly overhead on June 21 or 22.

q **10.** The time of year when the sun appears farthest north.

c **11.** The time of year when the sun is directly overhead at the equator.

i **12.** The time between two consecutive vernal equinoxes.

g **13.** Legal time within a certain region.

l **14.** Nearly the same size as the earth.

o **15.** Meteor fragments that reach the ground.

a. Tropic of Cancer
b. asteroids
c. equinox
d. solar system
e. Mercury
f. Olympus Mons
g. standard time
h. Pluto
i. tropical year
j. comet
k. Mars
l. Venus
m. Uranus and Neptune
n. Saturn and Uranus
o. meteorites
p. Jupiter
q. summer solstice
r. apparent solar time

QUESTIONS

Group A

Select the best term to complete the following statements. *Do not write in this book.*

1. The nine planets have one thing in common (a) they all move under the influence of the sun's gravity (b) they are the same size (c) the inner planets are larger than the outer planets (d) they are about the same temperature.

2. The most commonly accepted theory for the origin of the solar system is called the (a) colliding star theory (b) companion star theory (c) nebular theory (d) anti-matter theory.

3. The nebular theory begins with (a) a large star (b) a slowly rotating cloud of gas and dust (c) a small star (d) planets already formed.

4. The smallest of the inner planets is (a) Mercury (b) Venus (c) Earth (d) Mars.

5. The number of days required for Mercury to make one revolution around the sun is (a) 59 (b) 70 (c) 88 (d) 370.

6. When we use a telescope, most characteristics of Venus are concealed from us because it is (a) too near the sun (b) too far away from the earth (c) too far away from the sun (d) covered by heavy clouds.

7. The planets that have orbits smaller than the earth's and therefore make transits of the sun are (a) Mercury and Venus (b) Mercury and Mars (c) Mars and Venus (d) Jupiter and Mars.

8. The temperature on the surface of Venus is about (a) 200°C (b) 500°C (c) 1000°C (d) 2000°C.

9. Venus (a) has three moons (b) rotates in the same direction as the earth (c) rotates in the opposite direction from the earth (d) doesn't rotate.

10. The diameter of Mars is about half that of the earth while its mass is about (a) one half (b) one fourth (c) one fifth (d) one tenth.

11. The period of rotation of Mars on its axis (a) is about the same as that of the earth (b) is much less than that of the earth (c) is much greater than that of the earth (d) cannot be determined.

12. Asteroids vary in size from fewer than 16 km in diameter to greater than (a) 160 km (b) 1600 km (c) 16,000 km (d) 160,000 km.

13. The asteroid belt is between the orbits of (a) Earth and Mars (b) Mars and Jupiter (c) Jupiter and Saturn (d) Saturn and Uranus.

14. The number of outer planets (planets farther from the sun than Mars) is (a) 2 (b) 3 (c) 4 (d) 5.

15. Most of Jupiter is probably made of (a) ammonia and methane (b) ammonia and helium (c) methane and hydrogen (d) hydrogen and helium.

16. Jupiter's Red Spot (a) has recently been discovered (b) is a whirling column of colored gases (c) is a deep cavern on the surface (d) will always be present.

17. The rings of Saturn are made of particles that are (a) no bigger than sand grains (b) no bigger than dust particles (c) bigger than golf balls (d) bigger than basketballs.

18. In addition to its rings, how many satellites does Saturn have? (a) 5 (b) 7 (c) 10 (d) 11.

19. In size, the largest satellite of Saturn can be compared to (a) a basketball (b) a house (c) a city (d) a small planet.

20. One feature of Pluto that raises some doubt as to its origin is (a) the tilt of its orbit (b) its size (c) its atmosphere (d) its temperature.

21. The meteor shower that occurs in mid-August each year is known as the (a) aphelion (b) Perihelion (c) Orionids (d) Perseids.

22. The light from a comet is (a) from the head and tail of the comet (b) from the tail of the comet (c) from the head of the comet (d) never visible to the eye.

23. Seasons occur on earth because its axis is tilted. This inclination of the axis is (a) 3 degrees (b) 12.5 degrees (c) 23.5 degrees (d) 31.5 degrees.

24. At a point just north of the Arctic Circle, there are twenty-four hours of darkness at the time the sun is directly overhead at the (a) South Pole (b) Tropic of Capricorn (c) Equator (d) Tropic of Cancer.

25. The time it takes for the precession of the earth's axis to complete one turn is (a) 1 year (b) 100 years (c) 1000 years (d) 26,000 years.

26. A year measured by reference points among the stars is called (a) a calendar year (b) an arctic year (c) a tropical year (d) a sidereal year.

27. Apparent solar time is not used for clocks because it (a) never changes (b) changes constantly (c) changes occasionally (d) depends on the moon phase.

28. The difference between mean solar time and apparent solar time is never greater than (a) 4 minutes (b) 8 minutes (c) 12 minutes (d) 16 minutes.

29. The number of standard time zones in the entire world is (a) 12 (b) 24 (c) 36 (d) 360.

30. One reason for establishment of the International Date Line was to keep one location from having (a) two dates at one time (b) two times of day (c) too long a day (d) too short a day.

Group B

1. What are the steps in the development of the solar system as explained by the nebular theory?

2. Why is the planet Mercury visible only shortly before sunrise and shortly after sunset?

3. State the evidence that shows that the surface of Mars is changing.

4. List the planets in order of size starting with Jupiter. Also give the number of natural satellites of each.

5. Explain why some scientists think that an additional planet may have failed to develop when the solar system formed. Where would it be located if their theory were true?

6. Why do the rings of Saturn occasionally fail to show up in photographs?

7. What is a transit of the sun? Which planets viewed from earth exhibit this phenomena?

8. Describe the location of the sun, Earth, and Venus when Venus is in its new, crescent, gibbous and full phases as seen from Earth. Diagrams might help.

9. Why does apparent solar time change continuously?

10. At what time of year are apparent solar time and mean solar time nearly the same? Why?

11. Explain standard time zones and why we have them.

12. Explain the difference between a tropical year and a sidereal year.

13. Describe the apparent north-south motion of the sun during a single year. Give dates that help locate its position.

14. In your own words describe each of the following kinds of time. (a) apparent solar time (b) mean solar time (c) standard time.

15. Discuss the effect on seasons of the year if the earth's axis were not tilted to the plane of its orbit.

16. In what ways is Mars like earth? In what ways is it different?

17. Considering the rotation of the earth, the motion of the earth around the sun, and the speed of meteors, explain why meteor showers are more likely to be observed just before dawn than at any other time.

18. The earth travels around the sun at about 106,000 kilometers per hour. How many meters does it travel each second?

19. Calculate the circumference of the earth's orbit around the sun in kilometers (see problem 18).

20. The volume of a sphere is given by the formula $V = \frac{4}{3} \pi r^3$ where r is the radius of the sphere. From this, how would you account for the fact that Jupiter is 11 times as large as earth but more than 300 times its mass?

4
the moon

objectives

- Describe the stages in the theory of the origin of the moon.
- Describe the surface of the moon.
- Explain how the moon orbits the earth and the sun.
- Explain why the time between moon rises is greater than 24 hours.
- Draw a diagram to show the sun, earth, and moon positions for the various phases of the moon.
- Explain the reasons for solar and lunar eclipses.
- Describe moon and sun influence in producing tides.
- Compare the Julian and Gregorian calendars.

If you were on Mars or Venus studying the earth with a telescope, your attention might be drawn as much to the moon as to its parent planet Earth. You would see what appears to be a large and a small planet. The smaller body would move first ahead, then behind the larger as both travelled together around the sun. From a distance, you might logically decide that the earth and moon are a double planet system. Actually, they are a planet and satellite. Observers have been unable to find another planet in the solar system that has a companion so near its own size. Some of the satellites of Jupiter, Saturn, and Neptune are larger in diameter than the moon. But they are not so large compared to the size of the planets around which they revolve. The moon's diameter is 3476 km compared to the earth's diameter of 12,756 km.

PAST AND PRESENT ON THE MOON

History of the moon. Few things could add more to the understanding of the earth's past than firsthand exploration of another planet. Manned missions to the moon have provided such an opportunity for scientists. Exploration of the moon has changed it from merely a close neighbor to a familiar, small planet. Its rocky materials preserve a record of the moon's early history. The moon, unlike the earth, has undergone little change during the past four billion years. Although there is still much to be learned, the moon's history has become almost as well understood as that of the earth.

A great deal of scientific research has been done using the data and rocks gathered by astronauts on the moon.

75

FIG. 4–1. Almost all of our detailed knowledge of the moon has come from exploration of the moon by astronauts. (NASA)

FIG. 4–2. This lunar rock was brought back to earth for further study. (NASA)

As a result of this research scientists believe that the moon has passed through six separate stages in its development. The *first stage,* the period about which the least is known, is its origin. Study of lunar rocks seems to show that the earth and moon were formed in the same region of the solar system. This conclusion arrives from the presence of certain forms of oxygen found in lunar rocks. These forms of oxygen are the same as those found in earth rocks. Meteorites, formed elsewhere in the solar system, do not share this characteristic of lunar and earth rocks. It is believed that the moon was formed from material that surrounded the earth shortly after it was formed. According to this view, the moon was formed by the joining of small bodies orbiting the young earth.

However, other theories that account for the moon's existence cannot be ruled out. It is possible, for example, that the moon originally formed some distance from the earth and was later captured by the earth's gravity. If this happened, it is difficult to explain the moon's present orbit. Approaching the earth from a great distance should cause the moon to follow an orbit much farther from the earth than it does. Another theory proposes that the moon broke away from the earth. This theory also has flaws. If this had happened, the earth would have been spinning very fast. This fast spinning would cause both the earth and moon to move much differently than they do now.

The *second stage* in the moon's development began when its surface was entirely covered by an ocean of white-hot molten rock. The heat that caused the melting could have come from the energy carried by the bodies that collided and formed the moon. If the moon grew rapidly, the colliding bodies would create heat more rapidly than could be lost by cooling. The accumulated heat might have melted the entire moon. However, it seems certain that a molten layer at least several hundred kilometers deep was created.

Melting of the lunar material created the conditions found in the *third stage* of the moon's development. During this stage, the light and heavy materials in the molten rock were separated. Light materials floated up, while the heavier substances sank. The result was the formation of a thick, lunar crust made of lighter rocks. Below this was found a thick layer of denser rock material. The heaviest materials, like iron, probably sank to the center and formed a small, heavy core.

The *fourth stage* began when the molten surface of the moon cooled and formed a solid crust. However, in many places molten rock from below broke through the crust to form volcanoes. This volcanic activity created mountainous regions on the moon's surface. These regions are called the *highlands*. When you look at the moon, these highlands appear as the lighter parts on the surface. This is because the highlands reflect more light than the smoother parts of the moon's surface.

Following the stage of volcanic activity, the moon entered a *fifth stage*. This stage was marked by a period of intense bombardment by meteorites and larger bodies. The early solar system must have been filled with many loose fragments that never became part of the larger planets. These fragments probably bombarded the young planets until the solar system was nearly swept clean. In the case of the moon, there is no record of what collisions might have occurred in its earliest stages. However, there is ample evidence that about four billion years ago the moon, along with the other bodies in the solar system, went through a period of intense bombardment. During this time, gigantic impacts tore huge circular basins in the lunar crust. Materials thrown out by these violent collisions covered much of the original surface of the moon. See Figure 4–3. Smaller craters were also created as debris rained down on the moon. Most of the original crust of the moon was broken up by these impacts.

activity

Determine the diameter of the moon by using the apparatus shown below.

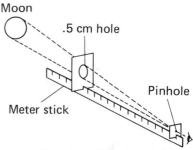

1. Place the card with a .5 cm hole at a distance from the pinhole card to just enclose the full moon. Record this distance and convert it to km.

2. The moon's diameter has the same relationship to its distance from the observer (382,000 km) as the diameter of the .5 cm opening has to its distance from the observer. Mathematically this is:

$$\frac{\text{moon's dia.}}{384,000 \text{ km}} = \frac{.5 \text{ cm}}{\text{dis. measured on meter stick (cm)}}$$

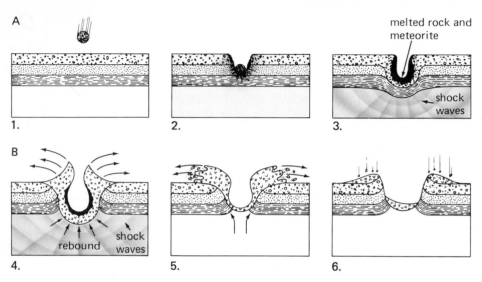

FIG. 4–3. These diagrams show how an impact crater is formed. A.) The energy of the rapidly moving meteorite is changed into heat and shock waves. B.) The rebound of the shock waves causes debris to be thrown out as solid and molten rock. Smaller "secondary" craters may form when material is splashed from the crater.

FIG. 4–4. A telescopic photo of a lunar mare or "sea." This is Mare Serenitatio or the "Sea" of Serenity, located in the Western Hemisphere of the moon. (Mt. Wilson and Palomar Observatory)

The *sixth stage* of the moon's history began when the bombardment by objects from space ended. During this stage, great floods of lava flowed onto the moon's surface. This lava came from the molten layer still lying beneath the solid crust. Much of the lava flowed into and filled the basins made by the earlier collisions. As the lava cooled and became solid, it formed the dark-colored lava plains visible on the moon. See Figure 4–4. These relatively smooth regions of the moon's surface are called *maria*, the Latin word for "seas." The maria got their name because long ago it was thought that they resembled oceans on the moon.

Volcanic activity continued for a billion years. Then the moon became quiet. For the past three billion years the moon has changed very little. This small planet, which once saw violent activity, is now dead.

The visible moon. The surface of the moon can be seen very clearly because there is no atmospheric layer masking its features. Early volcanic activity probably produced the gases that could have formed an atmosphere and oceans like those on earth. However, the moon's lower gravity allowed these gases to escape into space. The moon's mass is $1/81$ that of the earth. Its diameter is about one-quarter the size of the earth. These two factors combine to give the moon a surface gravity of only $1/6$ that of the earth. This means that a person weighing 82 kg would weigh only 14 kg on the moon. Lack of an atmosphere along with long periods of daylight and darkness (each lasts about two weeks) cause the moon's surface to become very hot then very cold. The lighted side of the moon has temperatures of over 100°C. The dark side reaches a low of −173°C.

The most noticeable features of the lunar surface are craters. They cover the moon so completely that each crater usually has smaller craters inside. The larger craters that can be seen from the earth are about 250 km in diameter. These larger craters often overlap one another. Some of the large craters, like the ones named Tycho and Copernicus, have bright streaks running through them. These streaks are called *rays*. They reach out for hundreds of kilometers. See Figure 4–5. The rays were made when material was splashed out by the impact that first formed the crater.

Most of the moon's surface is covered with crater-marked highlands. Almost all of the side of the moon never seen from the earth is made of these highlands. The smoother, dark maria are mostly found on the side of the moon that faces the earth. All the lunar surface is almost completely covered with a layer of loose, gray material. This material is debris that was thrown out when the craters were made. Billions of years of bombardment by tiny meteorites has turned some of this loose covering into very fine gray dust. See Figure 4–6. In the areas of the moon that have been explored, the depth of the layer of gray dust has been found to be about 3 m thick.

MOTIONS OF THE MOON

The moon's orbit. From our viewing point, the moon appears to revolve in an east to west orbit around the earth. At the same time, the earth travels in its own orbit around the sun. But the gravitational force acting between the earth and moon affects both bodies. To understand this, imagine that the earth and moon are connected by a weightless rod similar to the one shown in Figure 4–7. The balance point for this uneven, dumbbell-shaped system is closer to the earth's center than it is to the moon. The actual location of the balance point is about 1700 km below the earth's surface on the side facing the moon. The effects from the gravity of the earth and the moon force both bodies to circle around this point inside the earth. At the same time the moon circles the earth, the earth moves in a small orbit around the balance point. Only the center of the earth-moon system follows a smooth orbit around the sun. See Figure 4–8.

FIG. 4–5. A view of the full moon showing the dark maria and the lighter highlands. The crater Tycho with its bright rays is seen in the lower part. The numbered markings show the landing sites of the various Apollo missions. (Mt. Wilson and Palomar Observatory)

Interpret
What are rills on the moon? Galileo stated that the rills on the moon were the remains of ancient stream channels. Would present day information support this idea?

FIG. 4–6. This footprint of an astronaut on the moon shows that the lunar soil is composed of very fine particles. (NASA)

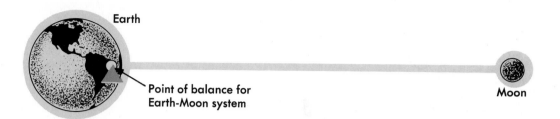

Earth

Point of balance for
Earth-Moon system

Moon

FIG. 4–7. The center of gravity around which both the earth and moon rotate is the balance point for an imaginary bar connecting the earth and moon.

FIG. 4–8. Only the center of the earth-moon system actually follows a smooth orbit around the sun. Both the earth and moon move in the kind of wavy paths shown for the earth in this diagram.

The moon revolves in the same direction as the earth rotates (from west to east). It follows an elliptical orbit around the earth at a speed of about 2200 miles per hour. As the moon revolves, it comes a little closer to the earth at one point in its orbit than at any other. When the moon is closest to the earth, it is said to be at *perigee* (pehr-i-jee), 360,000 km (225,000 mi) from center of the earth to center of the moon. At *apogee* (ap-o-gee), the farthest point, it is 404,800 km (253,000 mi) from center to center. The moon's average distance from the earth is almost 384,000 km (240,000 mi).

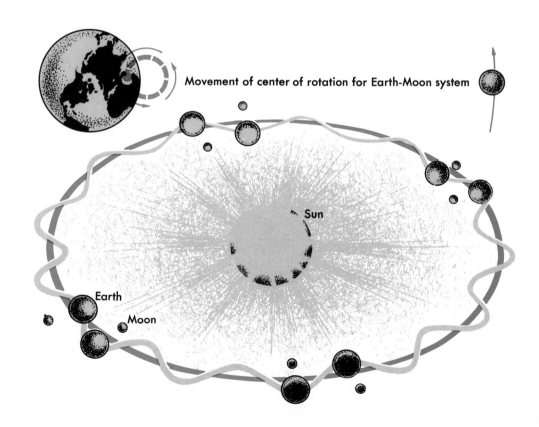

Movement of center of rotation for Earth-Moon system

Sun

Earth

Moon

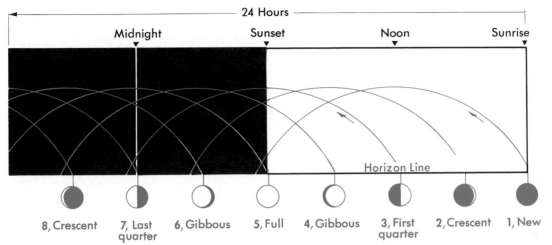

24 Hours

Midnight ▼   Sunset ▼   Noon ▼   Sunrise ▼

Horizon Line

8, Crescent   7, Last quarter   6, Gibbous   5, Full   4, Gibbous   3, First quarter   2, Crescent   1, New

Moonrise. If you watched the moon rise on successive nights, you would find that it rises later each night. You can only see the moon rise above the horizon when it is in positions 5 through 8. See Figure 4–9A. Can you explain why it is usually difficult to see the moon rise above the horizon in positions 1 through 4?

You have probably noticed that, after a few moonless nights, the moon appears as a thin crescent in the western sky just after sunset. A little later the next night it appears somewhat higher in the western sky. See Figure 4-9B, positions 2 and 3. Each successive night, even later after sunset, it continues to be seen still higher in the sky.

As the moon revolves around the earth from west to east, the earth rotates on its axis in the same direction. These separate motions create a time lag which results in the moon rising later each night. It is as though the earth needed to "catch up" by turning a little longer each day for the moon to be visible in its new position. Therefore the moon appears further toward the east each night. This extra turning takes the earth an additional 50 minutes. Thus the moon rises about 50 minutes later each night. But it also sets 50 minutes later.

When the moon is full, it rises in the east just about at sunset. The next night it does not begin to rise until after sunset. When the moon has almost completed its full-moon phase, it is often possible to see it in the western sky even after the sun has risen.

Phases of the moon. As the earth-moon system moves around the sun the moon's path takes it first between the sun and earth and then to the dark side of the earth. When

FIG. 4–9A. The phases of the moon appear as seen from the earth (1–8). The time scale at the top and orbital paths show how the moon rises about 50 minutes later each day as the phases progress from right to left.

FIG. 4–9B. The position of the moon in its orbit is shown as related to the numbered positions in Fig. 4–9A. One half of the moon is always lighted by the sun. The phases are a result of our changing view of the lighted portion facing the sun.

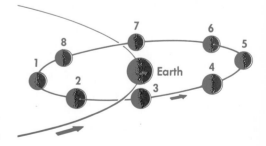

the moon is between the earth and sun, the part of the moon facing the earth does not receive light directly from the sun. Still, the moon is faintly visible because of sunlight reflected by the earth. The light reflected by the earth is called *earthshine*. After the moon has moved halfway around the earth, it is on the side of earth away from the sun. Then that half of the moon facing the earth appears fully lighted. But the lighted part of the moon again passes out of our sight as it moves back again between the sun and earth. It is this constantly changing view of the lighted half of the moon that is the reason for its phases.

Although we see little more than half of the moon as it passes through its phases, we always see the same portion. That is, the moon always keeps the same side turned toward the earth. To do this, the moon must rotate on its axis in about the same period as it appears to take in revolving completely around the earth. To measure the time needed for the moon to complete one revolution, we mark its position against the background of stars. At moonrise, the following night, the moon will appear slightly farther to the east among the stars. But in another 27⅓ days, the moon will return to its original position against the fixed background of stars. This means that the moon needs 27⅓ days to complete one revolution around the earth. Thus the moon must also rotate on its axis once every 27⅓ days to keep the same side toward the earth.

It is probably no coincidence that the moon rotates on its axis at the same rate that it travels around the earth. The moon is not perfectly round. In fact, it has a bulge on the side facing the earth. Such a bulge would result from the earth's greater gravitational attraction for the part of the moon which is closest. Once established, this bulge would continue to be attracted toward the earth. The rate at which the moon turns on its axis would then adjust itself to keep the bulge always toward the earth.

While the moon takes 27⅓ days to make one revolution around the earth, it takes slightly longer to go through a complete series of its phases. The period from one new moon to the next is 29½ days. The reason for this time difference can be seen in Figure 4-10. The moon at position A is in its new moon phase. After 27⅓ days it has moved into position B, completing one revolution (and one rotation). However, the earth has also moved farther around the sun during this time. Thus the moon must go on to

Explain

The term "dark side of the moon" refers to the portion of the moon's surface that faces away from the earth. What is wrong with this statement?

FIG. 4-10. For the moon to travel from new moon position (A) to the next new moon position (C), it must complete one revolution plus the distance from B to C.

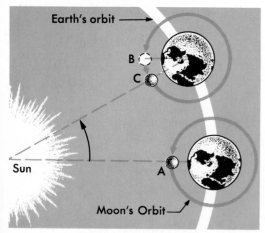

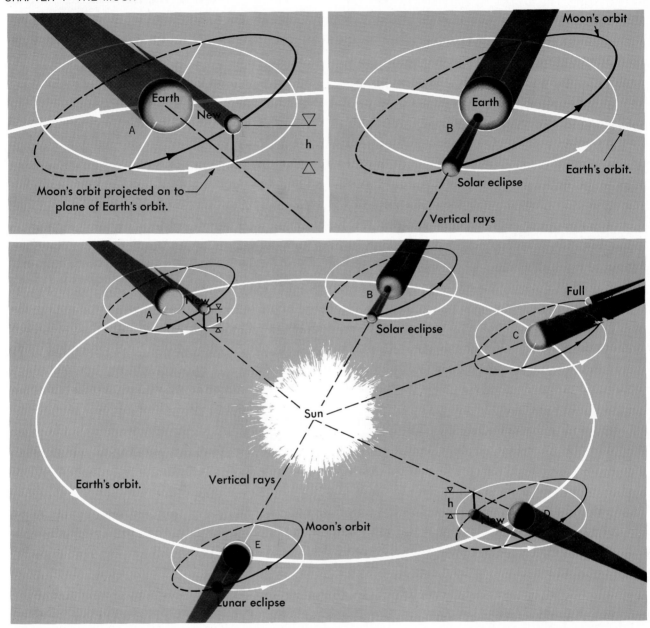

position C before it can return to its new moon phase again. This accounts for the difference in time between 27⅓ days for one revolution and 29½ days to go through one complete cycle of phases.

Eclipses of the sun and moon. All of the bodies moving around the sun, such as the earth and moon, cast long shadows into space. An eclipse occurs when either the

FIG. 4–11. An eclipse can occur only at positions B and E. In position A the moon arrives at the new moon phase as it crosses above the earth's plane. In position C the full moon phase occurs as the moon passes above the plane of the earth. The distance **h** is the elevation of the moon above and below the plane of the earth's orbit.

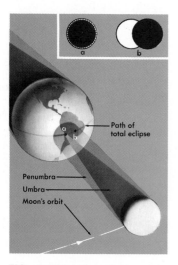

FIG. 4–12A. The diagram shows the position of the moon during a partial (b) and a total (a) eclipse of the sun.

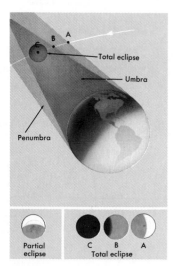

FIG. 4–12B. The moon enters the earth's penumbra at position **A**. At position **B** the moon is completely within the penumbra. A total eclipse of the moon occurs at position **C**. The moon is partially eclipsed as it passes just above the earth's penumbra.

earth or the moon passes through the other's shadow. Most of the time however, they move together around the sun without this happening. This is because the path of the moon around the earth is tilted about 5° to the plane of the earth's orbit around the sun. See Figure 4-11. This means that the moon usually crosses between the earth and the sun either too high or too low for its shadow to fall upon the earth. Likewise, when the moon passes on the side of the earth opposite the sun it usually misses the earth's shadow.

Shadows cast by the earth and moon are in two parts. One is a cone-shaped inner part completely cut off from the sun. This is called the *umbra.* The other is an outer part called the *penumbra,* in which sunlight is only partially blocked off. Sometimes the umbra of the moon's shadow falls upon the earth. All people within the umbra will see a total eclipse of the sun. Then the sun is completely covered by the moon. An eclipse of the sun occurs only during the new moon phase. This is when the shadow cast by the moon crosses the earth. See Figure 4–12A. Total eclipses of the sun do not occur very often. At any particular place a total solar eclipse is likely to be seen only once every several hundred years. Before the year 2000 only one total solar eclipse will be seen in the entire United States. This will be visible near the state of Washington in 1979.

Only a small part of the world is able to see any particular solar eclipse. This is because the moon's umbra never makes a large shadow spot on the earth's surface. The high speed of the earth's rotation sends the shadow sweeping across the earth's surface. For this reason, the total eclipse never lasts more than 7 minutes at any location. Those in the area outside the moon's umbra, but within the penumbra where a part of the sun is covered, see a *partial eclipse.*

An *eclipse of the moon* occurs during the full moon phase, when the earth's shadow may cross the lighted half of the moon. To produce a total eclipse, the moon must pass completely into the earth's umbra as is shown in Figure 4–12B. A partial eclipse of the moon occurs when the earth's shadow passes across one edge of the moon. Though eclipses of the moon occur only about as often as eclipses of the sun, the lunar eclipses are seen by more people. An eclipse of the moon can be seen everywhere on the dark side of the earth. But a total solar eclipse is seen only by those in the small shadow path of the moon's umbra as it glides across the earth's surface.

Sometimes the moon passes directly between the earth and the sun in a position that could produce an eclipse of the sun. However, the moon's umbra may be too short to reach the earth. This will only take place if the moon is at or near apogee when it comes between the earth and the sun. If the moon's umbra fails to reach the earth, an *annular* (ring-shaped) *eclipse* will occur. In an annular eclipse the sun is not completely blotted out. Instead it shows a thin ring of light around the outer edge as shown in Figure 4–12C (Stage c).

TIDES

The moon's influence on the oceans. The earth and the moon have a gravitational influence on each other. All parts of the moon are attracted toward the earth. In the same way, all parts of the earth are attracted toward the moon. It is this mutual attraction between the two bodies that causes the earth and moon to move as a single system. But the moon's gravity has one very obvious effect on the earth.

On the side of the moon facing the earth, the moon's gravity pulls out a bulge in the earth. The solid part of the earth bulges out only very slightly. But the water of the oceans moves more easily than the solid earth. Thus the bulge produced in the parts of the earth covered by water is very noticeable. A tidal bulge is produced in the part of the sea facing the moon. This is a clear example of the effect of the moon's gravity on the earth.

There is also a *tidal bulge* on the side of the earth opposite to the moon. See Figure 4–13. It is probably less clear why the moon's gravity can also be responsible for a tidal bulge on the opposite side of the earth. This can be explained as a result of smaller gravitational pull by the moon on the far side of the earth. The water there is farther from the moon and is affected less by its gravity. This allows the forces caused by the rotation of the earth and moon around a common center to push water away from the earth. Thus a tidal bulge is created there also. The bulge on the side facing the moon is called the *direct tide*. The bulge on the other side is called the *opposite tide*.

The two tidal bulges follow the moon as it moves around the earth. If the earth did not turn on its axis, the tidal

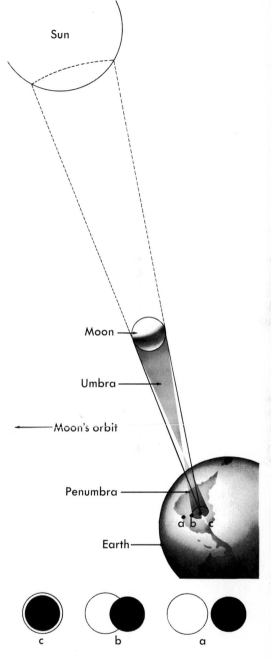

FIG. 4–12C. During an annular eclipse, the moon's umbra falls short of the earth. An observer at position c, instead of viewing a total eclipse of the sun, would be able to see a portion of the sun's photosphere.

FIG. 4-13. The moon's gravitational force is the main influence in creating tidal bulges on the earth. In this diagram the blue arrow indicates the direction of the earth's rotation.

FIG. 4-14. Due to the earth's rotation, the tidal bulges (A and B) do not occur directly between the earth and moon. This effect causes the moon to speed up in its orbit.

bulges would produce a high tide about every two weeks as they passed each location on the ocean shores. But the earth does rotate, once each day. Thus we should expect two high tides every 24 hours as the earth rotates and the tidal bulges pass by. However, the bulges also follow the moon's progress around the earth, so the earth's rotation has to "catch up" with the moving tides. It takes about 24 hours 50 minutes for both high tides to pass a given spot. The high tides actually do not occur when the moon is directly overhead but several hours later. This is because the rotation of the earth carries the tidal bulges ahead of their expected position in a direct line with the moon.

The tides in the earth's oceans also affect the speed of the earth's rotation. The tidal bulges are pulled around by the moon more slowly than the earth rotates. Friction against the sea floor from the moving masses of water in the tidal bulges tends to slow down the earth's rotation. The effect of this friction makes each day slightly longer than the previous one. In 1000 years the length of each day will be 0.2 seconds longer than it is now.

At the same time the tides are slowing down the rotation of the earth, they are causing the moon to revolve more rapidly. This is because the earth's rotation carries the tidal bulges forward from their expected position directly under the moon. Figure 4-14 shows this effect. As we see in the diagrams, bulge A which is closer to the moon, has a greater attraction for it than does bulge B. The attraction of bulge A for the moon tends to pull the moon ahead, speeding it up slightly. As the moon increases in speed, it also moves farther from the earth. During the earth's

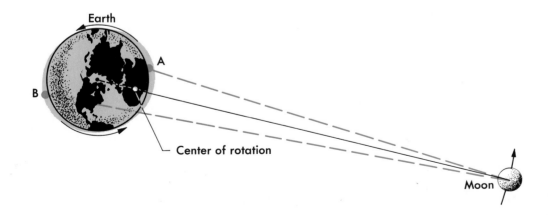

early history, if the moon did exist then, the moon was probably much closer to us than it is now.

The sun also influences the tides. The sun's gravitational force also raises tidal bulges on the earth. However, these tides are very small because of the sun's greater distance from the earth. Twice each month the sun and moon are in line with each other (at full moon and new moon). At such times the tidal effects of both the sun and moon are combined. The high tides produced as a result of this combined force are higher than usual and are known as *spring tides*. (The term is not related to the spring season.) See Figure 4–15.

In the first and last quarter phases of the moon, the sun and moon are at right angles to each other. This produces lower than normal tides called *neap tides*.

The influence of the moon on the tidal bulges also changes as its distance from the earth changes. At perigee, the moon is closer than at apogee, and its gravitational pull is greater. If the moon is at perigee at the time of its new or full moon phase (spring tides), the tides will be much higher than usual. See Figure 4–17. This event does not occur very often and can be predicted in advance. Coastal regions may have flooding problems during perigee spring tides, particularly if the weather is stormy.

The actual tides that are experienced at any location along the ocean shore are the result of many influences. Usually, the rise and fall of the tides at any particular place does not follow the expected pattern of two high and two low tides in 24 hours and 50 minutes. Some other factors that determine tidal action at different places are discussed in Chapter 17.

THE CALENDAR

The moon and the calendar. The calendar month was once based on the 29½-day cycle of the phases of the moon. This was a natural way of marking the passage of the year since the phases of the moon are very obvious and regular. But a year is 365¼ days, a number which cannot be evenly divided by 29½. This made it impossible to develop a calendar in which the length of the months is based on the phases of the moon.

The calendar of the Romans, which laid a foundation for our modern calendar, once had six months of 30 days

FIG. 4–15. Spring tides are shown during the new moon phase as a result of combined gravitational forces of the sun and moon.

FIG. 4–16. Neap tides occur when the gravitational forces of the sun and moon are at right angles during the quarter moon phase.

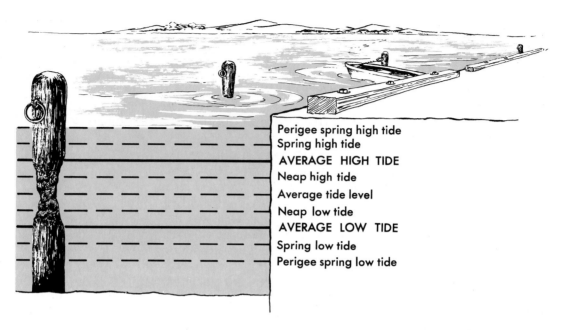

Perigee spring high tide
Spring high tide
AVERAGE HIGH TIDE
Neap high tide
Average tide level
Neap low tide
AVERAGE LOW TIDE
Spring low tide
Perigee spring low tide

FIG. 4–17. The various tide levels that might occur on a shoreline.

1582			**October**			1582
Sun.	**Mon.**	**Tue.**	**Wed.**	**Thu.**	**Fri.**	**Sat.**
	1	2	3	4	15	16
17	18	19	20	21	22	23
24	25	26	27	28	29	30
31						

FIG. 4–18. When most of Europe changed to the Gregorian calendar in 1582, ten days were dropped. On this calendar page, colored numbers represent Julian dates and black numbers Gregorian dates.

and six of 29 days. However, this still left each calendar year about eleven days short of the actual year. The Romans finally solved this problem by adding an extra month every few years to catch up again. During the reign of Julius Caesar, around 45 B.C., the Roman calendar had become so confused that Caesar ordered a complete change. The length of the months was increased to 30 or 31 days with one having only 28 days. This system ignored the phases of the moon and resulted in 12 calendar months out of a yearly total of 365 days. Since a year is actually about 365¼ days long it was necessary for the new calendar to put in an extra day every four years. Leap year, which has a whole extra day occurs whenever the last two digits of the calendar year are evenly divisible by four. This calendar became known as the Julian (after Julius Caesar) calendar.

Although it was a great improvement, the Julian calendar still was not quite correct. A year is actually 11 minutes and 14 seconds short of exactly 365¼ days. After the Julian calendar had been in use for hundreds of years, the months began to fall behind the seasons of the year. The vernal equinox had moved from March 21 back to March 11. In 1582, Pope Gregory ordered a change to bring the dates for church celebrations such as Easter back to their proper time of the year. The change was accomplished by taking out of the year 1582 the ten days that were gained. By official decree of the Pope, October 5 became October 15. See Figure 4–18.

At the same time, the century years (1800, 1900, etc.) were made leap years only if they could be evenly divided by 400. This omits three leap years every 400 years and satisfactorily keeps the months in agreement with the seasons. This form of the calendar is called the Gregorian calendar and is used by almost all countries.

A problem with the Gregorian calendar is that a normal year has 52 weeks and one extra day. A leap year has 52 weeks and two extra days. Thus a calendar date always falls one day later in the week each successive year. On leap year, the date "leaps" over a day. A more orderly calendar with respect to the dates and days of the week would have many advantages.

Many new kinds of calendars have been invented. For example, a Universal Calendar has been proposed which would have 13 months of 28 days each. This amounts to 364 days. An extra day without a date is therefore added at the end of each year. For leap year, two extra days are added. The chief objection to this calendar is that a year of 13 months cannot be evenly divided in halves or quarters.

A World Calendar with 12 months has also been proposed. The first month of each quarter (January, April, July and October) has 31 days. All the rest have 30 days. This system requires the addition of one extra day to the end of the year. The extra day would be a holiday known as "World Day." It would arrive between December 30 and January 1. In a leap year a second world day would be added between June 30 and July 1. See Figure 4–19.

Changes in the calendar are not easily made. For that reason, it seems likely that the calendar will remain as it is for some time.

JANUARY							FEBRUARY							MARCH							
S	M	T	W	T	F	S	S	M	T	W	T	F	S	S	M	T	W	T	F	S	
1	2	3	4	5	6	7				1	2	3	4						1	2	
8	9	10	11	12	13	14	5	6	7	8	9	10	11	3	4	5	6	7	8	9	
15	16	17	18	19	20	21	12	13	14	15	16	17	18	10	11	12	13	14	15	16	
22	23	24	25	26	27	28	19	20	21	22	23	24	25	17	18	19	20	21	22	23	
29	30	31					26	27	28	29	30			24	25	26	27	28	29	30	

APRIL							MAY							JUNE							
S	M	T	W	T	F	S	S	M	T	W	T	F	S	S	M	T	W	T	F	S	
1	2	3	4	5	6	7				1	2	3	4						1	2	
8	9	10	11	12	13	14	5	6	7	8	9	10	11	3	4	5	6	7	8	9	
15	16	17	18	19	20	21	12	13	14	15	16	17	18	10	11	12	13	14	15	16	
22	23	24	25	26	27	28	19	20	21	22	23	24	25	17	18	19	20	21	22	23	
29	30	31					26	27	28	29	30			24	25	26	27	28	29	30	W

JULY							AUGUST							SEPTEMBER							
S	M	T	W	T	F	S	S	M	T	W	T	F	S	S	M	T	W	T	F	S	
1	2	3	4	5	6	7				1	2	3	4						1	2	
8	9	10	11	12	13	14	5	6	7	8	9	10	11	3	4	5	6	7	8	9	
15	16	17	18	19	20	21	12	13	14	15	16	17	18	10	11	12	13	14	15	16	
22	23	24	25	26	27	28	19	20	21	22	23	24	25	17	18	19	20	21	22	23	
29	30	31					26	27	28	29	30			24	25	26	27	28	29	30	

OCTOBER							NOVEMBER							DECEMBER							
S	M	T	W	T	F	S	S	M	T	W	T	F	S	S	M	T	W	T	F	S	
1	2	3	4	5	6	7				1	2	3	4						1	2	
8	9	10	11	12	13	14	5	6	7	8	9	10	11	3	4	5	6	7	8	9	
15	16	17	18	19	20	21	12	13	14	15	16	17	18	10	11	12	13	14	15	16	
22	23	24	25	26	27	28	19	20	21	22	23	24	25	17	18	19	20	21	22	23	
29	22	31					26	27	28	29	30			24	25	26	27	28	29	30	W

W–World Day, December W (365th day), a world holiday, follows December 30th every year. Leap-year Day, June W, another world holiday, follows June 30th in leap years.
Easter Sunday, April 8th.

FIG. 4–19. Some of the suggested advantages of a World Calendar as shown above are:

1. A year of twelve months is easily divisible, since all dates regularly fall on the same days.
2. Each year begins on the first day of the week, Sunday, January 1.
3. Seasons are equalized into quarters, each having three months or 13 weeks.
4. Each quarter-year is arranged in the same pattern of three months of 31, 30, and 30 days respectively.
5. Each season begins on a Sunday and ends on a Saturday.
6. Holidays always fall on the same day and date. (International World Calendar Association)

VOCABULARY REVIEW

Match the word or words in the column on the right with the correct phrase in the column on the left. *Do not write in this book.*

1. Relatively smooth areas on the moon's surface.
2. Light streaks on the moon as if material had been thrown out.
3. Mountainous regions on the moon.
4. Moon is closest to earth.
5. Darkest part of shadow cone.
6. Part of shadow where light is only partly blocked.
7. Moon's umbra fails to reach the earth.
8. Changing view of lighted and dark parts of the moon.
9. Moon passes through the earth's shadow.
10. Tidal bulge on the side of the earth facing the moon.
11. Tidal bulge on the side of the earth away from the moon.
12. Occurs when the sun, earth, and moon are in line.
13. Tides of low range.
14. Contains 4 months of 31 days.

a. apogee
b. eclipse of moon
c. phases
d. opposite tide
e. rays
f. perigee
g. annular eclipse
h. highlands
i. maria
j. World calendar
k. umbra
l. Gregorian Calendar
m. penumbra
n. total eclipse
o. neap tides
p. spring tides
q. direct tides

QUESTIONS

Group A

Select the best term to complete the following statements. *Do not write in this book.*

1. Other than the earth, no planet in the solar system has (a) an atmosphere (b) a companion so nearly its own size (c) storms (d) gravity.

2. The moon's diameter is (a) 2160 km (b) 12,640 km (c) 3476 km (d) 7900 km.

3. Scientific evidence gathered on the moon suggests that its history consisted of (a) three (b) four (c) five (d) six stages of development.

4. Possible evidence for collisions of the moon with other bodies is (a) craters (b) rills (c) volcanoes (d) mountains.

5. In one of its early stages, the heat retained from its origin and from colliding bodies cause the moon's surface to be covered with (a) mountains (b) molten rock (c) dust (d) rills.

6. The lighted side of the moon has temperatures as high as (a) −173°C (b) −100°C (d) over 100°C.

7. On the moon, a boy weighing 66 kg would weigh (a) 18 kg (b) 66 kg (c) 11 kg (d) 33 kg.

8. The moon orbits around a point that is (a) at the earth's center (b) nearer the earth's surface than its center (c) half-way between the earth and moon (d) near the moon's surface.

9. At perigee, the moon is how many kilometers from the earth? (a) 360,000 (b) 404,800 (c) 225,000 (d) 253,000.

10. Each night the moon rises (a) a little earlier (b) about the same time (c) a little later (d) in the west.

11. The moon rises at a different time consecutive days because (a) it is slowing down (b) the earth is moving around the sun (c) the earth rotates from east to west (d) the moon moves from west to east around the earth.

12. We always see the same side of the moon because (a) the moon rotates on its axis at the same rate it moves around the earth (b) the earth rotates on its axis (c) the earth revolves around the sun (d) none of these.

13. The length of time it takes the moon to make one complete revolution around the earth is (a) 31 days (b) $29\frac{1}{2}$ days (c) $27\frac{1}{3}$ days (d) 30 days.

14. The time needed for the moon to go through a complete cycle of phases is (a) 31 days (b) $29\frac{1}{2}$ days (c) $27\frac{1}{3}$ days (d) 30 days.

15. The reason for the difference in the answers to questions 13 and 14 is (a) the earth's motion around the sun (b) the rotation of the earth on its axis (c) the rotation of the moon on its axis (d) the tilt of the earth's axis.

16. In each of its revolutions around the earth, the moon does not come directly between the sun and the earth because the (a) earth's axis tilts (b) moon's orbit tilts (c) moon does not rotate (d) earth's orbit tilts.

17. An annular eclipse could occur when the moon is (a) at apogee (b) full (c) at first quarter (d) at perigee.

18. The phase of the moon during an eclipse of the sun is (a) full (b) half (c) new (d) in any phase.

19. To produce a total eclipse of the moon (a) the earth must pass through the moon's umbra (b) the moon must pass through the earth's umbra (c) the moon must be at apogee (d) the moon must be at perigee.

20. The moon is most likely to be overhead (a) at high tide (b) after high tide (c) before high tide (d) at low tide.

21. Tides always (a) increase in height each day (b) decrease in height each day (c) speed up the earth's rotation (d) slow down the earth's rotation.

22. The tidal bulge tends to (a) speed up the moon (b) slow down the moon (c) bring the moon closer (d) have no effect on the moon.

23. Spring tides occur at (a) new moon and first quarter (b) new moon and last quarter (c) first quarter and last quarter (d) full and new moon.

24. Neap tides occur at (a) new moon and first quarter (b) new moon and last quarter (c) first and last quarter (d) full and new moon.

25. The problem of developing a calendar based on phases of the moon is that (a) the moon's motion is not regular (b) $29\frac{1}{2}$ is not evenly contained in $364\frac{1}{4}$ days (c) the months are too long (d) the moon's motion is too regular.

26. The Roman calendar fell short of a year by how many days? (a) 7 (b) 9 (c) 11 (d) 13.

27. The Julian calendar was not quite correct because it was (a) 11 minutes 14 seconds short each year (b) 11 minutes 14 seconds long each year (c) short by one day each 4 years (d) long by one day every 4 years.

28. The calendar used in almost all countries today is the (a) World calendar (b) Universal calendar (c) Julian calendar (d) Gregorian calendar.

29. The proposed Universal calendar has how many months? (a) 13 (b) 28 (c) 12 (d) 11.

30. The World calendar is probably preferable to the Universal calendar because it has (a) equal months (b) 365 days per year (c) 12 months (d) more quarters.

Group B

1. List the conditions that must exist for a total eclipse of the sun.

2. Give a brief description of the different ways the moon may have come into existence.

3. Why do we have leap years in the century years only when they are evenly divisible by 400?

4. Describe how the dust on the moon's surface was formed.

5. Suppose the moon rotated on its axis every 14 days instead of every 27. What effect would this have on the way we see the moon?

6. To a person on the moon, how would the earth appear at full moon, new moon, and during the first quarter?

7. Why does the moon have such a large temperature range?

8. Describe the six stages in the moon's development.

9. Why was the Julian calendar an improvement over the Roman calendar?

10. Why can everyone on the dark side of the earth see an eclipse of the moon?

11. Why can't there be an eclipse of the moon at first quarter?

12. Why can't everyone see a total eclipse of the sun?

13. Why do the spring tides occur at full moon?

14. If the moon turned on its axis at twice its present rate, what would be the probable effect on its temperature?

15. Light travels at 300,000 kilometers per second. How long does it take for light to travel the 400,000 km from the moon to the earth?

16. How does the rotation rate of the moon on its axis affect our knowledge of the moon's surface?

17. Give a simple explanation of opposite tides.

18. How do tides slow down the earth?

19. If the moon were farther from the earth than it is now, would high tides occur more often? Why?

20. The volume of a sphere is given by the formula $V = \frac{4}{3}\pi r^3$ where r is the radius. How many moons could fit into the volume of the earth?

probing the secrets of space

objectives

- [ ] Explain why the gravity barrier is a major obstacle to space study.
- [ ] Describe in simple terms how a rocket works.
- [ ] Describe the launching of an artificial satellite and its orbit.
- [ ] Explain the use of a minimum energy orbit in a flight to the inner planets.
- [ ] Identify the main reasons for sending spacecraft to the other planets.
- [ ] Describe the function of artificial earth satellites and the Space Shuttle.

We are in the age of space exploration. This is a time when machines and people can be sent from the earth to explore space. Behind these scientific explorations is the desire to find answers to difficult questions such as these: How did the earth and solar system begin? What can the other members of the solar system teach us about our environment on earth? Is life found only on earth?

If the future does provide any answers to these questions, even greater puzzles will still remain for us to solve. Are the earth and the other planets of the solar system unique among the stars? Or, does the universe contain swarms of planet systems? Are there many planets like the earth? Seeking answers to these questions will require pushing space exploration into the regions beyond our solar system. The things that are hardest to understand about the earth can be better understood by probing the secrets of the other planets and the space beyond.

GETTING INTO SPACE

The gravity barrier. Any object on the earth's surface is pulled toward the earth's center by a force due to gravity. The size of the gravitational force acting upon a body is determined by its mass and distance from the earth's center. See page 15. To move away from the earth, enough energy must be used to overcome the gravitational force pulling the object downward. For example, a rocket vehicle cannot start to move upward until the force delivered by its engine is greater than the gravitational force. The force the rocket engine must produce to lift the rocket straight up is determined by its total mass. But the gravi-

FIG. 5–1. A rocket cannot move upward until the thrust of its engines is greater than gravitational force. (NASA)

FIG. 5–2. The force needed to escape the earth's gravity can be compared to the force needed to climb out of a pit, shaped like the one shown here.

tational force pulling the rocket back toward the earth decreases as the rocket moves upward.

For a rocket vehicle trying to leave the earth, overcoming gravity can be compared to trying to climb out of a deep pit or hole. The walls of the pit would be very steep near the bottom but not nearly so steep at the top. See Figure 5–2. The slope of the pit's walls can be compared to the gravitational force pulling a rocket back to earth. At first, the gravitational force is very strong. The rocket engine must produce a powerful force to lift the vehicle off the ground. As the rocket moves farther from the earth, the force of gravity decreases. Less force is required to keep it from falling back. Finally, at a great distance from earth, the rocket is far enough out of the gravity "pit" so that it is not pulled back.

To escape the earth's gravity, a rocket engine must produce enough force to achieve a certain velocity. This is known as the escape velocity. If an object can achieve the escape velocity, it will continue to move away from the earth. At any speed less than the escape velocity, the object will eventually fall back to earth.

The escape velocity necessary for a vehicle to leave the earth is 40,000 km/hr. This value is determined by the amount of gravitational force at the earth's surface. Bodies with less mass than the earth, such as the moon, have less gravitational force. Thus, the moon has an escape velocity of only 8000 km/hr.

On the other hand, the escape velocity from Jupiter is 216,000 km/hr because Jupiter has a large mass.

A vehicle leaving earth must move away at a velocity at least as great as the escape velocity for earth. But once this velocity is reached, the space vehicle will need no additional force from its engines to continue moving away from the earth. According to the natural laws that describe moving bodies, the spaceship can turn off its engines and continue to move in a straight line at nearly a constant speed. Its path and speed will not change until its engine is turned on again, or is affected by the gravity of another planet, or some other body.

The rocket. Rocket engines are the only means now available that can furnish enough force to lift vehicles from the earth. In one way, a rocket engine is similar to an automobile engine. Both require a mixture of fuel and oxidizer to be burned. Hot gases are produced that expand and create the force necessary to move the vehicle. A basic difference between a rocket and an automobile engine is the speed with which the fuel is burned.

In a rocket engine, hot gases are produced by very rapid burning of fuel. As the hot gases expand, tremendous pressures push against the walls of the engine. If the engine had no opening to the outside, the gas pressure would be very nearly equal in all directions and the engine would quickly explode. However, a rocket engine is provided with an exhaust nozzle to permit the hot gases to escape. As the gases flow out of the exhaust, the force on the front of the engine is no longer balanced by a force to the rear. Unbalanced gas pressure pushing on the front of the engine thrusts the whole body of the rocket forward. This force is called its *thrust.* See Fig. 5–3. A rocket engine does not have to push against anything but itself. It operates even in the near vacuum of space.

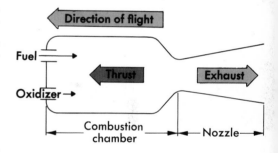

FIG. 5–3. The thrust of a rocket engine is produced by expansion of hot gases in the combustion chamber. Thrust is an unbalanced force produced by the exhaust. Its amount is determined by the speed and volume of the burnt gases leaving the nozzle.

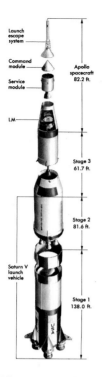

Launch
escape
system

Command
module

Service
module   Apollo
                      spacecraft
                      82.2 ft.

LM

Stage 3
61.7 ft.

Stage 2
81.6 ft.

Saturn V
launch
vehicle

Stage 1
138.0 ft.

FIG. 5–4. The *Saturn V* rocket vehicle at the time of blast-off weighs nearly 6 million pounds. Its total height including the *Apollo* spacecraft stands higher than a 30-story building.

As part of its load, a rocket vehicle must carry along both the fuel and oxidizer it needs. The fuel can be almost anything that is quickly and smoothly combustible. The most commonly used oxidizer is oxygen itself. But the oxygen used in rockets is in the liquid form because it occupies the least amount of space. At least 75 percent of the weight of a rocket vehicle at take-off is taken up by its fuel and oxidizer. Rockets that are intended to travel great distances must be very large in order to carry sufficient fuel and oxidizer. Such rockets must also be very powerful to get the entire load off the earth. See Fig. 5–4.

In the future, *nuclear rockets* may replace the fuel-oxidizer type. The source of heat in a nuclear rocket engine will be a device that obtains energy from atomic nuclear reactions. In a fueling process of this type, a substance is heated by the exhaust gas. Such a rocket will require much less fuel and no oxidizer. Thus a nuclear rocket of a given weight will be able to travel much farther. This means that nuclear rockets will be the most likely means for long space voyages in the future.

Earth satellites. If a rocket leaves the earth with a speed less than escape velocity, it will eventually fall back toward the earth. However, such a rocket could be given a sideways push somewhere near the top of the flight path. This would be done by tilting the rocket and firing its engine. Then the return path of the rocket as it falls back to earth will be more curved. If the curve of the return path can be made equal to or less than the curvature of the earth's surface, the rocket will never actually reach the earth. It will fall continuously in an orbit around the earth. See Fig. 5–5. The rocket will become a satellite of the earth.

FIG. 5–5. A rocket will remain in orbit if the curve of its flight path is equal to, or less than the curvature of the earth.

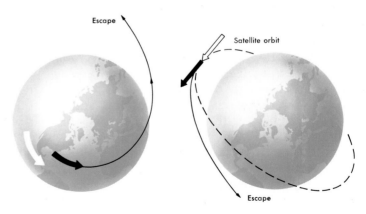

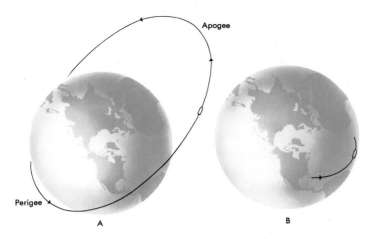

Apogee

Perigee

A   B

FIG. 5–6. The orbit shown in A, at the left, is not circular because either the launching angle was too steep or the speed of the rocket was too great. In B the angle of launching was too small or the speed was too slow to achieve an orbit.

The moment that the satellite is pushed into its orbit is very critical. If the angle at which it is aimed is too high or the speed too great, the satellite will go into a very elliptical orbit. See Fig. 5–6A. If the elliptical orbit has a perigee too close to the earth, the satellite will dip too deeply into the earth's atmosphere. There it will be vaporized by the heat generated from friction with the air. If the satellite is aimed at too low an angle, it will fall back to earth and disintegrate in the atmosphere before it can go into orbit. See Fig. 5–6B.

If the satellite is aimed correctly and has the right speed, it will enter a circular, or nearly circular, orbit. It is difficult to launch a satellite at exactly the correct angle and speed needed to achieve a circular orbit. Most satellites follow elliptical orbits that move them closer, then farther away from the earth. If their approach at perigee is close enough, friction with the upper atmosphere will slow the satellite's speed. With this loss of speed, the satellite begins to curve downward. As it moves closer to the earth, gravity forces the satellite to pick up speed again. It then moves away in a wider orbit. A satellite may repeat this in and out motion a number of times, dipping farther each time into the atmosphere. As it penetrates deeper, it will eventually slow down to such a point that it can no longer maintain any kind of orbit. The satellite then plunges toward the earth and is vaporized by its contact with the heavy air of the lower atmosphere.

To stay in an almost circular orbit at an altitude of 480 km, a satellite must maintain a speed of 27,520 km/hr. At a distance of 35,680 km, a satellite moves only about 10,720 km/hr. This speed is sufficient to take it

Discuss
The moon, like all earth-orbiting space vehicles, is an earth satellite. What would happen to the moon if it were suddenly to speed up; slow down; stop completely?

around its orbit once every 24 hours. Such a satellite would then appear to be stationary, if it moves in its orbit in the same direction as the earth rotates. The farther the satellite's orbit is from the earth, the slower it needs to revolve to maintain a circular orbit.

INTERPLANETARY TRAVEL

Flight to the inner planets. The problems encountered in flights to the inner planets are not much different than those involved in flights to a body as close as the moon. However, when traveling to a more distant planet such as Mars or Venus, the sun's gravity becomes a factor. A spacecraft can reach another planet and use less fuel if it becomes a temporary satellite of the sun. Such a path to the planets can be called a "minimum energy orbit." For example, to reach Venus, a spacecraft might be launched so that it achieves a velocity somewhat slower than the velocity the earth moves around the sun. The sun's gravity would then pull the craft in a direction toward the center of the solar system until it would finally cross the orbit of Venus. For a flight to Mars, a spacecraft is launched in the same direction as the earth moves around the sun. The earth's motion and the craft's velocity cause it to move in a path that carry it outward to intersect the orbit of Mars. See Figure 5–7.

The sun's gravity can supply part of the energy needed for these paths. As a result, flights between earth and its neighbor planets might not require any more power than flights to the moon. However, in flights between planets there are different problems in guiding the spacecraft to its target. An interplanetary spacecraft following a minimum energy path is a satellite of the sun with its own orbit. The rocket engines can be shut off except for brief firings to make small corrections in the path. This means that the craft's orbit must be carefully planned to

FIG. 5–7. Minimum Energy flight paths to Venus and Mars. Left, to reach Venus, the spacecraft might move in a direction opposite to the earth's motion around the sun. Right, a flight path to Mars that takes advantage of the earth's motion.

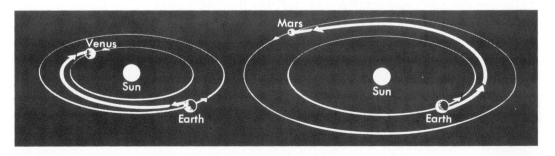

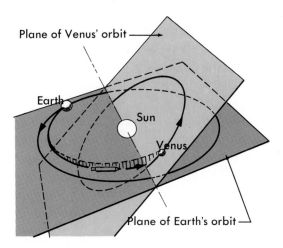

FIG. 5–8. Because the orbits of Earth and Venus are not on the same plane, a Venus-bound space vehicle launched in a minimum energy path would miss its target. The angular difference between the orbital planes of Earth and Venus is actually much less than shown in this diagram.

bring it and the target planet together at some point in the planet's orbit.

An example of the kind of difficulty met in guiding an interplanetary flight is this: The orbit of Venus is tilted about 3 degrees from the plane of the earth's orbit. Although this is a small difference, it means that Venus is sometimes as much as 4,800,000 km out of the plane of the earth's orbit. Unless this is taken into account, a vehicle launched in a minimum energy path toward Venus will miss its target. See Figure 5–8.

One solution to this problem would be to launch the vehicle when the earth crosses the plane of the orbit of Venus. At this time, there could be no error caused by differences between the orbital planes of Earth and Venus. For this reason, the flight path would not have to

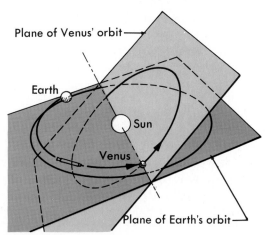

FIG. 5–9. Minimum correction needed for a flight path to Venus when the vehicle launching is timed to take place as Venus crosses the plane of the earth's orbit.

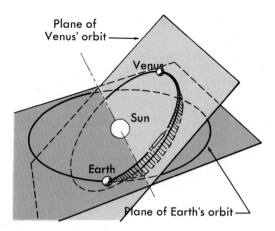

FIG. 5–10. The minimum energy path to Venus when Earth is not crossing the plane of Venus' orbit must be corrected as shown in the diagram.

be corrected during the journey. This would also hold true if the vehicle were launched when Venus crosses the plane of the earth's orbit. This situation is shown in Figure 5–9. If the Venus-bound vehicle were launched at any other time, it would have to be aimed at a slight angle to the plane of the earth's orbit. See Figure 5–10. This type of flight path would require greater speed at launching, and during any mid-flight adjustments. It would never be a minimum energy flight path. To reach the planets by using the least amount of energy, spacecraft can only be sent off during certain periods of time. These times are often called "launch windows." There is a different "launch window" for each planet that could be a destination.

In flights to the more distant outer planets, the gravity of the planets can be used to direct the flights. For example, a spacecraft could be sent off toward Jupiter following a minimum energy orbit. It would pass this giant planet in such a way that Jupiter's gravity would bend its path. It would then be flung toward Saturn as if hurled from a giant slingshot. See Figure 5–11. As the craft flies near Saturn, that planet would turn it toward Pluto. A similar plan could utilize the gravity force of Jupiter to guide a spacecraft to Uranus and Neptune.

Knowledge from the planets. Why do we spend huge amounts of money and use the efforts of thousands of people to send spacecraft to the other planets? There are four main reasons. First, the planets will probably help scientists discover how the earth came into existence. Each planet is in a different stage of development. Thus, they may show what the earth might have been like in the past, and how it might change in the future.

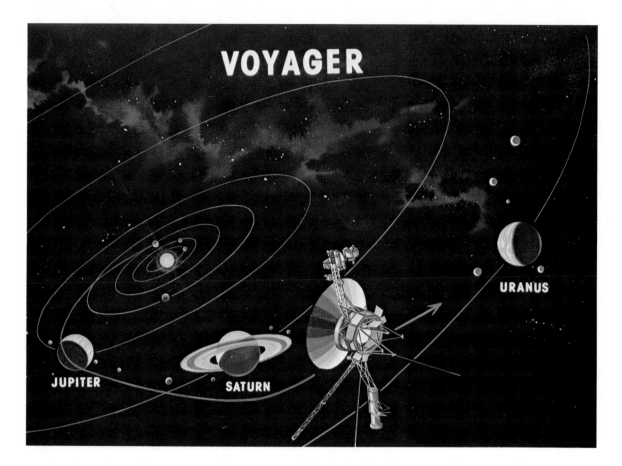

A second important reason for investigating the planets is to try to find evidence that life exists elsewhere. The discovery of life on other planets might mean that living things exist in many places in the universe.

Exploration of the other planets will also help prepare for the possibility that at some time the earth may become overpopulated. It may be necessary to establish colonies in other parts of the solar system to provide resources or living space.

By studying other planets at close range, scientists will also gain a better understanding of the earth. Mars, for example, has provided information that might explain what could cause our climate to pass into an ice age. An ice age on earth would mean disaster for the world's population. Scientists suspected that smoke, dust, and other kinds of pollution in our atmosphere could trigger a critical drop in the earth's temperature. But there was no direct way to test this theory. A spacecraft that orbited Mars for nearly a year was able to provide the answer. It

FIG. 5–11. A spacecraft is sent past Saturn by using the gravity of Jupiter. (NASA)

activity

Suppose you are on the moon. You and two members of your crew are returning from a mission when your ship crashlands 300 km from the base ship. The crash is on the sunlit side of the moon. Most of your equipment has been destroyed in the crash. Your survival depends on reaching the base ship. In addition to space suits, your crew is able to remove the following items from the wreck: 4 packages of food concentrate, 20 m nylon rope, 1 portable heater, 1 magnetic compass, 1 box matches, 1 first aid kit, 2-50 kg tanks of oxygen, 20 L water, 1 star chart, 1 case dehydrated milk, 1 solar powered receiver-transmitter, 3 signal flares, 1 large piece nylon fabric, 1 flashlight, 2 knives.

1. Rate each item in the list according to how important it will be to your survival.

2. Make a list of the items, in the order of their importance.

3. After each item, write what it will be used for that makes it important or unimportant.

4. Compare and discuss your list with other members of the class.

5. Compare your list with one prepared by astronauts for NASA. Your teacher will give you this list.

6. List revisions you would make in the order of the items if the crashlanding was on Mars instead of the moon.

observed Mars during a huge dust storm that lasted for months. The dust storm almost completely covered the planet. Temperatures on the Martian surface were measured as it went from a clear condition to a dust-filled atmosphere. The measurements showed that the Martian temperatures dropped an average of 20°C. The temperature drop was the result of dust blocking the heat from the sun. This provides strong evidence that a pollution filled atmosphere on the earth could set off an ice age.

Exploration of parts of the solar system has already given a clearer picture of how the earth and the other planets were probably formed and developed. The parts yet to be explored provide a giant scientific laboratory in which our knowledge of the earth can be expanded and tested.

IN ORBIT AROUND THE EARTH

Exploring the earth from space. The most important discovery since space exploration began has been expansion of our knowledge of the earth. From space, it has been possible to see the earth in new ways. Space exploration has also given us ways to discover the earth's valuable resources and use them in ways that will not harm the environment. Space technology is helping to meet the challenge of maintaining and improving the quality of life on our crowded planet. This is being done by con-

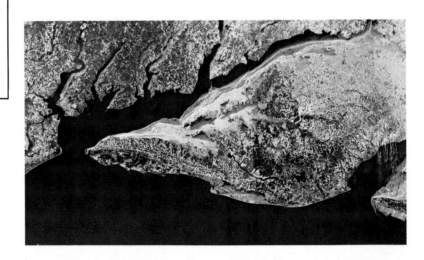

FIG. 5–12. A *LANDSAT* photograph of frozen Chesapeake Bay. (NASA)

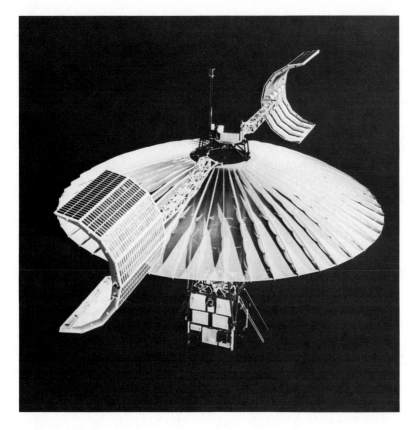

FIG. 5–13. A communications satellite. (NASA)

stantly studying the earth from space with a group of orbiting instruments called *applications satellites*.

An example of an applications satellite is one called *LANDSAT-1*. It is designed to investigate the earth's natural resources. Its sensors record the patterns made by land, water, plants, minerals and man-made structures. See Figure 5–12. *LANDSAT's* orbit lets it scan every part of the earth's surface once in 18 days. It has provided information that revealed previously unknown oil and mineral deposits in Oklahoma, Alaska, the Rocky Mountains, and the Brazilian jungles. Regions of the sea that might be productive new fishing grounds have been located. Changes in the environment like earthquakes, forest fires, and human activities have been examined.

One of the most important uses of *LANDSAT* involves its ability to survey the world's agriculture. In a matter of hours, the farmland of a large area can be examined and each kind of crop identified. Computers then use this information to print detailed maps showing the location of each type of crop. These maps become the main tool

FIG. 5-14. *Nimbus* meteorological satellite. (NASA)

for predicting areas of food shortages and surpluses. Such advance warning greatly improves the distribution of food throughout the world.

Applications satellites also make up a vital part of the world's communications systems. Systems on the earth's surface like telephone lines and radio stations connect between two places. Satellites connect all places in the region they cover. Three satellites placed in stationary orbits above the Atlantic, Pacific, and Indian Oceans can reach every point on the earth's surface. Satellites now in orbit provide telephone, telegraph, and television links between almost all parts of the world.

No scientific tool has done more to make accurate weather forecasts possible than the satellite. One such satellite is called *Nimbus.* It circles the earth in an orbit that carries it from pole to pole at an altitude of 1,120 km. Twice each day, it records temperatures and photographs weather patterns over the entire earth. Special sensors allow it to see through clouds and to locate and track icebergs in the sea near the polar regions. Mapping ice movements greatly increases the safety of ships moving through these waters. *Nimbus* also maps important ocean currents that have an effect on the world's weather. Two weather satellites set in stationary orbits above Brazil and the eastern Pacific also observe weather in the United States every 30 minutes. These satellites provide im-

mediate warning of threats from severe storms such as
hurricanes and tornadoes. They also provide full time
observation of less dangerous weather changes. These
space observations have given scientists the ability to
make weather forecasts with a degree of accuracy never
possible before.

In terms of effect on our daily lives, applications satellites are probably the most important product of space
exploration.

The Space Shuttle. The *Shuttle* has been developed as an
economical and simple means for reaching the earth's
orbit. Use of the *Space Shuttle* allows satellites to be put
into orbit. It also permits people to work in the space environment for long periods of time. The main part of the
Shuttle's system is the *Orbiter*. It is about the size of a
small jet air transport with short wings. It is powered by
rocket engines. See Figure 5–15. When launched from
the earth, *Orbiter* will mate with an extra outside fuel
tank and two solid-fuel rockets. At launch, both the *Orbiter's* engines and the solid rockets are used. When an
altitude of about 40 km is reached, the burned out solid
rockets detach and parachute into the ocean. They are

FIG. 5–15. The *Space Shuttle's Orbiter* in atmospheric flight. (NASA)

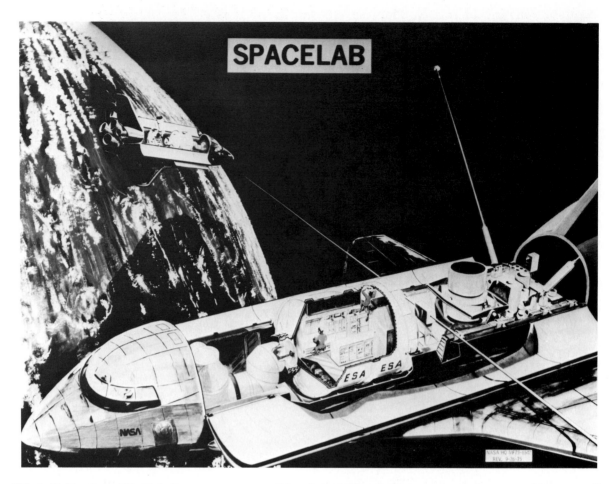

FIG. 5–16. The *Space Shuttle's Orbiter* can serve as a space laboratory in which scientists and engineers can work for periods up to 30 days in orbit around the earth. (NASA)

recovered by ships. The *Orbiter* proceeds into orbit around the earth. When the fuel in the extra outside tank is used up, it too is dropped off. This fuel tank is the only part of the system that is not used again.

Once in orbit, the *Shuttle* can launch, recover, and repair orbiting satellites. It can also serve as an orbiting space station for as long as 30 days. When its mission is completed, the *Orbiter* returns to earth, landing like an ordinary airplane. It can then be prepared for its next mission.

FIG. 5–17. NASA artist's rendering of the proposed *Jupiter Orbiter Probe* spacecraft observing one of Jupiter's satellites. Jupiter is in the background. The *Jupiter Orbiter* will be launched aboard the *Space Shuttle*. (NASA)

VOCABULARY REVIEW

Match the word or words in the column on the right with the correct phrase in the column on the left. *Do not write in this book.*

1. The initial obstacle that must be overcome to get into space.
2. The speed needed, near the earth's surface, to avoid falling back when the rocket engine stops firing.
3. Present day means of propulsion in outer space.
4. The force that makes a rocket move ahead.
5. The most likely propulsion for long space voyages in the future.
6. Follows a path whose curvature is equal to or less than that of the earth's surface.
7. Period of send-off time needed to reach a planet, using least energy.
8. Space program will give a better understanding of this.
9. Result of dust-filled atmosphere during storm on Mars.
10. Orbiting instruments for constant study of earth.
11. Scans the natural resources of the earth every 18 days.
12. A pole to pole weather satellite.
13. The time between observations of the U.S. weather by satellite.
14. The first economical way of reaching Earth's orbit.
15. The main part of the shuttle system.

a. nuclear rockets
b. applications satellites
c. *Nimbus*
d. 30 minutes
e. earth
f. *Orbiter*
g. gravity barrier
h. earth satellite
i. launch window
j. *Ranger*
k. fuel-oxidizer rocket engine
l. *Space Shuttle*
m. temperature drop
n. aeronautics
o. escape velocity
p. *Landsat 1*
q. thrust

QUESTIONS

Group A

Select the best term that completes the following statements. *Do not write in this book.*

1. Objects on the earth's surface are pulled toward the earth by a force due to (a) pressure (b) gravity (c) atmosphere (d) vacuum.

2. As an object moves away from the earth, the gravitational force acting on it (a) becomes less (b) remains constant (c) becomes greater (d) depends on the size of object.

3. Overcoming gravity can be compared to (a) reducing atmospheric pressure (b) climbing down a steep bank (c) climbing out of a pit (d) swimming.

4. The "escape velocity" needed to leave the earth is (a) 25,000 km/hr (b) 40,000 km/hr (c) 5000 km/hr (d) 8000 km/hr.

5. The moon has an "escape velocity" that is what fraction of the earth's "escape velocity?" (a) $5/8$ (b) $1/2$ (c) $1/4$ (d) $1/5$.

6. A rocket engine creates the force needed to drive a space vehicle. The hot gases within expand, putting pressure on the front of the engine that is (a) not balanced toward the rear (b) always balanced toward the rear (c) the only force acting (d) not important.

7. At blast-off, what part of the weight of a rocket vehicle is taken up by fuel and oxidizer? (a) 50% (b) 75% (c) 100% (d) 10%.

8. A nuclear engine operates by (a) throwing out radioactive matter as exhaust gases (b) absorbing energy (c) heating a substance that is given off as exhaust gases (d) using a fuel and oxidizer.

9. A rocket going straight up with less than escape velocity will (a) fall back to the earth (b) never reach the earth (c) become a satellite of the earth (d) become a moon satellite.

10. To stay in an almost circular orbit 700 km above the earth's surface, a satellite must have a speed of (a) less than 27,520 km/hr (b) 27,520 km/hr (c) greater than 27,520 km/hr (d) any value.

11. At what altitude will an orbiting satellite stay over the same spot on the earth? (a) 10,720 km (b) 6700 km (c) 22,300 km (d) 35,680 km.

12. In problem 11 the assumption is that the satellite (a) moves in a direction opposite to that of the earth's rotation (b) moves in the same direction as the earth's rotation (c) does not have any speed (d) changes its altitude constantly.

13. By becoming a temporary satellite of the sun and thus using the least fuel, a space vehicle can get to another planet (a) in the quickest time (b) only after centuries of travel (c) by the maximum energy orbit (d) by the minimum energy orbit.

14. To reach Venus by using the least energy, a vehicle must be launched so that it moves (a) around the earth (b) away from the sun (c) slower than the earth (d) faster than the earth.

15. Compared to moon-flight rockets, rockets that reach neighboring planets must be (a) much more powerful (b) no more powerful (c) much less powerful (d) too big to build at this time.

16. Sometimes Venus is out of the plane of the earth's orbit by as much as (a) 4,800,000 km (b) 48,000 km (c) 4,800 km (d) 48 km.

17. The reason for the answer to question 16 is (a) Venus' orbit is nearly a perfect circle (b) Venus' orbit has a peculiar bump in it (c) Venus' orbit crosses the earth's orbit (d) Venus' orbit plane is tilted 3° from the earth's orbit plane.

18. The periods of time when a spacecraft can be sent off and use the least energy to reach another planet are called (a) activation times (b) periodic movements (c) launch windows (d) frequency take-offs.

19. Jupiter's gravity could be used to (a) change the path of a space ship (b) aid in landing on the planet (c) help a craft off the planet (d) all of these.

20. One of the main reasons for using money and the efforts of thousands of people to explore other planets is to (a) make life more interesting (b) use excess energy (c) discover how the earth developed (d) do things never done before.

21. A second reason for using money and people to explore other planets is to (a) find evidence that life exists elsewhere (b) provide a source of resources (c) gain a better understanding of the earth (d) all of these reasons.

22. Evidence collected from observing a huge dust storm on Mars indicates that (a) it increases the surface temperature (b) dust doesn't affect the atmosphere (c) atmospheric pollution could set off an ice age (d) none of these.

23. The most important part of space exploration has been (a) getting to the moon (b) planning a space colony (c) getting close-up views of other planets (d) making it possible to see the earth in new ways.

24. The group of orbiting instruments used for constant study of the earth are called (a) moons (b) artifical instruments packages (c) applications satellites (d) none of these.

25. The artificial satellite designed to investigate the earth's natural resources is called (a) *Telestar* (b) *LANDSAT-1* (c) *COMPACT* (d) *Nimbus*.

26. *LANDSAT*'s orbit causes it to scan every part of the earth's surface every (a) 30 minutes (b) hour (c) 24 hours (d) 18 days.

27. *LANDSAT*-1 can (a) predict weather and reveal minerals (b) reveal minerals and survey the world's agriculture (c) survey the world's agriculture and provide telephone and television links (d) all of these.

28. The satellite used for helping make weather forecasts is called (a) *LANDSAT-1* (b) *COMPACT* (c) *Nimbus* (d) *Weasat*.

29. The *Orbiter* is a rocket used in the (a) *Nimbus* project (b) *LANDSAT* project (c) *COMPACT* project (d) *Space Shuttle* project.

Group B

1. Explain why the gravity barrier is a major obstacle to space study.
2. What is meant by escape velocity? Give three examples.
3. Explain the principle behind the operation of a rocket engine.
4. How can a rocket "fall" continuously in an orbit around the earth?
5. Define the terms apogee and perigee.
6. Explain how earth satellites can appear to remain stationary over one place on the earth.
7. What is the "minimum energy orbit" plan?
8. What is one of the main difficulties in guiding a flight to Venus?
9. Explain what is meant by "launch windows."
10. How could Jupiter's gravity be utilized in a minimum energy flight to Uranus and Saturn?
11. Identify four reasons for sending spacecraft to other planets.
12. What have we learned from the huge dust storms on Mars?
13. Describe the function of *LANDSAT-1*.
14. Explain how satellites aid the world's communications systems.
15. Describe the function of *Nimbus*.
16. State some of the purposes and benefits of the *Space Shuttle*.

pioneers in a new frontier

Ancient Indian Astronomers

Scattered through the mountains and plains of central North America are strange circles of rocks called "medicine wheels." There is evidence that some of these stone monuments were built a thousand or more years ago by Plains Indian tribes. The picture at the left shows an aerial view of the medicine wheel in Alberta, Canada. Notice the lines forming the spokes of the wheel. They were used to mark the position of the sun and certain stars during the cycle of the seasons. These astronomical sightings were probably used as a kind of calendar.

Below is a closer view of a medicine wheel in the Big Horn Mountains, Wyoming. These medicine wheels are similar to other structures found in many parts of the world. Stonehenge in western England, thought to be built by early Druids, shows the same basic circular pattern. Searching for astronomical knowledge has always been a vital part of human history.

A Laboratory in Space

Scientific research enters a new era with the Spacelab. Transported in NASA's Space Shuttle, Spacelab personnel enter the Spacelab for missions lasting from 7 to 30 days. Here the crew of scientists and technicians conduct experiments in space, performed in an environment free of the earth's gravity and atmosphere. The work area is basically a 4.2m diameter pressurized cabin with outside platforms for instruments that will be exposed to the space environment. After returning to Earth, the Spacelab can be fitted for a different group of experiments, then returned to space for another mission.

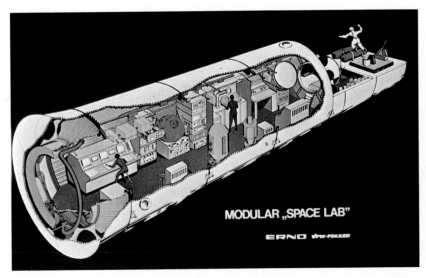

MODULAR "SPACE LAB"

ERNO VFW-FOKKER

People and Space Technology

The three NASA scientists in the photograph below are examining a readout on a new bacteria detector. It is hundreds of times faster than the standard technique used. This new detector was developed as a result of the earlier work done on developing a system to detect life on other planets.

In this photograph, an orthopedic researcher receives data on the walking patterns of a child with cerebral palsy. The information gathered will be important in prescribing physical therapy for individual leg muscles. Small electronic body sensors developed by NASA for space research are used to gather this data.

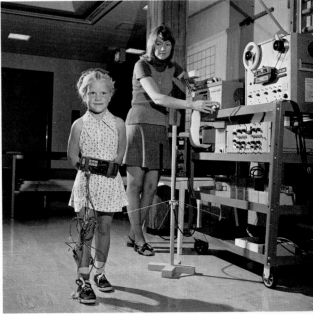

unit 2
the planet earth

6
models of the planet earth

objectives

- [ ] Compare ancient beliefs about the earth's place in the universe with what is believed today.
- [ ] List several reasons why the earth is believed to be a sphere.
- [ ] Describe a method of measuring the size of the earth.
- [ ] Describe how directions and locations can be determined on earth.
- [ ] Explain how map projections can be made.
- [ ] List the uses of maps.
- [ ] Describe how topographic maps can be interpreted.

Compared to the entire solar system, the earth is smaller than a speck of dust in your classroom. But to an insect crawling over the ground, the earth is a jungle of pebbles, fallen leaves, grass blades, twigs, and all sorts of giant barriers. However, if you walked over the earth's surface you would be in about the same position as the insect. Neither of you can see the earth as the ball-shaped planet it is, among a family of planets all moving around the sun. But unlike the insect, you are able to carry in your mind a picture or model that goes far beyond what your eyes and other senses tell you. Until recently, these models of the earth gave us the only clues to the earth's size, shape, and position in our solar system.

Humans probably began their search for an earth-model long before they learned to write. But their earliest recorded ideas of our planet did not correctly describe its size or shape. However, some early Greek teachers thought the earth was round. From these early beginnings, humans created many models to help describe the different features of the planet earth.

The most common models used to describe the surface features of the earth are called maps. A map is a visual model of the earth's surface features. Several of the ways maps can be made accurately, and then used to describe the earth's surface will be taken up in this chapter.

A CLOSER LOOK AT OUR PLANET

The earth's place in the universe. It was natural for ancient people to think of home as the center of all that was known to exist at that time. When wandering tribes settled down to form communities and nations, these too were thought of as centers of the entire area. It was believed that the earth was central to the surrounding heavens which carried the sun, moon and stars around the sky. Each of the ancient civilizations developed a model of the earth to fit its own beliefs. See Figure 6-1. The one thing that all of these models seemed to have in common was that they placed the earth, or at least a part of it, at the center of the universe.

Overcoming these early beliefs required thousands of years of effort. Thoughtful men observed and compared their findings with these ancient models. It was not until men like Galileo and Kepler gathered enough evidence to show that the earth was just one of several minor bodies circling around the sun, that finally moved the earth from its central place in the universe.

The shape of the earth. Evidence for the earth's actual shape can be gathered from several familiar observations:

1. A large body of water shows a curved surface. Ships sailing over the horizon on the sea seem to sink out of sight, the masts or taller parts disappearing last. A more exact way of showing the curve of a water surface would be to set three long poles into a lake bottom. The poles should be set about one kilometer (.6 miles) apart and be of the same length above the water surface. If an observer sights along the tops of these poles, he sees that the middle one is several centimeters higher than the two at either end.

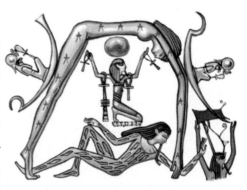

FIG. 6-1A. The universe as conceived by the Egyptians was a combination of gods. Keb, the earth-god, lay beneath the arching goddess of the heavens who was supported by the god of the atmosphere. Boats carrying the sun and moon gods sailed across the heavens.

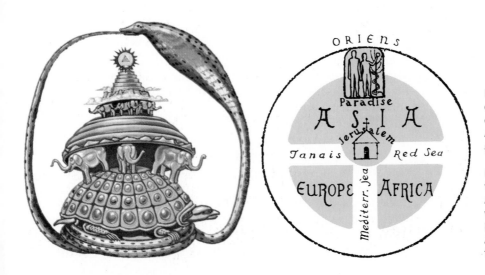

FIG. 6-1B. One Hindu idea of the earth had it supported on the backs of elephants who stood on a turtle. Around the entire arrangement was a cobra representing water.

FIG. 6-1C. One representation of the earth used during the Middle Ages was a disk. The continents were divided by the Red Sea, the Mediterranean, and the Don River. At the center was Jerusalem with the Garden of Eden in Asia.

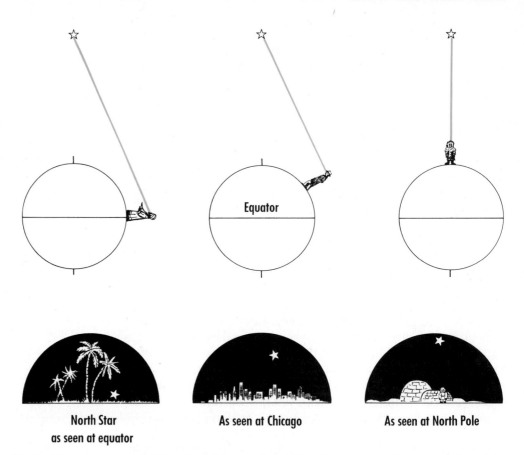

North Star

as seen at equator

As seen at Chicago

As seen at North Pole

FIG. 6–2. The elevation of the North Star (Polaris) above the horizon depends upon the position of the observer between the equator and North Pole.

2. The height or altitude of a particular star changes as an observer moves between the equator and poles. Figure 6–2 explains how the observer's viewing angle of the star is always changing as he moves over the rounded surface of the earth.

3. If the earth were flat, the horizon would always remain fixed along the edge of the earth. Then the horizon would never appear to move closer or farther away as an observer changes his viewpoint. Actually, the horizon does move away as an observer gains altitude and more of the earth's surface becomes visible. This could only happen on a rounded surface.

4. During a lunar eclipse the earth's shadow appears as the arc of a circle as it crosses between the sun and moon.

Other, more direct, evidence of the earth's shape is now available as a result of the exploration of space. Photographs taken by astronauts and from unmanned space vehicles reveal the earth's curved surface. See Figure 6–3.

Examine

How does the shape of the earth affect the number of miles in a degree of latitude as you move from the equator toward the poles?

The size of the earth. Determining the size of the earth has been a more difficult problem than finding its shape. However, more than two thousand years ago a Greek mathematician and astronomer named Eratosthenes (air uh TAHS thuh nees) measured the size of the earth with surprising accuracy. At the time he accomplished this, he was head of the library at Alexandria, the greatest institution of learning in the ancient world. He knew of a deep well in a city to the south of Alexandria where the sun's rays reached the bottom only once a year on about June 21. He reasoned that he could calculate the circumference of the earth if he knew the distance between

FIG. 6–3. Photo of the earth taken from the moon aboard *Apollo XVII*.

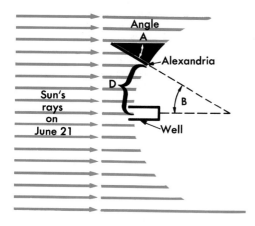

FIG. 6–4. Eratosthenes' method for measuring the size of the earth.

Discuss

How has our knowledge about the shape of the earth affected procedures for navigation over long distances?

Alexandria and the well, plus the angle of the noonday sun in Alexandria on June 21. His method is shown in Figure 6–4.

Angle A is equal to angle B. According to the principles of geometry, this angle has the same relation to 360° as the distance D does to the earth's circumference. Eratosthenes found that this angle was slightly more than 7 degrees. Since 7 degrees is about 1/50 th of a complete circle, the distance between Alexandria and the well must be about 1/50 th of the earth's circumference. Eratosthenes' calculation for the earth's circumference turned out to be about 46,250 kilometers (about 28,721 miles). This is not too far from the modern value of 40,000 km (25,000 miles). His final result was remarkably accurate, considering his crude measurements. Fortunately, the various errors almost canceled each other. Basically the same method used by Eratosthenes is still in use today.

During the seventeenth century, Sir Isaac Newton came to the conclusion that the earth could not be perfectly round. He reasoned that its rotation caused a slight flattening at the poles. His theory was proven correct in 1743 when scientists found that the earth's circumference at the equator was slightly larger than its circumference around the poles. This meant that the earth had the basic shape of an *oblate* (flattened) *spheroid.* That is, the outline of the earth is not a perfect circle.

However, the earth is so very close to a perfect sphere that it is rounder for its size than any bowling ball or basketball. Careful analysis of disturbances in the orbits of man-made earth satellites shows that the equatorial bulge is actually not at the equator as Newton had predicted. Instead it is a little south of the equator. This new location gives the earth an irregular shape that has been compared to the shape of a pear. However, this is not an accurate description. Since the earth's shape is so nearly perfectly round it is usually drawn as a regular circle. Figure 6–5 shows the generally accepted dimensions for the size and shape of the earth.

Directions and location on the earth. If the earth did not turn on its axis, it would be very difficult to describe any direction on its surface. Since the earth is very nearly a perfect sphere, it has no top, bottom, or sides to use in establishing direction. But the earth does rotate. Thus the ends of its axis of rotation, the north and south poles, provide the reference points needed. Based on the position of the poles, the four cardinal points of direction (N,E,S,W) can be located on the earth. The north-south direction runs along the axis line connecting the two poles. The east-west direction runs at right angles to the polar axis along a line parallel to the equator.

In describing a location on the earth's surface, we make use of imaginary lines. For example, to describe changes in location in a north-south direction, a system of imaginary lines drawn parallel to the equator is used. See Figure 6-6. The north-south location from the equator, of any place on the earth's surface, is called the *latitude* of that point. These imaginary lines are called *parallels of latitude*.

The latitude of a place is established in the following way: The distance from the equator to either of the poles is one-quarter of a full circle around the earth, or 90° of the full 360° circle. If the equator is taken as 0°, the location of any parallel of latitude can be described as a certain number of degrees north or south of the equator. For example, both the north and south poles have latitudes of 90°. Thus a point half way between the equator and one of the poles has a latitude of 45°. Of course, it is necessary to state whether the parallel lies north or south of the equator. Washington, D.C., for example, has a latitude of about 39°N. This fixes its position as 39° north of the equator.

To be more precise, each degree of latitude is divided into 60 equal parts called *minutes* (symbol '). A more precise latitude for Washington, D.C. is 38°53′N. We can obtain even greater precision by dividing minutes into 60 equal parts called *seconds* (symbol ″). In actual distance over the earth's surface, a degree of latitude (60 minutes) is equal to about 111 km (69 mi). A minute of latitude is 1.85 km (1.15 miles, which is one *nautical* mile.

Given the latitude of a particular place we can only determine our north-south location. That is, our distance in degrees north or south of the equator. To locate where we are along this line of latitude, we must also know our east-west position. To do this, another set of imaginary lines, called *meridians*, are drawn extending from pole to pole. Each meridian outlines a complete circle when drawn around the entire earth. See Figure 6-7. The east-west position of a place, called *longitude*, is established by use of these meridian lines. This is done by using degrees, as with latitude, to establish the position of a certain meridian. However, there is no natural starting point that is called 0° longitude. To solve this problem, a particular meridian has been selected by agreement among all na-

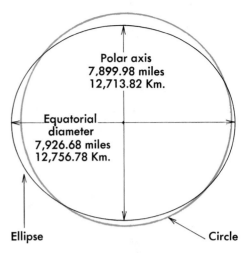

FIG. 6-5. For most purposes, the model of the earth with the shape shown in the diagram is satisfactory. The true shape of the earth is very slightly irregular, probably a little like a pear. However, the earth would appear as a perfect sphere if seen from a distance.

FIG. 6-6. Parallels of latitude are shown here at intervals of 15°. Note that each parallel forms a complete circle around the earth.

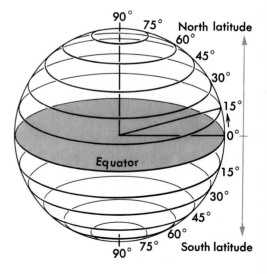

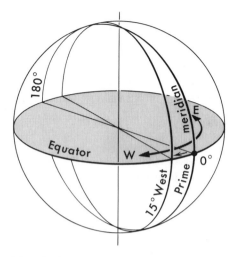

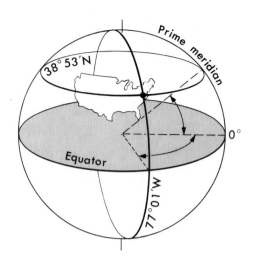

FIG. 6–7. Meridians of longitude are shown here at intervals of 15°, measured east or west from the zero degree (0°) meridian.

FIG. 6–8. The exact location of Washington, D.C. is determined by measuring longitude west of the prime meridian, and latitude north of the equator.

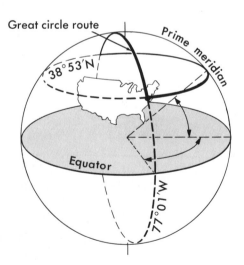

FIG. 6–9. A traveler flying a great-circle route from Washington, D.C. to Canton, China would have to pass over or near the North Pole.

tions. The meridian which passes through Greenwich, England, was selected as the 0° meridian. Greenwich is close to London and was originally the location of the Royal Observatory. The 0° meridian is often called the *prime meridian.*

Locations to the east of the prime meridian are said to have *east longitude;* those to the west have *west longitude.* The 180° meridian (directly opposite the prime meridian on the earth) also separates east and west longitude. Washington, D.C., is located west of the prime meridian; its longitude is about 77°W. to fix the position of Washington precisely, we say it is 38°53′N and 77°1′W. See Figure 6–8.

Another type of imaginary line often used on the earth is called a *great circle.* This is any line that completely circles the earth and divides the globe into two equal halves. The equator is a great circle, and each circle of longitude is a great circle. Great circles are useful in aviation because they indicate the shortest distance between any two points. See Figure 6–9.

MAPPING THE EARTH'S SURFACE

How are maps made? A map is a model of the earth's surface, or a part of that surface taken from a globe and reproduced on a flat area such as a piece of paper. Maps can never be pictures of the earth's surface in the same way that photographs are. Since the earth's curved surface is drawn on a map as if it were flat, a part of the surface shown will always be distorted. For a map to give accurate information, certain systems must be followed in making and using a particular map.

It is possible to show how a map distorts the earth's surface by using the skin of an orange. If a large piece of orange skin is flattened, its shape will be changed by sketching and tearing. The larger the piece of skin, the more it must be distorted to flatten it. Similarly, the larger the part of the earth's surface that is being shown on a map, the greater the distortion will be. A map of a smaller area such as a city will have very little curvature and will show only slight distortion due to flattening.

Since the earth is almost perfectly round, its surface can be correctly shown only on another round body such as a globe. However, globes that are large enough to show small details are expensive and difficult to handle. Thus to represent the planet earth, we must rely mostly on maps drawn on flat surfaces.

The various ways that the curved surface of the earth can be transferred onto a flat map with the least amount of distortion are called *map projections*. Most map projections are the result of mathematical calculations. However, with just a little effort it is possible to picture the method used to produce them. Imagine a transparent globe, lighted from inside. Surface markings cast shadows on a piece of paper that is held against the globe. Differences in the way the flat paper is held against the lighted globe will produce a variety of patterns in the shadows cast on the paper.

A commonly used projection is a map made by wrapping a sheet of paper into a cylinder around a globe. If the cylinder is unrolled, the map projection would appear as shown, in Figure 6–10. The meridians of longitude are shown as equally spaced, straight, parallel lines. Since we know that meridians on a globe converge at the poles and are farthest apart where they cross the equator, this type of projection introduces a distortion of land areas that becomes greater near the poles. See Figure 6–11. Because the cylinder is in contact with the earth's surface only at the equator, this projection represents most accurately only the area near the equator.

However, the advantages of this type projection make it a most valuable tool for navigation. Some of these advantages are: All compass directions are shown as straight lines. The four cardinal points of the compass are located on the four sides of the map. All lines of latitude and

Discover
Look up the latitude of your school in an atlas or almanac. The latitude of a nearby city would be close enough. From your latitude in degrees, find the distance of your school from the equator.

FIG. 6–10. A cylindrical map projection can be visualized as the result of projecting the lines of latitude and longitude from the center of a globe onto a cylinder.

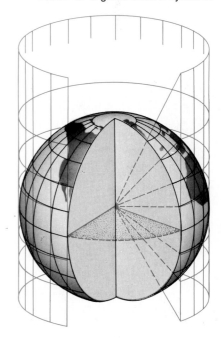

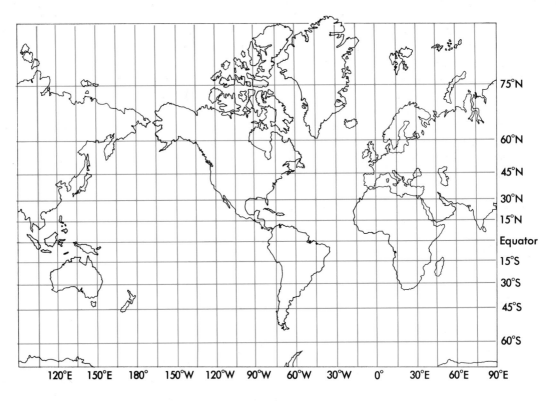

FIG. 6–11. This diagram illustrates how a cylindrical projection distorts high-latitude regions (above). The correct size is shown below.

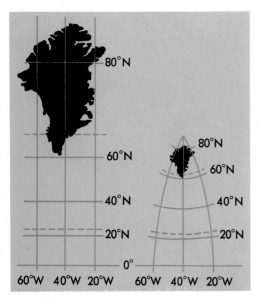

longitude are clearly shown. You are probably familiar with this projection since it is most commonly used in your textbooks.

A second type of projection can be constructed by placing a flat sheet of paper on a globe so that it touches the surface of the earth at only one point. Distortion on this map is due to the unequal spacing of the parallels of latitude as you move farther from the point in contact. See Figure 6–12.

The advantage of this type of projection for navigation is that all great circle routes are shown as straight lines. This is extremely helpful for plotting the shortest route between any two points on the earth's surface. This kind of map projection is often used by navigators to plot polar routes for air travel.

A third type of projection uses a cone as the basis for a map projection. The cone is arranged so that the axis of the globe is in line with the axis of the cone. The parallel of latitude where the cone and the globe are in contact will show the least distortion on the completed map. See Figure 6–13. This type of projection is often used to accurately map relatively small areas. These areas can be taken from the part of the projection showing the least distortion. To map a number of neighboring small areas of the earth's surface, a series of cone projections may be used. This is called a *polyconic* projection. Each cone comes in contact with the earth at a slightly different latitude. The different latitudes where each cone touches the globe are fitted together to form a continuous map as shown in Figure 6–13. The advantage of this type of projection is that the shape and size of relatively small areas on the map are very nearly the same as those on the globe.

Using a map. People use maps to find out many kinds of information about the earth's surface. *Political* maps show national and local boundaries clearly. They often indicate the relative sizes of towns and cities along with their political importance. Political maps are often combined with *relief maps*. Relief maps show elevation of the land surface by using different color keys. *Navigation charts* provide navigators with routes and distances. *Hydrographic maps* are useful in showing depths of water and the shape of the sea floor. *Weather* and *climate maps* also give information about the general conditions of the atmosphere. *Geologic maps* show the arrangement of rock formations and are particularly useful in the mining and petroleum industries.

To understand any map, the first thing you must know is its relation to compass directions. The most common method of showing compass directions on a map is to make the top of the map north. Looking at the map, then, right is east, left is west and the bottom is south. Meridians of longitude are usually drawn as lines running from top to

FIG. 6–12. Left, the principle of a map projection on a plane surface. Right, an actual map produced by this method.

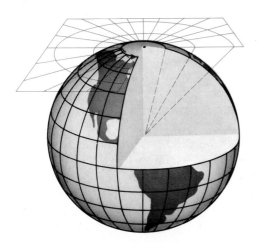

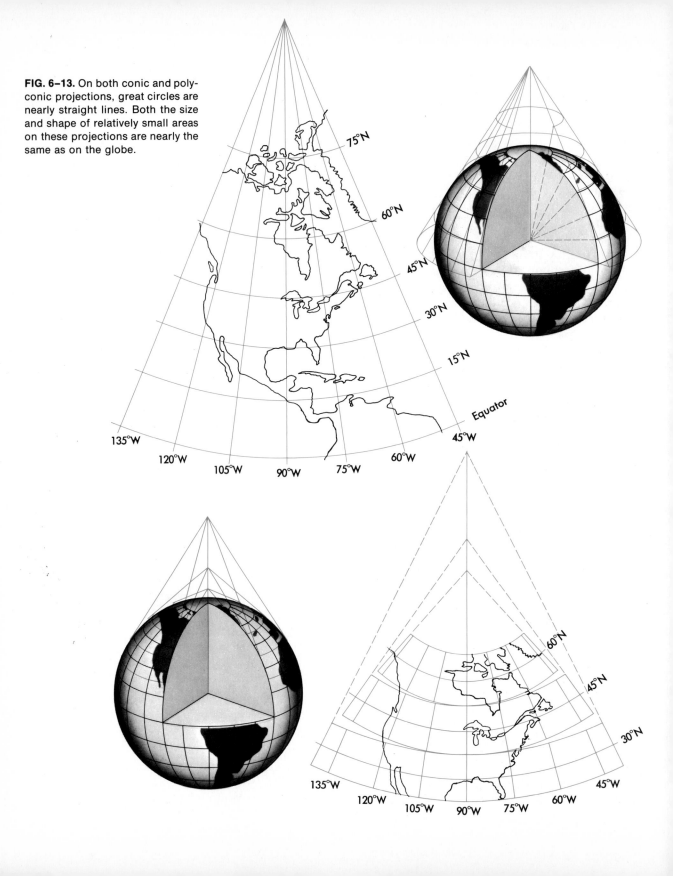

FIG. 6–13. On both conic and polyconic projections, great circles are nearly straight lines. Both the size and shape of relatively small areas on these projections are nearly the same as on the globe.

bottom. Parallels of latitude are lines running from side to side. Since these lines may either be straight or curved, depending upon the map projection, directions must always be read in relation to the parallels and meridians. This means that north is at the top of the map only if the meridians shown actually run from top to bottom.

In addition to knowing directions, the map user is usually interested in finding distances. All maps must indicate the relationship between actual distance on the earth and the same distance measured on the map. This relationship is called the *scale* of the map.

A map may be designed to show a large area of the earth. If this is the case, the scale selected will allow a short distance on the map to represent a large distance on the earth. For example, one inch on the map may represent 100 miles on earth. More detailed maps showing smaller areas use a larger scale, such as one inch to the mile. The scale of a map is commonly shown as a *graphic scale*. This is a line which is divided into parts marked to represent distances on the earth. Graphic scales are frequently used on maps covering smaller areas. Thus distances can be quickly found by directly comparing a measurement on the map to the number of divisions it covers on the graphic scale.

Another way of indicating the scale of a map is to use a fraction such as 1:62,500. This is called a *fractional scale*. It means that 1 unit of measurement on the map represents 62,500 of the same units on the earth. Occasionally the map scale may be given in a form such as "one inch equals one mile." This is known as a *verbal scale*.

TOPOGRAPHIC MAPS

A guide to the study of landforms. Some earth scientists use maps to study the landforms that give shape to the earth's surface. All the details that make up the surface features of the land are called its *topography*. A map made to show these details is known as a *topographic map*.

The governments of most countries make topographic maps of their territories for military, scientific and commercial use. In the United States, the Geological Survey, a branch of the Department of Interior, has mapped a large part of the country. The results are available in the form of detailed maps called *topographic sheets* or *quadrangles*. Most of these maps represent an area that covers 7.5 min-

Observe
Look at several road maps available at your local gasoline station. See if you can determine the type of projection used to draw these maps.

FIG. 6–14A. If the sea should rise by a number of equal increases, the new shorelines formed would correspond to contour lines showing the shape of the island.

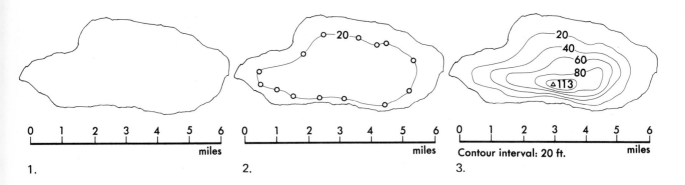

1.

2.

3.

Contour interval: 20 ft.

FIG. 6–14B. This sea island is 6 miles long, 3 miles wide, oval shaped, and 113 ft above sea level at its highest point. **1.** The map above shows only the shape of the land at sea level. The scale shows the length and width of the island. It does not show the *elevation*, the *steepness* of the slope, or the *shape* of the land above sea level.

2. A mapmaker surveys the island and proceeds to turn this map into a contour map. On the map he locates a series of points shown by his survey to be 20 ft above sea level. He then joins these points with a *contour line:* a line drawn through all points at the same height above sea level. Every point on this line is 20 ft higher (vertically) than any point at sea level, regardless of the uneven spacing between successive contour lines.

3. Now the mapmaker draws additional contour lines, showing where the island reaches the 40, 60, 80, and 100 ft elevations. Notice that the contour lines show the *elevation of the island, the steepness of the slopes* (which depends upon the closeness of the contour lines), and the *shape of the land above sea level.* Notice also that a symbol (△) called a bench mark is used to denote the elevation of any point which falls between the contour interval of 20 ft.

utes of latitude and 7.5 minutes of longitude. See pages 132–133. Topographic maps are also available for some areas which are 15 by 15 minutes; 30 by 30 minutes, and 1 degree by 1, 2, or 3 degrees.

Using topographic maps. Topographic maps almost always show the elevation and the shape of the land by using *contour lines.* A contour line is drawn on a map so that it connects all points on the ground of the same elevation. The shape of the contour lines gives us an idea of the shape of the land. The way these contour lines are able to show land features can be illustrated by a small island as shown in Figure 6–14A. Its shoreline could be one contour line connecting all places on the island at sea level elevation. Suppose the sea rose. Then a new shoreline with a new slope would be created. If the sea level should continue to rise, each time by the same amount, a series of new shorelines would be formed, each one outlining a new shape. If seen from above, each of the shorelines would represent a definite elevation above the original sea level.

The elevation of the land everywhere, no matter how far from the sea, is measured from *mean sea level.* This is a point midway between the highest and lowest tide. Mean sea level represents zero elevation or the starting point from which all elevations of the land are measured. Each new shoreline would at the same time also represent a contour line used to show the shape of the island. See Figure 6–14B. On an actual topographic map the contour lines are drawn from points of known elevation. The contour lines are drawn to connect points which are all at the same elevation above sea level.

The difference in elevation between any two contour lines is called the *contour interval.* A map maker chooses an interval suited to the size of the map and the topography of the region. The contour interval he chooses depends upon the *relief* of the land. Relief is defined as the difference in elevation between the highest and lowest points of the area being mapped. In maps of very mountainous areas where the relief is high, the contour interval may be as great as 50 or 100 feet to prevent crowding of the contour lines. In maps of low relief, some flat areas may have a contour interval of only one or two feet.

Interpreting a contour map. Just as printed words on a page transmit ideas, contour lines on a map are able to give a clear picture of the elevation, steepness and shape of the land surface. However, it takes some training and practice before contours on a topographic map can be interpreted. The following points should be kept in mind when working with topographic maps.

1. *A contour line connects all points having the same elevation.* The contour interval determines the elevation at which each contour line will be drawn. If a contour interval is 10 feet, contours will be shown for elevations of 10, 20, 30, 40, 50, 60 feet and so on. Usually, every fifth contour line is heavier and much darker. These are called *index contours* and are used to mark the actual elevation on the map.

Only the elevation of points directly on contour lines can be determined exactly. A point located between two contour lines has an elevation somewhere between that of the two lines. See Figure 6–15A.

Topographic maps almost always show a few points whose exact elevations are marked, although they are not located on contour lines. The elevations of these points are measured during the process of making the map. These known elevation points are marked with an X or BM (for "Bench Mark") printed in brown or black ink. Their exact elevation in feet is located near the mark.

2. The steepness of a given land surface can be easily determined from the contour lines. *Contours spaced widely apart show a gradual change in elevation and indicate nearly a flat surface. Closely spaced contours mean the elevation increases quickly and indicate a steep slope.* Contour lines that are almost touching indicate a very steep or vertical cliff. Where a map shows a vertical cliff grading into a more gentle slope the contours may appear to split. This, however, is only the effect of separate contour lines running together. See Figure 6–15B. A single contour line can never split. In the same way,

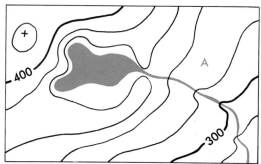

FIG. 6–15A. The elevation of point A is more than 320 feet but less than 340 feet. What is the contour interval?

FIG. 6–15B. A coastal valley shown in a side view or *profile* (top), and the same area represented by contour lines (bottom). The darker shading in the profile indicates the steepness of the slope also shown by the closeness of the contour lines. Notice that the 50-foot contour line intersects the stream channel at only one point. Explain.

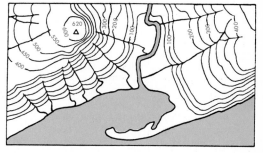

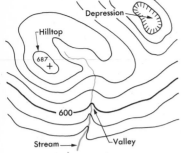

FIG. 6–16. Closed contours may indicate a hilltop as on left; with hachures they may signify a depression. Notice the heavier index contour and the contours bending in a V-shape with the point of the V directed up the stream valley.

Symbol	Color	Meaning
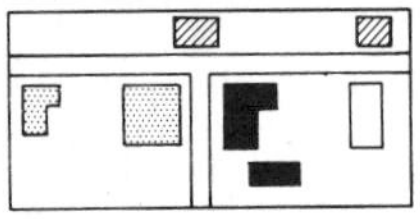	Black	Buildings
	Black	Church and school
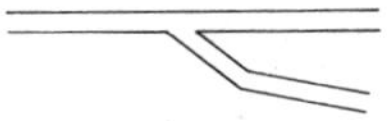	Red or Black	Road or highway
	Black	Railroad
	Blue	Stream
	Blue	Intermittent stream
	Blue	Lake or pond
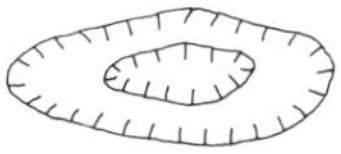	Brown	Depression
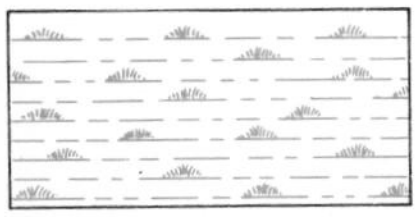	Blue or Green	Marsh or swamp

FIG. 6–17. The symbols shown here are some commonly used on topographic maps.

contour lines can never cross, since the point of crossing would have two elevations at the same place. The only exception would be a place where there is an overhanging cliff. But this is a very rare feature.

3. *When crossing a valley, contours bend to form a V shape.* The point of the V shows the upslope direction of the valley. If there is a stream flowing in the valley, the V in the contour lines will point upstream, or in the direction from which the water flows.

4. *Contour lines which form closed loops indicate a hilltop.* All contour lines join themselves again at some point. On a map of a small land area, however, most contours run off the map before closing on themselves.

5. A hole or depression is also indicated by using a closed contour loop. But, unless some indication is given that a depression is shown, the resulting contours might be mistaken for a hilltop. *A depression contour is indicated by use of hachure lines.* A hachure line is a short straight line pointing in the direction of the depression. See Figure 6–16.

Some of the colors and symbols used on topographic maps are shown in Figure 6–17.

Locating landforms. One of the most important uses of any map is to locate a particular point or place. With topographic maps the three methods commonly used are:

1. *Relation to easy-to-identify features.* The simplest and often most convenient method for locating a point is to give its distance and direction from any easily located feature on the map. The feature selected may be a mountain, lake, city, or any other feature that cannot be easily mistaken.

2. *By latitude and longitude.* On the topographic maps published by the United States Geological Survey, the eastern and western boundaries are marked as meridians of longitude. The longitude is marked at both the top and bottom of the maps in degrees (°) and minutes ('). Usually at least two other meridians are indicated on the maps between those at the edges.

The top and bottom boundaries on the map are marked as parallels of latitude. These parallels are marked at the left and right sides of the map. Usually at least two additional parallels are drawn or indicated by cross hairs (+) at 5 minute (') intervals.

3. *By township and range.* In the nineteenth century settlers began to move into the lands of the American mid-

west. At that time, a system of land boundaries was de-
vised to establish ownership. The basic plan divided the
land surface into squares six miles on a side. These areas
were called *townships.* Each township was subdivided
into thirty-six *sections,* each of which was usually one
mile on each side. Each section could be further subdivided
into half-sections, quarter-sections, or sixteenth sections.

On a map, townships are described by numbers and
compass directions from some selected point of latitude
and longitude. Vertical rows of a township square are
called ranges and are numbered from east to west. Figure
6–18 shows how a point is located by means of township
and range. The location of point X by this method would
be described, for example, as: in the southeast quarter
of the northeast one-fourth of Section 24, of the township
which is second in the south horizonal row, and second in
the west vertical row. States not included in this system of
land division are all the eastern coastal states (except
Florida), West Virginia, Kentucky, Tennessee, Texas, and
some parts of Ohio.

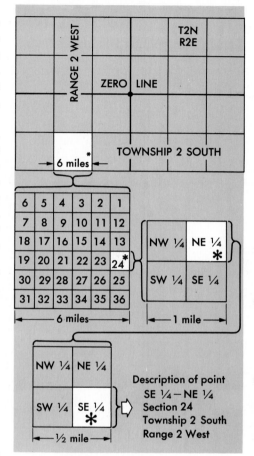

FIG. 6–18. The location of a point
by the township and range method.

activity

Studying a topographic map. Refer to the topographic
map reproduced on pages 132–133. Answer the following
questions that require the application of the rules for interpreting
contour maps.

1. In what general direction does the Neversink River flow?

2. Determine the elevation to the closest contour of the
   bench mark (△) shown in the south–central part of the
   map. (Just below the G in FORESTBURG).

3. What is the latitude and longitude of Wolf Reservoir to the
   nearest minute? (At the cross hairs).

4. Is the railway line entering the city of Monticello ascending
   or descending as it moves north?

5. Which of the following terms would best describe the entire
   area lying between Route 209 and the railroad in the lower
   southeast corner of the map: plain, plateau, valley,
   peneplane, river basin? Explain your answer.

6. Using the scale found on the bottom of the map, determine
   to the nearest half mile the number of miles along Route 17
   from Mastens Lake to Bridgeville.

Construct
Draw a simple topographic map
of your school playground.
Indicate those features that
would make it easy to identify. If
you assume the ground to be at
base level, determine the highest
point in the area.

MONTICELLO QUADRANGLE
15 MINUTE SERIES

WAR DEPARTMENT
CORPS OF ENGINEERS, U.S. ARMY

NEW YORK 1:62,500

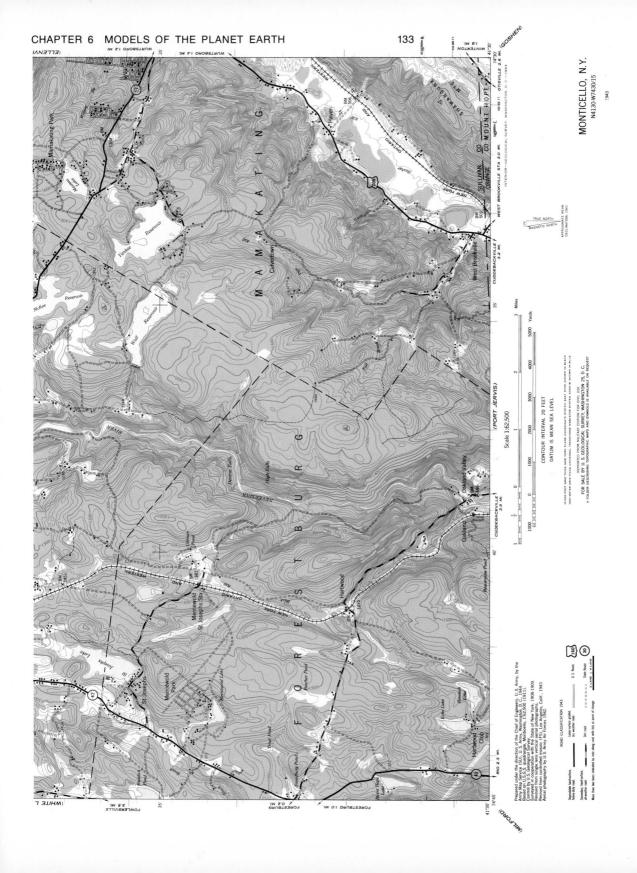

Scale 1:62,500

CONTOUR INTERVAL 20 FEET

DATUM IS MEAN SEA LEVEL

MONTICELLO, N.Y.

N4130-W7430/15

1943

TRUE NORTH

MAGNETIC NORTH

APPROXIMATE MEAN
DECLINATION, 1943

Prepared under the direction of the Chief of Engineers, U.S. Army, by the
Army Map Service (SU), U.S. Army, Washington, D.C., 1944.
Based on U.S.G.S. quadrangle, Monticello, 1:62,500 (1911).
Control by U.S. Geological Survey.
Surveyed in cooperation with the State of New York, 1908-1909.
Reprinted from single lens vertical aerial photographs.
Revised from controlled mosaic (FS), Los Angeles, Calif., 1943.
Aerial photography by U.S. Army Air Forces 1942.

REPRINTED FROM MILITARY EDITION FOR CIVIL USE
FOR SALE BY U.S. GEOLOGICAL SURVEY, WASHINGTON 25, D.C.
A FOLDER DESCRIBING TOPOGRAPHIC MAPS AND SYMBOLS IS AVAILABLE ON REQUEST

ROAD CLASSIFICATION 1943

Dependable hard-surface,
heavy-duty road.

Secondary, hard-surface,
all-weather road.

Loose surface graded,
dry weather road.

Dirt road.

U.S. Route

State Route

More than one have been indicated by note along road with tick at point of change.

INTERIOR—GEOLOGICAL SURVEY, WASHINGTON D.C.—1944

7. Locate McKee Reservoir along Route 17. Just to the north-west of the reservoir is a small lake. Is the stream connected to the lake an inlet or an outlet from the reservoir? How do you know?

8. Locate Anawana Lake due north of the town of Monticello. A second lake, connected by a main stream to Lake Anawana, lies just to the southeast. Which lake has a higher elevation? How do you know?

VOCABULARY REVIEW

Match the word or words in the column on the right with the correct phrase in the column on the left. *Do not write in this book.*

1. Measured the earth's circumference.
2. The distance around the earth.
3. The shape of the earth.
4. The most common models of the earth's surface.
5. Necessary to transfer curved surface of the earth to the flat surface of a map.
6. The north-south location of any place on the earth's surface.
7. 1/60th of a degree of latitude.
8. The east-west location of any place on the earth's surface.
9. The meridian which passes through Greenwich, England.
10. Any line drawn on the earth's surface which divides it into two equal parts.
11. Show national and local boundaries clearly.
12. Show depths of water and sea floor shape.
13. The relationship between a distance on the earth and the same distance measured on a map.
14. A map which shows the shape of the land surface.
15. A line drawn through all points with the same elevation.

a. circumference
b. maps
c. longitude
d. prime meridian
e. Eratosthenes
f. oblate spheroid
g. minute
h. political maps
i. latitude
j. contour line
k. map projection
l. geologic map
m. great circle
n. topography
o. scale
p. hydrographic maps
q. topographic map

QUESTIONS

Group A

Select the best term to complete the following statements. *Do not write in this book.*

1. Early models of the earth gave clues to the earth's (a) size (b) shape (c) position in the solar system (d) all of these.

2. The most common models used in scientific descriptions of the earth are (a) layered (b) spheres (c) maps (d) full scale.

3. The earliest models of the universe placed the earth (a) on a crystal sphere (b) at a point which moved around the sun (c) at the center (d) at the top.

4. Which of the following statements is evidence for the curved surface of the earth? (a) The horizon is always at the edge of the earth. (b) The horizon remains fixed as an observer climbs to a higher viewpoint. (c) Ships suddenly disappear when they sail away. (d) The horizon moves away as an observer gains altitude.

5. Which of the following statements is *not* evidence for the spherical shape of the earth? (a) Ships sail out of sight over the horizon. (b) The elevation of a star changes from one place to another. (c) The earth's shadow is curved. (d) The stars move in circles around the pole star.

6. One of the first measurements of the size of the earth was made by (a) Alexander (b) Eratosthenes (c) Astronauts (d) Sir Isaac Newton.

7. The 7 degree angle which was part of the early measurements of the earth's size is nearest what part of a complete circle? (a) 1/50 (b) 1/360 (c) 1/7 (d) 1/5.

8. Which of the following statements is true of Eratosthenes' measurement of the earth's circumference? (a) his calculation of the earth's circumference was about 46,250 kilometers (b) his calculations were remarkably accurate (c) he used geometric principles (d) all of the above.

9. The earth's circumference is nearest (a) 46,250 km (b) 26,660 km (c) 40,000 km (d) 25,000 km.

10. Sir Isaac Newton decided that the earth could not be perfectly round because (a) it is rotating (b) it has ice caps (c) rivers wear it away (d) rocks are heavier than water.

11. The earth has a shape which is (a) more like a pear than a sphere (b) more like an orange than a sphere (c) more like a sphere than a basketball (d) more like a sphere than the best sphere that can be made.

12. Locations north or south of the equator are described by a system of lines called (a) parallels of latitude (b) lines of longitude (c) meridians (d) great circles.

13. Which of the following is *not* the latitude of a place on the earth (a) 90°N latitude (b) 120°S latitude (c) 30° latitude (d) 0° latitude.

14. Zero degrees latitude is the location of (a) the North Pole (b) the South Pole (c) the Equator (d) a point halfway between the Equator and the North Pole.

15. Washington, D.C. has a latitude nearest (a) 38°53'N (b) 38°S (c) 39°E (d) 53°38'W.

16. A minute of latitude is most nearly (a) 100 km (b) 60 km (c) 2 km (d) 1 km.

17. A minute of latitude is sometimes called (a) 1 mile (b) 1 second (c) 1 kilometer (d) 1 nautical mile.

18. Meridian lines are used to determine (a) latitude (b) north-south position (c) distance from the North Pole (d) longitude.

19. A degree of latitude is equal in distance to a degree of longitude (a) nowhere on the earth (b) at the equator (c) at the poles (d) only at the prime meridian.

20. California is located at about 120 degrees (a) north latitude (b) south latitude (c) east longitude (d) west longitude.

21. Which of the following is *not* a great circle? (a) the 45° parallel of latitude (b) the equator (c) the 180° meridian combined with the prime meridian (d) any circle on the earth's surface which passes through both the North and South Pole.

22. A map projection has the least distortion at the (a) equator (b) poles (c) point it touches the globe (d) points far from the point it touches the globe.

23. The only undistorted representation of the earth's surface is (a) a cylindrical map projection (b) a globe (c) a conical map projection (d) a flat surface map projection.

24. If north is at the top of a map, east would be (a) at the bottom (b) to the left (c) to the right (d) at the right or left.

25. The scale is most likely to be correct for a whole map if it is the map of (a) a large area (b) a small area (c) an ocean area (d) a land area.

26. Close spacing of contour lines on a topographic map indicates (a) a depression (b) a steep slope (c) a gentle slope (d) a flat area.

27. Contour lines which cross (a) never occur (b) occur only on flat areas (c) occur only on steep slopes (d) occur only where there is an overhanging cliff.

28. A township is a square whose area is (a) 6 square miles (b) 12 square miles (c) 24 square miles (d) 36 square miles.

29. Index contours are always (a) even numbers of feet (b) multiples of 10 (c) marked for the elevation represented (d) drawn as dashed lines.

Group B

1. What is an oblate spheroid?

2. What is the latitude and longitude where the prime meridian crosses the equator? What is the latitude and longitude of the North Pole?

3. What is meant by the relief of the land?

4. Give another name for a topographic sheet and tell what it is.

5. The capital of which state in the United States is at 38°35′ north latitude and 121°30′ west longitude? Which at 46°48′N and 100°47′W?

6. To the nearest degree, what is the latitude and longitude of your city?

7. Why does the distance measured by a degree of latitude always stay the same while the distance for a degree of longitude varies?

8. If the earth did not rotate, how would you establish reference points for directions?

9. Is a degree of longitude the same distance on the earth as a degree of latitude? Explain.

10. Can contour lines cross each other? Explain.

11. How many kilometers are equal to one degree of latitude?

12. How many kilometers are equal to one minute of latitude?

13. How many meters are equal to one second of latitude?

14. Some map projections are made by placing a cylinder around a globe of the earth. What effect would it have on the map projection if the diameter of the cylinder were twice the diameter of the globe?

15. A mile is about 1600 meters. Which is larger, a mile or a nautical mile? How many meters larger?

16. Two cities north of the equator are at the same latitude. They are connected on a map by a line which is a great circle route between them. Where will the great circle lie in relation to the parallel of latitude for the two cities?

17. The scale of a map is 1:24,000. What distance on the map would represent a distance of 2.4 km on the earth? Express your answer in centimeters.

18. Explain Eratosthenes' method for measuring the earth's circumference.

19. Suppose that Eratosthenes had measured the distance between the well and a city 550 kilometers to the north. Suppose also that he found the angle to be 5.0 degrees. What value would he have obtained for the earth's circumference?

7
earth chemistry

objectives

- [ ] List the three fundamental particles found in an atom.
- [ ] Describe the characteristics of these particles.
- [ ] Explain the role of atoms in the makeup of elements.
- [ ] Define the terms atomic number and atomic mass.
- [ ] Determine the number of particles in an atom, when given the atomic number and atomic mass.
- [ ] Describe two ways electrons are involved in chemical bonding.
- [ ] Explain how forces that hold molecules together form a gas, liquid, or solid.

Moon rocks are not very different from any rocks found on the earth. Rocks found on Mars also strongly resemble earth rocks. Scientists believe that pieces of the solid surface of any of the inner planets will not differ greatly from the many rocks on the earth's surface. If there are other systems of planets around distant stars, fragments from many of those planets will probably be very similar to some kinds of earth rocks. Why is there such a similarity between earth rocks and rocks from the moon, Mars, or even other solar systems?

The reason we make this conclusion is based on our knowledge of how all matter is put together. These ideas are part of the *atomic theory*. This theory assumes that every substance, whether a part of the earth or part of Mars, is made up of atoms. In addition, the atomic theory states that there are only a certain number of the kinds of atoms in the universe. Thus, in all forms of matter there are only different arrangements of these same kinds of atoms. Thus, rocks found on any planet must contain the same kinds of atoms as those found on earth.

Although the atomic theory allows us to make predictions about the kind of matter found on other planets, its greatest importance is in understanding the composition of the earth. The atomic theory provides us with a tool for studying the materials that make up this planet.

ATOMS

Structure of atoms. One of the most important things to keep in mind about the atom is its size. A single atom is so small that it is difficult to even imagine. In fact, no one has

138

actually seen a single atom! It would take more than a million average atoms, side by side, to equal the thickness of the paper on which this page is printed. Anything so small cannot be studied individually. Again, it was necessary to invent a scientific model to describe the structure of a single atom.

Experiments carried out over hundreds of years have shown that atoms must have some connection with electricity. Scientists of the eighteenth century, such as Benjamin Franklin, demonstrated that matter contains two kinds of electricity. It was Ben Franklin who first called the two kinds of electrical charge "negative" and "positive."

FIG. 7–1. An area on earth seen from five increasingly closer viewpoints. (A) An astronaut's view of the Salton Sea in southern California. The area circled is shown from a distant view at ground level (B) and at close range (C). A mineral sample taken from the rock of mountains is shown in (D) along with its atomic make-up in (E). (NASA–7–1A)

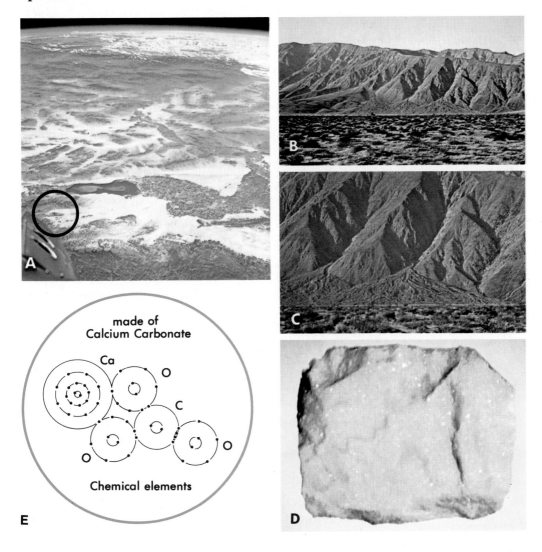

made of
Calcium Carbonate

Ca

O

C

O

O

Chemical elements

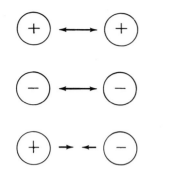

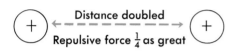

FIG. 7–2. The rules which explain attraction and repulsion of electrically charged bodies are illustrated. Notice in the lower diagram that increasing the distance separating the charged bodies greatly decreases the force acting between them.

Experiment
Use magnets to study attraction and repulsion. Use electrically charged objects for the same kind of investigation.

The model which best accounts for the various properties of individual atoms has three smaller parts to its structure. These fundamental atomic particles are *electrons, protons,* and *neutrons.* Models which describe the structure of atoms use different numbers and arrangements of these three basic atomic particles. One of the most important things to know about electrons and protons is that they carry electrical charges. Electrons carry a negative electrical charge; protons carry a positive charge. When any object has an electrical charge, it is able to influence another electrically charged body. If two objects have the same charge, such as two electrons, they will repel each other. Two positively charged bodies, such as two protons, will also repel each other. This rule is: *Like charged objects repel.*

On the other hand, charges that are not alike, attract one another. An electron and proton, for example, will be drawn toward each other because they carry opposite electrical charges. The rule here is: *Unlike charged objects attract.* The force of attraction or repulsion between charged bodies depends on the distance between them. The rules of electrical attraction and repulsion are shown in Figure 7–2. The third basic particle, the neutron, has no electrical charge. It is electrically neutral.

Besides the opposite electrical charges which make them attract each other, electrons, and protons in atoms have another important difference. All of the protons within a particular atom are packed together into a small region in the atom's center. This central core is called its *nucleus.* The atomic nucleus also contains any neutrons which the atom may have. The atom's electrons move in a certain region of space around the nucleus because they possess a certain amount of energy. Since the electrons are negatively charged, they are attracted toward the nucleus. Thus the electrons tend to remain relatively close to the nucleus of the atom to which they belong.

Electrons are not actually pulled into the body of the nucleus by the attraction of the protons. This is because the very rapid movements of the electrons around the nucleus give them enough energy to remain outside the nucleus. Keep in mind that neutrons in the nucleus play no part in the attraction of electrons since they are electrically neutral.

Kinds of atoms. The word "atom" comes from a Greek word which means "not able to be divided." This is an important feature of the atomic theory. An atom is the smallest complete part of any kind of matter. They are the smallest objects that can be recognized as ordinary matter. If any atom were separated into its parts, the result would be a collection of electrons, protons, and neutrons. These atomic particles bear no resemblance to ordinary matter, such as a piece of rock.

Careful investigation over a long period of time has shown that there are only a few kinds of matter that cannot be charged into a simpler form. This meant that these substances were made up of only one kind of atom. Since atoms cannot be changed into any simpler kind of matter, a substance composed of only one kind of atom cannot be simplified. Matter made up of only one kind of atom is called an *element*.

About ninety elements occur naturally on or in the earth. Around a dozen additional elements have been made artificially. This means that there is a total of about one hundred different kinds of atoms that are known to exist. Of the natural elements, a few very common ones make up most of the earth's crust. Table 7–1 shows the most abundant elements found in the earth's crust. Notice that in Table 7–1 a symbol is given for each element named. Each of the elements has a symbol of one or two letters. The symbol for an element is understood to represent a single atom of that element. Each kind of atom is also given an *atomic number*. This number is equal to the number of protons in the atom's nucleus. For example, the atomic number of an oxygen atom is eight. See Table 7–2. Since an atom normally contains the same number of protons and electrons, the atomic number also tells the number of electrons in that particular atom. Thus the oxygen atom has eight electrons moving around its nucleus along with the eight protons in the nucleus.

Thus the atom of each element has its own number of protons and electrons as a characteristic property that helps to identify it. But the atoms of different elements can be identified in another way. Each kind of atom has its own *atomic mass*. However, since atoms are so small, it is not practical to use ordinary units of mass such as grams. Instead, a special atomic mass scale is used on which the mass of a proton is assigned the value "1." Thus each of the protons in an atomic nucleus is said to have a mass of 1. Neutrons, which have nearly the same mass as protons, also have a mass of 1. Electrons are much lighter than either protons or neutrons. It takes

Table 7–1

Element	Symbol	Per cent by Weight in Crust
Oxygen	O	46.60
Silicon	Si	27.72
Aluminum	Al	8.13
Iron	Fe	5.00
Calcium	Ca	3.63
Sodium	Na	2.83
Potassium	K	2.59
Magnesium	Mg	2.09
Eighty other elements		1.41

Investigate
Examine Table 7–1. Compare the symbols used for the eight most abundant elements with their actual names. Find out why some symbols do not match the letters found in their names.

Table 7–2

Element	Atomic number (Protons)		Neutrons		Atomic Mass
Oxygen	8	+	8	=	16
Silicon	14	+	14	=	28
Iron	26	+	30	=	56
Lead	82	+	125	=	207
Uranium	92	+	146	=	238

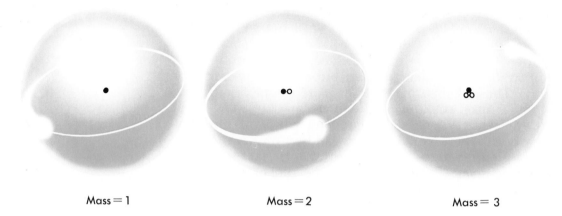

Mass = 1   Mass = 2   Mass = 3

FIG. 7–3. The isotopes of hydrogen. Solid dots represent protons and open circles represent neutrons.

1,840 electrons to equal the mass of one proton. Since electrons add so little to the total mass of an atom, it is usually ignored. This means that the atomic mass for any atom is calculated by the number of protons and neutrons in its nucleus. The total atomic mass for an atom is the sum of the masses of its protons and neutrons. For example, the atomic mass of oxygen is 16 as shown in Table 7–2. This is the sum of its protons and neutrons. The atomic number for oxygen is 8 which means that it has 8 protons. Therefore, an oxygen atom must have 8 neutrons in its nucleus.

Isotopes. Not all the atoms of a given element contain the same number of neutrons. This can be illustrated by the 3 different atoms of hydrogen. The atomic number for hydrogen is 1 which indicates that it is the simplest kind of atom. With an atomic number of 1, a hydrogen atom has only a single proton in its nucleus with one electron moving around it. This is illustrated in Figure 7–3, left diagram. Almost all hydrogen atoms have this arrangement.

But a few hydrogen atoms also have a single neutron in their nuclei. These hydrogen atoms still have only one proton and one electron but the neutron adds 1 unit to the atomic mass. These hydrogen atoms have an atomic mass of 2. A very rare form of hydrogen has two neutrons in its nucleus. It has an atomic mass of 3. All three forms of hydrogen differ from each other in the number of neutrons found in the nucleus. Each additional neutron gives each type of hydrogen atom an additional unit of atomic mass. When atoms of the same element differ in mass because of the different number of neutrons in their nuclei,

FIG. 7–4. A graph of the energy required to remove an electron from elements with atomic numbers 1–20.

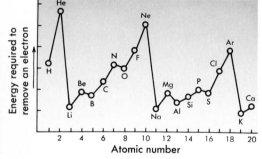

they are isotopes of that element. Hydrogen has three isotopes as shown in Figure 7–3.

MOLECULES

Electron arrangement in atoms. In the earth's crust,
atoms of a particular element are not usually found alone.
Almost all the elements which occur in the earth's crust
are chemically joined with other kinds of elements. When
two or more elements are chemically united, the resulting
new substance is called a *compound*. Although the smallest complete unit of an element is an atom, the smallest
complete unit of a compound is called a *molecule*. Each
compound is made up of only one kind of molecule just as
an element is made up of only one kind of atom.

Different kinds of atoms join together to form molecules because of the way their electrons are arranged. Remember that the electrons in an atom move around its
nucleus. For example, the single electron present in hydrogen atoms moves in a sphere-shaped region around
the nucleus. The electron remains in the neighborhood
of the nucleus because it is in a state of balance. Although
its motion tends to carry it away, its negative charge attracts it toward the positively charged nucleus.

However, it is possible to remove electrons from their
atoms. If energy in the form of heat or light is added to
hydrogen atoms, they may lose electrons. The same is
true for almost all other kinds of atoms. The correct addition of energy can result in the loss of one or more of the
atom's electrons.

In experiments performed to determine the amount
of energy needed to remove an electron from various
kinds of atoms, an important discovery was made. The
graph shown in Figure 7–4 shows the results of this kind
of experiment with some of the lighter elements. We can
see that there is a pattern in this graph. Certain kinds
of atoms require high energies to lose an electron. These
atoms, found where there are peaks on the graph, are
helium (He), neon (Ne) and argon (Ar). If the graph were
extended to include all the elements, similar peaks for
the elements krypton (Kr), xenon (Xe) and radon (Rn)
would appear.

A conclusion that we can draw from this information
is that these are elements which do not easily lose electrons. Thus they have a more stable electron arrangement
than other atoms. The atomic numbers of these elements

activity

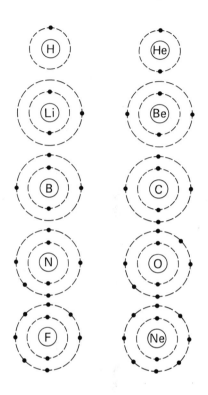

Use the electron structures above and
Fig. 7–4 to answer the following:

1. Do hydrogen (H) and lithium (Li)
   require less energy to remove an
   electron than helium (He)?

2. When an electron is removed from
   a hydrogen atom, is a pair of electrons broken up?

3. Is this also true for lithium?

4. Does boron (B) require less energy
   to remove an electron than beryl-
   lium (Be) and carbon (C)? Explain.

5. As electrons are added to an atom,
   what effect is there on the amount of
   energy needed to remove them?

6. Draw the electron structure of a
   sodium atom.

are: helium = 2, neon = 10, argon = 18, krypton = 36, xenon = 54 and radon = 86. We can further conclude that 2, 10, 18, 36, 54 and 86 electrons are favored numbers to be found around a nucleus.

Chemical bonds. For a molecule to be created, two or more different atoms must be held together by a *chemical bond.* Certain numbers of electrons indicate a stable atomic structure. This fact helps explain why some elements more readily form chemical bonds.

To illustrate the chemical bonding of elements into compounds, consider a common substance such as water. If we analyze water, we find that it is a compound made up of the elements hydrogen and oxygen. Hydrogen atoms (atomic number = 1) normally have only one electron. Since two electrons would be a more stable number, hydrogen will accept another electron if it is available. The element oxygen (atomic number = 8) holds eight electrons around its nucleus. Ten is the closest stable number of electrons. Therefore, we can predict that oxygen will accept two electrons to its structure.

If hydrogen atoms and oxygen are chemically combined, it is possible for both elements to acquire their stable number of electrons. Two hydrogen atoms can each share their single electrons with an oxygen atom, giving it a stable number of 10. At the same time, the oxygen atom can share two of its electrons; one with each hydrogen atom, resulting in a stable number of 2. The chemical bond which produces a water molecule is shown in Figure 7-5. Since the water molecule has a structure unlike either the hydrogen or oxygen atom, its properties are also unlike these elements.

Because of their electron arrangements, two hydrogen atoms will always join with a single oxygen atom to form a water molecule. This makes it possible to represent a water molecule by the formula, H_2O. A *chemical formula* represents one molecule of a certain compound. The formula for a compound tells what elements it contains and the number of atoms of each element present.

Some compounds are formed by chemical bonds that do not involve the sharing of electrons between atoms. An example is the compound sodium chloride, NaCl (common salt). Sodium atoms (atomic number = 11) have one more than the stable number 10 electrons. Chlorine atoms (atomic number = 17) have one less than the stable electron number of 18. Sodium atoms have a strong tendency

FIG. 7-5. A water molecule. Notice that the electrons of oxygen are arranged in two levels around its nucleus.

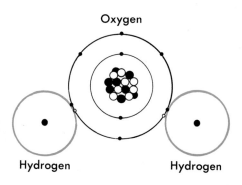

Oxygen

Hydrogen   Hydrogen

to lose one electron while chlorine atoms try to gain an electron. This means that one sodium atom will give up an electron to a chlorine atom as shown in Figure 7-6. Thus the chemical bond which is formed between the sodium and chlorine atom does not result from the sharing of electrons as in the water molecule.

When an electron is transferred from one atom to another, the atoms become electrically charged. Normally, atoms such as sodium and chlorine carry no overall electrical charge. This is because their charged particles, protons and electrons, are present in equal numbers. Their opposite electrical charges cancel each other exactly, thus neutralizing the atom. However, if an atom loses or gains electrons, it will take on either a positive or negative charge. Sodium, for example, becomes positively charged when it gives up an electron to chlorine. The loss of one electron still leaves a sodium atom with eleven protons in its nucleus. With only ten electrons circling the sodium atom's nucleus, there is then one excess proton not canceled by an electron. Thus, when the sodium atom loses an electron, it takes on a positive charge. It is then said to be a positively charged sodium *ion* whose symbol is Na$^+$. An ion is an atom or group of atoms which carry an electrical charge.

In the same way that sodium becomes an ion by losing an electron, chlorine becomes an ion by gaining one. The addition of an electron to a chlorine atom gives it eighteen electrons but only seventeen protons in its nucleus. The extra electron changes the neutral state of a chlorine atom to a negatively charged chloride ion (Cl$^-$).

Forces which hold molecules together. Two or more kinds of atoms may combine to form a chemical compound. And, in some cases, it is easy to see why the resulting compounds are usually solids. For example, you know that the compound sodium chloride, NaCl, or common table salt, is normally solid. The reason for this can be found in the way this compound is formed.

When sodium and chlorine atoms combine chemically to form sodium chloride, each one of these atoms becomes an ion. The sodium takes on a positive charge (Na$^+$) and the chlorine takes on a negative charge (Cl$^-$). When a number of these oppositely charged ions attract each other they arrange themselves in a definite pattern. Each positive (Na$^+$) ion is surrounded by negative (Cl$^-$) ions. See Figure 7-7. Each ion is strongly attracted to its op-

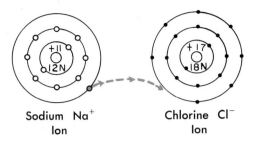

Sodium Na$^+$ Chlorine Cl$^-$
Ion   Ion

FIG. 7-6. The compound sodium chloride is formed by an electron transfer from the outer shell of sodium to chlorine, as indicated by the arrow.

FIG. 7–7. A crystal of sodium chloride is made of sodium ions (smaller spheres) and chloride ions held by their opposite electrical charges. Notice that each sodium ion is surrounded by oppositely charged chloride ions.

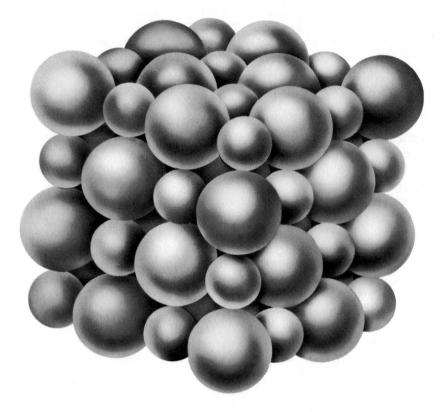

FIG. 7–8. Salt crystals have the shape of a cube because of the arrangement of their ions. (B.M. Shaub)

positely charged neighbor and can only fit into the pattern in a definite position. The sodium chloride structure that is formed will have a definite shape as shown in Figure 7–8.

Each piece of sodium chloride formed has the shape of a cube, because the ions from which it is made fit together in a cubical pattern. Any solid substance, such as sodium chloride, having a definite natural shape is called a *crystal*. Many compounds are able to form crystals, although the crystal arrangement is often difficult to see. The individual crystals are frequently so small that they can only be seen with a microscope. Also, the crystals often break after they are formed so that their original shape is not easily seen.

Other than its crystal shape, the ions which make up sodium chloride also give it another important property. It is hard to melt. To melt a solid material, heat energy must be added. This energy causes the particles (molecules or atoms) in the solid to move faster. When enough heat energy has been added, the individual particles move so rapidly that they break away from each other. In a

solid, the particles are locked tightly together and the
material holds its shape. But a liquid has no definite shape
and can flow freely because its particles move around
each other. If more heat is added to a liquid, some of its
particles will move so fast that they will escape from the
liquid. At this point the liquid begins to change to a gas.
The relative amount of motion of particles in solids, liquids
and gases is illustrated in Figure 7–9.

Sodium chloride is a difficult compound to melt because
of its strong *ionic bond*. Ionic bonds are attractions of unlike charged particles. These bonds form substances that
require a large amount of heat energy for them to melt
and become liquid. This is not the case for all substances.
Some substances may be composed of molecules which
are held together less strongly than ions. An example of
such a substance is water.

Unlike sodium chloride, water is made up of separate
molecules. Each molecule has the formula H_2O representing the two hydrogen atoms sharing electrons with an oxygen atom. But these electrons are not shared by equal attractions. The oxygen has a larger number of protons in

Experiment

Not all matter appears to pass
through a solid, liquid, and
gaseous state. Obtain some
camphor or moth balls and heat
them slowly in a closed beaker.
Observe what happens to the
solid. What appears to be
collecting on the sides and top of
the beaker?

FIG. 7–9. The relative motion of
molecules in solids, liquids, and
gases is represented in these diagrams.

its nucleus than does the hydrogen. Because of this, the oxygen attracts the electrons slightly more than the hydrogen does. This means that the water molecule has a slight negative charge around its oxygen atom. At the same time, there is a slight positive charge around the end of the molecule where the hydrogen atoms are located. Water molecules can then attract each other with their oppositely charged ends. See Figure 7–10.

If enough heat is lost, the attraction between water molecules can produce water in the solid form (ice). This form has a definite crystal pattern. If heat is then added to the water molecules, they begin moving farther apart. At a certain temperature the ice melts and becomes a liquid. Ice melts at a much lower temperature (0°C) than sodium chloride (801°C). This is because water molecules are held together by much weaker forces than the strongly attracted ions in salt. The relatively weak attraction between water molecules also allows them to become a gas at low temperature (100°C at ordinary atmospheric pressure).

Many of the materials found on the earth are liquids and gases. This is so because, as in the case of water, their molecules are not strongly attracted to each other. The substances which make up the gases of the atmosphere are another example of weakly attracted molecules. The most abundant gases in the atmosphere are nitrogen and oxygen. Each of these is made up of molecules which contain two identical atoms. Nitrogen gas has the formula N_2. Each of its molecules consists of two nitrogen atoms. The formula for oxygen gas is O_2. In both nitrogen and oxygen molecules, the electron arrangements of the atoms become stable by sharing electrons with an identical atom. The resulting molecules then have very little at-

FIG. 7–10. The arrangement of water molecules in (A) water vapor, (B) liquid water and (C) ice, is represented in these diagrams. Attraction of water molecules for each other causes them to form hollow rings in ice crystals.

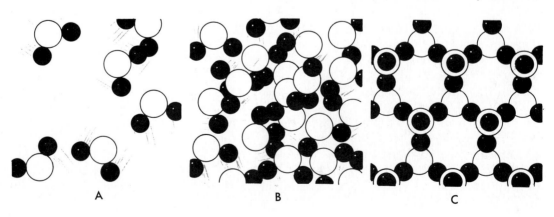

A   B   C

tractive force for each other. A similar situation exists among the molecules of the other gases found in the atmosphere.

The only liquid present in large amounts at or near the earth's surface is water. Temperatures at the earth's surface are such that water exists in the liquid state. If the earth were much colder or warmer, this would not be true. As already mentioned, water molecules are held to each other by relatively weak forces. Thus there is enough heat energy on most of the earth's surface to prevent water from becoming solid ice. See Figure 7–11.

Most of the matter in the earth is in the solid form. This means that most of the molecules of the earth are bound together by relatively strong forces. These forces are due to ion formation or electron sharing. These solid substances exist in a great many different forms. Just think of the variety of rocks and minerals that can be found in the crust.

FIG. 7–11. The sun's radiant energy allows water near the earth's surface to remain in a liquid state. (H.R.W. Photo by Russell Dian)

VOCABULARY REVIEW

Match the word or words in the column on the right with the correct phrase in the column on the left. *Do not write in this book.*

1. Assumes that every substance is made up of atoms.
2. Have only a negative electrical charge.
3. Part of the nucleus which carries a positive electrical charge.
4. Part of the nucleus which has no electrical charge.
5. Central core of an atom.
6. Means "not able to be divided."
7. Matter made up of only one kind of atom.
8. Equal to the number of protons in the nucleus of an atom.
9. Equals the sum of the number of protons and neutrons in the nucleus of an atom.
10. Atoms of the same element but with different numbers of neutrons in their nuclei.
11. The resulting substance when two or more kinds of atoms are chemically united.
12. Has the same relationship to a compound as an atom has to an element.
13. Holds a molecule together.
14. An atom or group of atoms which carries an electrical charge.
15. Any solid substance with a definite natural shape.

a. protons
b. nucleus
c. repulsion
d. atomic number
e. electrons
f. atom
g. neutrons
h. atomic theory
i. atomic mass
j. element
k. chemical bond
l. molecule
m. crystal
n. isotopes
o. chemical formula
p. compound
q. ion

QUESTIONS

Group A

Select the best term to complete the following statements. *Do not write in this book.*

1. The atomic theory states that (a) all atoms are identical (b) only light atoms are identical (c) there are only a certain number of kinds of atoms (d) only heavy atoms are identical.

2. Atoms cannot be studied individually because they (a) are too small (b) are not solid (c) move around too rapidly (d) are electrically charged.

3. Which of the following does *not* describe a kind of electrical charge? (a) positive (b) negative (c) neutral (d) nuclear.

4. Which of the following does *not* carry an electrical charge? (a) protons (b) neutrons (c) atomic nucleus (d) electrons.

5. Which of the following is *not* a part of every atom? (a) proton (b) neutron (c) nucleus (d) electron.

6. An atom which has lost an electron is brought near an atom which has gained an electron. The atoms will (a) attract (b) repel (c) sometimes attract and other times repel (d) have no electrical effect on each other.

7. Two atoms have charges which cause them to repel each other. If the atoms are moved so that their separation is twice as great, the repulsive force will be (a) zero (b) twice as great (c) one-half as large (d) one-fourth as large.

8. In an atom, neutrons will most likely be found (a) near the electrons (b) near the protons (c) in the outer regions of the atom (d) only if it is a heavy atom.

9. A substance which is composed of only one kind of atom is called (a) an element (b) a compound (c) a solid (d) an ion.

10. Of the elements known to exist, the number which occur naturally is (a) very few (b) about 12 (c) about 90 (d) about 100.

11. The most common element by weight in the earth's crust is identified by the symbol (a) Al (b) Fe (c) O (d) Si.

12. The atomic number of an atom is equal to the number of (a) electrons in the nucleus (b) neutrons in the nucleus (c) protons plus neutrons in the nucleus (d) protons in the nucleus.

13. The atomic number of oxygen is (a) 16 (b) 12 (c) 8 (d) 4.

14. The atomic number of uranium is (a) 238 (b) 146 (c) 92 (d) 82.

15. Which of the following elements has atoms whose atomic number and atomic mass are numerically equal? (a) oxygen (b) silicon (c) iron (d) hydrogen.

16. The atomic mass of oxygen is (a) 16 (b) 12 (c) 8 (d) 4.

17. The atomic mass of uranium is (a) 238 (b) 146 (c) 92 (d) 82.

18. The atomic number of argon is 18. Its atomic mass is 40. The number of neutrons in the nucleus of argon is (a) 18 (b) 22 (c) 40 (d) 58.

19. From the information in problem 18, the number of electrons in argon atoms is (a) 18 (b) 22 (c) 40 (d) 58.

20. How many electrons does it take to equal the mass of one proton? (a) 16 (b) 32 (c) 920 (d) 1840.

21. Hydrogen atoms have atomic masses of any of the following except (a) 1 (b) 2 (c) 3 (d) 4.

22. Most hydrogen atoms have an atomic mass of (a) 1 (b) 2 (c) 3 (d) 4.

23. Most argon atoms have an atomic mass of 40 but natural argon is given in most references with an atomic mass of 39.94. This means that (a) some argon atoms have an atomic mass greater than 40 (b) all argon atoms have an atomic mass of 39.94 (c) some argon atoms have an atomic mass less than 40 (d) some argon atoms have an atomic mass of 39.94.

24. The smallest complete unit of a compound is called (a) an atom (b) a molecule (c) a proton (d) a neutron.

25. A water molecule is made when two hydrogen atoms combine with (a) one oxygen atom (b) one oxygen molecule (c) two oxygen molecules (d) two oxygen atoms.

26. The chemical bond in a water molecule is produced by (a) hydrogen giving up electrons (b) oxygen giving up electrons (c) the sharing of electrons between hydrogen and oxygen (d) the taking on of electrons by oxygen.

27. Sodium atoms combine with chlorine atoms by (a) sodium atoms taking on electrons (b) sodium atoms giving up electrons (c) chlorine atoms giving up electrons (d) sodium atoms sharing electrons with chlorine atoms.

28. A crystal of sodium chloride is held together by forces (a) due to sharing of electrons (b) due to gravity (c) that are weak (d) caused when ions of unlike charge come near one another.

29. Water exists as a liquid at room temperature. If heated to 100°C, the forces holding it together may be broken converting it to (a) its elements (b) a solid (c) a gas (d) ions.

30. Most of the matter in the earth is in the solid form, which means that most of the molecules of the earth are (a) ionic (b) giving up electrons (c) held together by relatively strong bonds (d) held together by relatively weak bonds.

Group B

1. List the four most abundant elements in the earth's crust. What are their chemical symbols?

2. Where in atoms do scientists find protons? neutrons? electrons?

3. Why are electrons neglected in determining atomic mass?

4. What is an ion?

5. Define what is meant by isotope, atomic number and atomic mass.

6. Why aren't atomic masses expressed in grams?

7. Atoms of the element aluminum have 13 protons and 14 neutrons. What is the atomic number and the atomic mass of aluminum?

8. Atoms of the element potassium have 19 electrons and an atomic mass of 39. What is the atomic number of potassium? How many neutrons are in each atom?

9. Of what importance is it that the numbers of electrons and protons in atoms are equal?

10. What role does electricity play in the structure of atoms?

11. Draw a diagram of a hydrogen atom and an oxygen atom. Show locations for neutrons, protons and electrons in the diagram.

12. When the atomic mass for a naturally occurring element is given it is seldom a whole number. Why?

13. Why does hydrogen gas usually occur as molecules of two atoms?

14. What is the difference in the way atoms combine to form sodium chloride and water molecules?

15. Tritium is the name for an atom which has one proton and two neutrons in its nucleus. Deuterium has one proton and one neutron. Protium has just one proton. What is the common name for all three of these atoms?

16. Explain the following statement briefly. "The reason for different kinds of atoms joining together to form molecules is found in their electron arrangement."

17. The element silicon (Si) has atomic number 14 and oxygen (O) has atomic number 8. Suppose silicon and oxygen combine so that the oxygen is grouped around one silicon atom. What is the chemical formula that will give all atoms a stable number of electrons?

18. Water may exist as a solid, liquid or gas depending on its temperature. Explain why this is so.

objectives

☐ Describe several ways silicon and oxygen combine to form silicates.

☐ List several examples of silicate minerals.

☐ List and describe the composition of several non-silicate minerals.

☐ Describe several properties and tests that are used to identify a mineral.

☐ Explain how rocks can be classified according to how they were formed.

Almost any rock that you pick up off the ground was originally part of the earth's crust. The crust forms the outer layer of the earth's surface on which you live. Since it is an actual piece of the crust, the rock is very likely to be made up of about 50 percent of the element oxygen. The other 50 percent of the materials in the rocks of the crust could probably be accounted for by about eight other kinds of elements. See Figure 8–1A. The properties of the rock, and almost the entire earth's crust, depend upon the way in which these relatively few, but commonly found elements join together.

Each element has its own chemical properties. It can become part of the molecules in compounds that form the materials of the earth. How do these elements join together to produce the many different kinds of rocks and other materials that form the crust? This chapter will serve as a brief answer to this difficult question.

MINERALS

Basic mineral types. A *mineral* is a single chemical compound or element that is found naturally. *Rocks* are a combination of different minerals found in the earth's crust. Since oxygen and silicon are the most abundant elements in the earth's crust, the most common mineral compounds must include these elements. These minerals are known as *silicates*. There are five main groups of silicate minerals.

The key to identifying the silicate minerals is understanding the structure of silicon atoms. An atom of silicon (atomic number = 14) has a total of fourteen electrons.

FIG. 8–1B. The basic structure of many minerals is a tetrahedron made up of four oxygen atoms and one silicon atom.

FIG. 8–2. Quartz is made up of silicon and oxygen atoms arranged in a continuous network, as shown below. (American Museum of Natural History)

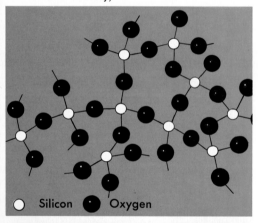

Silicon ○　Oxygen ●

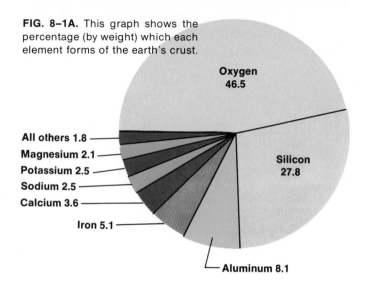

FIG. 8–1A. This graph shows the percentage (by weight) which each element forms of the earth's crust.

Oxygen 46.5

Silicon 27.8

All others 1.8
Magnesium 2.1
Potassium 2.5
Sodium 2.5
Calcium 3.6
Iron 5.1
Aluminum 8.1

This number of electrons is mid-way between the stable electron numbers of ten and eighteen. One way a silicon atom can obtain a stable electron number is to share its four outer electrons with other atoms. A common arrangement is a silicon atom surrounded by four oxygen atoms. Each of these atoms shares an electron with the silicon. See Figure 8–1B. This arrangement forms a four-sided pyramid called a *tetrahedron*. But an oxygen atom (atomic number = 8) is able to accept two more electrons before reaching a stable electron number (10). By sharing only one of the four available electrons in a silicon atom, each oxygen atom is able to form a second chemical bond with still another silicon atom.

Thus each silicon atom is bonded to four oxygen atoms in a continuing pattern, to form a network as shown in Figure 8–2. The resulting mineral is *quartz*. Quartz has the chemical formula SiO_2. Quartz is one of the hardest of all minerals because all of its atoms are tightly joined together. See Figure 8–3.

A group of minerals called *feldspars* have a structure similar to quartz. In the feldspars, however, aluminum atoms replace silicon in some of the tetrahedrons. Sodium (Na), calcium (Ca), or potassium (K) atoms may also be present in the feldspar crystal. These atoms form weaker bonds than those between silicon and oxygen. Thus the feldspars are softer than quartz. Some kinds of feldspar are shown in Figure 8–4.

Another group of silicate minerals is produced when only one oxygen in the silicon-oxygen tetrahedron fails to

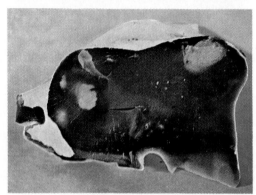

FIG. 8–3A. Top to bottom: Smoky quartz is a variety of colored quartz crystals. Agate is fine crystalline quartz. Tiger's Eye quartz is often cut into stones with a curved surface that reflects a band of light resembling a cat's eye.

FIG. 8–3B. Top to bottom: Chalcedony is a general name for fine-grained varieties of quartz having a waxy appearance. Flint is also a fine grained variety of quartz usually dark gray in color. Jasper is similar to agate, but has no color bands.

FIG. 8–4A. Left to right: Orthoclase feldspar is a potassium-aluminum silicate. Labradorite is a form of plagioclase feldspar, a sodium-calcium aluminum silicate.

FIG. 8–4B. Left to right: Microcline is a variety of orthoclase feldspar. Albite is a type of plagioclase feldspar which contains large amounts of sodium.

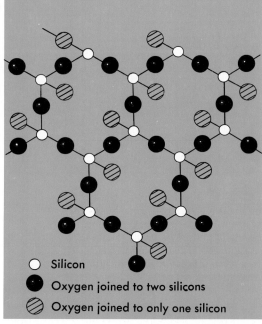

FIG. 8–5. Mica splits into thin sheets because its atomic arrangement is the one shown in the diagram at the left. Above is a photograph of the mineral "muscovite," a type of mica.

○ Silicon
● Oxygen joined to two silicons
⊘ Oxygen joined to only one silicon

attach itself to another silicon atom. One oxygen may be joined to an atom of potassium or aluminum instead of silicon. See Figure 8–5. Minerals having this type of structure are called *micas.* The micas separate easily into thin layers because the silicon-oxygen bonds going in three directions are stronger than the fourth bond involving another kind of atom.

If only two oxygens in the silicon-oxygen tetrahedron are joined to other silicons, a fourth group of silicate minerals is produced. The oxygens which are not joined to silicon may be bonded to atoms such as magnesium (Mg), iron (Fe), calcium (Ca), and aluminum (Al). See Figure 8–6. Minerals with this structure are known as *hornblende* and *pyroxene.* Both of these minerals have a similar appearance and are difficult to tell apart.

If none of the oxygens in the tetrahedron are joined to other silicons the mineral *olivine* is formed. In olivine, the silicon-oxygen tetrahedrons are separate but are linked by magnesium and iron atoms.

In addition to the five basic groups of silicate minerals already mentioned, there are many kinds which are less common. Some are shown in Figure 8–7 to illustrate their variety in color and form.

Examine
Look at the clear window in an electric house fuse. Try to break out a part of the window. Observe how the material can be peeled into thin layers. What do you think this material is?

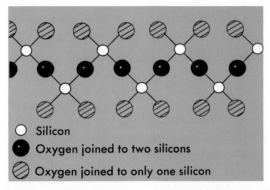

○ Silicon
● Oxygen joined to two silicons
⊘ Oxygen joined to only one silicon

FIG. 8–6. The structure of pyroxene minerals is in a crystal pattern, based on the arrangement of atoms shown in the diagram at the right. The photographs are of two varieties of pyroxene minerals. Hornblende (left) and augite (right).

FIG. 8-7. All of these minerals are silicates. They are, top left, biotite mica; right, stilbite, a mineral belonging to a general group of silicates called "zeolites." Bottom left, chrysotile, a type of asbestos and right, serpentine, an ornamental stone often used in the same way as marble.

Other mineral types. Almost all common rocks are mixtures of the various kinds of silicate minerals. However, there are many other kinds of minerals which do not contain silicon and oxygen. These minerals can be put into three groups:

1. *Rock forming minerals other than silicates.* These include a large number of compounds as well as a few native elements such as sulfur. See Figure 8-8.

2. *Metal ore minerals.* An ore is a rock which is an important source of some useful metal. To be considered a valuable ore a rock must contain enough of a metal to make its removal cheap and practical. See Figure 8-9.

3. *Gem minerals.* Most gems are mineral crystals of unusual color and brilliance. They are usually cut to increase their brilliance and color. The value of a gem is determined by its size, lack of flaws, hardness, color and brilliance. See Figure 8-10.

Identification of minerals. Laboratory tests with special equipment are needed for the complete identification of a

FIG. 8-8A. Non-silicate minerals. Top, two varieties of calcite, $CaCO_3$. Bottom, two varieties of gypsum, $CaSO_4$.

FIG. 8-8B. Non-silicate minerals. Left, top to bottom, halite (NaCl) and fluorite. Right, top, kernite which yields borax and, bottom, apatite which is mainly calcium phosphate.

FIG. 8-8C. These two non-silicate minerals are elements. To the left is graphite, which is a form of carbon. At the right is sulfur.

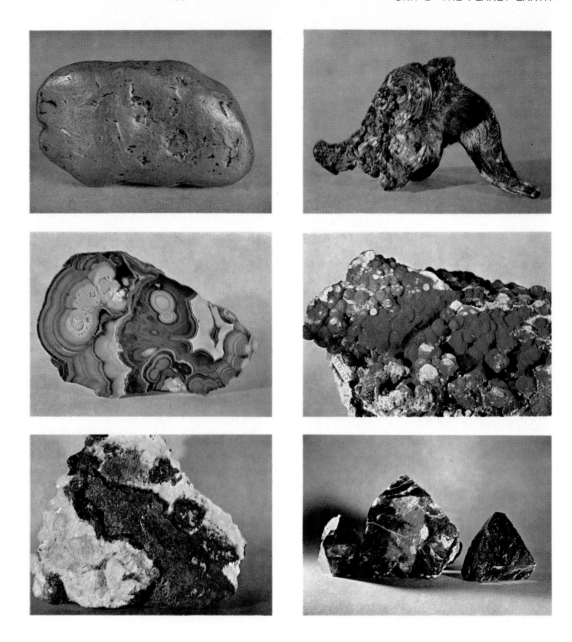

FIG. 8–9
NATIVE GOLD Bolivia   **NATIVE SILVER** Norway
MALACHITE (COPPER) Urals, U.S.S.R.   **AZURITE (COPPER)** Arizona
CUPRITE (COPPER) Arizona   **SPHALERITE (ZINC)** Spain

FIG. 8–9 (cont.)
HEMATITE (IRON) England
LIMONITE (IRON) New Hampshire
SIDERITE (IRON) Austria

CINNABAR (MERCURY) California
BAUXITE (ALUMINUM) Arkansas
URANINITE (Black) (URANIUM) India

FIG. 8–10
OPAL Australia
TOPAZ Burma and Brazil
TOURMALINE California

GARNET Alaska and Connecticut
BERYL, Aquamarine S.W. Africa
CORUNDUM, Ruby Madagascar

mineral. However, there are some simple tests by which many common minerals can be identified. Some of these tests are described below:

1. *Color.* One of the more easily seen characteristics of a mineral is its color. Look at the color photographs in the chapter and note the differences among the many minerals shown. The color of the surface tarnish of those minerals which look like metal should be considered. For example, the iron mineral pyrite is the color of gold on a fresh surface. But it is a much darker yellow when tarnished. Other minerals contain different elements which affect their color. The colorless mineral called corundum becomes a ruby when colored red by traces of chromium. Sapphires are produced from corundum colored bluish by traces of iron and titanium. Such differences must be kept in mind when identifying unknown minerals according to color.

2. *Streak* is the color left by the mineral when rubbed against a streak plate. The back of a piece of building tile, or a piece of porcelain with a dull surface may be used for a *streak plate.* Again it must be kept in mind that the color of the streak may not be the same as the mineral. For example, the streak of gold-colored pyrite is black For most minerals, however, the streak is either colorless or very light.

3. *Luster* is the ability of the mineral to reflect, bend or absorb light. Many terms are used to describe luster: *dull, pearly, waxy, metallic, glassy, brilliant* (diamond-like).

4. *Crystal form.* The internal structure of a mineral can be understood by studying its crystal form. In most minerals, the atoms, ions and molecules are arranged in a particular pattern. This pattern gives the crystal its characteristic shape. Some mineral crystals are so perfectly shaped that they seem to have been cut and polished artificially rather than created naturally. The chemical properties of these atoms make it possible for them to fit together in a definite pattern.

There are six basic shapes of mineral crystals. See Figure 8–11. Think of these shapes as having imaginary internal lines called axes. The mineral molecules are arranged along the axes giving a crystal its shape which is different from that of other mineral crystals.

Other combined forms exist besides the basic forms shown in Figure 8–11. Most minerals are not found as separate large crystals. They are more commonly masses of small crystals which can be seen only with a micro-

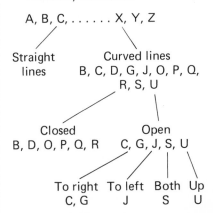

The Six Basic Crystal Systems

Garnet   Sulfur

Isometric or Cubic System: Three axes of equal length intersect at 90° angles. Ex. galena, pyrite, halite.

Orthorhombic System: Three axes of different lengths intersect at 90° angles. Ex. olivine, topaz.

Zircon   Orthoclase   Amphibole

Tetragonal System: Three axes intersect at 90° angles. The two horizontal axes are of equal length. The vertical axis is longer or shorter than the horizontal axes. Ex. rutile, cassiterite, chalcopyrite.

Monoclinic System: Of three axes of different lengths, two intersect at 90° angles. The third axis is oblique to the others. Ex. gypsum, micas, augite, cryolite, kaolinite.

Tourmaline   Rhodonite   Chalcanthite

Hexagonal System: Of the four axes, the three horizontal axes intersect at 60° angles. The vertical axis is longer or shorter. Ex. quartz, calcite, apatite, hematite.

Triclinic System: The three axes are of unequal length and are oblique to one another. Ex. plagioclase feldspars, including albite and labradorite, turquoise.

FIG. 8–11

scope. Sometimes only x-ray will show it. A few minerals are not made up of crystals. There is no orderly arrangement of their molecules. These are called *amorphous* or *massive*.

5. *Cleavage and fracture.* These terms tell how a mineral splits or breaks. Some minerals split along certain directions leaving flat surfaces. The micas for example, split easily along one direction. This means that mica has one cleavage plane. The different kinds of cleavage are described according to the number and direction of the cleavage planes. See Figure 8–12.

Most minerals do not break along cleavage planes, but split unevenly in one of several ways. Some types of fracture are *conchoidal* (shell-like), *splintery, irregular* and *earthy*. See Figure 8–13.

6. *Hardness.* The hardness of a mineral is determined by how easily it can be scratched. Certain common minerals are used as standards of comparison. The scale of mineral hardness, called *Moh's Scale*, is given in Table 8–1. The range is from talc, one of the softest minerals, to diamond, the hardest of all minerals.

To test an unknown mineral for hardness find out which mineral on the scale it can scratch and which it cannot. For example, gypsum is harder than talc, but softer than calcite. Any mineral which scratches gypsum but not calcite has a hardness on Moh's Scale between 2 and 3.

The following standards of hardness are useful additions to the scale:

Fingernail	2.5
Copper penny	3.0
Glass or knife-blade	5.5
Steel file	6.5

Care must be taken in testing hardness. Talc, which is a soft mineral, may only appear to scratch an unknown mineral. A soft mineral can make a mark on a harder one that can be mistaken for a scratch. Such a mark can be rubbed off, a true scratch cannot. It is wise to test both ways. Try the known mineral on the unknown as well as the reverse.

7. *Specific gravity.* You have noticed that some minerals are heavier than others. The relative weights of two or more minerals can be compared. This can be estimated by weighing or by handling (hefting) equal sized pieces of different minerals.

FIG. 8–12. Two types of cleavage in three directions are shown in the top diagrams. Lower diagram illustrates cleavage in one direction.

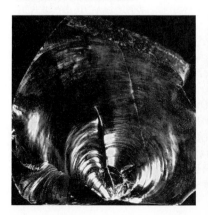

FIG. 8–13. Fracture. Left, conchoidal fracture in obsidian. Right, earthy fracture in yellow ocher, an impure form of limonite. (American Museum of Natural History)

Table 8–1

Moh's Scale of Mineral Hardness

1.	Talc
2.	Gypsum
3.	Calcite
4.	Fluorite
5.	Apatite
6.	Orthoclase
7.	Quartz
8.	Topaz
9.	Corundum
10.	Diamond

Compare
What is the relationship between the terms "specific gravity" and "density?"

A much more exact method is to express the relative weight of a mineral by comparing its weight to the weight of an equal volume of water.

This can be determined by first weighing the mineral sample. It is then suspended by a fine thread and weighed while it is completely under water. Its weight in water is then subtracted from the dry weight. This difference is the weight of a volume of water equal to the volume of the mineral sample. Specific gravity can be obtained by dividing the weight of an equal volume of water into a dry weight.

For example, a sample of mineral is found to weigh 40 grams dry and 30 grams when submerged in water. Then:

$$\text{Specific gravity of the mineral} = \frac{\text{weight of sample in air}}{\text{weight of equal volume of water}}$$

or

$$\frac{40 \text{ grams}}{(40-30) \text{ grams}} = \frac{40}{10} = 4$$

Perhaps you already have some idea of the specific gravity of common minerals. You know from experience how a certain size of the most common minerals should feel (heft). When you pick up a pebble it will probably consist of quartz (Sp. gr. = 2.65), feldspars (Sp. gr. = 2.60 to 2.75) and calcite (Sp. gr. = 2.72). A pebble made of the mineral pyrite (Sp. gr. = 5.02) would feel too heavy for its size and you would probably consider it an unusual pebble. With a little practice it will be possible for you to handle a mineral specimen and judge whether its specific gravity is average, high or low.

Special properties of minerals. Following is a list of some other properties of certain minerals which will help you identify them.

1. *Magnetism.* Some minerals are attracted to a magnet. *Magnetite* and some samples of *pyrrhotite* (PIR-oh-tyt), both iron ores, are attracted to a magnet. Lodestone, a form of magnetite, acts as a magnet. Its magnetic properties have been known since ancient times.

2. *Fluorescence.* In some minerals certain atoms absorb ultraviolet light and give off visible light rays. These minerals glow under ultraviolet (black) light and are called fluorescent. See Figure 8–14. A few minerals continue to glow after the ultraviolet light is cut off. They are called *phosphorescent.*

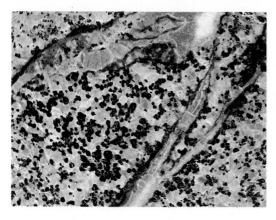 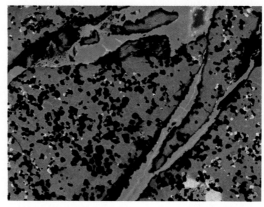

3. *Radioactivity.* Minerals which contain uranium and radium are radioactive, and are usually fluorescent. These atoms can be detected with a Geiger counter.

4. *Optical properties.* The way that light is changed as it passes through minerals is often useful in identifying them. For example, when looking at an object through calcite, a double image will be seen. See Figure 8–15. Some optical properties of minerals cannot be seen without the use of special equipment. A special kind of microscope is needed to measure the degree that light rays are bent as they pass through a mineral crystal. This is one of the most dependable methods used to identify a mineral.

Simple testing methods. Many tests have been developed for identifying minerals. Some that need only simple equipment are described below.

1. *Simple chemical test.* The common mineral calcite ($CaCO_3$) can be identified by an acid test. A drop of cold, dilute hydrochloric acid placed on calcite will cause bubbles of carbon dioxide (CO_2) gas to form. The same will happen to dolomite if hot acid is used, but will act more slowly if cold acid is used.

2. *Flame tests.* A Bunsen burner flame can be used to test minerals.

If you blow air through a metal tube (blowpipe) into the flame, the flame can be directed against a mineral sample. See Figure 8–16, top. When an oxidizing flame is used, oxygen reacts with the mineral to produce colors that indicate the presence of certain metals. In a reducing flame, hot gases remove oxygen from the mineral. This may cause a color change.

One way to use the blowpipe and flame is in bead tests. This is done by dipping a loop of platinum wire into

FIG. 8–14. Above, left, the minerals calcite (greenish white), willemite (light brown), and franklinite (dark brown), as they appear in ordinary light. Right, under ultraviolet light, calcite and willemite fluoresce red and green, respectively. Franklinite does not fluoresce.

FIG. 8–15. The mineral calcite, in a clear form, produces a double image of objects viewed through it.

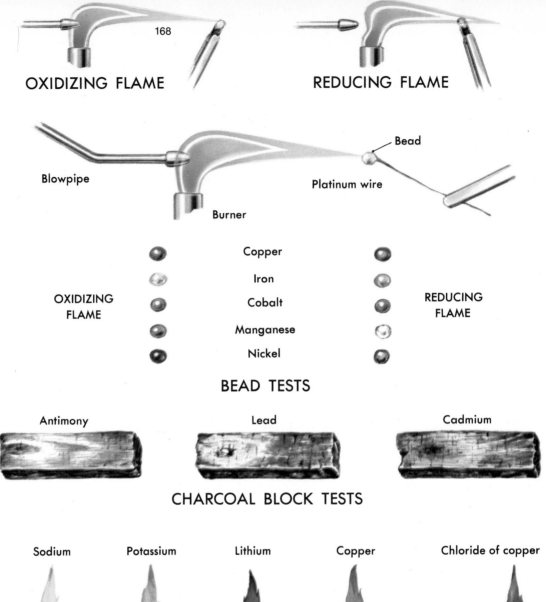

OXIDIZING FLAME

REDUCING FLAME

168

Bead

Blowpipe

Platinum wire

Burner

OXIDIZING
FLAME

Copper

Iron

Cobalt

Manganese

Nickel

REDUCING
FLAME

BEAD TESTS

Antimony

Lead

Cadmium

CHARCOAL BLOCK TESTS

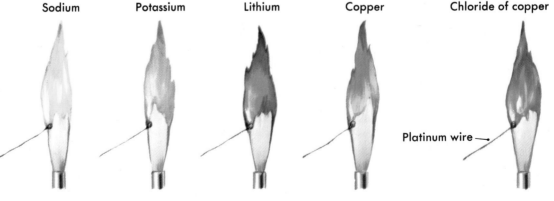

Sodium

Potassium

Lithium

Copper

Chloride of copper

Platinum wire

FLAME TESTS

FIG. 8–16. These color tests are frequently used in identifying minerals.

powdered borax. The loop is then held in the burner flame while the borax melts and forms a colorless bead. Then the bead is applied to the powdered mineral. When the bead is again held to the flame, the borax and metals in the mineral may react to produce a color. Different colors may be produced by using the oxidizing and reducing flames.

Another common test is to heat a small amount of the mineral on a block of charcoal. The appearance of the material left on the block after heating may help to identify some of the elements in the mineral. Characteristic fumes or a bead of metal may also be produced.

A bit of the powdered mineral may also be moistened with acid and held on a clean platinum wire. The wire is then put into a burner flame. A change in the color of the flame shows the presence of certain elements. See Figure 8–16, bottom.

ROCKS

Types of rocks. Suppose that you decide to start a rock collection. It is easy to collect rocks since they are found everywhere. They have many different forms and colors. But how will you organize your collection? Rocks can be grouped by their color, roughness or smoothness, or their mineral composition. But there is another way which will probably work better. It will include all rocks and is based on how they were formed.

Any rock that you would pick up for your collection is only a small piece broken off a larger rock. The earth's crust is made up of great bodies of rock which were formed in different ways. Scientists have discovered after careful study, that there are three ways that rocks are produced. See Figure 8–17.

One group is called *igneous*. The word "igneous" means "from fire." Igneous rocks are formed by the cooling and hardening of hot melted (molten) rock below the earth's crust. All rocks now in the earth's crust are believed to have first been igneous.

When igneous rocks are at the surface of the earth, conditions there help to break up the large masses. Small pieces are produced by the wearing away of solid rock. These pieces are called *sediments*. Sediments can be joined together again to form solid rock. Any rock made from sediments is called *sedimentary rock*.

Explore
Take some pictures along road cuts or of outcrops. If you live in the city examine the walls of buildings or park statues and try to identify the type of rock.

FIG. 8–17. These masses of rock in Zion National Park came into existence as layer upon layer of sediment was deposited. (Ramsey)

The third way that rocks are formed calls for the changing of igneous or sedimentary rocks. When these are buried very deep in the earth, they are under great pressure and become very hot. The heat and the pressure cause the rock to change into another form. This is called *metamorphic* rock. Only the conditions deep in the earth can produce metamorphic rock.

By examining a small piece of rock carefully we can often find signs that tell us something about the large rock from which it came. But to do this we need more detailed knowledge of the properties of the three types of rock.

Igneous rocks. The molten rock from which igneous rocks are formed is called *magma*. Magma may cool either at or below the earth's surface. When magma reaches the surface it is called lava. Igneous rocks can be divided into two groups according to where the magma cools and hardens.

How quickly or slowly the magma cools determines the appearance of the igneous rock being formed. When magma cools slowly mineral crystals are able to grow to a large size. The slow cooling of magma below the surface produces *intrusive* igneous rocks. The slow loss of heat allows atoms enough time to join together in well-organized patterns. Because of this the silicon-oxygen tetrahedrons form well-developed crystals. This results in intrusive igneous rocks having a coarse texture due to the larger crystals present. See Figure 8–18A. Granite is an example of a coarse-grained intrusive igneous rock. Such rocks are often called *plutonic* rocks. See Figure 8–18B.

Rapid cooling of lava at the surface produces *extrusive* igneous rocks. The rapid cooling may prevent large crystals from forming. This makes the rock look glassy like in *obsidian*. See Figure 8–19. Gases may be trapped in the lava as it hardens forming *pumice*. Slower cooling allows the growth of small mineral crystals. The resulting rocks such as *basalt* have a fine texture. See Figure 8–20.

Differences in the rate of cooling of magma affect the mineral composition of igneous rocks as well as their texture. Probably all magma begins with the same composition. It is thought to be dark, rich in iron and magnesium, very like basalt. If magma is forced out without any opportunity to cool before reaching the surface it will harden as *basaltic* rock. Large areas of the earth's surface are covered with such rocks formed by basaltic lava from volcanoes.

When magma cools below the surface, the first crystals to form will contain much of the iron and magnesium. These crystals tend to sink into the lower part of the still liquid magma. Thus the upper layers of magma will contain larger amounts of silicon and potassium than were originally there. The upper layers form a light-colored *granitic* rock. Granite type rocks are found here. The dark material below forms basaltic rock.

Sedimentary rocks. There are three types of sedimentary rock. One is called *fragmental*. The rock fragments are carried away from their source by water, wind, or ice. They are then left as deposits. The fragments are cemented into rocks by great pressure and the cementing action of other materials. One example of a fragmental sedimentary rock is *conglomerate*. This is a coarse-grained rock made up of gravel, pebbles or even boulders.

Cemented grains of quartz sand make up another group of sedimentary rocks. These are the *sandstones*. Gaps between the sand grains leave spaces which liquids move through easily.

QUARTZ   HORNBLENDE
FELDSPAR   MICA

FIG. 8–18A. A piece of granite is composed of minerals representing four mineral groups.

FIG. 8–18B. Plutonic rocks. Top, pink granite. Bottom, hornblende syenite.

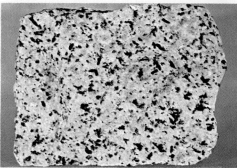

FIG. 8–19. Massive obsidian formations. (Ramsey)

FIG. 8–20A. Natural salt formations found near the Sierra Nevada Mts. form evaporites of the mineral Halite. (Shostal)

Shale is a clay deposit that has become rock mainly as a result of pressure. The flaky clay particles are usually pressed into flat layers. Because of this shale tends to split into flat pieces.

A second group of sedimentary rocks is *chemical* in origin. These are rocks which are formed of materials that were dissolved in water. When the water evaporates it leaves materials behind, which become solid again. These materials are called *evaporites. Halite* and *gypsum* are part of this group. See Figure 8–20A.

The smallest group of sedimentary rocks are the *organic* rocks. These rocks are made from materials that once were living things. *Limestone* is a sedimentary rock made up of calcium carbonate ($CaCO_3$). Animals and plants take calcium carbonate from water. Their skeletons settle to form organic limestone. In *coquina,* an organic limestone, pieces of shells are often found. *Coal* is a sedimentary rock of *organic* origin. It is made up of plant material that is partly decayed, and then buried. The carbon from these plants forms most of the material in coal. See Figure 8–21A.

Metamorphic rocks. The third large division of rocks are those called *metamorphic.* These rocks are formed deep in the earth's crust. Great pressure and heat from the weight of the rock above, intrusion by magma, and movement of the crust cause the rocks to change. Minerals can be melted to produce new compounds. Pressure can mash

FIG. 8–20.

FINE-GRAINED IGNEOUS ROCKS
PYROXENITE Ontario
ANDESITE PORPHYRY Nevada
RHYOLITE BRECCIA Mexico

TRACHYTE TUFF West Germany
VESICULAR BASALT Colorado
RED OBSIDIAN Oregon

FIG 8-21A

CONGLOMERATE Colorado
SHALE Texas
LIMESTONE Massachusetts

SANDSTONE New York
ROCK SALT New York
DOLOMITE Minnesota

FIG. 8–21B

SLATE Pennsylvania
QUARTZITE South Dakota
MICA SCHIST Vermont

MARBLE Georgia
PHYLLITE Massachusetts
GNEISS Massachusetts

FIG. 8–22. Stratified sedimentary rocks. (Ramsey)

FIG. 8–23. A cliff formed of massive igneous rock in Yosemite National Park. (Ramsey)

rocks into new textures, and chemical changes can be brought about by gases and liquids escaping from magma. See Figure 8–21B.

All metamorphic rocks are made from sedimentary, igneous or other metamorphic rocks. For example, *slate* is a metamorphic form of the sedimentary rock shale. The tendency of shale to split into flat sheets can still be seen in slate.

Greater heat and pressure can change slate into another metamorphic stage called *phyllite*. And still greater heat and pressure can change phyllite into *schist*.

Marble is formed from limestone. The limestone crystals grow larger and many impurities are driven off. In general, marble is white, though it may have some dark streaks containing other minerals.

Quartzite is a metamorphic rock formed from sandstone. It is usually more compact and firmly cemented, so that the spaces between the sandstone particles disappear. Because quartzite is very hard and durable it forms the large part of many hills and mountains from which weaker rocks have been worn away.

Graphite is almost pure carbon, formed from the metamorphic changes of coal. It is a soft black solid widely used in pencils and as a lubricant for machinery.

Rock structures. Large masses of the three kinds of rocks are put together in different ways. Some are arranged in layers called *beds* or *strata*. The strata may be seen because of differences in color, texture or composition. See Figure 8–22. Sedimentary rocks are almost always *stratified*.

Those rocks which show no strata are said to be *massive* or unstratified. Igneous rocks usually are arranged in massive formations. See Figure 8–23. In some lava flows, however, strata is shown by beds of volcanic dust or fragments lying between layers of lava.

Metamorphic rocks can be either stratified or massive depending on their composition, and on the conditions under which they are formed. See Figure 8–24(A & B).

One thing that all three kinds of rock have in common are cracks or joints in fairly regular patterns. In igneous rocks the *joints* are probably formed during the cooling process. A column-like formation may result from jointing in the fine-grained igneous rock. See Figure 8–25.

In sedimentary rocks, jointing may be the result of the drying and cracking of the beds. It may also be caused by movements of the crust which may slowly bend and twist the beds. Joints in sedimentary rocks are usually at right angles to the direction of the bed. This often separates the rocks into large blocks. When these blocks fall, cliffs are formed. See Figure 8–26.

Jointing in metamorphic rocks is similar to that of the rock from which it originated. For example, in a rock such as slate, joints may be similar to those found in the sedimentary rock called shale.

FIG. 8–24A. A formation of massive metamorphic rock in Zion National Park. (Ramsey)

FIG. 8–24B. Stratified metamorphic rock formation. (Ramsey)

FIG. 8–25. Devil's Postpile National Monument is a large cliff of basaltic rocks with vertical joints which form separate columns. (Ramsey)

FIG. 8–26. These strange formations in Bryce Canyon National Park were formed from a cliff made of layers of weak sedimentary rocks. (Union Pacific Railroad)

VOCABULARY REVIEW

Match the word or words in the column on the right with the correct phrase in the column on the left. *Do not write in this book.*

1. A single chemical compound found naturally as part of the earth.
2. Silicon and oxygen joined to form a network of SiO_2 molecules.
3. Similar to quartz with aluminum atoms in place of some silicon atoms.
4. Separates easily into thin layers.
5. Only two oxygens in the silicon-oxygen tetrahedron are joined to other silicons.
6. The color of a thin layer of the finely powdered mineral.
7. Ability of a mineral to split with a flat surface along certain directions.
8. A volcanic rock light enough to float on water.
9. A form of magnetite.
10. Rocks formed directly from magma.
11. Formed from rock fragments produced by the wearing of solid rock at the surface.
12. Rocks formed from igneous or sedimentary rocks under great heat and pressure.
13. Igneous rock with a glassy appearance.
14. Clay that has become rock due mainly to pressure.
15. Formed from phyllite under great heat and pressure.

a. feldspars
b. streak
c. pumice
d. basaltic
e. mineral
f. micas
g. lodestone
h. quartz
i. hornblende
j. igneous
k. cleavage
l. metamorphic
m. shale
n. graphite
o. obsidian
p. schist
q. sedimentary rocks

QUESTIONS

Group A

Select the best term to complete the following statements. *Do not write in this book.*

1. A majority of the minerals in the earth's crust contain (a) oxygen and silicon (b) silicon and hydrogen (c) hydrogen and aluminum (d) aluminum and silicon.
2. Minerals which contain oxygen and silicon are known as (a) oxidation (b) silicates (c) silicons (d) hydrates.
3. A mineral which contains only silicon and oxygen is (a) hornblende (b) mica (c) feldspar (d) quartz.
4. Micas can be separated into thin layers because one oxygen in the silicon-oxygen tetrahedron has a (a) strong bond with a silicon (b) weak bond with a silicon (c) strong bond with another kind of atom (d) weak bond with another kind of atom.
5. A mineral whose structure is closest to that of hornblende is (a) mica (b) olivine (c) pyroxene (d) quartz.

6. Which of the following is a group of minerals, all of which contain silicon and oxygen? (a) rock forming minerals other than silicates (b) gem minerals (c) metal ore minerals (d) micas.

7. The most obvious characteristic used in the identification of minerals is (a) streak (b) luster (c) color (c) specific gravity.

8. Which of the following is *not* a term used to describe the luster of minerals? (a) bright (b) dull (c) silky (d) adamantine.

9. Which of the following is *not* a term used to describe fracture? (a) irregular (b) conchoidal (c) flat (d) earthy.

10. Moh's scale is used in describing the mineral characteristic of (a) cleavage (b) specific gravity (c) magnetism (d) hardness.

11. Any mineral which can be scratched by a fingernail has a hardness (a) less than calcite (b) greater than gypsum but less than calcite (c) greater than calcite (d) less than gypsum but greater than talc.

12. A common substance which could be used in place of calcite on Moh's scale is a (a) fingernail (b) copper penny (c) piece of glass (d) steel file.

13. The specific gravity of a mineral sample can be found by dividing the weight of the sample by (a) the weight of the dry sample (b) the volume of the sample (c) the volume of an equal weight of water (d) the weight of an equal volume of water.

14. Which of the following is not attracted by a magnet? (a) magnetite (b) pyrite (c) lodestone (d) pyrrohotite.

15. Fluorescence is the process in which substances absorb (a) visible light and also give off visible light (b) visible light and give off ultraviolet light (c) ultraviolet light and give off ultraviolet light (d) ultraviolet light and give off visible light.

16. Which of the following mineral tests may not require a flame? (a) borax bead (b) charcoal block (c) colored flame (d) all of these require flames.

17. To organize a rock collection, you may start by separating them by (a) color (b) texture (c) how they were formed (d) any method you wish.

18. Which of the following does *not* refer to one of the three basic ways in which rocks are formed? (a) igneous (b) evaporation (c) metamorphic (d) sedimentary.

19. Which one of the following rocks is not related to the others in the way it was formed? (a) basalt (b) obsidian (c) dolomite (d) pumice.

20. Granitic rocks belong to which of the following rock types? (a) igneous (b) metamorphic (c) sedimentary (d) basaltic.

21. All of the following are similar, in that they are made up of individual grains joined together except (a) shale (b) conglomerate (c) sandstone (d) halite.

22. Halite and gypsum are examples of (a) igneous rock (b) plutonic rocks (c) evaporite rocks (d) fragmental rocks.

23. Limestone and dolomite are similar except that dolomite (a) is not as old (b) is older (c) contains magnesium (d) has been exposed to the atmosphere longer.

24. Which of the processes listed below does not produce metamorphic rocks (a) precipitation (b) heat (c) pressure (d) chemical changes.

25. All metamorphic rocks are produced from existing (a) igneous rocks only (b) sedimentary rocks only (c) igneous or sedimentary rocks (d) plutonic rocks only.

26. Which rock listed first does not come from the rock listed second as a result of metamorphism? (a) marble from limestone (b) quartzite from sandstone (c) slate from shale (d) phyllite from schist.

27. If the following rocks had a common origin, which would be the last to form? (a) slate (b) schist (c) shale (d) phyllite.

28. Diamonds are a form of pure carbon. Which of the following is most closely related to diamonds? (a) graphite (b) coal (c) slate (d) shale.

29. Stratified rocks are rocks that are (a) massive (b) in layers (c) all the same color (d) all the same composition.

30. Joints in rocks are (a) only found in igneous rocks (b) places where rocks bend (c) cracks in the rocks (d) only found in sedimentary rocks.

Group B

1. How is a borax bead made and how is it used in identifying minerals?
2. Give a brief description of the three general ways in which rocks are formed.
3. Which mineral groups do not contain silicon and oxygen?
4. Explain why it is difficult to identify a mineral simply by its color.
5. What is the special optical property of calcite?
6. Name at least eight tests which can be used in identifying minerals.
7. What is the streak of a mineral and how is it obtained?
8. Explain how evaporites are formed and give at least one example.
9. A mineral sample has a dry weight of 24 grams and weighs 16 grams when submerged in water. What is the specific gravity of the mineral?
10. How do metamorphic rocks differ from the rocks from which they were formed?
11. Name at least three igneous rocks, three sedimentary rocks and three metamorphic rocks.
12. Describe the steps in the formation of a schist.
13. What is Moh's scale and how is it used?
14. Describe three types of mineral identification tests which require flames.
15. Explain the difference between cleavage and fracture.
16. Draw a diagram of a silicon-oxygen tetrahedron.
17. How do rocks produced by intrusive and extrusive volcanic activity differ?
18. How olivine different from quartz, mica and hornblende?
19. What factors determine the properties of silicate minerals?
20. When an object is submerged in water, it appears to lose one gram for each cubic centimeter of its volume. The density of an object is the number of grams of mass for each cubic centimeter of its volume. Density has units of $\frac{grams}{cm^3}$. What is the specific gravity of a mineral whose density is 3.5 $\frac{grams}{cm^3}$?

9

the restless earth

objectives

- [ ] Describe the theory of continental drift.

- [ ] Describe the theory of plate tectonics.

- [ ] Identify the main layers of the earth and state the evidence of their existence.

- [ ] List and describe three kinds of crustal plate boundaries.

- [ ] Compare the earth's magnetic field to a bar magnet.

- [ ] Describe the major causes, processes, and features of volcanic activity.

- [ ] Explain how volcanism is related to plate tectonics.

In 1912 a theory was proposed that most scientists thought was contrary to common sense. The theory stated that the continents were not stationary, but slowly drifted along the earth's surface. This was called the theory of *continental drift* and was proposed by Alfred Wegener, a German scientist. Most of Wegener's fellow scientists did not believe that his theory of continental drift was possible. They did not understand how the huge continents could drift and plow through the solid earth that surrounds them. Others before Wegener noticed that the continents showed evidence that they may have once been joined together and then somehow separated. The outlines of western Africa and eastern South America, for example, seem to fit together like the pieces of a jigsaw puzzle. Some of the continents now separated by wide oceans have similar rock formations. Remains of living things also seem to indicate that the land masses were once connected. Wegener's theory of continental drift explained these observations. It proposed that there was once only a single super continent called *Pangaea* (Pan-*gee*-ā). Then a few hundred million years ago, Pangaea was thought to have broken up and the continents then drifted apart.

Wegener's fellow scientists found his theory unacceptable. They could see no way that continents could move over the earth's surface like rafts on a lake. They were right. It did not happen that way. Yet today, most scientists accept the general principle behind the theory of continental drift. This is because new discoveries made during the past few decades have revealed that the continents do move. However, they do not drift like individual rafts. The continents move because they are passengers on a number of rigid plates of rock that move along the earth's surface. Discovery of the existence of these moving plates has given rise to the theory of *plate tectonics*. As a result of this theory, there has been an explosion of new knowledge about the forces that shape the earth.

PLATE TECTONICS: A THEORY

The earth's interior. According to the modern theory of plate tectonics, the earth's crust is made up of a number of separate rigid plates. The plates are exposed to forces that cause them to move. This movement creates the familiar landscape of the earth's surface. "Tectonics" refers to the forces that move and shape the crustal plates. These forces seem to come from deep inside the earth. The foundation of the theory of plate tectonics comes from knowledge of the earth's interior.

The deepest holes drilled into the earth in the hunt for petroleum reach to a depth of about 10 km. It may be possible to drill about 10 times deeper than this in the future. This is only about 1/2000 of the distance to the earth's center. The deep interior of the earth can never be reached by any direct method like drilling. However, the earth itself provides a method for investigating its internal structure. When an earthquake occurs, waves are sent out that pass through the earth. See Figure 9–1. These

FIG. 9–1. This scientist is examining a record of earthquake waves. (U.S. Geological Survey)

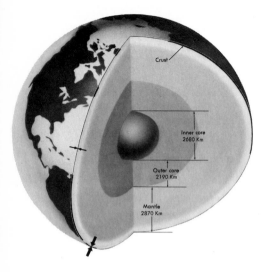

FIG. 9–2. The layers that make up the earth. The relative thickness of the crust is exaggerated. The crust is about 11 km thick in the ocean basins (small arrows), and about 32 km thick in continental areas (large arrows).

Examine
Use a fresh hard-boiled egg to help demonstrate the layered earth. Do not remove the shell. Cut the egg in half and compare what you see with Fig. 9–2.

Illustrate
Make a drawing that includes all of the following dimensions to scale. Let 1 cm = 500 km. Earth-radius = 6400 km, Core-radius = 1340 km, Mantle-radius = 6389, Satellite orbit = 250 km, Ocean deep = 11 km, Mt. Everest = 9 km. Explain any problems you may have.

waves can be detected and measured far away from the actual location of the earthquake. The waves move through the rock at a speed that is determined by how dense or rigid the rock is. Measurement of the behavior of earthquake waves reveals the density and other general properties of the earth's interior.

Analysis of earthquake waves shows that the earth is divided into four separate layers. See Figure 9–2. There is a thin outer layer called the *crust.* The crust is thicker beneath the continents than beneath the oceans. Below the crust is the *mantle.* The mantle is 2870 km thick. At the earth's center is the *core,* a sphere with a diameter of 7060 km. The core is made up of two parts, the inner core and outer core.

The way that earthquake waves are bent as they move through the earth is shown in Figure 9–3. Each of the earth's different layers can be determined by the effect it has on these waves. For example, waves bend very sharply as they enter the denser core. This creates a region on the surface where the waves are not felt. This region is called the *shadow zone.* Since certain kinds of earthquake waves do not pass through the outer core, scientists believe that it is made of liquid. The inner core, however, seems to be solid. Behavior of earthquake waves, as well as the high density of the earth itself, indicates that the core is made of a very heavy material. Evidence seems to indicate that the core is made up mostly of iron mixed with some nickel. Apparently when the earth was molten during its early history, the heavy materials like iron sank to form a dense core. The lighter rock materials floated upward to form the mantle and crust.

The crust is separated from the mantle by a sharp boundary called the *Mohorovicic discontinuity.* This boundary is named after the Yugoslavian scientist who first discovered its existence. It is called the "Moho" for simplicity. The Moho apparently marks a boundary where the rocks of the crust meet the rocks of the mantle. The rocks of these two layers differ in mineral composition, in crystal structure, or in both. The crust and upper part of the mantle together make up a hard shell called the *lithosphere.* Unlike the crust, the lithosphere has no sharp boundary separating it from the remainder of the mantle. Rather, it gradually becomes a softer region called the *asthenosphere.* The mantle rock of the asthenosphere begins at a depth of 100 km and extends to a depth of 250 km. The rock within the asethenosphere is partly

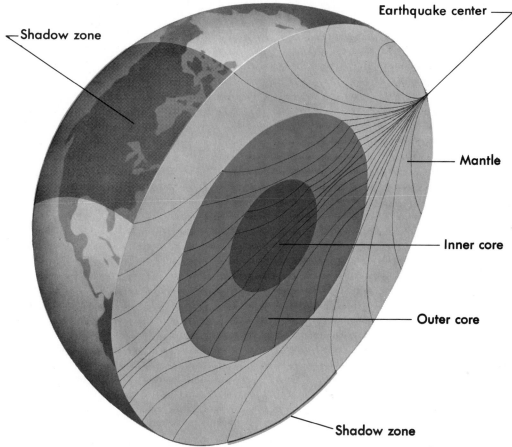

Shadow zone

Earthquake center

Mantle

Inner core

Outer core

Shadow zone

FIG. 9–3. The shadow zone is a region on the earth's surface where waves from a given earthquake are rarely felt. It results from the bending of some of the waves and the blocking of others by the earth's core.

melted and flows like a very thick liquid. The ability of this rock to flow is an important part of the theory of plate tectonics. The motion of the crustal plates is believed to be caused by slow movements of the mantle rock within the asthenosphere.

Plate boundaries. According to the theory of plate tectonics, the lithosphere is divided into separate plates. See Figure 9–4. These plates move since they can slide slowly over the partly molten zone of the upper mantle. The moving plates carry continents along with them like logs frozen in moving ice on a river. Where the plates come together are the zones of active change in the lithosphere. Study of the changes that take place along these plate boundaries has caused a revolution in our understanding of the forces that shape the earth's surface.

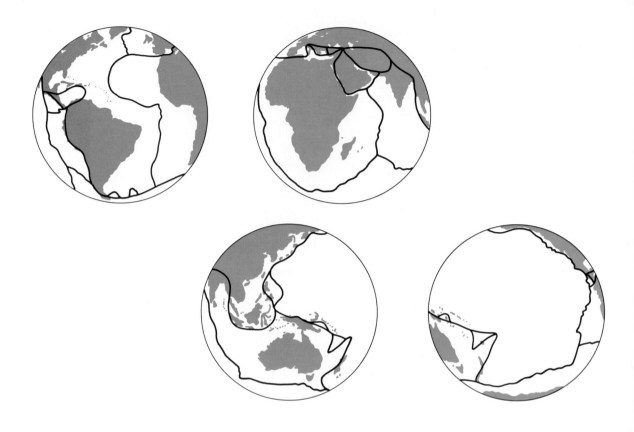

FIG. 9–4. The lithosphere is broken into a number of crustal plates that make up the outer covering of the planet.

There are three kinds of plate boundaries that differ according to the way each plate moves in relation to the other. For example, one kind of plate boundary occurs where both plates are moving away from each other. The gap created as the plates separate is filled with molten rock. The molten rock rises up from the asthenosphere. See Figure 9–5A. This type of plate boundary is found only on the sea floor. The locations of these spreading boundaries are marked by *mid-ocean ridges.* These ridges run around the entire globe along the floor of every ocean of the world. They represent the boundaries of various plates that are moving apart. As the plates separate and molten material fills in from below, new crust is created. This process, which continuously produces new sea floor moving out in opposite directions from the mid-ocean ridges, is called *sea-floor spreading.* A typical rate at which the plates move apart during sea-floor spreading is 5 centimeters per year.

All plates that meet along the mid-ocean ridges are
being pushed apart by sea-floor spreading. This means
that the plates have other boundaries where they are
being pushed against still other plates. When two plates
collide, one plate may be pushed down while the other
rides up. See Figure 9–5B. The region along plate boundaries where one plate is forced under another is called a
subduction zone. In some cases when plates collide, they
slide past each other. See Figure 9–5C. Each of the individual plates has some combination of these three kinds
of boundaries.

The forces that move the plates are not completely
understood. But they seem to result from the upward
movement of the molten material from the mantle. It may
be that this heated material rises to the surface to form
the mid-ocean ridges. At other plate boundaries, the edges
of colliding plates are pushed down and melted again.
This creates new crust in one place and destroys it in
another in an endless cycle. See Figure 9–6. Each plate
slides along as if carried on a very slow moving conveyer
belt. It starts at a mid-ocean ridge and moves toward
another of its boundaries. Here it is forced beneath
another plate and back into the earth's interior.

FIG. 9–5. A) Molten rock rises up
from the asthenosphere into the
gap created by the separating
plates; B) When two plates collide
one is pushed down while the
other rides up; C) In some cases
when plates collide, they slide past
each other.

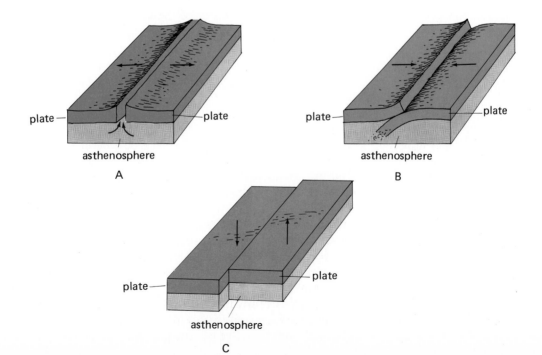

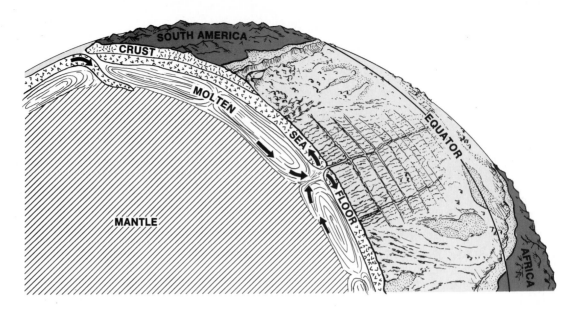

FIG. 9–6. Material is shown rising at the mid-ocean ridges, forcing the crustal plates to move apart. Where a plate meets the spreading sea floor, as along the west coast of South America, the sea floor is pushed beneath the plate.

FIG. 9–7. The geomagnetic poles are presently located at an angle of 11.5° from the axis of rotation.

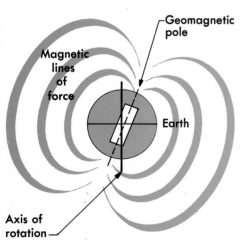

Plate tectonics and the earth's magnetism. Sea-floor spreading was one of the most important discoveries that led to the theory of plate tectonics. Proof of sea-floor spreading came from study of the earth's magnetism. Anyone who has used a compass to determine directions knows that the earth is like a giant magnet. Some features of the earth's magnetism can be described in the following way. Imagine a very powerful bar-shaped magnet buried near the earth's center. This imaginary magnet would be tilted at an angle to the geographic pole. See Figure 9–7. The points on the surface just above the poles of the imaginary magnet are called the *geomagnetic poles*. The magnetic axis is inclined in relation to the earth's geographic axis (axis of rotation). As a result, a compass needle does not always point in the direction of the earth's geographic pole. If a compass needle does not point to the geographic north pole, where does it point? Recent observations have located the north magnetic pole at 78.5°N. latitude and 69°W. longitude. The south magnetic pole is presently located at 78.5°S. latitude, 111°E. longitude. By connecting these two points with an imaginary line drawn through the earth, it would be apparent that the geomagnetic axis does not pass through the center of the earth. See Figure 9–7.

The angle between the direction of the geographic pole and the direction in which the compass needle points is called *magnetic declination*. Declination is measured in degrees east or west of the geographic north pole. In Figure 9–8, a compass needle at point A in the Northern Hemisphere points west of the geographic pole. At point C, the compass needle lines up with the geographic axis so that there is no declination. At point B, the compass indicates a declination east of geographic north. Because of its importance in navigation, magnetic declination has been determined for points over the earth. Charts and tables are available that give the declinations for most regions of the world. The pattern of magnetic declination for the United States is shown in Figure 9–9.

Lines drawn through all points having equal declination are called *isogonic lines*. The line of zero declination is called the *agonic line*.

As yet, there is no final explanation to the origin of the earth's magnetism. Acceptable explanation seems to lie in the properties and motions of the earth's outer core. Earlier in this chapter we learned that the earth's outer core is believed to be a liquid when under ordinary conditions. The entire core (inner and outer) is probably composed of heavy metallic elements, including iron and nickel. Slow movements of the liquid metal in the outer core may take place independently of the earth's rotation. These movements generate electrical currents in the same way an electrical generator or dynamo works. Elec-

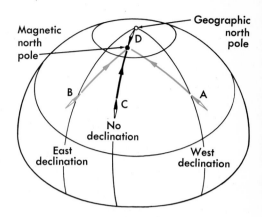

FIG. 9–8. Compass declination depends upon one's position with respect to magnetic and geographic north poles.

FIG. 9–9. The lines on this map connect points having the same magnetic declination. Their unevenness indicates that the source of the earth's magnetism is not uniform.

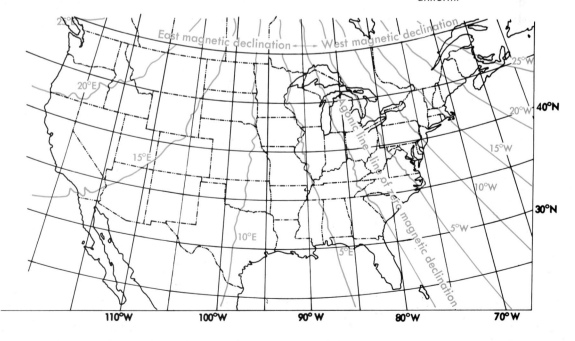

trical currents created in the outer core turn the solid inner core into a magnet. It is no different from the way that an iron bar becomes magnetized if it is wrapped with wires carrying an electrical current. See Figure 9–10.

Like a compass, some kinds of minerals are affected by the earth's magnetism. In molten rock, individual crystals of these minerals line up in the direction of the earth's magnetic poles. When the rock cools and solidifies, these crystals are frozen in a position that shows the direction of the magnetic poles. Rocks formed millions of years ago contain a record of the earth's magnetism at the time they were formed.

Study of the ancient magnetic records found in rocks has revealed two surprising facts. First, the earth's magnetic poles seem to have shifted position many times during the last billion years. However, the magnetic poles must always stay close to the geographic poles since the rotation of the earth helps create the magnetic poles. The apparent wandering of the magnetic poles must actually be the result of motion of the continents. Thus, ancient evidence of magnetics in rocks can help scientists to trace the motion of the crustal plates in the past.

Another surprising discovery was that the earth's magnetic field reverses itself from time to time. That is, the magnetic north and south poles change places. Magnetic evidence in rocks shows that at least nine such reversals have taken place during the past 3.5 million years. Reversals of the earth's magnetic field have become part of the evidence that supports the idea of sea-floor spreading. Reversals are also used as a way of measuring the speed with which the plates move apart.

When lava flows from a mid-ocean ridge, it hardens and moves away on either side of the ridge. The lava becomes split into two narrow stripes. One stripe lies on each side of the ridge. The rocks in these stripes carry a record of the direction of the earth's magnetic poles at the time they were formed. Each reversal of the magnetic poles will be shown by parallel striped magnetic patterns on each side of the ridge. See Figure 9–11. In time, the stripes of lava move away from the mid-ocean ridge carrying their magnetic record.

The discovery of these magnetic stripes means that sea-floor spreading almost certainly does occur. The rate at which new sea floor is spreading can be determined from the times of the magnetic reversals and the distance of the stripes from the mid-ocean ridges.

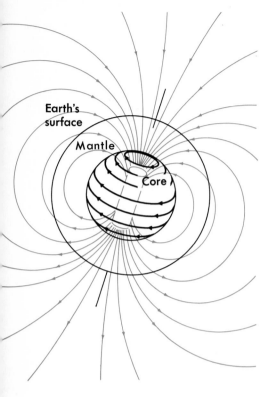

FIG. 9–10. This diagram illustrates how the earth's core might be able to produce the magnetic field as a result of its motion.

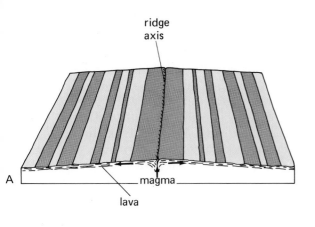

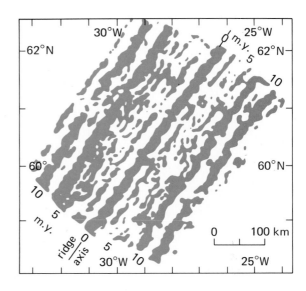

FIG. 9–11A. Reversals of the earth's magnetic field leave a record in the form of magnetic stripes in the lava flowing out of the mid-ocean ridges.

FIG. 9–11B. The pattern of magnetic stripes as they are found on the sea floor along the mid-Atlantic ridge southwest of Iceland. Ages of the rock are given in millions of years.

VOLCANISM

Origin of magma. Wherever deep holes have been dug into the earth's crust, temperature has been found to increase with depth. Once beyond the shallow depths where the rocks are affected by seasonal temperatures on the surface, temperature increases at a steady rate. Measurement of this temperature rise in many deep wells and mines shows an average increase of 1°C for each 30 meters of depth.

Since holes have not been drilled to any great depth into the crust, it is not known if the temperature keeps increasing 1°C for every 30 meters. However, it seems very unlikely that the temperature keeps increasing at this rate all the way to the earth's center. If this were true, the temperature of the core would be several hundred thousand degrees celsius. The temperature probably increases very slowly below about 100 km. An estimate of the temperature and pressure of the earth's interior is given in Figure 9–12 on page 192.

It is believed that radioactive substances are the source of heat within the earth. Most of this heat remains trapped in the rocks that contain the radioactive materials. Very little escapes to the surface. Most of the energy on the earth's surface is received from the sun. The lighter rocks of the crust and upper mantle contain the largest amount of radioactive substances. This means that the earth's

Interpret

From the graph in Figure 9–12, determine the temperature and pressure at the boundary between the mantle and the core.

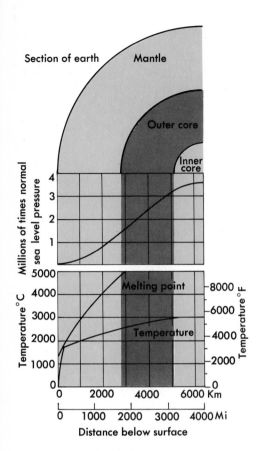

FIG. 9–12. These graphs show how temperature and pressure increase beneath the earth's surface. Notice that there is a depth in the upper mantle at which the melting temperature is reached.

Construct

Make a model of intrusive volcanic structures as shown in Figure 9–13.

internal heat is concentrated in these regions and does not increase very much toward the center of the planet.

If rocks melted at certain temperatures, regardless of other conditions, the earth would be a ball of molten rock covered by a thin solid shell of crust. However, other conditions, chiefly pressure, also affect the temperature at which rocks melt. In general, as the pressure increases, the melting temperature of rock also increases. This means that most of the rocks in the mantle, although very hot, are not molten because they are under high pressure. Only the upper mantle rocks in the asthenosphere are in the molten state. In addition to pressure, the amount of water contained in the minerals of the rock affects the way it melts. Rock with a high water content will melt at a lower temperature when under pressure than similar rock containing less water.

Some magma is believed to come directly from the asthenosphere at a depth greater than 200 km. It might be expected that magma coming from such great depths would cool and solidify before reaching the surface. However, the pressure drops as the magma moves upward. This reduces the temperature at which the rock melts. If magma moves up fast enough it can still be molten on reaching the surface.

Magma can also be created when one crustal plate is pushed beneath another along a plate boundary. The rock of the plate that is pushed down is heated as it is thrust toward the mantle. The plate being pushed down is often part of the sea floor. Thus, its rock has a high water content. This "wet" rock melts easily and becomes less dense. It then moves up and pushes into the rocks of the overriding plate above, producing volcanic activity.

Magma always moves upward because molten rock is lighter than solid rock. It works its way upward through the cracks and weaker parts of the rocks above. Sometimes magma cools and becomes solid before it reaches the surface. This results in *intrusive volcanism*. The main effects of intrusive volcanism are found in the rocks beneath the surface. If liquid magma reaches the surface it is called *lava*. The effects produced by lava on the earth's surface are called *extrusive volcanism*. It is extrusive volcanism that produces volcanic eruptions.

Results of volcanic activity beneath the surface. If bodies of magma solidify beneath the surface, many different types of rock structures can be produced. The principal structures formed beneath the surface are:

1. **Batholiths.** These are the largest of all the structures formed by intrusive bodies of magma. Batholiths are very large masses of solidified magma covering hundreds of square kilometers. The name "batholith" means "deep" rock. Batholiths extend downward to unknown depths. They form the cores of many major mountain ranges. A small batholith, less than 100 square kilometers, is usually called a *stock*.

2. **Laccoliths.** Magma can move between existing rock layers and spread outward toward the surface. The magma pushes the rock layers above into a dome-shaped mass. This results in small mountains being formed on the surface of the earth. These dome-shaped structures are called *laccoliths*. "Laccolith" means "lake" of rock. Laccoliths, unlike either batholiths or stocks, have a definite floor. They are frequently found in groups.

3. **Sills.** A sheet of fluid magma sometimes flows between the layers of existing rock. This raises the rocks just enough to make room for the intruding magma. The magma then hardens to form a sill. The walls of a sill are roughly parallel to the surrounding rock layers.

4. **Dikes.** A mass of magma can fill and then solidify in vertical cracks in rocks. This structure is called a *dike*. Dikes are often outgrowths of the large pockets of magma that form batholiths. See Figure 9–15 on page 194.

Volcanoes. A volcano is an opening in the earth's crust, through which hot gases, liquids, and solid materials reach the surface. The material poured out during a volcanic eruption includes large amounts of gases, as well

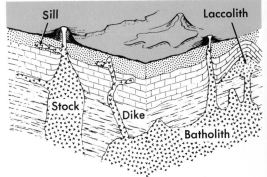

FIG. 9–13. Intrusive volcanic structures.

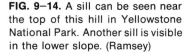

FIG. 9–14. A sill can be seen near the top of this hill in Yellowstone National Park. Another sill is visible in the lower slope. (Ramsey)

FIG. 9–15. Igneous dikes shown cutting diagonally across layered rock formation. (Ramsey)

FIG. 9–16. A quiet eruption being observed at the crater of Kilauea volcano, Hawaii. (Rowan-Photo Researchers)

as lava. Volcanoes can form on the surface of continents or on the sea floor.

When a volcano is actually erupting, it is said to be *active.* There are various kinds of volcanic eruptions. The type of eruption that will take place seems to depend on the pressure that forces the lava onto the surface. For example, in Hawaii the volcanoes produce thin lava. This lava gives off its gases with very little pressure and spreads out quickly. The lava pours out in more or less steady streams, often from large fountains. See Figure 9–16. This type of eruption is called *quiet* since it is fairly calm.

Some volcanoes explode with great violence. Before the eruption, there are often loud rumblings and earthquakes. An eruption of this type is called *explosive* because it usually happens suddenly and violently. One of the largest explosive eruptions in history took place in 1883 on the small island of Krakatoa. See Figure 9–17. This small island was located between Sumatra and Java in Indonesia. Suddenly, in August of 1883, a series of gigantic explosions blasted away about two thirds of the

island. About five cubic miles of material were thrown into the air. A dust cloud from the explosion completely circled the earth. The explosion and resulting giant waves produced in the surrounding sea killed between 30,000 and 40,000 people. Many deaths were reported more than 1,610 km away.

Another famous explosive eruption took place in the West Indies. Mount Pelée, on the island of Martinique, had been inactive since 1851. Suddenly in 1902, it blew off the top of its crater. A cloud of dust and hot gases raced down the mountain toward the city of St. Pierre. About 30,000 people lost their lives from breathing the hot dust and gases. After the first explosive eruption, an immense rock plug formed in the throat of the volcano. Pushed up by the pressures from below, the gigantic rock plug slowly rose 300 meters above the rim of the broken crater. See Figure 9–18. But its weak rock soon crumbled.

Many volcanoes are a combination of both quiet and explosive eruptions. These *intermediate volcanoes* sometimes erupt explosively, sometimes quietly, or in a combination of both. Stromboli (*strom*-boh-lee), located north of Sicily in the Mediterranean Sea, is a volcano of this type. It erupts every fifteen minutes or so with minor explosions throwing fragments of partly solidified lava into the air. Clouds of steam produced during the eruptions reflect light from the glowing lava below. Stromboli is known as the "lighthouse of the Mediterranean" because this effect is visible from a great distance.

Vesuvius, near Naples, Italy is another intermediate volcano. The first known eruption of this volcano took place in A.D. 79. This explosive eruption buried the ancient

FIG. 9–17. An old engraving shows the island of Krakatoa before the eruption of 1883. (Culver Pictures, Inc.)

FIG. 9–18. This great rock plug rose during the 1902 eruption of Mount Pelée. (American Museum of Natural History)

FIG. 9–19. A deeply eroded portion of a great lava plateau, the Columbia Plateau in Washington. (U.S. Geological Survey)

FIG. 9–20. Pahoehoe lava eventually decomposes to produce a very fertile soil. (Ramsey)

cities of Pompeii and Herculaneum. Since then it has had both explosive and quiet eruptions every 50 to 100 years.

A fourth type of eruption is called a *fissure flow*. A fissure is an open crack in an otherwise solid surface. Large amounts of lava flowing from many fissures in the surface build vast areas of lava plains or plateaus. See Figure 9–19.

Products of eruptions. Three types of materials are thrown out by volcanoes. First, there is molten lava. The composition of lava varies in different volcanoes. It may be thin and flow easily, or very thick and flow slowly. The temperature of lava usually reaches 1000–1200°C before it begins to cool. The cooled lava then takes many forms. It may cool rapidly on the surface, though the part below is still very hot. The crusted lava on the top then breaks into jagged floating chunks as the liquid portion slowly flows on. This type of lava flow is called *block lava* or, in Hawaii, *aa* (*ah*-ah).

The lava often solidifies. This gives its surface a ropy appearance. The Hawaiian word *pahoehoe* (pah-*hoy*-ay-*hoy*-ay) is used to describe this type of lava. See Figure 9–20. Sometimes the outer parts of a lava flow cool

rapidly, while the interior remains liquid. This liquid material may flow out later, leaving tunnels in the solid lava. See Figure 9–21.

Gases are a second type of material produced during volcanic eruptions. Steam is by far the most abundant gas that escapes from volcanoes. There are two ways that steam is formed during a volcanic eruption. Steam is formed when water beneath the ground meets the hot magma. However, much of the steam probably comes from the water vapor that is dissolved in the magma itself. Many other kinds of gas are also given off, but in much smaller amounts than the steam. Around or near volcanoes, there are often fissures or holes from which steam and other gases escape. These are called *fumaroles*. Some volcanic gases such as carbon monoxide, sulfur dioxide, and hydrogen sulfide are poisonous.

A third kind of material thrown out by volcanoes consists of solid fragments of varying sizes. These are actually fragments of solidified lava flung up by the force of escaping gases. The smallest of these rock particles are *dust* and *volcanic ash*. Many eruptions send out great clouds of dust and ash that settle on the surrounding land. Larger particles of rock that can be carried some distance are called *cinders*. Various sized rock fragments with a round shape, and often with twisted ends, are called *volcanic bombs*. See Figure 9–22. Very large chunks of rock thrown out in an eruption are known as *blocks*. These can weigh several tons such as the one shown in Figure 9–23.

Volcanic features formed on the earth's surface. The lava and solid materials that are produced during volcanic eruptions can create massive rock piles around the volcano. These are called *cones* and are classified into three main types. *Shield cones* are broad at the base with gentle slopes. They are made from layer upon layer of lava flowing out around the volcanic opening. The lava flow is fluid enough to cover a wide area. Shield cones are generally formed as a result of quiet eruptions. The Hawaiian Islands are a cluster of shield cones built up from the sea floor.

Explosive eruptions will form *cinder cones*. These have fairly narrow bases and steep sides. A cinder cone is a pile of solid fragments thrown out during an explosive eruption. Because their materials are loosely arranged, cinder cones are seldom very high. Figure 9–24 illustrates a shield and cinder cone.

FIG. 9–21. Lava tunnels or tubes are formed when molten lava drains from partly solidified flows. (Bob & Ira Spring)

FIG. 9–22. The largest volcanic bombs may measure several feet across. (Ramsey)

FIG. 9–23. This large volcanic block was thrown out during the last eruption of Mount Lassen in 1917. (Ramsey)

FIG. 9–24. The figure on the left shows a typical shield volcano. The one on the right shows a steeper sloping cinder cone. (Ramsey)

Composite cones are a combination of lava and rock fragments. They are built of alternating layers of solidified lava. Volcanoes with this layered or stratified make-up are sometimes called *stratovolcanoes*. Some of the world's best known volcanic mountains, such as Fujiyama in Japan, and Mounts Rainier, Hood and Shasta in the United States, have this type of structure.

Openings commonly develop in cones some distance below the main opening. Materials flowing from these openings can build up smaller volcanoes called *parasitic cones*. Other types of cones are very small. These are *spatter cones* that form in the lava fields away from the main vent. As the lava cools, a crust forms. Hot gases then force lava out in spattering fountains through openings in the crust.

The funnel-shaped pit at the top of a volcanic cone is known as the *crater*. The crater usually widens as the very hot magma melts and breaks down the crater walls. The rim can also collapse when explosions occur within

FIG. 9–25. Mount Shasta in California is an example of a stratovolcano. Notice the smaller cone that developed on the side of the original cone. (Ramsey)

the crater. Often a smaller cone is built within the crater by materials erupting from the vent.

Occasionally a large, basin-shaped depression can be formed when a volcanic cone collapses. A cone can collapse into a basin when magma pours out across its surface or drains downward again. Explosions can also completely destroy the upper part of the cone. When the cone collapses or is blown away, the immense pit or depression formed is called a *caldera*. See Figure 9–26.

When a volcano becomes extinct, no more material is added to the cone. When it is active, a volcano continues to build its cone much faster than weather and erosion can wear it away. When it stops erupting for a long period, valleys and gullies are cut deeply into its slopes by erosion. Eventually the softer parts of the dormant volcano are worn away and only the hard rock formed by the solidified

FIG. 9–26. The waters of Crater Lake, Oregon, fill the caldera of ancient Mount Mazama. The circular form of the crater and the symmetry of the inner cone, called Wizard Island, are visible in this view. (HRW Photo by Richard Weiss)

FIG. 9–27. A volcano in old age, Shiprock near Farmington, New Mexico. The neck exposed by erosion is all that remains of this ancient volcano. (Shostal)

activity

Most volcanic activity occurs in certain regions of the earth's surface. In this activity you will locate several volcanoes on a map of the world. Obtain a map of the world that has the crustal plates outlined. Your teacher may provide you with one.

Volcano	Name	Longitude	Latitude
A	Aconcagua	70W	35S
B	Tungurahua	80W	0N
C	Pelee	61W	15N
D	Tajumulco	90W	15N
E	Popocatepetl	100W	20N
F	Lassen	122W	40N
G	Rainier	122W	47N
H	Katmai	155W	60N
I	Fujiyama	139E	35N
J	Tambora	120E	10S
K	Krakatoa	108E	5S
L	Mauna Loa	155W	20N
M	Kilimanjaro	37E	3S
N	Etna	15E	38N
O	Vesuvius	14E	41N
P	Teide	16W	28N
Q	Laki	20W	65N

Find the location of Volcano "A" on the map by reading the longitude across the top or bottom of the map and the latitude along the side. Write the letter "A" at that location. Do the same thing for each of the other volcanoes. Show each location by writing the proper letter.

When finished locating the volcanoes, answer the following questions:

1. Which ocean has a ring of volcanoes around it?

2. Are most of the volcanoes located on the edge of crustal plates?

3. Which volcanoes are not located on the edge of a crustal plate?

magma remains. Often the hard central neck or plug is left standing as an isolated shaft of rock. Dikes running out from the central plug can also be exposed. See Figure 9–27. At this stage, intrusive formations such as laccoliths, associated with the original source of the magma, are frequently exposed on the surface. See Figure 9–28.

Volcanism and plate tectonics. Most of the world's volcanoes are found along the boundaries of the great crustal plates. The mid-ocean ridges are made up of an almost continuous line of active volcanoes. As the plates move apart along the mid-ocean ridges, magma moves up from the asthenosphere to fill the gap. Some of the magma that reaches the sea floor is still molten. A volcanic eruption is produced. Most of these eruptions are hidden beneath the sea. However, some eruptions may produce a volcanic cone that rises above the surface of the sea. An example of this kind of eruption, along the mid-ocean ridge, was the volcano called Surtsey. Surtsey rose from the sea in

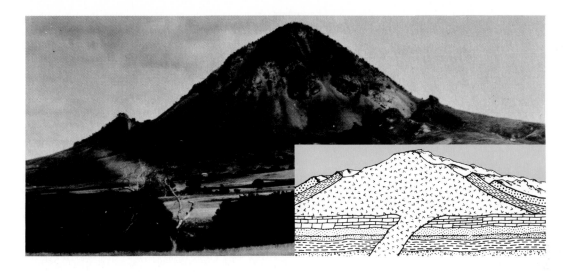

the Atlantic Ocean just south of Iceland in 1963. See Figure 9–29. Iceland itself is part of the mid-Atlantic ocean ridge that is raised above sea level. Such islands are formed when lava pours out of an opening along a mid-ocean ridge faster than the moving plates can carry it away.

FIG. 9–28. This is an exposed laccolith, showing outcrops of the upturned sedimentary rocks around its base. (U.S. Geological Survey)

FIG. 9–29. This volcano, called Surtsey, emerged from the sea just south of Iceland in 1963. (Vance Henry/Taurus Photos)

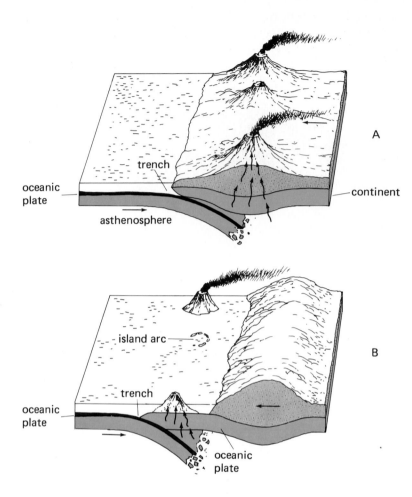

FIG. 9–30. A) The oceanic plate pushed down produces a trench. The plate melts and forms magma that rises to form volcanoes. B) When one oceanic plate is pushed beneath another plate, island arcs are formed along the trench.

Volcanic activity also occurs when one plate is pushed beneath another. When the edge of a plate that is part of the sea floor meets another plate carrying a continent, the oceanic plate is pushed beneath the continent. As the oceanic plate is pushed down, it bends and produces a deep *trench* along the edge of the continent. The oceanic plate melts and forms a light magma that rises to produce volcanoes along the edge of the continents. See Figure 9–30A.

When one part of the sea floor is pushed beneath another oceanic plate, a different kind of volcanic activity results. A chain of volcanoes called an *island arc* grows along the trench marking the boundary of the plates. See Figure 9–30B. The earliest stages of an island arc produce a series of small volcanic islands such as the Aleutian Islands stretching across the north Pacific Ocean. Older island arcs grow larger, like the group of volcanic islands that make up Japan.

Not all volcanoes are found along plate boundaries. The cause of volcanoes that occur away from plate boundaries is not known. For some reason, magma is able to work its way upward at some locations within the interiors of the crustal plates. It may be that some of these volcanoes are the beginnings of a spreading center that can split apart an existing continent.

VOCABULARY REVIEW

Match the word or words from the column on the right with the correct phrase in the column on the left. *Do not write in this book.*

1. The super continent as proposed by the continental drift theory.
2. The modern theory stating that the earth's crust is made up of several rigid parts.
3. The thin outer layer of the earth.
4. The layer of the earth between the crust and the core.
5. The Mohorovicic discontinuity.
6. The hard shell of the earth.
7. The location of the spreading boundaries of the crustal plate.
8. Where one crustal plate is forced beneath another.
9. The angle between the direction of the geographic pole and the direction the compass needle points.
10. Magma cools and becomes solid before it reaches the surface.
11. Very large body of solidified magma extending to unknown depths.
12. Thin sheet-like intrusion between rock layers.
13. Massive rock piles that form around volcanoes.
14. A deep region in the ocean where the oceanic plate has been pushed beneath the continent.
15. A chain of volcanoes that grows along the boundary of a plate.

a. mid-ocean ridges
b. core
c. mantle
d. batholiths
e. trench
f. Pangaea
g. island arc
h. crust
i. sills
j. lithosphere
k. agonic lines
l. cones
m. fumaroles
n. intrusive volcanism
o. plate tectonics
p. magnetic declination
q. subduction zone
r. Moho

QUESTIONS

Group A

Select the best term to complete the following statements. *Do not write in this book.*

1. The theory of continental drift is supported by the (a) size of continents (b) the equal spacing of mountains (c) shapes of the continents (d) thickness of the mantle.
2. The theory of continental drift is (a) entirely acceptable (b) partly acceptable (c) completely unacceptable.
3. According to the theory of plate tectonics, the earth is made up of (a) several rubbery plates (b) thousands of small pieces (c) a single flexible plate (d) several rigid plates.
4. The term that refers to forces that move and shape the crust is (a) plates (b) regulators (c) tectonics (d) volcanism.
5. Drilling to the center of the earth (a) is possible today (b) will be possible soon (c) is being considered for the year 2000 (d) can not be done with known materials.

6. Starting at the surface, the order of the layers of earth is (a) core, mantle, crust (b) core, crust, mantle (c) crust, core, mantle (d) crust, mantle, core.

7. The layer of the earth that we know best is the (a) inner core (b) crust (c) outer core (d) mantle.

8. Together, the crust and upper mantle make up the (a) outer core (b) asthenosphere (c) atmosphere (d) lithosphere.

9. The ring-shaped region where waves are not felt for a particular earthquake is called the (a) thermal zone (b) density zone (c) shadow zone (d) pressure zone.

10. The Mohorovicic discontinuity apparently separates the earth's (a) mantle and crust (b) core and mantle (c) inner and outer cores (d) crust and core.

11. Where plates move apart is a region of (a) mid-ocean ridges (b) subduction (c) no activity (d) all of these.

12. Where one plate is forced beneath another is a region of (a) mid-ocean ridges (b) subduction (c) no activity (d) all of these.

13. Proof of sea-floor spreading is found in a study of the earth's (a) life forms (b) earthquakes (c) magnetism (d) mountains.

14. The magnetic field of the earth can be compared to that of a (a) bar magnet (b) horseshoe magnet (c) coiled spring (d) donut.

15. Compared to the geographic poles of the earth, the magnetic poles are (a) near the equator (b) at the same location (c) located in different places (d) all of these.

16. The angle between the directions of true north and magnetic north is called magnetic (a) inclination (b) declination (c) geographic (d) azimuth.

17. Magnetic records on the sea floor show that the earth's magnetic field (a) has always been the same direction (b) has reversed (c) has changed only slightly (d) none of these.

18. Temperatures taken deep in wells and mines show that with depth, temperature (a) decreases (b) remains about the same (c) increases (d) changes unpredictably.

19. Even though much of the rock found in the deep crust and mantle is very hot, it is kept solid because of (a) low pressures (b) high pressures (c) radioactivity (d) solar energy.

20. Molten rock found below the surface of the earth is called (a) lava (b) a batholith (c) a laccolith (d) magma.

21. The effects of lava on the earth's surface are referred to as (a) intrusive volcanism (b) extrusive volcanism (c) quiet volcanism (d) intermediate volcanism.

22. The cores of many major mountain ranges are formed from (a) batholiths (b) sills (c) dikes (d) block lava.

23. When magma solidifies in vertical cracks, it forms (a) sills (b) block lava (c) dikes (d) pahoehoe lava.

24. Most volcanoes are found (a) along the boundaries of crustal plates (b) in the middle of crustal plates (c) in a line across crustal plates (d) in Europe.

Group B

1. Briefly describe the theory of continental drift.
2. How is the theory of plate tectonics similar to the theory of continental drift?

3. How does the theory of plate tectonics differ from the theory of continental drift?

4. Draw a diagram showing the main layers of the earth and label them with their correct names.

5. What evidence do we have supporting the existence of the core of the earth?

6. Explain the relationship between mid-ocean ridges and sea-floor spreading.

7. Describe the three kinds of crustal plate boundaries.

8. How is the earth's magnetic field like the field of a bar magnet?

9. Explain the relationship between the geographic and the geomagnetic poles.

10. How is magnetic declination related to isogonic lines?

11. Describe the nature of the evidence that supports sea-floor spreading.

12. How is intrusive volcanism similar to extrusive volcanism? How does it differ?

13. Describe four features that are the result of intrusive volcanism.

14. Name and briefly describe three types of cones associated with volcanoes..

15. How is volcanism related to plate tectonics?

16. How are island arcs and ocean trenches related?

10
movement of the earth's crust

objectives

- [ ] Describe the effects of the forces that bend rock.
- [ ] Describe the origin of the forces that deform the crust.
- [ ] Explain earthquakes and how they are studied.
- [ ] Describe how most earthquakes are believed to be related to crustal plates.
- [ ] Compare different kinds of mountains and the processes that formed them.

The scientific study of the earth is called *geology*. Geological discoveries indicate that the rocks on the earth's surface are not as solid and permanent as they appear to be. There is evidence of huge forces in the crust that are able to break, bend, and lift even the strongest rocks. Seashells and other remains of sea animals are found buried in the rocks of high mountains. This could happen only if rocks made from sediments once on the sea floor have been lifted far above sea level. In many places, sedimentary rocks are tilted and bent into arches and valleys. However, sedimentary rocks are always formed in horizontal layers. In order to be tilted into steep angles, or bent into uneven shapes, they must be pushed by a great force.

The modern approach to studying changes in the rocks of the earth's crust was clearly expressed by James Hutton, a Scottish geologist, in 1785. His theory was that *the present is the key to the past*. This principle is called *uniformitarianism*. It means that the same forces that changed the rocks in the past are still operating today, and causing the same kinds of changes. Before Hutton, it was commonly believed that the earth was shaped by great and violent events in the past, like volcanoes. After that, the planet was thought to have settled down and remained the same. Hutton's principle of uniformitarianism can only be correct if the earth is very old. The forces that work now to change the earth's crust operate very slowly. Therefore, it must have taken many millions of years to create the features that today make up the landscape.

The features found on the earth's crust are the result of slow but extremely powerful forces that have been operating for hundreds of millions of years. How rocks

FIG. 10–1. Fossils of sea animals have been found near the top of these mountains in the Canadian Rockies. (Atkinson)

FIG. 10–2. Solid rock is plastic under intense pressure. The limestone cylinder on the left was distorted into the form at the right by a vertical pressure of 125,100 pounds per square inch; simultaneously a pressure of 22,100 pounds per square inch was applied from the sides. (U.S. Dept. of the Interior)

react to these forces, and the source of the forces, is one of the most important parts of geology.

CHANGING THE SHAPE OF THE CRUST

Deforming rocks. When force is applied to some kinds of materials, such as an iron bar, they may respond by bending. Other materials, like glass, are more likely to break when force is applied. Rocks can respond to forces in both these ways. Forces can act within the crust to squeeze rocks together or pull them apart. Other forces cause layers of rock to slip past one another. When these forces are applied slowly, rocks will usually first deform like a piece of rubber. Then they return to their original shape when the force is removed. However, all rocks have a certain limit beyond which they cannot be deformed without permanently changing their shape. When this limit is reached, the rock either continues to bend, or it breaks. Generally, rocks that are under high pressure and temperature deep within the crust bend a great deal before finally breaking. See Figure 10–2. Rocks near the surface also bend, but they reach their limit sooner and then break.

Folding. The process by which rocks respond to forces by moving into new positions without breaking up is called *folding*. When rocks are folded, the layers are often pushed into shapes that resemble waves on water. These wave-like folds in rock can be small, or they can be large enough to raise mountains. Whatever its size, a fold usually has an upturned and downturned section. The section that is raised is called an *anticline*. The corresponding lowered section is a *syncline*. See Figure 10–3.

If the earth's surface was not constantly worn down, large folds in rocks would usually form ridges from the anticlines. Valleys would form from the synclines. See Figure 10–4A. However, the folds are often worn off as they develop. This action keeps the surface almost flat. See Figure 10–4B. When this happens, the only evidence of folding is the tilt of the rock layers beneath the surface. The anticline areas can also be worn down more rapidly because they are the first to be exposed. Wearing down of the anticline areas then leaves the original syncline as a mountain or ridge. See Figure 10–4C and D. Ridges

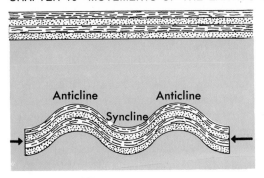

formed from synclines and valleys that were originally anticlines are common in the Appalachian Mountains.

Rocks exposed at the earth's surface are constantly changing. As a result, it is not always easy to find evidence that shows folding and other features of the rock structure in a region. Useful clues that help identify the rock structure of an area are found in sedimentary layers. All sedimentary rock layers are nearly flat or horizontal when formed. This is because most sedimentary rocks are formed from material deposited on the level surfaces at the bottom of bodies of water. A sedimentary rock layer that is not level was probably tilted by movements of the rocks in the crust. The angle at which a rock layer tilts from the horizontal is called its *dip*. For example, a certain bed of rock might have a dip of 30°. The direction of dip is also usually given using compass directions. Thus, the dip of a rock layer might be 30° south. It is also useful to indicate whether the rock layer runs in a general north-south or east-west direction. This is called the *strike* of the rock layer. The strike is always measured at right

FIG. 10–3. Rocks can respond to pressure by becoming folded, as shown in the diagram. Strongly folded metamorphic rocks near Calico, California, are shown in the photograph. (Ramsey)

FIG. 10–4. Synclinal mountains develop as weaker beds are worn down; A) leveling the initial anticlines and synclines; B) weak beds in the anticlines are exposed; C) uplift brings about further erosion, and D) the resistant rocks of the syncline remain as a ridge.

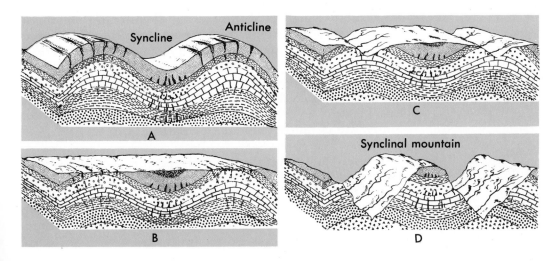

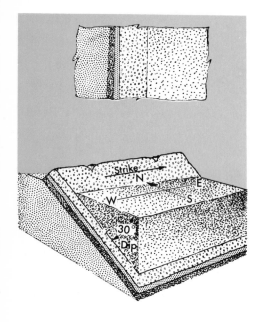

FIG. 10–5. The dip and strike of sedimentary rocks. What is the angle of the dip shown? The sketch at the top shows how the beds might appear on the surface.

FIG. 10–6. Rock relationships in a pitching fold.

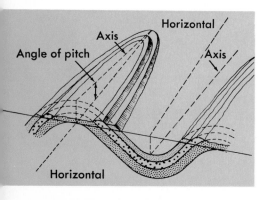

angles to the direction of its dip. A rock layer might be described as having a dip of 30° south and an east-west strike. See Figure 10–5.

Once the dip and strike of the rock layers in an area are known, the rock structure beneath the surface can usually be determined. For example, the dip of a group of layers may show that they are part of a fold whose *axis* is not horizontal. The axis of a fold is an imaginary line running along the top of an anticline or syncline. A fold whose axis is not horizontal is said to be a *pitching fold*. See Figure 10–6. The surface appearance of a typical pitching fold is clearly illustrated in Sheep Mountain in Figure 10–7.

Faulting. Rocks do not always respond to changes by folding. Sudden or abrupt crustal movements can fracture or crack rocks. This is most likely to occur on rocks near the surface since they are under the least amount of pressure. If there is no movement between the sides of a fracture, it is called a *joint*. Joints may also form in igneous rocks when cooling takes place. (See page 177.) These joints are always at right angles to the surface at which cooling occurs.

Often, the same pressures that produce rock fractures and joints, also force the fracture surfaces to slip against each other. If there is such movement of the rock along the sides of the fracture, it is called a *fault*. When faulting occurs, the rocks can slip against each other along the fracture in several ways. The rocks along one side of the fault can move horizontally. See Figure 10–8. This is called a *strike-slip fault*. A more common kind of movement along faults is an upward or downward movement of the rocks. This type of movement produces *vertical faults*, since the movement of the rocks is mainly vertical in direction. There are several kinds of vertical faults. They differ in the way the rock blocks on each side of the fault move in relation to each other. Figures 10–9, 10, and 11 show several ways that vertical faulting can occur. Many faults show evidence of both lateral and vertical motion.

In a few places on the earth, huge *parallel faults* have developed. These faults can cause the blocks of rock between them to sink. The result of this movement has been the formation of broad valleys with steep cliffs on both sides. This kind of valley is called a *rift valley* or *graben*. Graben is a German word meaning "trench." See Figure

FIG. 10-7. Sheep Mountain, Wyoming, shows how the anticline of a pitching fold can appear after it is eroded. Can you imagine what the appearance of the various layers would be if they were not eroded? (Barnum Brown)

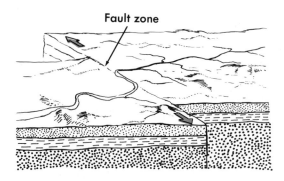

FIG. 10-8. Diagram of a strike-slip fault showing streams that have been offset by movement along the fault plane.

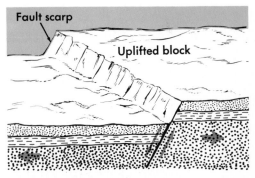

FIG. 10-9. When a fault is tilted away from the uplifted block, a "normal" fault is produced. This type of faulting often forms a steep cliff or "fault scarp." Arrows indicate direction of crustal movement that caused the fault.

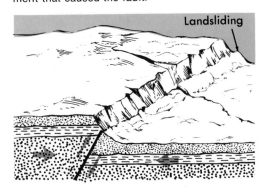

FIG. 10-10. A fault tilted toward the uplifted block produces a "reverse" fault. The fault scarp then becomes an overhanging cliff that usually collapses.

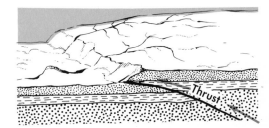

FIG. 10-11. If a fault is tilted very slightly, one block of rock may be thrust or pushed up over another to form a "thrust" fault. This action can produce a tightening of the crust.

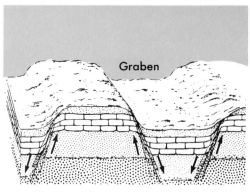

FIG. 10–12. A rift valley or graben is formed by faulting, as shown in the diagram. Left, the Crag Lake graben in Idaho. (U.S. Geological Survey)

10–12. The Dead Sea occupies a rift valley as does the upper part of the Rhine River. The center of the mid-ocean ridge is occupied by large rift valleys formed by spreading of the sea floor.

Origin of forces that deform the crust. According to the theory of plate tectonics, the earth's solid outer crust is made of a series of rigid moving plates. As the plates rub together, collide, or spread apart, a *stress* is created in the rocks of the lithosphere. Stress is a force that tends to change the shape or size of the rock. Up to a certain point, the rocks making up the plates can be deformed and then return to their original shapes. This is like a stretched rubber band. However, stress that builds up quickly can cause the rock to fracture. These breaks commonly occur along existing or old plate boundaries. Stresses can build up more slowly over many thousands of years. When this occurs, the rocks are more likely to be folded without breaking.

Not all the changes of the rocks in the crust are caused by movement of the crustal plates. Stress can also be created because the solid lithosphere floats on the partly molten asthenosphere below. Just as a ship floats high or low on water, depending on its cargo, the crust can move up or down if its weight changes. The principle that states that the solid crust floats in a state of balance is called *isostasy* (eye-*sos*-tuh-see). The word isostasy derives from Greek and means "equal standing." Isostasy means that the heavier parts of the crust sink deeper toward the mantle. Mountains, for example, have deep roots that make them heavy and cause them to sink. They could be compared to a heavy piece of wood that floats deeper in water than a lighter piece. As the top of a mountain is worn off, the rest of the mountain will continue to rise until its roots are level with the crust. The material that is worn off the mountain will be deposited as sediment in a nearby body of water. The added weight of the sediment will cause that region to sink. See Figure 10-13. When the crust rises and falls due to the effects of isostasy, the rocks are exposed to stress that may result in folding and faulting.

Isostasy also explains what happens when very thick deposits accumulate on the floors of large bodies of water. Sediments from the higher parts of the land wash into the ocean basins near the shore. The weight of this sediment, pressing down on the floor of the basins, forces them to sink and more sediments can be deposited. As a result of the sinking, the basin remains at a lower level than the surrounding land. Accumulation of a great thickness of sediment formed this way produces an area called a *geosyncline.*

Studies of the effects glaciers have on the crust also seem to show the principle of isostasy in operation. During the earth's past, glaciers covered much of the continent's surface. The weight of the ice pressed down on the land. Now that the glaciers have almost disappeared, the land that was covered with ice is slowly rising. See the map in Figure 10-14.

Gravity also creates stress in rocks. Large sheets of sediment can slide downhill, folding and breaking as they move. This happens when a large amount of magma in a batholith pushes up from below. Pushing up the crust causes thick layers of sediment to slide down the gentle slopes that are created. Parts of the Appalachian Mountains may have been folded in this way.

EARTHQUAKES

Earthquakes. Most earthquakes occur when there is movement of the masses of rock on the sides of a fault. The walls of a fault are usually pressed very closely together. Because they are close, it means that the rocks along a fault do not usually slide smoothly against each other. Instead, pressure slowly builds up until the rocks bend and then finally break. Due to an abrupt release of this pressure, there is a sudden slipping of great blocks

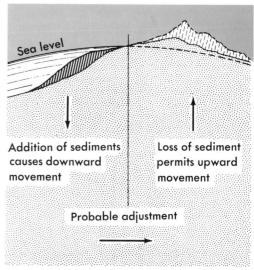

FIG. 10-13. Diagram illustrating the theory of isostasy.

FIG. 10-14. The darker areas on the map are rising, while the lighter areas are sinking. The numbers indicate the rising (+) or sinking (−) in millimeters per year. The lifting is probably the result of weight being removed after the glaciers melted. Sinking areas can be caused by the formation of geosynclines.

Pre-existing fault plane

Under stress

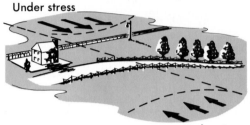

Under stress

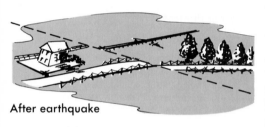

After earthquake

FIG. 10–15. According to the elastic rebound theory of earthquakes, stress builds up slowly and is then suddenly released.

of rock along the fault zone. The result is the trembling and vibration of solid rock called an *earthquake.*

This explanation of the way that earthquakes are produced is usually called the *elastic rebound theory.* It is commonly accepted as the explanation for the occurrence of most earthquakes. Measurements made along faults over many years show that very slow movement of rock on both sides of large faults does take place. This movement builds up strain in the rocks that causes them to bend and stretch like rubber bands. When the strain becomes too great, the rocks break and spring back (rebound). As the rocks along the fault snap into new positions, they also release energy in the form of earthquake waves. The actual point along a fault where the slippage occurs and causes the earthquake is called the *focus.* The way that elastic rebound actually works along a fault is shown in Figure 10–15. Small earthquakes can also be created by volcanic eruptions, or by land slides.

Some earthquakes are caused by faults very deep beneath the earth's surface. These earthquakes can probably not be explained by the elastic rebound theory. The temperatures and pressures at those depths are much different from the conditions near the surface. However, it is the sudden release of energy stored in rocks under great strain that is responsible for all large earthquakes.

Detecting earthquakes. When an earthquake occurs, vibrations in the form of waves move out in all directions through the surrounding rock. The effect of these waves is felt most strongly at the surface, where there are loose or water-soaked sediments instead of solid rock. Most earthquake damage to buildings is likely to occur where the structures are not built on solid rock. Observe what happens to a bowl of gelatin when the bowl is struck quickly. The bowl shakes very little but the gelatin vibrates in a series of waves.

Since most earthquakes occur along fault lines, the focus is usually a line rather than a point. The point, or line, on the earth's surface directly above the focus is called the *epicenter.* During an earthquake shock, the waves travel out from the focus in widening circles. Their movement is similar to the ripples made when a rock is thrown into quiet water.

Earthquake waves can be detected by an instrument called a *seismograph* (*syz*-muh-graf). The principle behind the working of a seismograph is illustrated in Figure

10–16. Analysis of seismograph records shows that there are three types of earthquake waves. The seismograph record shown in Figure 10–17A, indicates the three kinds of waves. The straight line beginning at the left is the record shown in Figure 10–17A indicates the three kinds line suddenly takes on a wavy appearance at the letter P. At the letter S the wave also changes and continues to letter L, where it changes again. The first disturbance that appears on the seismograph is the *primary wave*, the second is the *secondary wave*, the third is the *surface-long wave*.

The primary and secondary waves travel from the focus. See Figure 10–17B. They are often called *body waves* because they travel through the body of the earth. Primary waves are the result of a back-and-forth vibration of rock. Secondary waves are caused by an up-and-down (or side-to-side) motion of the rock. Notice in Figure 10–17B that secondary waves are not detected on the side of the earth opposite an earthquake. Surface waves are created when either type of body wave reaches the surface.

Another important difference between the types of earthquake waves is the speed at which they move. Primary waves are the fastest, secondary waves are slower, and surface waves are the slowest of the three. Although primary and secondary waves travel at different speeds,

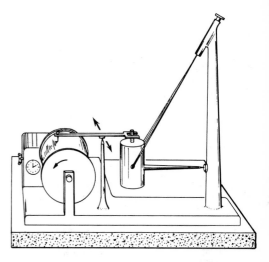

FIG. 10–16. A simple seismograph like the one shown in the diagram can be constructed. When the earth vibrates, the suspended weight remains stationary, allowing the attached pen to make recordings on the revolving drum. This kind of seismograph is best able to record side-to-side waves.

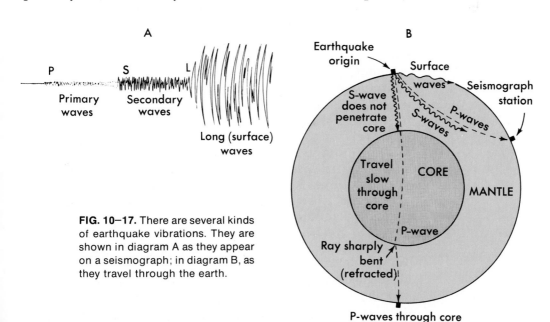

FIG. 10–17. There are several kinds of earthquake vibrations. They are shown in diagram A as they appear on a seismograph; in diagram B, as they travel through the earth.

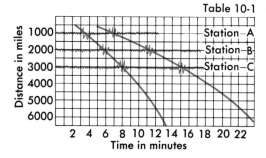

Table 10-1

Table 10-2

Distance miles	Time of Travel					
	P wave		S wave		S-P interval	
	M	S	M	S	M	S
1000	3	22	6	03	2	41
2000	5	56	10	48	4	52
3000	8	01	14	28	6	27
4000	9	50	17	50	8	00
5000	11	26	20	51	9	25
6000	12	43	23	27	10	44
7000	13	50	25	39	11	49

FIG. 10-18. The epicenter of an earthquake can be located by means of seismograph recordings taken by at least three different stations.

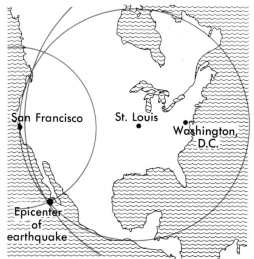

the differences in speed between them always remains the same. Primary waves travel about 1.7 times faster than secondary waves. Knowing this makes it possible to use a seismograph record to find the distance to an earthquake focus and its epicenter. Secondary waves will fall farther behind as the distance from the focus increases. Think of it as similar to a race between a fast horse and a slow one. The slow horse will fall farther and farther behind as the distance away from the starting gate increases. The time lag between the arrival time of the two kinds of waves can be plotted on a graph similar to the one in Table 10-1. Data from the graph can then be translated into a distance in Table 10-2.

For example, if secondary waves arrive 2 minutes and 41 seconds after the primary waves at seismograph station A, the earthquake focus is 1610 km away. If several kinds of seismograph records are made at one station, the direction from which the waves came can also be determined. It is possible then to locate the general position of the epicenter of the earthquake. However, a more accurate way of locating the epicenter makes use of the records from several stations. In this method, circles are drawn for at least three stations. Each station is used as a center, and the distance from it to the focus is the radius of a circle. The epicenter is very near the point where the three circles meet. See Figure 10-18.

The thousands of earthquakes that occur every year have many different intensities. On the average, only two earthquakes each year actually cause destruction to life and property. A limited amount of damage is caused by an additional twelve. About one hundred more would be damaging if they occurred in populated regions. The intensity of an earthquake is measured in terms of its ability to cause damage to structures and change the earth's surface.

The first attempts to measure the strength of earthquakes were based on descriptions of what happened during the quake. But the people who experienced the earthquake described their observations differently. In order to provide a reliable description, a scale was developed by Mercalli in 1902. A modified form of this scale is still often used today to describe the intensity of an earthquake. The scale starts with the number I intensity described as so slight that it can barely be felt. The numbers go as high as XII, which indicates total destruction. However, many things influence the amount of damage done by an earthquake. The Mercalli scale is not useful in comparing the intensity of earthquakes. A scale is needed to measure the *magnitude* or total energy released by an earthquake.

The scale that measures earthquake magnitude is called the *Richter scale*. Earthquake magnitude is determined on this scale by using seismograph records. Numbers 1 to 8.6 are used to describe the magnitude. Each number indicates a greater release of energy. Each higher number indicates an energy about 30 times greater than the preceding number. Thus, an earthquake with a magnitude of 5.5 releases about 30 times more energy than one of magnitude 4.5. The largest earthquakes measured have magnitudes near 8.6. Those with magnitudes less than 2.5 are usually not felt by people in the area. See Table 10–3.

Locations of earthquakes. We know that the depth of the focus of an earthquake can be measured. Most earthquakes occur at shallow depths within 55 km of the surface. However, some have a focus with intermediate depth that is between 55 to 240 km. A few deep-focus earthquakes come from a depth of 300 to 650 km. At a depth greater than this, the rock is probably too soft to accumulate the strain needed to produce an earthquake.

Locating the epicenters of shallow, intermediate, and deep-focus earthquakes on a map can begin to help show the cause of many severe earthquakes. Often, the epicenters of the three types of earthquakes are found to be

Table 10–3

Richter Magnitudes	Earthquake effects
0	Smallest detectable quake.
2.5–3	Generally not felt but recorded. About 100,000 such earthquakes of this magnitude occur each year.
4.5	Can cause local damage.
6.0	Destructive in a populated region.
7.0	Called a major earthquake.
7.8	San Francisco earthquake of 1906.
8.0 or greater	Great earthquakes. Cause total destruction to close population centers.

FIG. 10–19. Epicenters of earthquakes of various depths along the western edge of South America.

nearly parallel lines. See Figure 10–19. Investigation shows that these earthquakes mark the location where one crustal plate is diving beneath another. The source of the earthquake follows the edge of the descending plate. See Figure 10–20. It appears that earthquakes are one of the phenomena resulting from movement of boundaries of the crustal plates.

FIG. 10–20. A sloping fault zone along the west coast of South America accounts for the various depths of earthquakes.

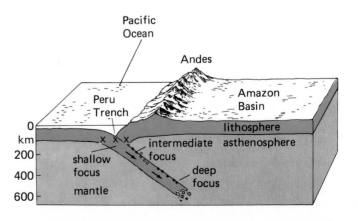

FIG. 10–21. Aerial photo of San Andreas fault near San Francisco. (U.S. Geological Survey)

An example of earthquake activity that occurs along plate boundaries is found on the San Andreas fault in California. See Figure 10–21. Movement along this fault was the cause of the San Francisco earthquake of 1906. Many other less severe earthquakes have occurred in the area since then. The San Andreas fault is the boundary where the north moving Pacific plate brushes against the west moving North American plate.

Earthquakes can occur anywhere, although they are rare in regions not near plate boundaries. One of the strongest earthquakes known to have occurred in North America happened near New Madrid, Missouri in 1812. A major earthquake also occurred in Charleston, South Carolina in 1886. These earthquakes are thought to be the result of isostatic adjustments to the extra weight of accumulated sediments in the areas. Occasional earthquakes occur in New England. These earthquakes may be the result of adjustments of the crust to weight lost after the ice of the last glacial age melted.

Predicting and controlling earthquakes. In the future, earthquake warnings may be given just as serious storms are forecast now. Scientists are discovering how to detect changes in the crust that can signal an approaching earthquake. Most faults that are near centers of population have been located and mapped. Instruments are placed along these faults to measure small changes in the zone around the fault. Built-up strain can be detected. The slight tilting of the ground that precedes many earthquakes can also be observed. Small changes in the earth's magnetic field can also be useful information for predicting a coming earthquake. A severe earthquake is often preceded by a decrease in the speed of P-waves generated by milder disturbances. Just before the severe earthquake, the P-wave speed suddenly increases. Studying the waves produced by mild earthquakes can then give advance warning of the more severe quakes.

Scientists are also concerned about the problem of controlling earthquakes. It is known that earthquakes have been artificially set off when water was forced into deep wells. The weight of water stored in reservoirs and the explosion of nuclear bombs underground have also set off earthquakes. When the processes involved in the formation of earthquakes are better understood, it may be possible to develop a system for earthquake control. This could be done by releasing the strains along a fault and producing a slow, steady motion. Rapid movements that cause the elastic rebound characteristic of earthquakes could thus be avoided.

MOUNTAIN BUILDING

Kinds of mountains. Mountains are more than simple elevated parts of the earth's crust. They are complicated features whose rock structures give evidence of their formation. Mountains are grouped according to the most characteristic process that created them. There are four main groups. For example, the largest mountain ranges of the world are made by *folding*. The Alps, Rockies, Himalayas, Appalachians, and Urals are all basically made up of very large and complex folds. However, these giant mountain systems also show much faulting and intrusive igneous activity.

Other mountains have been formed by faults that broke parts of the crust into large blocks. These blocks were then tilted to create *fault-block* mountains. See Figure 10–22. The Sierra Nevadas of California and the Teton Moun-

Investigate
Would alluvial fans be more likely to form on the sides of folded mountains or block mountains?

FIG. 10–22. The basic structure of a typical fault-block mountain.

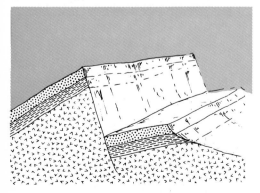

FIG. 10–23. The Sierra Nevadas are fault-block mountains. (Ramsey)

tains of Wyoming are examples of fault-block mountains. Much of Nevada, Arizona, western Utah, southern Oregon, northern New Mexico, and southeastern California are covered by fault-block mountains. The blocks form nearly parallel mountain ranges averaging 80 km in length. The Grand Canyon has been cut into one of these upturned blocks. See Figure 10–24.

Volcanic mountains are formed from piles of lava and other volcanic materials that have been forced from volcanic openings. The Cascade Mountains of Washington and Oregon are examples of mountains made up of a series of volcanic cones. Most volcanic activity occurs on the sea floor along the mid-ocean ridges. The sea covers some of the largest volcanic mountains on the earth.

An unusual kind of mountain can be produced from old igneous or metamorphic rocks that have been covered with layers of sediments. Some force, such as a batholith or laccolith rising from below, raises a circular dome like a giant blister on the earth's surface. See Figure 10–25A on page 222. As the sedimentary layers are worn off the dome, *domed* mountains are formed. See Figure 10–25B on page 222. The Black Hills of South Dakota and the

FIG. 10–24. The Grand Canyon has been cut into one of the tilted blocks that is part of the system of fault-block mountains of southern Utah and northern Arizona.

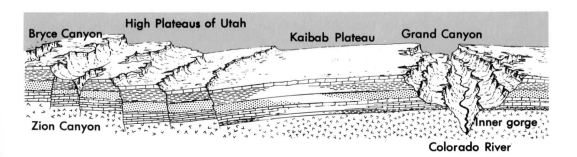

FIG. 10–25. *A)* Domed mountains are formed when bedrock with a sedimentary covering is uplifted. *B)* Later the surface is worn away, creating a more or less circular mountainous area.

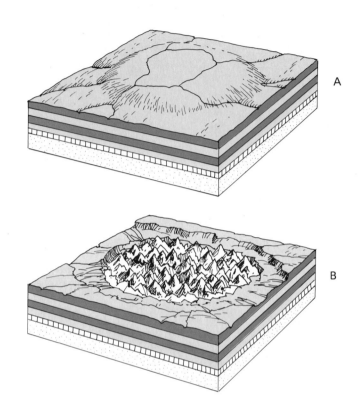

activity

In this activity, you will locate some of the world's major earthquakes and mountain ranges. You will then see how they are related to other features on the earth. To do this, you will need a map of the earth that shows the crustal plate boundaries and volcanoes.

Each of the positions listed in the chart is a place where a major earthquake has occurred. On your map, write the letter identifying each earthquake at the longitude and latitude stated.

The position stated for each of the major mountain ranges is near the center of the range. On your map, write the name of each major mountain range, as near as you can, to the proper longitude and latitude. Then answer these questions:

1. Describe the general relationship between the earthquake regions and the crustal plates.

2. Describe the general relationship between the mountain ranges listed and the crustal plates.

3. How are the positions of earthquakes, volcanoes, and mountain ranges related?

4. Are all the earthquakes listed found along the edges of crustal plates?

5. Are all the mountain ranges listed found along the edges of crustal plates?

Adirondack Mountains of New York are examples of domed mountains.

Plate tectonics and mountains. One kind of boundary between crustal plates forms when a part of the sea floor is pushed under the edge of a continent. Such a boundary is found where the western edge of South America meets the Pacific sea floor. This boundary was probably formed when South America separated from Africa, moved westward and was thrust up over the Pacific Ocean floor. The action at this kind of plate boundary is believed to be responsible for creating great mountain systems. It may also have built the continents themselves. Plate tectonics and mountain building are not completely understood. However, scientists believe that the following series of events happen repeatedly during the earth's long history. First, materials wash off the land and accumulate as sediment on the sea floor near the edge of a continent. Because of the weight of sediment, the lithosphere breaks near the edge of the continent. A plate boundary is then formed. The lighter continental plate is pushed up over the oceanic plate. The oceanic plate plunges down where it, and the sediments, become molten. Melting of the oceanic plate results in volcanic activity. This activity creates an island arc of volcanoes. Sediments that continue to accumulate between the island arc and the edge of the continent form a geosyncline. See Figure 10–26A. Continued movement of the plates pushes the old island arc toward the continental edge. The sediments of the geosyncline are crumpled and pushed upward, forming folded layers. Magma from below joins with these folded layers. The magma becomes the core of a growing mountain range that is forming along the margin of the continent. See Figure 10–26B. Continuing uplift and volcanic activity finally build a large mountain system. The Andes Mountains of South America are examples of the kind of mountain range that is still growing.

The continents might have grown to their present size by repeated episodes of mountain building. Each time a thin band of mountains grew up, they were added to where the continent's edge met the sea floor. If this theory is correct, the continents were once made up of mountains that have since worn away. Evidence of these ancient

EARTHQUAKE	LONGITUDE	LATITUDE
A – China	110E	35N
B – India	88E	22N
C – Pakistan	65E	25N
D – Syria	36E	34N
E – Italy	16E	38N
F – Portugal	9W	38N
G – Chile	72W	33S
H – Chile	75W	50S
I – Equador	78W	0
J – Nicaragua	85W	13N
K – Guatemala	91W	15N
L – California	118W	34N
M – California	122W	37N
N – Alaska	150W	61N
O – Japan	139E	36N
P – Japan	143E	43N

MOUNTAIN RANGE	LONGITUDE	LATITUDE
Himalayas	75E	30N
Alps	10E	45N
Atlas	0	30N
Appalachian	80W	40N
Andes	70W	30S
Coast	120W	40N

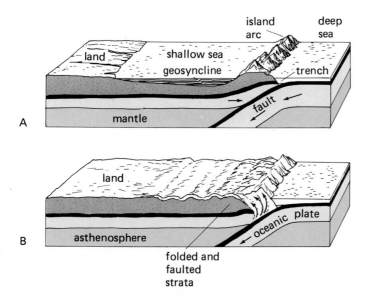

FIG. 10–26. These diagrams show how mountains can be formed along the edges of continents and increase the size of the land area.

mountains is found in rock structures of the land that remains.

Some mountains also formed when two continents collided head-on. When this happened, one continent was not pushed beneath the other because both were made up of light rock. Rather, the colliding plates were crumpled and deformed to create a mountain range. An example of a mountain system formed this way is the Himalayas. These mountains were created about 45 million years ago, when India collided with Asia.

The Appalachian Mountains also exhibit evidence of continental collision. More than 200 million years ago, North America collided with North Africa. The two continents have since separated. But evidence of that collision still remains within the Appalachians. Continental collision has also played a part in the formation of the Alps in Europe.

VOCABULARY REVIEW

Match the word or words in the column on the right with the correct phrase in the column on the left. *Do not write in this book.*

1. Scientific study of the earth.
2. Theory that the present is the key to the past.
3. Rocks respond to forces by moving into new positions without breaking.
4. Raised section of a fold.
5. Movement of rock along the side of a fracture.
6. Present in the center of a mid-ocean ridge.
7. Principle that the solid crust floats in a state of balance.
8. Accumulation of great thickness of sediments in the ocean near shore.
9. Results from abrupt release of pressure along a fault zone.
10. Actual point along a fault where slippage occurs.
11. Point on the earth's surface directly above the focus.
12. Instrument used to detect earthquake waves.
13. Causes the first disturbance that appears on the seismograph.
14. Scale used to measure earthquake magnitude.
15. Results from a force such as a laccolith rising from below.

a. folding
b. isostasy
c. anticline
d. rift valley
e. geology
f. earthquake
g. uniformitarianism
h. focus
i. fault
j. seismograph
k. secondary wave
l. geosyncline
m. stress
n. epicenter
o. dome mountain
p. strike
q. Richter scale
r. primary wave

QUESTIONS

Group A

Select the best term to complete the following statements. *Do not write in this book.*

1. Evidence that the earth's crust has been lifted in places is found in (a) remains of sea animals high on mountains (b) heavy rocks in the deep ocean (c) remains of sea animals deep in the ocean (d) heavy rocks on mountain tops.
2. The theory that "the present is the key to the past" was first stated by (a) Albert Einstein (b) Johanne Kepler (c) Nicolaus Copernicus (d) James Hutton.
3. Forces can change the features of the earth because (a) the forces are abrupt (b) the forces are small (c) the forces act over a long period of time (d) rocks of the crust were molten at one time.
4. Which of the following is not closely related to the others? (a) syncline (b) anticline (c) geosyncline (d) folding.
5. Sedimentary layers that are tilted show movements of the rock, because sedimentary layers are originally deposited in (a) a horizontal position (b) a vertical position (c) a slanted position (d) dry weather only.
6. The tilt of a rock layer from the horizontal is called its (a) axis (b) strike (c) slant (d) dip.

7. Movement of rock along the sides of a fracture is called a (a) joint (b) strike (c) slip (d) fault.

8. Stress in rocks might be due to (a) motion of crustal plates (b) joints (c) a dip (d) a strike.

9. The center of a mid-ocean ridge is a (a) high mountain (b) fold (c) rift valley (d) riverbed.

10. The principle that says the heavier parts of the crust sink deeper toward the mantle is called (a) float (b) isostasy (c) uplift (d) thrust.

11. Sediments washed into the ocean near shore can result in the accumulation of a great thickness that produces an area called (a) a geosyncline (b) an anticline (c) a syncline (d) a plate.

12. The actual point along a fault where slippage occurs and causes the earthquake is called that earthquake's (a) circumcenter (b) epicenter (c) orthocenter (d) focus.

13. The point on the earth's surface directly above the source or center of an earthquake is called the (a) circumcenter (b) epicenter (c) orthocenter (d) focus.

14. Body waves are made up of (a) primary waves only (b) secondary waves only (c) primary and secondary waves (d) long waves.

15. The earthquake waves that travel fastest through the earth are (a) primary waves (b) secondary waves (c) surface waves (d) long waves.

16. Using the Richter scale of earthquake magnitudes, each higher number represents how many more times as much energy than the preceding number? (a) 5 (b) 10 (c) 15 (d) 30.

17. Most earthquakes seem to be caused by movements (a) in the earth's core (b) deep in the earth's mantle (c) at shallow depths in the earth's crust (d) on the earth's surface.

18. Most earthquakes occur (a) near the edges of crustal plates (b) in the centers of crustal plates (c) in the middle of an ocean (d) at the earth's poles.

19. In addition to predicting earthquakes, it is hoped that we can control them by (a) increasing their magnitudes and having fewer of them (b) decreasing their magnitudes and having fewer of them (c) decreasing their magnitudes and having more of them (d) increasing their magnitudes and having more of them.

20. Mountains formed when the crust breaks into large blocks that are then tilted are called (a) folded mountains (b) fault-block mountains (c) volcanic mountains (d) dome mountains.

21. The greatest volcanic mountains of the earth are (a) in southern Europe (b) in North America (c) in South America (d) in the ocean.

22. When part of the sea floor is pushed under the edge of a continent (a) the continent rises (b) the ocean rises (c) the continent sinks (d) a desert forms.

Group B

1. Describe the principle of uniformitarianism.

2. What is the difference between a syncline and an anticline?

3. Describe the formation of a rift valley. What is another name for a rift valley?

4. In Figure 10–12 on page 000, what evidence shows that these are faults and not joints in the rocks?

5. How does the theory of plate tectonics account for stress in the earth's crust?

6. Give a brief description of the principle of isostasy.

7. What is the elastic rebound theory for earthquakes?

8. What is believed to be the principle cause of most earthquakes?

9. How are scientists able to tell how far away an earthquake has occurred?

10. How are the focus and the epicenter of an earthquake related?

11. How do scientists plan to control earthquakes in the future?

12. What relation exists between the location of most earthquakes and most volcanoes?

13. What do the origins of the Alps, Rocky, Himalaya, Appalachian, and Ural mountains have in common?

14. How are fault-block mountains formed?

15. Briefly explain how mountain building is related to crustal plates.

our dynamic planet

Hot Spots

Narrow columns of hot material called "plumes" flow upward from deep in the mantle and create hot spots. Volcanic activity occurs where the plume pushes upward through the crust. While most of the earth's volcanoes are located along the edges of the crustal plates, there are about 100 places around the globe with volcanic activity that are not plate boundaries. These are hot spots.

Two hot spots are known to be currently active in the continental United States. One is located in the area of Yellowstone National Park. At the left is a picture of Yellowstone Park's Old Faithful, a geyser that erupts every 66 minutes. It is evidence of the molten plumes forcing their way up to the crust and heating the water trapped below the surface. Here steam builds up and erupts at regular intervals, relieving the pressure. Another hot spot is in the northeastern corner of New Mexico.

A much larger and more active hot spot on the floor of the Pacific Ocean has formed the Hawaiian Islands. Pictured here is a volcanic eruption on the Hawaii island. Bursting forth from the hot interior, molten lava pours down the side of the mountain, destroying everything in its path.

The cause of plumes within the mantle is not known. They must be part of the activity within the earth that is responsible for the motion of the plates that make up the lithosphere. Further study of hot spots may lead to a better understanding of the forces that have repeatedly broken apart the continents and created ocean basins during the earth's long history.

Can Animals Predict Earthquakes?

For many years observers all over the world have reported unusual behavior by animals before severe earthquakes. Some of the strange behavior has taken place weeks before the earthquake occurred. In other cases, the animals became disturbed only hours or minutes before. There have been reports of horses trembling, neighing, and running wildly, of dogs howling, of cattle bellowing and refusing to graze, of rats leaving their holes, and even of fish leaping out of the water. All reports of strange animal behavior before an earthquake agree on restless and noisy responses. In China, the observations of such animal behavior have been used as part of a system to make a prediction of major earthquakes.

How animals are able to anticipate earthquakes is not known. It may be that they are able to sense slight changes in earth conditions that are known to precede large earthquakes. Among these slight changes are shifts in the earth's magnetic field and tilting of the land as the crust is lifted slightly. Animals may also be able to detect the very small shocks that occur before many large earthquakes. Additional research on the way animals seem to be able to predict damaging earthquakes may lead to development of instruments that can duplicate their apparent abilities. Such instruments would be a valuable part of an earthquake warning system.

People Who Study the Earth in Change

A surveyor pictured at left measures the topography of the land. From these measurements come various maps that give us a detailed picture of the earth. Because of changes constantly occurring on our dynamic planet, a surveyor's work is never done.

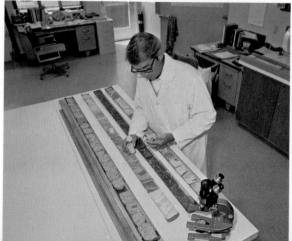

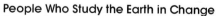

A geologist examines cores from the crust of the earth. Each layer tells a story of hundreds of years in earth's history. The earth builds up and wears away in ever-changing cycles. Through the cores, the geologist interprets this story.

11
weathering and erosion

objectives

- ☐ Name and describe two ways that weathering occurs.

- ☐ Explain how the nature of rock, climate, and topographic conditions affect the rate of weathering.

- ☐ List and describe the layers of a mature soil.

- ☐ Explain how soil is produced.

- ☐ Describe the roles gravity and wind play in erosion.

- ☐ Describe several methods currently being used to conserve soil.

- ☐ Describe the life cycle of mountains, plains, and plateaus.

Rocks appear to us to be solid and permanent. But they are almost always undergoing change. When rocks are exposed to air and water at the earth's surface, they meet different conditions from those under which they are formed. Gases of the atmosphere, moisture, and temperature changes bring about changes in the appearance and composition of the rock. Most rocks will slowly crumble into small pieces as they react to conditions at the surface. All changes that destroy the original rock structure are called *weathering*. When rocks are exposed to weathering processes, the formation of soil is one important result. *Soil* is a mixture of loose rock fragments and materials produced by living things growing in the soil.

Weathering of large masses of rocks allows movement of resulting smaller fragments. Before the breaking up of rocks by the processes of weathering, little movement can take place. Movement of rock materials over the earth's surface produced by various natural agents is called *erosion*. The smaller fragments produced by weathering can fall down a slope as a direct result of gravity. They can also be easily carried off by erosional agents, such as wind and running water. Once weathering has taken place, the agents of erosion work to move the products. It is the combined action of weathering and erosion that accounts for much of the present shape of the land surface.

WEATHERING

Types of weathering. The effects of weathering may be seen wherever rock structures are exposed. Small loose

232

chips are easily pulled off, and some of the fragments may even crumble at a touch. When the rock is broken open, the inner parts are found to be firm and often different in color from their surface appearance. Air and water have been able to penetrate into cracks and openings near the surface. Here, too, the exposed rock is often stained and softened. See Figure 11-1.

The weathering process that produces these changes takes place in two ways. The rock may be broken up without any change in its mineral composition. This is called *mechanical weathering*. At the same time, processes are at work which change the individual mineral crystals. This is called *chemical weathering*.

An example of mechanical weathering is the effect of freezing water on exposed rock surfaces. When water freezes, it expands by about ten percent of its volume. If water seeping into cracks in rocks freezes, pressure is created by its expansion. This can force pieces of the rock to break off from its original structure. The breaking off of great blocks of rock when water freezes and expands is called *frost action*. See Figure 11-2. However, the greatest effect of frost action is the constant chipping away of small grains of rock. Frost action is a common cause of rock weathering at high altitudes and other places where temperatures often fall below freezing.

FIG. 11-2. Frost action is breaking the rock on this mountainside into smaller pieces. (Ramsey)

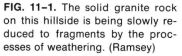

FIG. 11-1. The solid granite rock on this hillside is being slowly reduced to fragments by the processes of weathering. (Ramsey)

FIG. 11–3. Mud cracks form when wet soil containing large amounts of clay or silt particles is dried. (Ramsey)

Examine
Collect a number of different rock specimens from your local area. Have them soak in a pail overnight. Place them in a plastic bag and put them in your freezer overnight. Observe any changes in their appearance.

A similar process takes place in drier climates. Over long periods during which water evaporates from rock surfaces, salt crystals are likely to grow in small openings. These salt crystals originate from minerals dissolved by the water. Their growth as the water evaporates can produce enough pressure to break off fragments of the rock.

Rocks which contain clay particles can absorb water at their surface. This absorption of water can create a swelling in the rocks. Later, drying and shrinking will then produce cracks. This process is also responsible for the familiar pattern of cracks found in dried mud rich in clay particles. See Figure 11–3.

Working together with mechanical weathering processes are the chemical changes that cause rock to crumble. In chemical weathering, reactions take place between minerals in the rock, carbon dioxide, oxygen, and water. Dry carbon dioxide (CO_2) gas, which is always present in the atmosphere, has no effect whatever on rock. But many minerals are affected when carbon dioxide from the air dissolves in water. Carbon dioxide gas dissolves in water to produce a weak solution of carbonic acid. Carbonic acid readily reacts with certain minerals, such as the feldspars and calcite, which are commonly found in rocks. This action of carbonic acid is an example of chemical weathering. The rocks containing minerals which react with carbonic acid, weather very rapidly in moist climates. This process is called *carbonation*.

Like carbon dioxide, oxygen gas in the atmosphere can also affect certain rocks when dissolved in water. This kind of weathering is called *oxidation*. Iron-bearing minerals will quickly combine with dissolved oxygen to form reddish-brown iron oxide. The red color of certain rocks and soils is almost always due to chemical weathering, which accounts for the presence of iron oxide.

Besides aiding the action of carbon dioxide and oxygen, water also changes some minerals in other ways. Water can dissolve or combine directly with some minerals. This is known as the process of *hydration*. Most of the minerals affected by hydration dissolve in water. In this way, they are removed from the rocks. Often water seeping down from the surface dissolves minerals as it passes through the upper rock layers. These dissolved minerals are then carried down to greater depths and finally deposited. This is called *leaching*. In leaching, minerals are transported from the upper layers of soil and rock to the lower parts. Valuable mineral ore deposits may be created when leaching concentrates minerals in a small region beneath the surface.

In addition to mechanical and chemical weathering, plants and animals also play a part in the decomposition of rocks. Roots of trees and shrubs can work their way into the cracks of rock and wedge it apart. See Figure 11–4. Burrowing animals, such as gophers and prairie dogs, are also a factor in weathering. Their digging activities constantly expose new rock surfaces to the process of weathering. Even such a small creature as the earthworm plays a role in weathering. Earthworms bring fine rock particles to the surface where they are exposed to the weathering action of the atmosphere. Their burrows also allow space in the ground for water and air to penetrate the upper layers more easily.

The results of weathering. The various processes of mechanical and chemical weathering that have been described are relatively slow. But they have been at work for millions of years. The combined effects of these processes have broken up and decomposed almost all the rocks on or near the surface of the crust. For this reason, the land surfaces of the earth have accumulated a layer of rock fragments of various types. The solid unchanged

activity

Punch about 10 small holes in the bottom of a paper cup. Cover these holes, on the inside, with a piece of facial tissue. Half fill the cup with sand. Add one teaspoonful of salt. Now place the cup inside a cup that has no holes. Support the sides in several places with pieces of paper, so that the inner cup is about one centimeter off the bottom of the outer cup. Add 5 teaspoons of water, one at a time. Answer these questions:

1. Describe the appearance of the sand and salt before the water was added.

2. Describe the appearance of the sand and salt after the water drained through it.

3. Where has the salt gone?

4. What simple test can you use to check your answer to 3?

FIG. 11–4. Trees growing in cracks can force rocks apart. (Ramsey)

FIG. 11–5. Weathering takes place most rapidly along existing cracks in the rock. This has resulted in the deep grooves following vertical cracks in this resistant sandstone. (Ramsey)

FIG. 11–6. This California mountain is covered with residual boulders. (Doug Wilson-Black Star)

rock beneath the layers of looser material is often called *bedrock.*

The appearance of weathered rock fragments depends upon the nature of the original rock and the way in which weathering has taken place. Coarse-grained igneous and some metamorphic rocks usually separate into their individual mineral grains. The result is a coarse sand or gravel.

Usually joints or cracks are already present in bedrock. Weathering can then take place fairly rapidly. As a result, rocks often tend to separate into a series of blocks as weathering takes place along parallel sets of joints. See Figure 11–5. An unbroken pattern of weathering along a series of joints can eventually leave separate blocks with rounded off edges. Smaller fragments are usually carried away, leaving large *residual boulders* in their place, similar to those shown in Figure 11–6.

Some rocks tend to break apart in shells or plates. This is called *exfoliation,* and gives the weathered rock surface a rounded appearance. See Figure 11–7. Exfoliation seems to be the result of a combination of both mechanical and chemical weathering. It may also be the result of pressure released when rocks buried under a great weight are brought to the surface.

The various shapes and forms created by weathering in the bedrock mean that new rock is being exposed. As the surface rock is being reduced to small fragments, erosion continues removing the old material. Fresh bedrock then becomes exposed to the forces of weathering. The speed with which rock is weathered away depends upon three main factors:

1. *The nature of the rock.* The rate at which rocks weather varies according to their composition. Igneous and metamorphic rocks in general, react slowly to weathering by mechanical processes. They are affected largely by chemical processes. Although their decomposition is slow, they eventually crumble as the minerals change and allow the grains to separate. Quartz is the least affected mineral commonly present in igneous and metamorphic rocks. It remains as individual grains; quartz makes up most of the mineral matter in ordinary sand.

Among the sedimentary rocks, limestone and some rocks containing calcite are the most rapidly weathered. They are affected by carbonation and decay rapidly in a moist climate. See Figure 11–8. Many other sedimentary rocks are attacked mainly by mechanical weathering processes. The rate of weathering of most sedimentary rocks depends upon the material that holds the fragments of the sediment together. Shales and sandstones that are not firmly cemented together gradually return to their original condition of separate particles of clay and sand. On the other hand, conglomerates and sandstones that are strongly cemented by silicates, last even longer than most igneous rocks.

2. *Climate.* In dry climates, weathering takes place slowly. The lack of water slows the rate of the chemical weathering processes. The same is true of cold climates where extremely low temperatures reduce chemical weathering. In humid, warm climates, weathering is

FIG. 11–8. These old tombstones on Cape Cod show how various kinds of rock weather at different rates. The marble slab on the right is dated 1854 while the slate marker on the left is from 1835. Which shows the greater evidence of weathering? (Ramsey)

FIG. 11–7. Exfoliation results in the peeling and flaking of rock surfaces, as shown in these granite outcrops. The picture was taken at the summit of the Sierra Nevadas in California. (U.S. Geological Survey)

FIG. 11–9. These two photographs show how the weathering processes in dry and humid regions produce different landscapes. (Ramsey)

Discuss

A farmer would be very concerned about the type of soil on his land. Explain why a soil having either a high sand content or high clay content would be very poor farming soil.

fairly rapid. In a changeable climate, weathering is generally more rapid than in a more constant one. Temperature changes affect mechanical weathering. The difference in landforms between parts of the earth is mainly due to the effects of weathering in different climates. See Figure 11–9.

3. *Topographic conditions.* Altitude, the slope of the land, and exposure to sun and rain all influence weathering. With greater altitude, temperatures decrease and rainfall increases. Steep slopes allow quick removal of weathered rock. New surfaces are thus exposed more quickly to the action of weathering.

Soil. Once a layer of loose rock fragments has been created by weathering, the upper exposed rock material continues to decompose. The lower parts are partly protected and therefore do not weather as rapidly. Eventually, a layer of very fine particles is created at the surface. This layer of small rock fragments forms the basis for the production of soil.

When soil is first formed, it consists of grains of mineral matter. The minerals present in the original soil are those of the rocks which have been weathered. A variety of feldspar minerals are the most plentiful of the rock forming minerals. When the feldspars are subjected to weathering, fine grains of a silicate mineral containing aluminum and water are formed. This material makes up *clay*. Other minerals containing aluminum may also form clay when they weather.

Granite and other rocks containing a large percentage of quartz can form *sandy soils*. Such soils contain large amounts of the relatively unaffected quartz grains. These are left after the other minerals in the rocks have decomposed. In addition to clay and sand, soils may also contain a number of other rock particles. Their sizes range from the microscopic clay particles to the larger sand grains. These particles are classed as *silt*.

The weathered mineral and rock grains that form soil very often do not remain in the same location as the bedrock of their origin. Erosion transports them to other locations. Thus soil may have a composition different from the rock on which it rests.

The weathering of rock is only the first step in the formation of soil. The second step begins when plants grow in the young soil and start to add organic matter. Plants help the growth of animals. As the plants and animals die, their remains form *humus*, a dark organic material which is a part of fully developed soils.

Several well formed layers make up the *soil profile* of a mature soil. A typical soil profile is shown in Figure 11–10. The layers themselves are known as *horizons*. Soils generally consist of three principal horizons labeled by the letters A, B, and C. See Figure 11–11.

The top, or A horizon, of a mature soil is the topsoil. The longest period of the soil's life occurs in this layer. During this time it nourishes the growth of plants, bacteria, fungi, and small burrowing animals. It is also the zone from which surface water *leaches* the most soluble minerals. Leaching is the process by which minerals are removed from the soil. Immediately below the A horizon is the subsoil or B horizon. In most soils the B horizon contains the dissolved materials leached from the surface layer. In dry climates, the B horizon may also contain minerals brought up from below by the rapid evaporation of water on the soil surface. The lowermost or C horizon consists of partly weathered bedrock. This bottom layer is in the first stages of mechanical and chemical change. These changes eventually produce the B and A horizons.

The type of soil found at a particular place is determined mostly by the climate of the region. Where it is humid, large amounts of water pass down through the upper soil horizons. Much of the mineral matter in the A horizon is leached and then concentrated in the B horizon. The presence of these substances gives the B horizon a dense and hard composition. Where rainfall is abundant, the B horizon is often very thick.

In drier climates the soil horizons are frequently distinct. This is due to the lack of water passing downward through the top layers to leach minerals in the lower soil layers. The soils of drier regions are often very fertile because there has been little opportunity for leaching to remove needed minerals.

A mature soil reacts to the particular climate of its location. Therefore, there are hundreds of different sets of conditions that influenced its weathering and the slow organic processes that brought it to maturity.

EROSION

Gravity and erosion. When rock fragments are separated from a mass of solid rock by weathering, they fall to the

FIG. 11–10. A photograph of a mature soil showing the separate layers which have been formed. (Ramsey)

FIG. 11–11. A diagram of a typical mature soil with each horizon labeled.

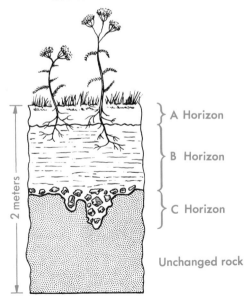

FIG. 11–12. Rock material falling to the base of this cliff has formed a talus slope. (Ramsey)

FIG. 11–13. The soil on this hillside is moving down as shown by its appearance on the lower slope. Near the top, movement is in the form of soil creep which is not so obvious. (Ramsey)

lowest part of the surrounding surface. If the weathered rock lies on a steep slope or on the face of a cliff, the rock fragments tumble down until they reach the base of the slope. A pile of rock fragments of various sizes begins to accumulate at the base of most cliffs and steeper slopes. Such accumulation of broken rock is called *talus*. See Figure 11–12. On gentle slopes the rock fragments move down so slowly that the talus weathers sufficiently to form soil. In this case, the talus slope becomes covered with vegetation.

The formation of talus is only one of the more obvious results of gravity-produced erosion. Less obvious is the slow downhill movement of loose, weathered rock material pulled along by its own weight. This process is called *soil creep*. Usually it is not noticed unless buildings, fences or other objects on the surface are moved along with it. See Figure 11–13. Soil creep is generally most rapid when the ground is wet. Water lubricates the rock particles and allows them to move more freely. Absorption of water by the soil also increases its weight.

The movement of large amounts of loose material down slopes may be much more sudden and dramatic than soil creep. In a landslide, masses of loose rock and soil on a slope abruptly break loose and come tearing down. At the beginning a landslide may be only a small amount of loose rock breaking loose from the top of a slope. As the loose rock moves down the slope, more material breaks loose. The moving rock mass gains momentum until it ends as a great avalanche of boulders, rocks and soil. Such a landslide destroys everything in its path until it reaches ground-level and comes to a halt.

Landslides are common in mountainous regions where steep rock cliffs have been carved by glaciers. Landslides are most likely to happen without warning. It is not obvious what sets them off. Heavy rainfall or melting snow may be responsible in some cases. Earthquakes also have been known to trigger large landslides.

Usually landslides take place in stages, with a small amount of land slipping, first in one part of the slope, then in another. These minor landslides are said to be *slumping* and give many slopes a hilly, bumpy appearance. See Figure 11–14. Slumping may take place on a large scale along some cliffs which have a weak rock base. Large blocks may slump down the cliff-side as erosion removes rock from its base.

In mountainous regions that are ordinarily dry, heavy rains may produce a flowing movement of a mixture of fine rock particles and water called a *mudflow*. When a mudflow occurs, masses of mud move through valleys, frequently spreading out in a sheet at the base of the mountains.

The kinds of erosion already described are those which are caused directly by gravity acting upon individual fragments of rock. All such processes of erosion directly controlled by gravity are called *mass-wasting*. However, gravity also acts indirectly to cause erosion. The most important of all the causes of erosion is water falling on, and running over the earth's surface. Running water moves because gravity pulls it down to a lower level. Ice, mainly in the form of great glaciers, is also a powerful

FIG. 11–14. A large block of this slope has slumped, probably as a result of material removed when the highway was built. (Ramsey)

FIG. 11–15A. An area of desert pavement in the Colorado desert of California. (Ramsey)

FIG. 11–15B. The unusual rock formation was the result of the sand-blast action of wind-blown mineral grains of quartz. (Black Star-Shulthess)

agent of erosion. A more detailed study of the effects of running water and glacial ice as agents of erosion will be taken up in Chapters 12 and 13. However, a third agent of erosion, the wind, is not as closely controlled by gravity.

Wind erosion. Wind which blows with enough speed is able to pick up loose particles of rock. However, winds are usually able to move only the smallest rock fragments, such as silt, dust, and sand grains. In regions where rainfall is abundant, there are usually enough plants to protect the ground against the full force of wind. But in dry areas the plant cover is spotty and much of the surface of weathered rock is exposed. Under these conditions, wind becomes an important agent of erosion.

The most common form of wind erosion is *deflation* (from the Latin, *deflare*, meaning to blow away). This refers to the lifting and movement of silt and dust particles. Usually deflation tends to remove a layer of the finer soil particles, leaving behind fragments too large to be lifted. The remaining pebbles and gravel often form a sheet which protects the materials below from further erosion. Such a layer of closely fitted small stones is called *desert pavement*. See Figure 11–15A. Deflation may also cause the formation of shallow depressions. These are produced when bare soil is exposed to wind in only a limited area. This can happen where, for some reason, the natural plant cover has disappeared. Once the protective plants are gone, the wind strips off a layer of material to form a *blowout* or *deflation hollow*. Running water carries fine sediments into this depression. Later, wind action blows these sediments away. A deflation hollow tends to expand by this method. They may grow to several kilometers wide and 5 to 20 meters deep.

Sand grains carried by the wind increase its ability to carry on erosion. In this case, the erosion is caused by the hard sand grains, which are usually quartz, striking softer rock. Sand grains are seldom lifted more than a short distance off the ground. This means that erosion takes place only close to the ground surface. See Figure 11–15B.

Various rock shapes such as natural bridges, rock pinnacles, rocks perched on pedestals, and even large desert basins are believed to have been caused by the eroding effects of wind-driven sand. However, it is very doubtful that such large features could be produced by wind action. Erosion of large masses of rock by wind-blown particles takes place very slowly. It is effective only close to the ground where heavier sand grains can be lifted. Only in the few locations that have strong and steady winds, large amounts of loose sand and relatively soft rocks, can there be much erosion by wind-blown sand.

An exception is where pebbles and larger stones have been exposed to the wind in deserts and along beaches. Stones that develop one or more smooth polished faces are called *ventifacts* (from the Latin, *ventus*, meaning wind). See Figure 11–15C.

Deposits made by wind. All the material moved by wind eventually settles and is deposited. The ability of wind to move material diminishes as its speed decreases. Thus deposits are formed where a decrease in wind speed drops a part of the load. Such deposits are often temporary and are carried away by the next strong wind. Just as often, however, the material becomes covered over. Pressure from the overlying sediments then forces the fragments together, and perhaps substances are deposited between the grains to cement them together. The deposit of wind-blown material then becomes sedimentary rock, a relatively permanent part of the earth's crust.

The most common of all wind deposits are *dunes*, or hills of wind-blown sand. Dunes are formed where there is a supply of dry, unprotected soil and winds strong enough to move it. Large areas of dunes are common in deserts. But sand dunes are also common along the shorelines of the sea and larger lakes. A dune is started when an obstruction breaks the speed of the wind. With the reduction

FIG. 11–15C. Ventifacts when developed to perfection have three curved surfaces intersecting in three sharp edges. (U.S. Geological Survey)

FIG. 11–16. Movement of sand across the surface of a dune creates a gentle slope facing the wind. Notice the size relationship in the photo between the dune and the man. (Photo Researchers–Fran Hall)

FIG. 11–17. This deposit of loess in Linn County, Iowa, is about 11 meters thick. (U.S. Geological Survey)

Examine

Obtain a piece of soft rock like limestone and a piece of hard rock like granite. Rub them both 100 times with a piece of coarse sandpaper. Record your observations.

in wind speed, a small deposit of material accumulates on the sheltered or lee side of the obstacle. As the small mound of material grows, it acts as a larger windbreak. More material is then deposited so the pile grows larger.

The gentlest slope of a dune is typically on the side facing the wind. It is in this location because the force of the wind tends to flatten the side of the dune against which it blows. The sand that the wind pushes over the crest tumbles down on the lee side, giving it a steeper slope. The wind sweeping around the sides often builds two long pointed extensions which give the dune a crescent shape. See Figure 11–16. These crescent-shaped dunes are called *barchans* (bahr-kans). They are most common in areas where there is a limited supply of sand. Most dunes have a complicated shape, a result of one dune being piled on top of the other.

The wind, blowing continuously, moves sand grains up the windward slope and over the crest of dunes to the leeward side. Generally if the wind blows from the same direction, it will force the dunes to move in the direction of the wind. In fairly level areas this dune migration or movement continues unless there are enough plants to hold the sand in place. Sometimes it is necessary to plant grasses, trees, or shrubs to prevent dunes from drifting over highways, railroads, farmland, or buildings. Protective fences are often used for the same purpose.

Dunes are formed of the heavier sand particles carried by the wind for fairly short distances. The finer particles of dust are carried to greater heights by the wind and travel much farther. The fine wind-blown material is probably finally deposited in such thin layers that it is not noticed. However, in certain places throughout the world, thick, unlayered deposits of a yellowish, fine-grained sediment have been formed by the accumulation of fine wind-blown dust. This material is known by its German name of *loess* (luhs). Although *loess* is soft and easily eroded, it sometimes forms steep bluffs because of its tendency to break into vertical slabs. See Figure 11–17.

A large area in northern China is covered entirely with loess. Apparently the deposited material was carried by the wind from the interior deserts of Asia. Larger deposits of loess are also found in Central Europe. In North America loess appears in the north central states along the eastern border of the Mississippi valley, and in eastern Oregon and Washington. These deposits were probably built up by dust from the dried beds of lakes and streams after the last ice age.

FIG. 11–18. Gullying has ruined great areas of valuable farmland. This field is practically useless for farming because of heavy erosion. (Soil Conservation Service)

Soil erosion and conservation. Erosion is a natural process. It is part of the chain of natural events that constantly change the shape of the earth's solid surface. Erosion of the soil which covers bedrock is a part of this wearing down of the crust. It would occur even if people were not present on the planet. However, natural soil erosion is a slow process. It is usually kept in balance by the accumulation of new soil. If human activities did not disturb this balance, new soil would be formed about as fast as the existing soil was eroded. However, we use the land to grow crops and to pasture their animals. The natural balance between soil erosion and soil formation can be upset by the unwise use of the land.

Rapid soil erosion is encouraged when the natural plant protection is removed. This happens when land is cleared of trees and small plant cover in preparation for farming. Grazing animals can also destroy the natural grasses and low-level plants. These activities expose the upper soil layers to the full effect of the erosion processes. Plowed land becomes cut through with deep gullies. Furrows in plowed land, particularly in rows running up and down slopes, allow water to run swiftly over the bare soil. Each furrow becomes a small gulley which expands with each rain. Finally the land is filled with miniature canyons. See Figure 11–18. This type of destructive soil erosion is called *gullying*.

Less obvious but equally destructive is *sheet erosion*. This type of erosion occurs when water strips away ex-

FIG. 11–19. Dust storms cause tremendous damage, as is illustrated by this Colorado farmyard after repeated storms. (Wide World Photo)

FIG. 11–20. Contour plowing helps to prevent gullying by causing water to flow along the plowed furrows. Another soil conservation method is shown in this photograph. Can you find it? (U.S.D.A. Photo)

FIG. 11–21A. Strip cropping helps to hold the topsoil in place on gentle slopes. Runoff is reduced by the strips of cover crops. (U.S.D.A. Photo)

FIG. 11–21B. Terracing builds up low ridges that slow down runoff water, resulting in greater absorption by the soil, thus causing relatively little erosion. Photo Researchers (Charbonnier/ Realities)

posed topsoil slowly and evenly. Finally only a smooth surface of subsoil or rock is left in a once fertile field. In dry periods wind replaces water as the means of rapid erosion. The loose topsoil is blown away as clouds of dust and drifting sand. See Figure 11–19.

Continuing erosion by water and wind reduces the fertility of the soil by removing the upper layers, thus exposing the lower horizons which are lacking in organic matter and are difficult to cultivate. The non-fertile subsoil prevents plants from growing which might otherwise protect it from further erosion. The cycle ends with complete removal of all the soil layers during a few years' time.

Rapid and destructive soil erosion can be controlled. Cover plants can be laid out on the bare soil as protection. But the most important methods of conservation are those which allow the soil to be cultivated and crops grown. Some of the methods used are:

1. *Contour plowing:* plowing in such a way that the furrows follow the contours of the land. This prevents direct flow of water down slopes which might cause gullying. See Figure 11–20.

2. *Strip cropping:* arrangement of crops in alternate bands, such as corn, and cover crops which help to hold the soil. See Figure 11–21A.

3. *Terracing:* construction of step-like ridges following the contours of the field. These hold or slow down the run-

off water to prevent rapid erosion on the slopes. See Figure 11–21B.

4. *Crop rotation:* alternation of row crops which expose the soil one year with cover crops which protect the soil the next year. This stops erosion in the early stages and allows small gullies to fill with soil.

Much of the soil over the land surface of the world has been damaged by erosion. To protect the soil while using it is a never-ending struggle that required knowledge and skill. Once soil has been swept away by erosion, no amount of knowledge or work can repair the damage quickly. Thus loss of soil resources through careless and unwise use is a problem for everyone. Care of the soil is a national and international problem, for all of our lives depend upon its fertility.

EROSION SHAPES THE LAND

Origin of the minor landforms. All land forms, regardless of the forces that shaped them, have one thing in common. They are always temporary results of the action of two opposite influences on the earth's crust. One of these forces is diastrophism, the movements that bend, break, and lift the crust into elevated landforms such as mountains. Associated with diastrophism in raising parts of the crust are the processes of volcanism. Opposing the building actions of diastrophism and volcanism are the powerful agents of weathering and erosion. The wearing action of these two forces tends to level off the land surface.

However, the landscape that exists at any particular place always has a more ancient chapter in its history. A mountain exists because it was built up by diastrophism or volcanic action that raised it above the surrounding land. But a mountain is always being worn down by weathering and erosion. Its present shape is partly a product of these forces, which are at work now and have been for millions of years.

The rocks which make up a mountain have a much older history than the mountain itself. They were created long before any landform of which they are now a part. The building and wearing forces which constantly carve the land surface are also controlled by the structure of the ancient rocks on which they must work. Thus all landforms have their present appearance as an outgrowth of

FIG. 11–22. These landscapes illustrate the three factors that control land forms. Yosemite Valley (A) is a land form which was formed by a combination of glacial processes and the resistant rock structure of the region. The landscape in (B) is entirely the result of its basic rock structure. The mountain shown in (C) is in a mature stage in its life history and has a form typical of its age. (Ramsey)

FIG. 11–23. Closest in this photograph are the Alabama Hills which are mature mountains that contrast with the youthful Sierra Nevadas in the background. (Ramsey)

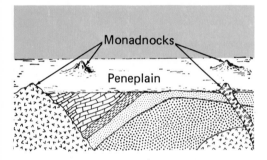

FIG. 11–24. Monadnocks in a peneplain developed on a mountain region are remnants of folded and faulted underlying rock formations.

FIG. 11–25. The Catskills of New York are the remains of a mature plateau. The ridges are relatively flat-topped and have horizontal terraces along their slopes. (Litton Industries, Aero Service Division)

a complicated, ancient, and fairly recent history. This must be taken into account when thinking about the way in which any landscape has come into being.

The weathering and erosion of the major landforms (mountains, plains, and plateaus) result in the formation of a variety of minor landforms. All of the influences which give these *minor landforms* their various shapes can be grouped into one of three categories:

1. *Structure* is a term which refers to the rocks from which a minor land form may be carved. Two different features of rocks are included in their structure. One feature is the *type* of rock, such as sedimentary, igneous or metamorphic. This is important and will determine how well the rock is able to resist the forces acting to wear it away. A second feature of rock structure is *formation*. This is the way they have been faulted, folded, or lifted by crustal movements. Formation also controls how the landform will be carved by erosion, since bending and breaking generally make rock less resistant.

2. *Processes* are agents of weathering and erosion which attack the rock structure. Streams, glaciers, wind, and waves are the most important forces which produce landforms with characteristic shapes. Usually one process is most important at a particular place. For example, waves are the most powerful influence on the shape of most shorelines.

3. *Stage* is a term which describes how long a particular process has been at work and how effective its work has been. A stream, for example, in its early stages may barely affect the surface over which it flows. In time the stream can cut a deep channel and create a major landform. However, the main factors in producing landforms are the rock structure and the various processes which shape it. Stages in the development of a landform are often interrupted and changed. An old river can be made youthful if the underlying strata undergo uplift. Such changes in

rock structure often make it difficult to establish the stage of a given landform.

The three factors that control the origin of minor landforms are illustrated in Figure 11–22.

Life cycle of mountains. No matter how they have come into existence, mountains usually follow a cycle of development as they are exposed to the forces of erosion. Mountains are created by the forces of uplift. As long as these forces continue, the mountain is usually raised faster than it is worn away. These mountains are said to be *youthful.* They are rugged with sharp peaks and deep narrow valleys. However, mountain growth eventually stops. Then the rugged scenery becomes worn down to form *mature* mountains. These are recognized by their rounded tops and gentler slopes. See Figure 11–23. As erosion continues over a very long period of time, mountains may become worn down until the area is almost level. This period of the mountain cycle is called *old age.* The final appearance of the surface at this stage is called a *peneplain,* which means "almost a plain."

A peneplain usually has low rolling hills with occasional resistant knobs called *monadnocks.* See Figure 11–24. A peneplain might be mistaken for a true plain. However, it hides under its surface the folded, twisted or tilted rocks that are unlike the horizontal layers of true plains. Southern New England is a raised peneplain with typical monadnocks in New Hampshire and Massachusetts.

Some scientists believe that continents do not remain in a stable condition long enough for streams to form peneplains by erosion. If this is true, the way these wide level surfaces are formed is still an unsolved problem.

Life cycle of plains and plateaus. Both plains and plateaus usually pass through a series of stages as erosion wears away their surfaces. In the case of a plain, the changes are not great because the streams have little of the land surface to cut away before they reach base level. A mature or old plain is different from one in youth mainly because it possesses well developed streams. A plateau, on the other hand, is exposed to much erosion because it is at a higher elevation. The effect of the streams in attacking the rock of a plateau depends upon the climate and the type of rock. A young plateau usually has deep stream valleys with broad, flat regions separating them. Mature

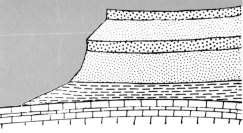

FIG. 11–26. Note the structures of the mesa above and the buttes below. (Santa Fe Railway Photo; Union Pacific Railway Photo)

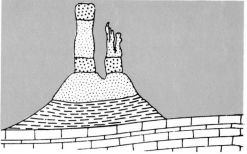

plateaus, such as the Catskill region in New York State, are rugged areas and are generally called mountains. See Figure 11–25. Many streams have cut wide valleys through the broad surfaces of the original plateau. In old age, plateaus are worn almost level, with traces of the original plateau left here and there. In dry regions these remnants have steep walls with flat tops. If they have broad tops they are called *mesas*. The smaller ones with narrow tops are called *buttes*. See Figure 11–26. In humid regions the last remnants of old plateaus are more rounded.

VOCABULARY REVIEW

Match the word or words in the column on the right with the correct phrase in the column on the left. *Do not write in this book.*

1. Changes that destroy the original rock structure.
2. The action of CO_2 when dissolved in water.
3. Water combining with minerals in rock.
4. Breaking apart in sheets or shells.
5. Dark organic material in soil.
6. Accumulation of broken rock at the base of a cliff.
7. The layers which make up soil.
8. All types of erosion caused directly by gravity.
9. Stones or pebbles which have been smoothed by wind erosion.
10. Large accumulations of fine wind-blown dust.
11. Furrows that follow the contour of the land to prevent erosion.
12. Movement of rock materials over the earth's surface by various natural agents.
13. Refers to a period of time during which a particular process has been at work.
14. The surface in the mountain cycle when it is almost a plain.
15. An occasional resistant knob on a peneplain.
16. Remnants of plateaus.

a. loess
b. mesa
c. humus
d. monadnocks
e. exfoliation
f. stage
g. deflation
h. peneplain
i. contour plowing
j. soil profile
k. blowout
l. ventifacts
m. weathering
n. erosion
o. carbonation
p. terracing
q. hydration
r. mass wasting
s. talus

QUESTIONS

Group A

Select the best term to complete the following statements. *Do not write in this book.*

1. The most important result of the weathering of rock as far as living things are concerned is the (a) release of oxygen by frost action (b) formation of soil (c) erosion which occurs (d) leaching which occurs.

2. Rock in which a change in mineral content is taking place is said to be undergoing (a) erosion (b) mechanical weathering (c) frost action (d) chemical weathering.

3. Which one of the following involves mechanical weathering? (a) exfoliation (b) hydration (c) carbonation (d) oxidation.

4. The breaking up of rock by freezing water is called (a) hydration (b) leaching (c) frost action (d) mass wasting.

5. Which one of the following ingredients of air is *not* involved in weathering? (a) nitrogen (b) oxygen (c) carbon dioxide (d) water.

6. A process similar to frost action which causes rocks to break apart but which is found in drier climates is (a) exfoliation (b) growth of salt crystals (c) blowout (d) leaching.

7. Rocks containing clay undergo a weathering very similar to (a) hydration (b) oxidation (c) frost action (d) carbonation.

8. Large caves in limestone rock are formed when carbon dioxide dissolves in water to produce (a) quartz (b) oxides (c) hydrates (d) carbonic acid.

9. Sand or gravel is formed from the weathering of rocks such as (a) limestone (b) shale (c) granite (d) calcite.

10. Large unweathered rocks left behind after weathering of bedrock are called (a) ventifacts (b) talus (c) residual boulders (d) barchans.

11. Weathering takes place slowly in climates that are (a) dry or very cold (b) humid (c) moderate (d) warm.

12. Soil composed mostly of a mixture of particles ranging from microscopic clay particles to larger sand grains is called (a) humus (b) talus (c) loess (d) silt.

13. The fine grains often formed from weathered feldspars are called (a) sand (b) clay (c) humus (d) talus.

14. A mature soil commonly has a soil profile consisting of (a) one layer (b) two layers (c) three layers (d) four layers.

15. Subsoil is the name given to the layer of the soil profile which is generally labeled (a) A (b) B (c) C (d) D.

16. The layers of the soil profile are called (a) horizons (b) ventifacts (c) barchans (d) sublayers.

17. The layer of soil where most of the animal and plant life is found is labeled (a) A (b) B (c) C (d) D.

18. The dissolved materials of the soil generally leach into the (a) A horizon (b) B horizon (c) C horizon (d) bedrock.

19. Soils in drier climates are often very fertile, mainly because (a) no plants can grow to use up the minerals (b) they have well defined horizons (c) much leaching occurs (d) very little leaching occurs.

20. Directly or indirectly, the cause of most erosion is (a) water (b) wind (c) gravity (d) the earth's rotation.

21. Slow, downhill movement of weathered rock is called (a) a landslide (b) slumping (c) soil creep (d) deflation.

22. Slumping is considered an early stage of (a) soil creep (b) landslide (c) mudflow (d) deflation.

23. The most important agent of erosion is (a) water (b) wind (c) heat (d) man.

24. The most common form of wind erosion is (a) oxidation (b) exfoliation (c) mass wasting (d) deflation.

26. Humans can increase the weathering and erosion of soil by (a) grazing animals on it (b) clearing the plant life (c) plowing the soil (d) all of these.

27. Once soil has been swept away by erosion (a) no amount of work can quickly repair the damage (b) crop rotation should be started (c) contour plowing should be done (d) terracing will repair the damage.

28. The life cycles of plains and plateaus are very similar except that plateaus generally pass through the series of erosion stages faster since they (a) are at a higher elevation (b) have more streams (c) are flatter on top (d) are made of more durable materials.

Group B

1. Why does weathering generally occur before erosion?

2. Describe how you can tell whether a given rock sample has undergone weathering.

3. Contrast mechanical weathering and chemical weathering.

4. Give three different examples of how water produces mechanical weathering of rock.

5. Why are limestone and calcite caves found only in areas with an ample water supply?

6. How can you tell whether iron-bearing minerals have been weathered?

7. Give at least three examples of how living things cause weathering.

8. What three main factors determine the speed of rock weathering?

9. Describe the two steps involved in the formation of soil.

10. Describe a typical soil profile.

11. What relationship exists between soil creep, slumping and landslides?

12. Why is gravity the indirect cause of all erosion regardless of the eroding agent?

13. What fact seems to make it doubtful that such large features as natural bridges, rock pinnacles, rocks perched on pedestals, and the like were formed by wind erosion?

14. Describe the process by which a sand dune moves along the ground.

15. How are sand dunes and loess related?

16. Although natural erosion is constantly changing the earth's solid surface, it would not generally cause much concern if it were not for the activities of man. Why is this so?

17. What is the principal means by which igneous, sedimentary, and metamorphic rocks weather?

18. Describe how the four methods used by farmers aid in controlling erosion.

19. Explain the formation of monadnocks.

20. What do buttes and mesas have in common? How do they differ?

21. What features of a peneplain are like those of a true plain? What features are different?

Water is the most powerful of all erosion agents. It is more effective in moving weathered rock fragments than all other forces of erosion combined. Each drop of water that falls on the land does its share in wearing away and moving the rocks that make up the earth's crust.

The earth has not always had its present supply of liquid water. There were no oceans, lakes, or rivers during the earth's earliest history. These bodies of water could form only after the earth's crust cooled. It is likely that much of the earth's water appeared originally in liquid form beneath the earth's surface. It may have been given off as a gas in the violent volcanic eruptions that were common in the earth's early stages.

Heavy layers of clouds probably surrounded the young planet. Eventually as the cloud cover began to drop its accumulated moisture, rain fell and filled the ocean basins. At this time, an endless water cycle was started. Solar energy absorbed by water in the sea caused evaporation. This water vapor again formed clouds, then rain. Raindrops fell to become running streams which used the energy from gravity to carve and move rock materials. It is this continuous cycle of water movement which gives water its power as an agent of erosion.

THE WATER CYCLE

The hydrologic cycle. Part of the earth's water supply is bound up as water molecules in certain sedimentary rocks. The remaining water on earth follows a never-ending path, leading from evaporation to water vapor, then condensing back to clouds to produce precipitation

objectives

- [ ] Explain the operation and importance of the hydrologic cycle.

- [ ] Describe the importance of water conservation.

- [ ] Trace the steps in the development of a river system.

- [ ] Describe erosion caused by moving water.

- [ ] Explain how the water table is related to the water in the ground.

- [ ] Describe the formation of springs and geysers.

- [ ] Describe the formation of sinks and caverns.

253

in the form of rain or snow. Then run-off flows over the surface or sinks into the ground. Finally, it is stored until evaporation again takes place. This series of events is known as the *hydrologic cycle,* or water cycle.

Each year about 95,000 cubic miles of water enters the air as water vapor. Most of this, about 80,000 cubic miles, comes from the sea. This water evaporates when the sea absorbs the sun's heat energy. However, an additional 15,000 cubic miles of water evaporates each year from other sources. An important source of this moisture is the process of *transpiration,* a means by which plants give off water vapor. Water is also evaporated from lakes, streams and from the upper layers of the soil.

Most of the water that evaporates into the air falls back again into the oceans. About 71,000 cubic miles of water is returned annually to the oceans as rainfall. This leaves 24,000 cubic miles of water to fall on the land surfaces. Of this, 9,000 cubic miles is run-off in streams and rivers. This run-off water moves over the land surface and returns to the sea within a few weeks. But the remaining 15,000 cubic miles of water soaks into the land to become *ground water.* Some of this ground water fills the spaces between soil particles, providing an available water supply to plants for use in their vital life processes. A great deal of ground water sinks far into the pores and cracks of the deeper bedrock. This water slowly meanders by various routes back to the oceans. The complete hydrologic cycle is illustrated in Figure 12–1.

We can say that this continuous hydrologic cycle gives the earth a *water budget.* The income part of the budget is the water received by precipitation, rain, and snow. Outgo is represented by the ways that water evaporates and enters the air. The water budget is probably in balance for the whole earth. For example, the amount of water lost by evaporation over the complete surface of the earth equals the amount of moisture which falls to the earth as precipitation.

The water budget, however, much like the heat budget, is not in balance at every particular place on the earth's surface. Precipitation is affected by the patterns of weather and the shape of the earth's surface. One side of a mountain range may receive large amounts of precipitation while the other side remains a dry desert. Evaporation also differs greatly from one location to another. Far more solar energy is received in regions near the equator than in areas further away from it. Evaporation is also affected by winds which sweep up a great deal of moisture.

The amount of evaporation and precipitation at a particular place also changes with the seasons of the year. Some locations receive more moisture every month of the year than is lost by evaporation. In these places, surplus water constantly runs off in streams or becomes ground water. Other very dry locations consistently receive less water than is lost by evaporation. Under these conditions, plant life is scarce and run-off occurs only after occasional, heavy rains.

In places which are neither close to the equator nor to the poles, the situation varies throughout the year. During the cooler months, temperatures remain low enough for evaporation to use less water than the amount that is received by precipitation. During these months a surplus of moisture exists. The soil becomes saturated and excess water runs off in streams. During the warmer months, however, an increase in the rate of evaporation causes an overall loss of moisture. In the warmer period, the amount of moisture in the soil drops to a low level. Streams can continue to flow only if they are supplied from distant sources or from ground water.

Investigate

Blow up three balloons to nearly the same size. Place one balloon outside in sunlight; the second balloon in a shady spot; and the third balloon on some crushed ice. After 15 minutes observe the size of the three balloons. In which balloon would evaporation take place fastest?

FIG. 12–1. A summary of the hydrologic cycle.

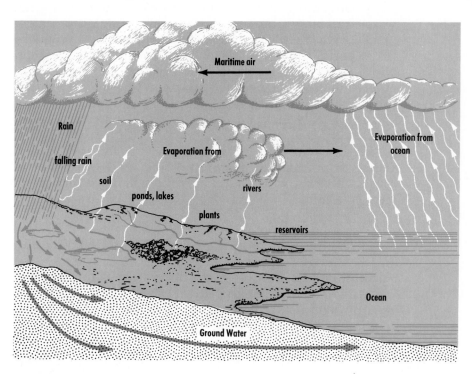

Table 12.1 — Water in Industry

Quantity	How Consumed
5 gallons	To process 1 gallon of milk.
10 gallons	To produce 1 gallon of gasoline.
80 gallons	To generate 1 kilowatt-hour of electricity.
300 gallons	To manufacture 1 pound of synthetic rubber.
65,000 gallons	To produce 1 ton of steel.

FIG. 12–2. The surface of the ground shown in this photograph has been eroded by falling raindrops, except where it was protected by stones. Notice the coin at the left for scale. (Ramsey)

Water conservation. It is estimated that each person in the United States drinks an average of 200 gallons of water per year. Each individual uses up another 15,000 gallons each year for washing, laundry, cooking, and the operation of heating and air conditioning equipment. This personal use of water is unimportant, however, when compared with 160,000 gallons per person which are used each year by industry. Table 12–1 will give you some idea of the vast amounts of water consumed in America's industrial processes.

To these figures must be added the billions of gallons of water needed to irrigate crops. All of this vital water must come from the water which runs off in streams or enters ground water supplies. This amounts to about 1300 billion gallons daily in the United States. It represents the entire supply available which can be obtained at reasonable cost. As the population of the country grows, the demand for water also grows. Conservation of this limited resource becomes essential.

On the average 90 percent of the water used by cities and industries is returned to rivers or the sea as waste. The untreated sewage of many cities is disposed of by dumping. Much waste water discharged into rivers by industries contains harmful materials. In some cases, the water can never be used again even after all practical purification methods have been tried. Such practices leave parts of our limited water supply unfit for human consumption. To insure a future supply of water large enough to meet the needs of an expanding population, we must achieve better control of water pollution.

It is possible that the supply of fresh water can be increased by treatment of sea water. This subject is discussed in Chapter 16. However, our best hope in the near future to provide an adequate supply of fresh water at reasonable cost is by conserving the water which is presently available.

RUN-OFF OF WATER OVER THE LAND SURFACE

Formation of a river system. Erosion by running water begins when each raindrop strikes the earth. Energy from the falling rain is delivered against the surface at the moment of impact. See Figure 12–2. Even after reaching the solid surface, however, a raindrop still does not lose all its energy until it reaches sea level. Gravity forces water to move or try to move down slopes until it reaches the level of the oceans. Thus the water that falls and remains on land surfaces must run off in established streams or else erode new channels. When a new path is carved, the first step in the formation of a river system begins as the water tumbles in broad sheets down the slopes.

The run-off from several slopes collects in low places. The energy from the moving water is then concentrated on a smaller surface and the rate of erosion increases. Soon a gully is formed. The gully serves as a collection channel for water from all nearby slopes. The increasing volume of water further enlarges the gully. These are the earliest stages in the development of a river. A beginning stream is not likely to flow except immediately following a rain. Even so, each time the water flows, it enlarges the gully and collects more water. Eventually this process results in a fully developed river valley containing a permanent stream.

In addition to the growth of the beginning river and its valley, branch gullies are produced by streams flowing in from the sides. These streams become the *tributaries* for the growing river system. See Figure 12–3. Erosion at the head of the gullies, where the water enters the stream, lengthens them and continues to cut back the existing slopes. This is called *headward erosion*. Chiefly it is the process by which a beginning river system extends its main branches back towards its source to collect the run-off water from a larger area of land.

Compare
Determine the conditions necessary for the formation of each of the following drainage patterns: deranged, dendrite, trellis, rectangular, and radial.

FIG. 12–3. Stages in the development of rivers and streams. (Top left) Gullies appear on the steeper slopes. (Top right) Gullies expand by downcutting and headward erosion. (Bottom right) Streams establish separate drainage basins. (Bottom left)

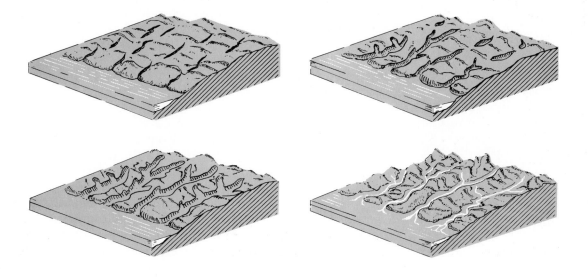

FIG. 12–4. Sharp ridges shown in this photograph are divides separating desert streams which flow during periods of heavy rainfall. Can you find an example of headward erosion which will probably cause stream piracy in the future? (Ramsey)

FIG. 12–5. The valley of the Yellowstone River shows the characteristics of a young stream. (Ramsey)

With headward erosion, the main stream and its tributaries eventually form a branching system of channels. These channels drain a series of slopes that make up the *drainage basin* or *watershed* of the river system. The ridges or regions of high ground that separate the various watershed areas from each other are called *divides*. See Figure 12–3.

Divides between neighboring streams gradually become lower as erosion continues its work from both sides. The rate of erosion is generally not equal on both sides of the divides. This is due to differences in the kind of material through which the stream flows or the speed of stream flow. Occasionally, a speed-up of erosion of one stream cuts through a divide. When this happens, all the water then flows in one of the stream channels. This process is called stream capture or *stream piracy*. See Figure 12–4. An established river system is often enlarged by stream piracy.

A stream's ability to cut down and widen its channel by erosion depends largely upon the swiftness of its water flow. In turn, the velocity of a stream depends upon the amount of water it carries and on its slope or *gradient*. A stream's gradient is a term expressing the difference in elevation between its head and mouth. Rivers with a steep gradient generally have a high velocity, allowing the river to rapidly erode its channel. However, as the river grows older, its gradient becomes less steep and the channel is cut down toward the level of its mouth.

In its early stages, when the gradient is still steep, a river usually deepens its channel more rapidly than it can

FIG. 12–6. This aerial photograph of Niagara Falls shows how it has been worn back over the past few hundred years. Like all waterfalls, it will eventually disappear from the river channel. (Niagara Frontier State Commission)

cut into the sides. This produces a *V-shaped* valley such as that shown in Figure 12–5. A river in this stage of development is said to be in its youth. Besides a V-shaped valley, *waterfalls* and *rapids* are very common features of youthful streams. Both waterfalls and rapids frequently appear where there is unusually hard rock in the stream channel. Hardened rock resists erosion by the stream and tends to remain as a barrier. However, such barriers are usually only temporary, and both waterfalls and rapids disappear as the stream grows older. See Figure 12–6.

Young rivers usually have relatively few tributaries. This is because there has not been time for a large system of feeder streams to develop. For this reason, a young river usually carries a small volume of water. Much of the water falling in the watershed of a young river system does not reach the main stream. It remains in the higher lands to form lakes and swampy areas.

On the other hand, a *mature* river has well-established tributaries. It effectively drains off the water falling into its drainage system. Because of its many tributaries and good drainage, a river carries its largest volume of water at the mature stage. In its later stages, a stream has less

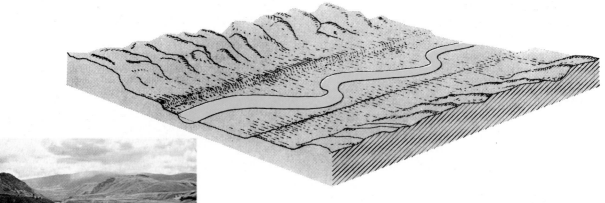

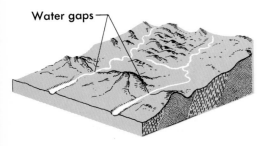

FIG. 12–7. Above right, a diagram of a stream with terraces. Above, a photograph of a river in British Columbia with well-developed terraces. (Ramsey)

FIG. 12–8. The process by which a water gap may become a wind gap following stream capture.

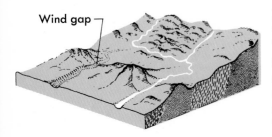

tendency to deepen its channel. Its main erosion is on the sides of its valley walls. Thus an *older* river usually loses the V-shape of its youthful channel and creates a wider valley with a relatively flat floor. The waterfalls and rapids of the youthful phase start to disappear and the channel becomes flat. Then the river channel normally occupies only a small part of the valley floor.

Movements of the crust can raise or lower the surface of the land. Any movement that increases the slope of the land will change the gradients of existing streams. A stream whose gradient has been increased in this way is said to be *rejuvenated*.

The steeper gradient of a rejuvenated stream allows it to cut deeper into the valley floor. Rejuvenation often results in the formation of step-like *terraces* in stream valleys. See Figure 12–7. When the land surface rises very slowly, the existing streams are cut down only as fast as they are elevated. When this happens, the mountains raised across the channel of an existing river will be deeply notched. This notch is called a *water gap*. The river that cuts the water gap may later be captured by a stream running parallel to the mountains. Thus the water gap is abandoned and it becomes a *wind gap*. See Figure 12–8.

Erosion by moving water. Running water carries on its work of carving the land surface by a series of processes. The first event that produces a movement of rock fragments is the splash of a raindrop. The force of raindrops on bare soil tends to seal off the surface. The impact of the drops shifts the particles around on the surface and the finer soil grains are thus fitted between larger ones. The pores and channels through which water might enter the soil are now plugged up. Soil particles are also pressed together by energy from the falling drops, closing the spaces between the particles. The effects of these actions waterproof the soil surface so that most water falling on it will be lost as run off.

At first, the run-off appears as shallow layers of water running over the surface. This brings about *sheet erosion*. This type of erosion removes the lighter soil particles and substances which dissolve. When the run-off water enters lower places on the surface, its energy becomes more concentrated. At the same time, its power to erode is directed against smaller areas of the surface, allowing the larger rock fragments more freedom of movement. Small particles are carried along, suspended in the water. Heavier fragments of rock are rolled or bounced along the stream bed. Each contact between the loose rock fragments carried by the water and the rocks of the stream bed wears off a small amount of both. The moving rocks become rounded and smooth, while the stream bed is worn away. Rocks are often swirled around for some time in a whirlpool motion over the same area on the stream bed. This creates a bowl-shaped cavity in the rock of the stream bed. Such a cavity is called a *pothole*. See Figure 12–9.

As the load carried by the stream increases, its rate of flow slows down and there is a tendency for bends to develop. Once a slight bend develops in a stream bed, the channel curve in that place usually grows larger. The reason for this is that water usually flows fastest around the outside edge of the curve. The faster flowing water erodes the outside bank more readily than it does the inner bank. This results in an ever-increasing enlargement of the curve, as shown in Figure 12–10. Slower moving water on the inside of the curve allows sediments to settle and form a bar. Older streams often develop a series of looping curves called *meanders* (mee-*an*-durz). See Figure 12–11. Frequently these meanders become so curved that the river cuts across the narrow neck of land

FIG. 12–9. A pothole formed in the solid rock of a stream channel. The continuous motion of the stones swirled along in the water helped to grind and enlarge the hole. (Ramsey)

Explain
Part of the Rio Grande river along its older stream bed, acts as a boundary between Texas and Mexico. Why is this a poor physical boundary line?

FIG. 12–10. A stream tends to develop larger curves because the outside bank of a small curve is eroded faster than the inner bank.

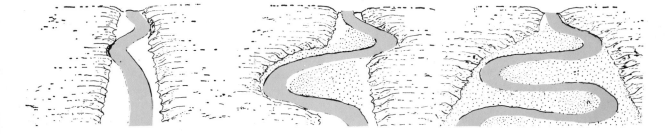

FIG. 12–11. A meandering river showing deposits which form bars along the inside curve of each meander. (National Air Photo Library)

FIG. 12–12. A stage in the formation of an oxbow lake; the cutting off of a meander loop. The river cuts deeply into the bank along the outside of the meander's curve.

at both ends of the curve, isolating the meander from the river. If the water remains in the isolated meander, it will form an *oxbow lake*. The process by which a meander becomes an oxbow lake is shown in Figure 12–12.

The combination of dissolved substances, suspended particles, and rock fragments rolled along by the water make up the *load* of a stream. Under certain conditions, a stream will decrease its load, generally when for some reason the water slows down. Because stream velocities are constantly changing, the loads of streams are also increasing and decreasing. As a result, deposits of solid material carried by the stream are formed, then picked up again by moving water. In some cases, stream deposits remain in place only a short time. In others, they become a more or less permanent feature of the land.

Most of the load carried by a stream is deposited when the stream reaches a larger body of water. As a stream empties into the relatively quiet water of an ocean, gulf, or into a lake, all of the remaining load is deposited at its

1   2   3   4

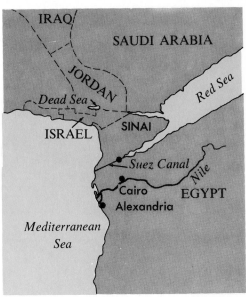

FIG. 12–13A. Delta of the Nile River as seen during an orbit of Gemini 4. The small diagram shows the area covered by the photograph. (NASA)

mouth. The sediment deposited usually takes the form of a triangle with its apex turned directly upstream. These fan-shaped deposits at the mouth of a stream are called *deltas*. See Figure 12–13A.

When streams flow down a steep slope from high land onto a level surface, a deposit resembling a delta is formed. As the stream flows swiftly down the steep mountain slope, it gathers a heavy load. When the stream reaches the level surface, it deposits its load in a delta-shaped form called an *alluvial fan*. The word "alluvial" refers to any deposit laid down by running water. See Figure 12–13B.

Floods. A stream adjusts itself to carry the amount of water flowing within its channel. If a stream always flowed with the same volume, its channel would change very little. But the flow of water (in nearly all streams) changes constantly. Most noticeable are the times when the stream overflows its banks. When this happens, the

FIG. 12–13B. The major difference between an alluvial fan and a delta deposit is their place of deposition. A fan is an alluvial *land* deposit. (Photo Researchers-Russ Kinne)

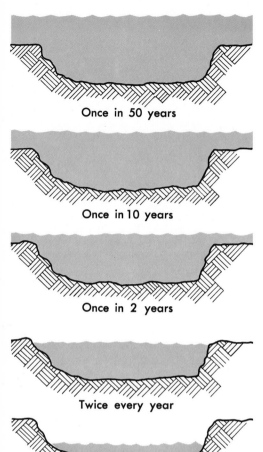

Once in 50 years

Once in 10 years

Once in 2 years

Twice every year

Ninety days each year

FIG. 12–14. The water-flow in a typical river varies greatly over a period of time.

stream is said to be in *flood stage*. Changes in water flow which produce floods are illustrated in Figure 12–14.

Spring floods are common in regions where the winters are harsh. Since rainfall and water released by melting snow cannot be absorbed by the frozen ground, it appears mostly as surface run-off in streams. Ice jams also increase the chance of spring flooding by blocking stream channels.

In many areas human activity has increased the size and number of floods. The natural ground cover of plants which helps protect the surface from heavy run-off has been removed. Forest fires and destructive logging operations, or the clearing of land for cultivation in watershed regions increase the run-off load in streams.

When a stream overflows its banks and spreads out over the nearby land, it deposits part of its load along the borders of the channel. Accumulation of these deposits along the edges of the stream eventually produces raised banks. Such elevated banks are called natural *levees*. These long, raised banks may be quite high and prominent along mature river channels.

However, not all the load deposited by a stream in flood stage goes into the formation of levees. Large amounts of sediment are left behind in fairly even layers after the water level has gone down. A series of floods produces a thick layer of deposited material on the level areas on both sides of the stream channel. Since this accumulation is generally very even, the part of the valley floor usually covered in the flood becomes a relatively flat flood plain. A *flood plain* with levees is shown in Figure 12–15. Swampy areas are common on flood plains because drainage is usually poor in the area between the levees and the outer walls of the valley.

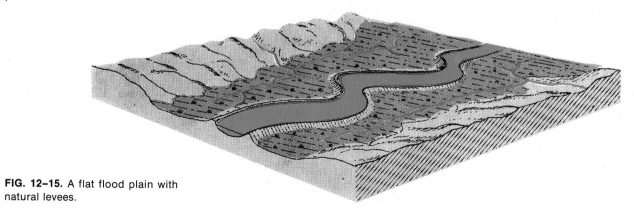

FIG. 12–15. A flat flood plain with natural levees.

Floods are a natural part of the development of streams and will continue to occur. They are a problem because many cities and farms are located on the fertile flood plains of rivers. To prevent loss of life and property in these areas, some control over floods becomes necessary. There are direct and indirect ways of controlling floods.

Indirect methods of flood control include forest and soil conservation measures which prevent heavy run-off during periods of high rainfall. The most common example of direct flood control methods is a dam. Dams help to control floods by creating artificial lakes which absorb excess run-off. The stored water can then be used for generating electrical power and for irrigating cropland during dry periods. Artificial levees are also built to directly control floods. However, these levees offer only temporary protection. As a river deposits sediment along its bed, the height of the levees must constantly be raised. They also require protection against erosion by the river. In some cases, a permanent overflow channel or floodway can be an effective means of *flood control*. During flood stage, a floodway helps carry excess water to prevent the main stream from overflowing.

WATER BENEATH THE EARTH'S SURFACE

The water table. Water can seep into the earth's crust because much of the land surface is covered by loose, weathered material. In many places, the outer layer of fragments lies over porous rock, through which water can penetrate. Below this is rock which may be completely solid. As water seeps down from the surface, it first enters the zone of aeration, a region where the spaces between the rocks contain both water and air. Most of the water in this zone is found as a thin film clinging to the surface of the rock and soil particles. Some of the water in the *zone of aeration* moves down to lower levels; some of it evaporates and some is absorbed by the roots of plants. The depth of the zone of aeration and the amount of water it contains varies greatly from one time or place to another. During a dry period this zone would increase in depth. In mountainous or hilly regions, it may be hundreds of feet thick. At other locations where the soil is saturated with water, the zone of aeration will be only a few inches thick or else completely absent.

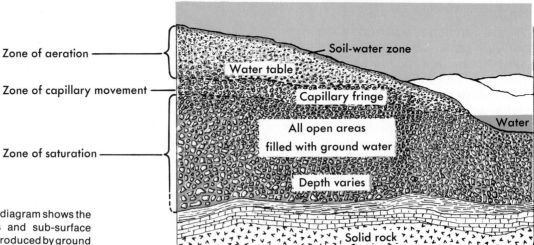

Zone of aeration

Soil-water zone

Water table

Zone of capillary movement

Capillary fringe

Water

Zone of saturation

All open areas
filled with ground water

Depth varies

Solid rock

FIG. 12–16. This diagram shows the various features and sub-surface moisture zones produced by ground water.

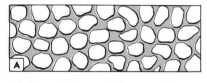
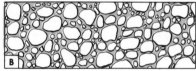
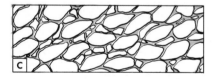

FIG. 12–17. (A) Rock composed of particles of nearly the same size has high permeability because of many pores or open spaces; (B) Rock with particles of many different sizes has lower permeability; (C) Mineral deposits between the rock particles reduce permeability to nearly zero; (D) Fine-grained rock which dissolves in water has high permeability because water dissolves cavities in it.

At the bottom of the zone of aeration is a region where water is drawn up from below by capillary action. Capillary water is water drawn upward against the force of gravity by a force known as *capillary tension*. In a soil profile, the top of the *zone of capillary movement* is called the *capillary fringe*. The depth of this zone depends on the size of the rock particles and is seldom greater than a few feet. In gravels and coarse sands the rise of capillary water is only a fraction of an inch, but in the smaller soil particles the grains are closer together and the rise may be several feet.

Below the zone of capillary movement lies the *zone of saturation* in which all spaces are filled with water. Water found in the zone of saturation is referred to as *ground water*. The top of the zone of saturation that marks the depth needed to reach ground water is usually called the *water table*. See Figure 12–16.

The depth of the water table below the ground surface at any particular location depends upon several things. One important influence is the amount of rainfall in the recent past. Another factor is how easily water can penetrate into the rock. The ability of rock to allow entrance of liquids is called *permeability* (per-mee-uh-*bil*-i-tee). The most permeable rocks are those composed of very coarse grains. These rocks usually have large pores between the grains. An example of this type of rock is sandstone. Very fine grained rocks, such as limestone, may become permeable if cracks are produced after the rock is formed. Various ways in which rocks become permeable are shown in Figure 12–17. Ground water ordinarily sinks deeper until it strikes a completely solid rock layer. The rocks above this layer then become completely saturated up to the water table.

Generally speaking, the water table follows the contours of the land surface. It slopes down where the surface slopes and rises in higher ground. However, the general slope of the water table is usually not as sharp as that of the land surface. This is true because ground water flows very slowly as it works its way through small cavities in the rock.

In many places depressions on the land surface dip below the water table. When this happens, ground water may flow out onto the surface to join with surface streams. Water may also collect on the surface to form lakes or swamps. See Figure 12–18. Many streams continue to flow in periods of low rainfall because they are fed by ground water.

Surface streams flow because they have a gradient which is determined by the slope of their beds. Ground water also flows because the gradient of its bed corresponds in a general way to the slope of the surface

FIG. 12–18. Above, a small stream channel permanently fills with water where it first cuts below the water table. Below, the dry upper zone has been lifted to show the surface of the water table.

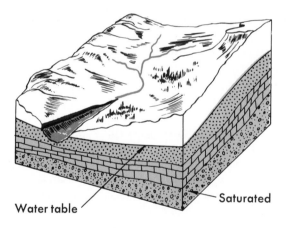

Water table   Saturated

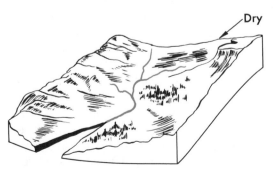

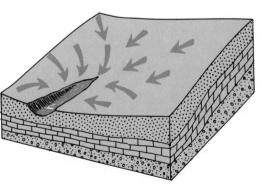

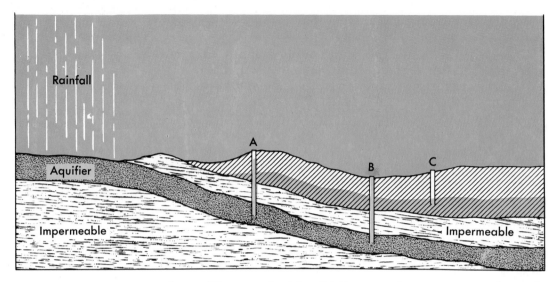

FIG. 12–19. The type of rock formations that result in artesian wells. At A, the artesian well does not flow. At B, the well flows because it is below the origin of the water, where the aquifer is exposed at the surface. Well C collects only ground water from the local area.

above. The speed with which ground water moves through the rock depends upon the gradient and the permeability of the rock structure.

Any hole below the water table will fill with water and form a well. The rate of water flow into a well depends upon the amount of ground water available at that particular place. An *artesian well* obtains water from a porous rock layer exposed at one end and sandwiched between an upper and lower layer of solid rock. The porous layer, called an *aquifer* (*ak*-wi-fer), slants downward from its exposed surface. See Figure 12–19.

Springs. A *spring* is found where ground water comes naturally to the surface. There are many types of springs; any exposed surface between the water table and the ground may result in the formation of a spring. Hillsides are common locations for springs because a slope on the surface often drops below the level of the water table. Such springs usually do not flow continuously, since the water table may drop below the slope in dry periods. However, one type of hillside spring is likely to flow continuously. This is a spring formed at the zone of contact between permeable and impermeable rocks. Water filters down through the permeable rock and comes to the surface on a hillside when it meets the impermeable rock layer. See Figure 12–20.

FIG. 12–20. Springs often appear where the water table intersects the surface above a layer of impermeable rock.

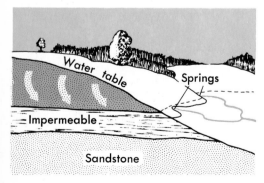

Hot springs and geysers. In some areas ground water is heated by volcanic activity. The water then finds its way

to the surface before cooling, and produces a *hot spring.* The water from a hot spring is heated by passing close to a body of magma. It may also be heated by mixing with steam and hot gases escaping from bodies of magma deep in the crust. Hot water is better able to dissolve minerals than cold water. Thus the water of hot springs is likely to contain unusually large amounts of mineral matter. Much of this dissolved mineral material is deposited when the water cools after reaching the surface. It often accumulates around the mouths of hot springs and builds up layers or terraces. See Figure 12–21. The chief mineral in these deposits is usually *travertine,* a form of calcite. Travertine is mostly white when freshly deposited but turns gray upon weathering. It may be colored red, brown, or yellow by impurities, such as iron compounds.

Geysers are a type of hot spring that, at times, throw steam and water into the air. Some geysers erupt from open pools and throw up sheets of water and steam. Others erupt through a small surface opening sending a column of water and steam high in the air. Both types tend to build deposits of a silicate mineral called *geyserite.*

FIG. 12–21. A hot spring in Yellowstone National Park. The water has deposited layers of travertine around the opening of the spring. (Ramsey)

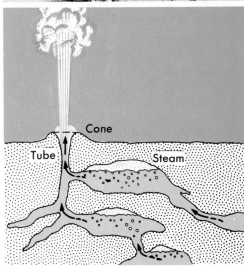

FIG. 12–22. (A) a geyser in Yellowstone National Park. (B) a deposit of geyserite built around an extinct geyser in Yellowstone National Park. (C) a diagram showing the underground structure of a typical geyser. (Shostal Photos—A) (Ramsey—B)

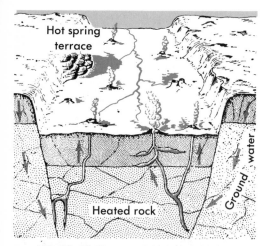

FIG. 12-23. Geysers and hot springs are generally found only in trenches formed by parallel cracks. Typical features and their origins are shown in this diagram.

FIG. 12-24. A limestone sink in Weston County, Wyoming. Sinks are a feature of erosion by ground water. (U.S. Geological Survey)

The underground structure of a geyser consists of a crooked tube leading to the surface. The tube usually connects several underground chambers, as shown in Figure 12-22. Heated ground water fills both the tube and chambers. Steam trapped in the rock cavities creates pressure that forces some water out onto the surface. This produces a small surge of water which signals all geyser eruptions. The water of a geyser boils explosively when the escaping water reduces pressure on the very hot water below. A roaring column of steam and boiling water is driven up and erupts on the surface. The eruption continues until most of the water and steam are emptied from the tube and storage chambers.

Afterwards, the ground water begins to collect again for a repetition of the process. The period between every geyser's eruption is determined by its supply of ground water, heat, and the shape of its tube and chambers. Geysers are found in special valleys called *geyser basins*. See Figure 12-23. Almost all of the geysers of the world are found in Yellowstone National Park, New Zealand, Iceland and Japan.

Ground water and rock. Pure water does not easily dissolve any but a few minerals. If ground water were pure, it would not have much effect on the rocks through which it moves. However, ground water does contain carbon dioxide, oxygen, and other dissolved materials. These are picked up during its movement on the surface, before moving down into the rock. Thus rocks beneath the ground can be affected chemically by ground water. An example is the process of *carbonation*. In this process carbon dioxide dissolved in ground water forms an acid which attacks certain minerals. This reaction is identical to the one that takes place in chemical weathering on the surface.

Rocks composed mainly of the mineral calcite, such as limestone, dolomite, and marble, are especially vulnerable to attack by carbonation. They dissolve slowly in water containing carbon dioxide. Thus ground water penetrating into openings in these rocks enlarges the rock cavities. If this dissolving action enlarges a number of connecting cracks, a *cavern* will be formed in the rock. Often the cavern will grow quite large with many large and small connecting chambers. This is particularly true in areas where large deposits of limestone rock exist beneath the surface. Examples of such caverns formed in

FIG. 12–25. Karst topography in a limestone region, showing some of the features of underground drainage.

limestone are the Carlsbad Caverns in New Mexico and Mammoth Cave in Kentucky. Many other places in North America have limestone regions with caverns.

When part of the roof of a cavern collapses, it often creates a roughly circular hole on the ground surface. Such holes are called *sinks*. See Figure 12–24. Sinks frequently become filled with water, forming ponds or lakes.

Cavern growth in limestone regions often produces a particular type of landscape where sink holes and similar depressions on the surface are very common. Ponds and lakes are also common, but there are almost no surface streams. The absence of streams means that there are no valleys in the region. This type of landscape is called a *karst plain*. See Figure 12–25. Karst plains are found in the limestone regions of Kentucky, Tennessee, and southern Indiana.

Caverns that lie above the water table cannot fill with water. Their appearance is marked by stone formations made of calcite deposited by water dripping into the spaces inside the cavern. Water dripping from the same spot on a cavern roof builds a hanging deposit called a *stalactite* (stuh-*lak*-tyt). At the point on the cavern floor where drops of water fall, another deposit of calcite called a *stalagmite* (stuh-*lag*-myt), is formed. Frequently the two formations grow until they meet forming a continuous *column*. See Figure 12–26.

FIG. 12–26. The limestone deposit that forms stalactites and stalagmites is known as *travertine* or *calcareous tufa*. (Luray Cavern)

VOCABULARY REVIEW

Match the word or words in the column on the right with the correct phrase in the column on the left. *Do not write in this book.*

1. The series of events in which water is evaporated and finally returned to storage.
2. Water that soaks into land to fill available space.
3. The probable balance between evaporation and precipitation.
4. Streams flowing into a river from the sides.
5. The series of slopes which make up the drainage basin.
6. Ridges or regions which separate the various watershed areas.
7. The notch cut in a mountain raised across a river channel.
8. Formed when water remains in an isolated meander.
9. The part of the load which is deposited at the mouth of a stream.
10. The region in which the spaces between the rock contain both water and air.
11. The name usually given to the top of the zone of saturation.
12. Obtains water from an aquifer located between two impermeable layers.
13. Occurs where the ground water comes naturally to the surface.
14. A hot spring which throws steam and water into the air.
15. Sometimes formed when the roof of a cavern collapses.

a. watershed
b. oxbow lake
c. hydrologic cycle
d. water gap
e. tributaries
f. meanders
g. ground water
h. delta
i. divides
j. water budget
k. spring
l. zone of aeration
m. cavern
n. geyser
o. artesian well
p. sink
q. water table

QUESTIONS

Group A

Select the best term to complete the following statements. *Do not write in this book.*

1. The earth's present supply of liquid water (a) has always been present in the oceans (b) probably first appeared as liquid water beneath the earth's surface (c) is the result of solar winds (d) was first present as ice.
2. Of the water which enters the atmosphere as vapor each year, what part comes from the sea? (a) all of it (b) about 15% (c) about 80% (d) about 95%.
3. The process by which plants give off water vapor is called (a) dew (b) precipitation (c) evaporation (d) transpiration.
4. Of the water that falls on the land surface each year, the part which becomes ground water is about (a) 50% (b) 40% (c) 20% (d) all of it.

5. Which of the following would play the smallest part in balancing the water budget for a particular area? (a) evaporation (b) winds (c) run-off (d) precipitation.

6. Which industrial process requires the greatest amount of water? To produce or process (a) a gallon of milk (b) a pound of synthetic rubber (c) a gallon of gasoline (d) a pound of steel.

7. Which process uses the greatest amount of water per person in the United States? (a) washing and laundry (b) cooking (c) drinking water (d) industry.

8. The amount of water which runs off in streams or enters the ground water supplies in the United States is about (a) 1300 gallons daily (b) 1,300,000,000,000 gallons yearly (c) 1,300,000,000,000 gallons daily (d) 1300 gallons yearly.

9. Water loses all its energy due to gravity only after it (a) falls as rain (b) reaches a river (c) is in a reservoir (d) reaches sea level.

10. The wearing of the land at the head of gullies which causes them to lengthen and reach up the slopes is called (a) headward erosion (b) tributaries (c) watershed (d) dividing.

11. Stream piracy involves (a) one county stealing another's streams (b) conservation (c) erosion of the divide between two streams (d) building dams.

12. Which of the following factors least affects the rate of stream flow? (a) slope of a stream bed (b) gradient of a stream bed (c) load of a stream (d) steepness of a stream bed.

13. As a stream grows older its gradient usually (a) becomes greater (b) becomes less (c) remains the same (d) becomes a channel.

14. Common features of a youthful stream usually *do not* include (a) large amounts of water (b) a V-shaped valley (c) waterfalls (d) many tributaries.

15. A stream whose gradient has been increased due to movements of the earth's crust is said to be (a) youthful (b) mature (c) rejuvenated (d) terraced.

16. When the stream running in a water gap is captured by another stream running parallel to the mountains, the water gap becomes a (a) terrace (b) wind gap (c) youthful gap (d) mature gap.

17. A bowl-shaped cavity worn in a stream bed by rocks swirling at that location for some time is called a (a) pothole (b) sink hole (c) rock hole (d) pithole.

18. Which of the following is *not* primarily related to looping curves in streams? (a) oxbow lakes (b) meanders (c) load (d) bars.

19. The deposit built when a stream reaches a level surface at the foot of a mountain is called (a) a stream delta (b) an alluvial delta (c) an alluvial fan (d) a stream fan.

20. Which of these is built in an attempt to control flooding by rivers? (a) flood stages (b) flood plains (c) deltas (d) artificial levees.

21. The region in the ground where all of the spaces between rock particles are filled with water is called the (a) zone of saturation (b) zone of aeration (c) capillary fringe (d) flood zone.

22. The ability of rock to allow liquids to enter is called (a) aeration (b) permeability (c) capillarity (d) saturation.

23. A porous water layer from which supplies of underground water may be drawn is called (a) a spring (b) a well zone (c) an artesian zone (d) an aquifer.

24. A form of calcite deposited by hot springs is (a) geyserite (b) travertine (c) pyrite (d) quartz.

25. A silicate mineral deposited by hot springs is (a) geyserite (b) travertine (c) pyrite (d) quartz.

26. Which of the following is *not* commonly found on a karst plain? (a) ponds (b) sink holes (c) surface streams (d) lakes.

27. A deposit of calcite hanging from the roof of a cavern is called a (a) stalagmite (b) stalactite (c) column (d) limestone icicle.

Group B

1. Describe the hydrologic cycle.

2. Name two factors which determine how fast water flows into a well.

3. What might cause the rejuvenation of a stream?

4. Explain how stalactites and stalagmites are formed.

5. What are several characteristics of a mature river?

6. Why do people inhabit flood plains when there is so much danger from floods?

7. At what time of year would you drill a well to a depth that would most likely provide a year-round supply of water?

8. What is meant by the water budget? Why might it not be balanced for a particular place on the earth's surface?

9. Explain how meanders are formed.

10. What are hot springs and how are geysers related to them?

11. Name several features of a youthful stream and describe how each changes with maturity.

12. How has man increased the possibility of floods? How has he decreased the possibility of floods?

13. Name several factors which affect the rate at which a stream channel is widened and deepened.

14. Would it be possible for a tributary to form a delta? Explain.

15. Explain briefly the relationship among the zone of aeration, the zone of saturation, the capillary fringe, and the water table.

16. How are oxbow lakes formed?

17. Why is long term prediction of stream flow difficult?

18. Would alluvial fans or deltas be composed of coarser materials? Explain.

19. How could you as an individual conserve water? How much water would be conserved by 3 billion people (the earth's population) if they all practiced this conservation idea? Do not suggest anything that would damage your health.

About ten thousand years ago early humans witnessed the end of an important chapter in Earth's history. Great sheets of ice that had covered about one-third of the earth's surface began to melt. Land that had been buried under thick ice layers for thousands of years began to be exposed. The blankets of ice had covered much of the continents of the Northern Hemisphere. In North America an ice sheet had reached as far south as what is now New Jersey, Ohio, Illinois, Kansas, Montana, and Washington.

Today the landscape throughout much of the world, and particularly in the northern United States and Canada, shows the effects of these great bodies of ice. Evidence contained in older rocks indicates that the earth was partly covered by ice at least several times in its ancient past. The ice sheets that retreated only a few thousand years ago were the last of a series of ice ages the earth has experienced. This chapter is concerned with the way the ice left its marks on the rocks.

ORIGIN AND MOVEMENT OF GLACIERS

Formation of glaciers. Water which remains on the earth's surface as a liquid quickly runs off into lakes, streams, and rivers, and finally back to the sea. However, if the average temperature is near or below the freezing point, the water falling on the surface freezes and remains as solid ice. Water accumulates from one year to the next in thick layers of snow. The great weight of the snow presses against the layers beneath. This pressure, along with some melting and freezing again, changes the snow into small grains of ice. The grainy ice thus produced is

objectives

- [ ] Explain how glaciers form.
- [ ] Describe the two kinds of glaciers presently found on earth.
- [ ] Describe how glaciers erode the land.
- [ ] List the two general kinds of glacial deposits.
- [ ] List the conditions necessary for a lake to form.
- [ ] Describe the life history of a lake.
- [ ] Describe the conditions that existed during the ice ages.
- [ ] List several possible explanations for the cause of the ice ages.

275

called *firn* or *névé* (nay-vay). In the lowest layers of the accumulated snow, the pressure is so great that the firn becomes a solid mass of ice. Each year more snow is added to the top of the ice layers. Eventually the weight becomes great enough to cause parts of the body of ice to move slowly over the earth's surface. The moving ice is called a *glacier.*

Today relatively few places have low enough average temperatures along with sufficient snowfall for glaciers to form. Only in the polar regions can ice survive all year at sea level. Another area is above the snow line in high mountains near equatorial, middle, and northern latitudes. The *snow line,* the elevation above sea level where snow remains all year, is higher near the equator and lower near the poles. If the amount of snow and ice that accumulates during a year is greater than the amount that melts during the warm season, an *ice field* (or snow field) will be formed. Ice fields cover all lands very near the poles and are found in mountains at lower latitudes. See Figure 13–1.

Interpret

Does the elevation of the snowline vary directly or indirectly with the latitude of the place in question? Explain.

FIG. 13–1. A part of the Columbia ice field in the Canadian Rockies is visible as the layer of snow and ice across the distant mountains. (Ramsey)

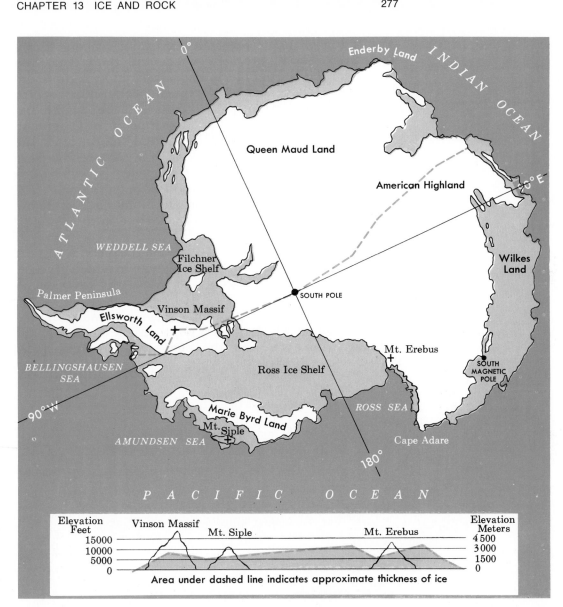

When ice moves out from a mountain ice field, an *alpine* or *valley* glacier is produced. These can be small glaciers or huge rivers of ice that fill an entire valley.

A second type of glacier found only in the polar regions is a continuous sheet of ice called a *continental glacier*. This type of glacier covers the entire land surface except for the tops of the highest mountains. Antarctica is covered by the largest of all existing continental glaciers. It is one and a half times as large as the United States (not including Alaska and Hawaii). Over much of its area, the

FIG. 13–2. Antarctica. The darker color represents the oceans surrounding the continent. The lighter color shows sea areas covered by ice. The white areas are land masses, also covered by ice. The dotted line in the drawing above shows the thickness of ice at varying elevations.

Antarctic ice cap is more than 3000 meters (about 10,000 feet) thick. See Figure 13–2. Greenland, the largest land body near the North Pole, is also buried under a continental glacier. It covers about 80 percent of the entire surface of this large island. Only small fringes of the land around the coast are exposed. At its thickest point the Greenland glacier is about 3000 meters in depth. If all the ice in the two great glaciers of Greenland and Antarctica were melted, it is estimated that the water released would raise the level of the sea by more than 60 meters (about 200 feet). This would be high enough to submerge completely all the major cities along every coastline in the world!

Movement of glaciers. Glacial motion has been a subject of much scientific study. The exact way in which glaciers move is still not completely understood; however, some principles are clear.

The advancing front wall of a glacier is called the *ice front*. In alpine glaciers the ice front moves as a result of pressure from above and from being on a sloping surface.

FIG. 13–3A. Movement of ice in an alpine glacier.

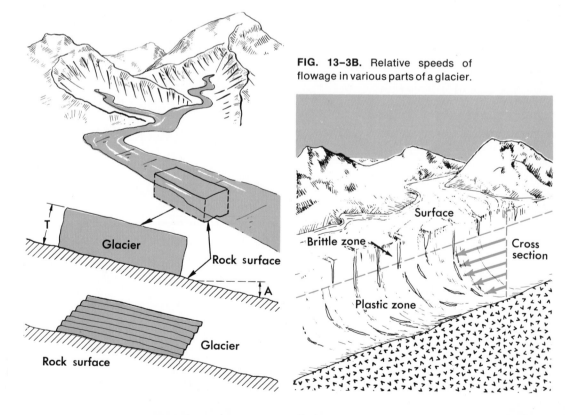

FIG. 13–3B. Relative speeds of flowage in various parts of a glacier.

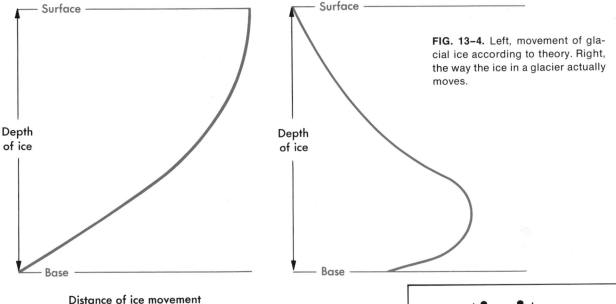

FIG. 13-4. Left, movement of glacial ice according to theory. Right, the way the ice in a glacier actually moves.

Think of a block of ice frozen to a sloping surface. The block of ice represents a glacier and the sloping surface, a valley floor. If the angle of the slope in Figure 13-3A is large enough (angle A) and the ice is thick enough (T), the ice should begin to move down the slope. The force that acts to pull it down is regulated by the angle of the slope. In glaciers, motion may also result from layers of ice sliding over each other.

Another explanation suggests that the bottom ice melts under pressure from the overlying weight, rolls forward on a thin film of water, and then refreezes. Still another possible way to explain glacial movement is that ice in the upper zone of the glacier is quite brittle. Below this brittle zone it is suggested that the ice may flow like melted plastic. See Figure 13-3B.

In the type of movements just described, the rate of ice flow plotted against depth would ordinarily produce a graph like that shown in the left part of Figure 13-4. But the actual measurement of the way glacial ice moves results in a different picture. See the right graph in Figure 13-4. The most rapid movement occurs at two-thirds of the distance down from the top of the glacier.

We can see then that a glacier does not just slide down a slope. Its movement is a combination of a sliding of upper

activity

The motion of silicone putty is very much like that of a glacier. Prepare a V-shaped trough by folding a 4 x 6 index card lengthwise and taping it to another index card. See the diagram below. Press the putty into the top of the trough and against the end card. Mold the putty so that it resembles a glacier. Press 5 small beads in a straight line across the putty as shown in the diagram. Prepare a data table to record how far each of the beads moved each hour for 5 or 6 hours. Compare your results to Fig. 13-5, page 280.

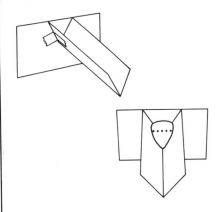

FIG. 13-5. The movement of different parts of an alpine glacier can be measured if stakes (dots) are driven into the ice. The crosses show the positions of the stakes at a later time.

ice layers and a plastic-like flow of deep layers. How rapidly any particular part of the glacier moves is determined by the slope, ice thickness, and temperature at that point. Thus not all parts of a glacier move at the same speed. By driving rows of stakes across the surface of an alpine glacier, geologists have determined that surface motion is most rapid near the center, decreasing toward the sides. The rate of movement is often one meter or more a day. See Figure 13-5.

The uneven surface movements of alpine glaciers cause *pressure ridges* to form in ice near the surface. This uneven pressure results in large cracks, called *crevasses* (kreh-*vas*-ez), appearing on the top and sides of the glacier. See Figure 13-6. Crevasses generally form across the width of the glacier. However, near the lower end they may develop in a lengthwise direction. See Figure 13-7A. Crevasses often extend more than 30 meters (100 ft) below the ice surface. The actual surface opening may be

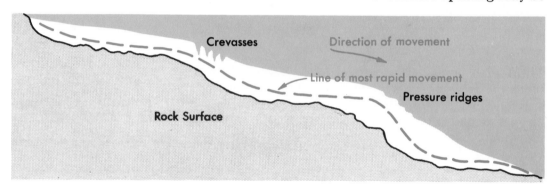

FIG. 13-6. Ridges and cracks (crevasses) appear on the surface of a glacier as a result of uneven movement.

FIG. 13-7A. A small glacier with many crevasses and pressure ridges. This type of glacier on a steep slope is often called a "hanging" glacier. (Ramsey)

hidden by a thin crust of snow which breaks at the slightest weight. Thus traveling over the top of a glacier can be
very dangerous, requiring experience and great caution.
See Figure 13–7B. Continental glaciers move as a result
of the continuous pressure of additional snow which is
added to their surface.

In an ice sheet there are usually centers from which
glaciers move outward in all directions. The ice sheets
which now cover Antarctica and Greenland, for instance,
have a general movement outward toward the shoreline.
Along the coast of Antarctica, the glacier has moved out
over the sea in places to form wide ice shelves. The rise
and fall of the tides are able to break off large sections of
this leading front of ice. These large blocks of ice then
float away as *icebergs*.

Glacial movement depends upon the balance between
ice being added by snowfall and ice lost by melting and
evaporation. As long as new snow is added faster than it
melts, the glacier will continue to advance. The glacier
will finally stop its forward movement and become stationary at the level where the ice melts as quickly as the
snow is added. Very slight differences in average yearly
temperatures and snowfall may upset this balance. The
glacier will then advance or retreat accordingly. Thus the
movement of a glacier is a sensitive indicator of climatic
temperature changes.

FIG. 13–7B. A mountain climber
shown scaling a treacherous glacial crevasse. (Springs)

THE WORK OF GLACIERS

FIG. 13–7C. A small round lake or
tarn is commonly found in the bowl
of a cirque. (U.S. Geological
Survey)

Glacial erosion. The formation of alpine glaciers cause
the landscape to become even more rugged. The glacial
processes which change the shape of the mountains begin
at the heads of the valleys in which the glaciers form.
Frost action breaks off rock from the valley walls, causing
them to become steeper. The glacier pulls loose blocks of
rock from the floor of the upper valley as it moves. These
actions create a bowl-shaped depression called a *cirque*
(surk). This French word, meaning "circus," refers to the
resemblance of the depression to a circus theater. See
Figure 13–7C.

When a number of cirques are formed close together,
the dividing ridges between become very sharp and jagged.
These ridges are called *aretes* (a-*rayts*) which means
"spines." See Figure 13–8. Several aretes may join to
form a sharp peak called a *horn*.

FIG. 13–8. The cycle of glacier formation and the development of glacial features: cirques, aretes and horn peaks.

FIG. 13–9. An outcrop of glacially polished rock photographed north of Lake Superior. A layer of till still covers the rock at the right. (Ramsey)

As the mass of ice grows in the cirque, the glacier slowly begins to move to lower levels along the general path of the existing valley. Large amounts of rock become mixed in the ice. The rock may range in size from microscopic pieces, up to large blocks broken off by frost action or by the force of the moving ice. These rock fragments are evidence that help explain how:

1. A glacier can erode even very hard rock over which it moves.

2. The rocks buried in the ice, in turn, act like the teeth of a file as they are dragged over the rock floor.

Solid rock over which a glacier has moved often shows a polished effect from the action of the tiny rock particles in the ice. At the same time deep scratches or grooves are left by the larger rocks. See Figure 13–9. A large rock projection may be smoothed into a glaciated rock *knob*. Such a knob usually has a smoothly sloping side in the direction from which the glacier came. The other side is steeper as a result of rock being plucked away by the ice. See Figure 13–10.

As a glacier moves through a valley, the valley walls and floor are ground away. This changes the original V-shape of the valley into a U-shape. See Figure 13–11. U-shaped valleys are considered direct evidence of glacial erosion. Glaciers are the only means by which a valley can acquire a U-shape. The movement of a glacier through a valley also straightens its sharp bends.

Glaciers in smaller, adjacent valleys may often join the main glacier. As the glaciers melt, these side valleys end in mid-air high above the main valley floor. They are called *hanging valleys*. See Figure 13–8. Streams flowing through them form steep waterfalls that drop to the main valley below.

Because of their great thickness and mass, continental glaciers completely override all existing land features except for the largest mountains. The landscape that results from erosion by continental glaciers is the opposite of the sharp and rugged features produced by alpine glaciers. Continental glaciers produce relatively smooth landscapes by grinding down mountains and all other relief features. See Figure 13–12. In a few cases existing valleys may be gouged out and deepened, especially if the direction of ice movement is parallel to the valley. In most places, however, the land surface is smoothed.

Glacial deposits. The ice of a glacier may melt as a result of higher temperatures. This can happen when an alpine glacier reaches lower altitudes or when a change in climate melts the ice sheets of continental glaciers. When a glacier melts, all of the material accumulated by the ice is deposited. This forms various kinds of features which fall under the general name of *glacial drift*. It is found in two completely different deposits:

FIG. 13–10. Lembert Dome is a granite glacial knob in Yosemite National Park. The glacier flowed from the right, producing a smoothly rounded surface, then plucked off the rock at the left to create a steeper slope. (National Park Service)

FIG. 13–11. A glaciated valley in the Sierra Nevada Mountains. (Ramsey)

FIG. 13–12. These two photographs show the difference in the kind of landscapes produced by alpine and continental glaciers. One is a view of the Canadian Rockies; the other was taken in northern New York. (Ramsey)

FIG. 13-13A. Glacial till exposed along a road in New England. (Ramsey)

FIG. 13-13B. A large glacial erratic boulder deposited on top of Mt. Mansfield, Vermont. (Mary S. Shaub)

FIG. 13-14. The joining of lateral moraines to form medial moraines is clearly seen in this view of an Alaskan glacier. (Rapho-Guillumette, Lowry)

1. *Till* is the unsorted material deposited directly from the ice. It is either scraped from the bottom of the glacier or left behind when melting takes place. See Figure 13-13A.

2. *Stratified drift* is material that has been sorted and deposited in layers by the action of melt-water flowing out from subglacial streams.

Usually it is easy to spot glacial deposits. The visible rock surfaces are polished and scratched. Large boulders, called *erratics*, may be present. See Figure 13-13B. These are carried by the ice and have an entirely different composition from the bedrock.

To distinguish between till and stratified drift, it is necessary to dig into a deposit. Till is characterized by small rocks, gravel, sand, and clay all mixed together. Stratified drift is characterized by layers of rounded rocks, pebbles, and clay. When highways are cut through hills in glaciated regions, the nature of the deposits is often exposed.

Glacial deposits of till and stratified drift are responsible for many characteristic landscape features. Some of the most common of these features are described in the following sections.

Features formed by till. The material carried along by the glacier often forms ridges or mounds on the ground or on the glacier itself. These deposits are called *moraines*. Accumulations of rocky debris along the edges of a valley glacier in a string of long ridges are called *lateral moraines*. Joining of two or more valley glaciers brings the

lateral moraines together to form *medial moraines*. See Figure 13–14.

The bottom of a glacier may become so loaded with rock material that all of it can no longer be carried. When this happens, part of this load is deposited below the ice. It is then passed over as the ice moves along. All the material left beneath the glacier, along with that deposited directly from the ice when melting takes place, makes up the *ground moraine*. After the glacier has disappeared, the surface of the ground moraine is left with a gentle, hilly appearance. Much of the landscape from eastern Ohio west to the Rockies and north into Canada was formed by ground moraine.

At the melting edge of the glacier, a large moraine is usually formed. It is produced by material pushed along ahead of the moving ice or dumped by melting ice at the glacier's edge. This end deposit is called a *terminal moraine*. See Figure 13–15. In glaciated regions old terminal moraines are often seen as belts of small hills with many hollows that may contain lakes or ponds. Large terminal moraines are found in Minnesota, the Dakotas, Wisconsin, northern Illinois, Indiana, and Ohio.

A common feature of areas covered with ground moraine are *drumlins*. These are long, low mounds of till. They are rounded in shape and are often found in clusters lying with their long axis parallel to the direction of glacial movement. See Figure 13–16.

FIG. 13–15. An alpine glacier in Alaska with its terminal moraine and outwash plain. (U.S. Geological Survey)

Illustrate
Draw a diagram which shows the location of each type of moraine as they would be found in a valley glacier.

FIG. 13–16. A series of drumlins near Rochester, New York. (E. R. Degginger)

Investigate
Check your local library for books on the geology of your area. See if you can find any evidence of local glacial activity. Look for examples of glacial deposits that you will recognize from your text illustrations.

Features formed by glacial meltwater. The melting of a glacier is an almost continuous process. Streams of meltwater flow from the edges and surface of the glacier, particularly during the warmer parts of the year. The glacial meltwater usually has a milky appearance, caused by the presence of very fine rock particles. This powdered rock is produced by the grinding action of rock against rock as the glacier moves. Along with the small rock particles, the meltwater carries enough drift to deposit a large *outwash plain* in front of the glacier. The plain is usually crossed by many streams carrying the meltwater.

The outwash plain is often pitted with many depressions called *kettles*. These are formed when pieces of ice break off from the glacier and are covered with drift. As the ice melts, the drift sinks and produces a depression. Kettles often fill with water to make ponds.

When the continental glaciers disappear, long, winding ridges of gravels and coarse sand may be left behind. These are called *eskers*. They are ridges of stratified drift deposited by a stream of meltwater flowing in tunnels within or beneath the glaciers. Eskers may extend for some distance over flat glaciated country, looking like a winding raised roadway. Glacial features such as eskers, produced by meltwater, are especially common where continental glaciers were melting and shrinking at the same time.

LAKES

Origin of Lakes. Any erosion process or other event that causes a depression in the earth's surface can produce a lake. The only other requirement is a plentiful supply of water.

Glaciers, a common source of basins, created lakes in several ways. As they moved over the surface, they gouged out depressions in low level areas. These filled with water and became lakes.

Hundreds of the lake basins in New England and New York were dug from solid rock by the continental glacier which lay over the region. The Finger Lakes of central New York were formed from existing stream valleys. These valleys were deepened by the ice, then dammed by glacial deposits. Almost every glacial cirque which is now exposed is occupied by a small lake called a *tarn*.

However, most of the lake basins were formed by deposits from glaciers rather than by glacial erosion. Many such lake basins were left in the uneven surface of the ground moraine that covers much of northern Europe and upper North America. The lakes were filled with water from melting ice. If they were below the level of the water table, rainfall and ground water helped to fill them.

Other glacial lakes came from terminal and lateral moraines damming existing streams. The belts of terminal moraines across Minnesota and the Dakotas produced many such lakes. Similar belts of moraines and associated lakes are found in Wisconsin, Indiana, Ohio and northern Illinois.

The Great Lakes are the direct result of a combination of glacial actions. These lakes were formed from existing broad river valleys covered by ice. Glacial erosion widened and deepened them. As the glacier melted, the meltwater flowed into these basins. The water, trapped by moraines to the south, formed dams. In their early stages, the lakes had outlets to the south through the Wabash and Illinois rivers, then to the Mississippi. However, with further melting of ice, the lakes grew in size. This caused them to drain to the Atlantic through the Susquehanna River of Pennsylvania and western New York. Later the lakes also drained through the Mohawk and Hudson valleys.

The Great Lakes grew in size until they were slightly larger than they are now. Then, as the glaciers retreated and the weight of ice lessened, the land surface was pushed up as the pressure decreased. The lake-beds were

FIG. 13–17. Steps in the formation of the Great Lakes. Note how the drainage pattern changed as the glacier retreated.

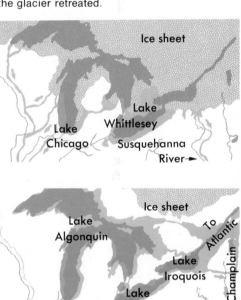

FIG. 13–18. The former glacial Lake Agassiz, the extent of which is outlined in light blue, once covered a very large area in Canada, North Dakota, and Minnesota.

Investigate

Our earliest evidence concerning Ice Age Man, has been found in bog sediment of ancient glacial lakes. Refer to the book "Bog People: Ice Age Man Preserved" by R. L. Bruse-Milford.

also uplifted, reducing them to their present size. This land shift also established final drainage to the north through the St. Lawrence River. This brief summary of the complicated history of the Great Lakes is shown in Figure 13–17.

A large glacial lake that formed at about the same time as the Great Lakes, covered parts of North Dakota, Minnesota, and the Canadian province of Manitoba. This was called Lake Agassiz (*Ah*-guh-see). It developed when the meltwaters from the continental glacier were blocked from their northward flow by an ice dam. When the ice dam melted, the lake disappeared. This body of water once covered an area greater than all the Great Lakes combined. See Figure 13–18. Lake Winnipeg in Canada is the chief remnant of the former giant Lake Agassiz.

The life history of lakes. Most lakes are relatively temporary features of a landscape. Even the largest is likely to exist for a much shorter time than many other features. There are two reasons for the rapid disappearance of lakes. First, the lake may lose its water, leaving the basin empty. Second, the lake basin itself may be destroyed.

Lakes often disappear because their water drains away. The most common cause is erosion by an outlet stream. If the outlet stream cuts its bed down below the level of the lake floor, all of the water in the lake will drain out. A lake may also lose its water because a change in climate brings dry conditions to the region. Evaporation of the water then causes the lake to disappear.

Lake basins may be destroyed by an overload of sediment. This sediment is carried in by the streams feeding the lake. The inflowing streams build deltas which grow toward the center of the lake. See Figure 13–19. Some of the sediment is washed down from neighboring slopes or by the action of waves cutting into the lake basins. Wind blows dust into the water and landslides add their share of soil. Plants growing around the lake's edges advance farther and farther into the lake. Then the lake becomes a bog or marsh. Finally the basin fills completely and becomes a meadow. See Figure 13–20.

If a lake has no outlet stream, it may become a *salt lake*. Salt lakes are similar to the sea. Since water can leave only by evaporation, the dissolved substances are left behind, making the water increasingly salty. Any lake will become salty if it is naturally without an outlet or is deprived of its outlet. If it is also located in a dry region

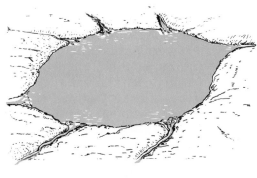

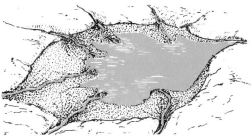

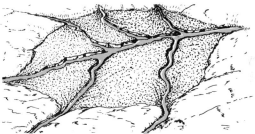

FIG. 13–19. Three stages in the filling of a lake by deposits of stream sediments.

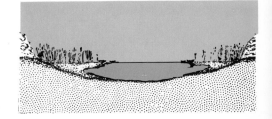

FIG. 13–20. A small lake or pond may fill slowly or be converted into a bog by the growth of vegetation around its edges.

where evaporation is rapid, it has more of a chance of becoming salty. The composition of the water in salt lakes depends upon dissolved minerals brought in by streams and by the chemical and biological processes which take place in the lake. Substances commonly found dissolved in the waters of salt lakes include sodium chloride, and the chlorides of calcium, magnesium and potassium as well as the sulfates of sodium, calcium, and magnesium. The salt deposits left in the beds of extinct salt lakes often contain valuable minerals. For example, borax is mined from such deposits.

THE ICE AGES

Glacial periods. For unknown reasons, the earth passes through periods when the average temperature everywhere on the surface becomes lower. These periods are the ice ages or glacial periods when great continental ice sheets spread outward from the poles. There is evidence that there have been at least several ice ages during the earth's long history. An ice age seems to have taken place about 600 million years ago and another about 275 million years ago. But the most recent ice age occurred during the last one million years. Its great ice sheets retreated only during the last ten to fifteen thousand years. The areas of the Northern Hemisphere covered by glaciers during this period are shown in Figure 13–21. All of Canada and the mountainous regions of Alaska were buried under the ice sheets. In the western mountains of the United States, alpine glaciers gathered into a single ice sheet that flowed west to the Pacific and eastward to the foothills of the Rocky Mountains. A great continental glacier with its center in the Hudson Bay region spread as far south as the Missouri and Ohio rivers.

Europe was almost covered by an ice sheet centered over the Baltic Sea. It spread south to Germany and the Low Countries, and west to the British Isles. In the east the glaciers reached Poland and Russia. Alpine glaciers grew in the Alps in southern Europe and in the mountains and highlands of Siberia. Large glaciers were also formed in the Southern Hemisphere. The Andes Mountains of South America supported a large ice sheet as did the South Island of New Zealand.

The features built on the land surface by the glaciers of the most recent ice age are still fresh, and easily found. Studies of the landforms show four separate cycles of growth and retreat of glaciers during this period. These cycles seem to have been the result of brief periods of extreme cold when the glaciers advanced. Relatively long periods of warmer climate followed when the ice sheets retreated.

Possible causes of the ice ages. There is no clear explanation of the climate changes which cause ice ages. For glaciers to grow, temperatures must drop, or the amount of moisture falling as precipitation must increase, or both must happen at the same time. A number of theories have been proposed to account for the conditions necessary to produce an ice age. A brief description of some of these theories will show the general thought and the problems involved:

1. *Changes in the amount of energy from the sun.* The amount of solar energy received by the earth might change due to several factors. One would be a small change in the tilt of the earth's axis and a slight difference in the earth's orbit around the sun. Changes in the earth's motions in space could cause a slight decrease in average temperature on the earth's surface. Such changes in the tilt of the earth's axis and shape of its orbit are known to occur in 21,000- to 90,000-year cycles. However, the glacial periods have come at irregular intervals and are

much more widely spaced in time. More than 300,000 years have gone by between some stages of the last ice age.

The amount of energy given off by the sun might vary. While the evidence is not certain, there is reason to think that the amount of solar energy produced does change slightly from time to time. However, a great deal more information about the behavior of the sun needs to be collected over a long period of time. A change in the sun's energy output, combined with other factors, might cause periods of glaciation on the earth.

2. *Topographical factors.* It may be that the ice ages are caused entirely by events which take place on the earth itself. For example, at least two of the glacial periods have occurred during times of mountain building. High mountains above the freezing level (snow line) act as snow-collecting areas and that might gradually help to reduce

FIG. 13-21. At its peak the last great glacial period covered about 30 per cent of the earth's land area. The glaciers (1) and ice pack covering the sea (2) are shown at their height. The solid lines mark present day sea coasts and lake shores.

the earth's temperature. This in turn would produce more snow and ice eventually leading to a glacial period. In addition, volcanic activity connected with mountain building might help to bring about cooler temperatures. Volcanic dust, thrown into the air, would shade the earth's surface and reflect solar radiation. Elevation of the earth's crust, particularly near the poles, might also help a glacier to form.

However, all of these topographic changes in the land surface require long periods of time. The glaciers seem to form and disappear relatively quickly. Further, it is difficult in this theory to account for the increased snowfall needed to create the large continental glaciers.

3. *Melting of the arctic ice.* Part of the Arctic Ocean is now covered by a thick layer of ice. Under present climatic conditions, this ice remains frozen. However, calculations show that if a large amount of warmer water from the North Atlantic were to enter the Arctic Ocean, the ice would melt. This does not happen now because a ridge on the sea floor keeps the two oceans from completely mixing. However, if the level of the Atlantic were to rise, water could pass over this barrier. Theoretically this is exactly what would be needed to produce an ice age.

If the Arctic Ocean were free of ice, more water would evaporate into the polar atmosphere. This would cause heavier snowfall on the continents bordering the north polar regions. The increased snowfall could possibly result in new glacier formations. But the trapped water in the glaciers would eventually cause the ocean levels to drop again. This would again result in separating the Arctic and Atlantic oceans. The Arctic Ocean would then freeze over again. The glaciers would be cut off from their supply of moisture and would begin to disappear. So the cycle of advance and retreat of the ice sheets would be complete.

The advantage of this theory is that there is no need to explain any basic change in the earth's position or motion to account for the ice ages. However, there is no evidence that removal of the ice in the Arctic Ocean would cause formation of the great glaciers. Also, the circulation of water in the oceans is not well understood. It would be difficult to establish whether mixing of Arctic and Atlantic water could actually take place on a large scale.

Many more questions need to be answered before a satisfactory explanation of the advances and retreats of the glaciers can be found.

VOCABULARY REVIEW

Match the word or words in the column on the right with the correct phrase in the column on the left. *Do not write in this book.*

1. Grainy ice formed from snow.
2. A large body of moving ice.
3. Large cracks in a glacier.
4. A chunk from a glacier "floating" in the sea.
5. Bowl shaped depression in which a glacier grows.
6. Material deposited directly from a glacier as it melts.
7. Large boulders left by a melting glacier.
8. Ridges or mounds formed on a glacier by the material it carries.
9. An elongated mound usually formed by continental glaciers.
10. Depressions found in an outwash plain.
11. A cirque filled with water.
12. Formed by continuous evaporation from a stationary body of water.

a. drumlin
b. salt lake
c. till
d. firn
e. tarn
f. cirque
g. eskers
h. kettles
i. crevasses
j. aretes
k. glacier
l. moraines
m. iceberg
n. erratics

QUESTIONS

Group A

Select the best term to complete the following statements. *Do not write in this book.*

1. For water falling on the land to remain for long periods of time very near where it falls, it needs to (a) fall as snow (b) fall as hail (c) fall where the temperature on the ground is at or below freezing (d) be absorbed by a plant.

2. Portions of ice fields become a glacier when (a) movement of the ice begins (b) the spring thaws occur (c) snow continues to fall (d) the temperature remains below freezing.

3. The elevation in mountains above which snow remains all year is called the (a) frost line (b) ice line (c) snow line (d) glacier line.

4. An ice field forms when (a) snow falls at least once each year (b) more snow falls than melts each year (c) an equal amount of snow falls and melts each year (d) a crust of ice forms on deep snow.

5. When ice moves out from a mountain ice field, (a) a continental glacier is formed (b) an iceberg is formed (c) an esker is formed (d) an alpine glacier is formed.

6. At the present continental glaciers are found only (a) on the North American continent (b) in the Southern Hemisphere (c) in the Antarctic (d) in the polar regions.

7. The movement of an alpine glacier is best described as (a) a combination of sliding motion for top layers with a flowing motion for lower layers (b) only a sliding motion (c) only a flowing motion (d) a combination of flowing of upper layers with sliding of lower layers.

8. A direct evidence of the uneven movement of glaciers is the formation of crevasses and (a) icebergs (b) pressure ridges (c) terminal moraines (d) drift.

9. Glaciers are a sensitive indicator of climate changes since they are (a) found where most weather occurs (b) found high in the mountains (c) in delicate balance between addition of ice by snowfall and loss of ice by melting and evaporation (d) easily checked for average depth which is an indicator of temperature change.

10. A sharp peak formed by the joining of several aretes is called a (a) cirque (b) horn (c) till (d) tarn.

11. Hanging valleys, such as those found surrounding Yosemite Valley in Yosemite National Park in California, were formed (a) by water erosion (b) by wind erosion (c) by meltwater from a continental glacier (d) by several small glaciers joining one larger glacier.

12. The erosional features caused by alpine and continental glaciers (a) are different in many ways (b) are very similar (c) occur only while melting (d) are caused by the ice itself.

13. The material deposited by a glacier is called (a) drift (b) arete (c) cirque (d) kettle.

14. Material deposited in layers by meltwater is called (a) erratic (b) moraine (c) stratified drift (d) till.

15. Glacial deposits can frequently be recognized since visible rocks are (a) jagged and sharp (b) weathered and dark in color (c) often polished and scratched (d) rounded and smooth.

16. An accumulation of rocky debris along the edges of a valley glacier often forms long ridges called (a) lateral moraines (b) medial moraines (c) ground moraines (d) terminal moraines.

17. Material deposited beneath a glacier is called a (a) lateral moraine (b) medial moraine (c) ground moraine (d) terminal moraine.

18. Long ridges made of gravels and coarse sand exposed when a continental glacier melts are called (a) kettles (b) lateral moraines (c) medial moraines (d) eskers.

19. Glacial features produced by meltwater are especially common (a) where valley glaciers are melting (b) where continental glaciers melted (c) near where valley glaciers formed (d) in the polar regions.

20. Which of the following is a feature most commonly produced by glaciers? (a) waterfalls (b) V-shaped valleys (c) lakes (d) plateaus.

21. Any erosion process or other event that causes a depression can produce a lake if (a) rainfall is plentiful (b) a glacier melts to give it water (c) the water table of the surrounding land is higher than the depression (d) any of these may produce the lake.

22. Almost every glacial cirque that is now exposed is occupied by a small lake called a (a) tarn (b) till (c) finger lake (d) basin.

23. Final drainage of the Great Lakes to the north happened when (a) dams were erected on the Wabash and Illinois Rivers (b) the continental glacier eroded a new path (c) the continental glacier melted, allowing the surface to bulge upward (d) the Great Lakes took on extra water.

24. Lake Winnipeg in Canada is the remnant of the former giant (a) Lake Superior (b) Lake Agassiz (c) Lake Erie (d) Lake Iroquois.

25. Most lakes are relatively temporary features of a landscape, they generally disappear when (a) an outlet river cuts a channel draining the lake (b) the lake basin is destroyed by becoming filled with sediment from incoming streams (c) landslides fill the basin (d) all of these are ways a lake is destroyed.

26. A salt lake forms when (a) water drains from mountains rich in minerals (b) not enough rain falls (c) the lake has no outlet stream (d) the lake has no inlet streams.

27. The most recent ice age occurred about (a) 600 million years ago (b) 275 million years ago (c) 1 million years ago (d) 1 thousand years ago.

28. The last ice age produced a polar ice sheet that (a) covered all of Canada and Alaska (b) extended south to Germany (c) covered the British Isles (d) all of these are correct.

29. In order for glaciers to grow (a) temperatures must drop (b) more precipitation must form (c) temperatures must drop or more moisture must precipitate (d) any one of these will cause a glacier to grow.

30. The ice ages were caused by (a) a small change in the tilt of the earth (b) excessive mountain building (c) there is no correct explanation at this time (d) melting of the Arctic ice pack.

Group B

1. Describe how a glacier is formed.
2. In what general locations would you expect to find glaciers at present?
3. What are the two main types of glaciers?
4. Describe the kinds of motion shown by the parts of a typical glacier.
5. What causes crevasses and pressure ridges in glaciers?
6. How do rocks become mixed with glacial ice?
7. How does the movement of a glacier through a valley change the shape of the valley?
8. How does the erosion caused by continental glaciers differ from that caused by alpine glaciers?
9. How can one tell the difference between till and stratified drift?
10. Name the kinds of moraines and describe how each is formed.
11. Describe three features formed by glacial meltwater.
12. In what ways are lakes formed by glaciers?
13. What reason can you give for the main drainage of the Great Lakes being north through the St. Lawrence River?
14. In what ways might a lake lose its water?
15. What processes aid in the filling of a lake bed with sediment?
16. What areas were covered by glaciers in North America during the last ice age?
17. What are three popular theories about the causes of ice ages?
18. Why is it said that no present theory can really explain what causes ice ages?

14
shorelines

objectives

- [ ] Describe four ways that rock is weathered by sea waves.

- [ ] List several examples of formations made by wave erosion.

- [ ] Describe the makeup of a typical beach.

- [ ] Describe the movement of materials along a shore.

- [ ] Describe the origin of shoreline features.

- [ ] Explain how the ice ages have caused changes in the shoreline.

All parts of the earth's surface are constantly exposed to natural forces. These forces are responsible for the processes that create change. These processes generally carry out their work of reshaping the earth's landscape slowly and quietly. But this is not the case where land and sea come together. The world's shorelines form a boundary zone separating land and sea. Here change is often very rapid. The reason for the unsettled conditions in this area is clearly visible. The energy stored in waves, accumulates far out at sea. This energy is released by the constant flinging of water against the shore. The margins of the land never escape the continual pounding. Rocks must bear the brunt of this energy contained in waves. The result is that the rocks are eventually reduced to fragments. Wave action then shifts the small rock fragments around to produce features which are characteristic of shorelines. A shoreline is always a very temporary result of the ability of waves to erode the rock, then transport and rearrange the fragments.

THE ATTACK ON THE SHORE

Wave erosion. Mechanical weathering resulting from sea water being tossed against rock formation by wave action, will attack the rock in four ways. (1) Water and air forced into cracks in the rock exert enough pressure to break off whole blocks; (2) fragments of rock thrown against the shore cut into the solid rock; (3) as waves cut back into the shore, rocks fall and grind together, and (4) small fragments of rock rub against each other when they

296

are suspended in the tumbling water. Chemical weathering of rock along the shore above the zone of wave action may also be rapid due to the presence of salt water.

When waves strike directly against the rock of the land, erosion can produce a *wave-cut cliff*, such as that shown in Figure 14–1. Wave erosion continues to attack the base of the cliff until a notch is finally cut. Eventually the undercutting of the cliff causes the overhanging rock to break away. Rock debris accumulates at the base of the cliff. This material is slowly ground up by wave action. As this process continues, the cliff is gradually worn back.

How rapidly rocks are worn away depends upon the nature of the rock material. Old maps are evidence that parts of the coast of England have retreated several kilometers since the time of the Romans. The shore along the coast of Cape Cod retreats inland at the rate of several meters each year. Both these coasts are composed of weak rock material. Shorelines made of harder rocks can slow the effects of wave action.

As a wave-cut cliff is worn back, a platform is usually left extending out beneath the water at the base of the

FIG. 14–1. Cliffs along the Oregon shoreline being cut back by wave action. A shelf or wave-cut terrace is partially exposed in the foreground. (Atkenson)

FIG. 14–2. The formation of a wave-cut terrace by wave erosion, and a wave-built terrace by deposition at the base of a cliff.

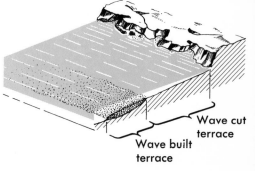

Wave cut terrace

Wave built terrace

FIG. 14–3A. As waves cut back into a section of the Oregon coast, these stacks were left standing. (Ramsey)

FIG. 14–3B. Sea-arches are visible near the shoreline off the coast of Portugal. (Mercier)

cliff. This platform is called a *wave-cut terrace*. See Figure 14–2. As wave action cuts into the terrace, much of the ground-up rock is carried out to sea. It may then be dropped some distance from shore, building an extension to the wave-cut terrace called a *wave-built terrace*. The finest particles are probably carried beyond the wave-built terrace and scattered over the sea floor by ocean currents.

After a time, the width of the terraces becomes very great. The waves must then lose most of their energy in the shallow water some distance out from the base of the cliff. When this occurs, the rate of wave erosion of the cliff will be greatly reduced. Wave erosion may even be halted unless wave action or currents erode the terraces and again expose the cliff to the force of the waves. A rise in sea level may also allow wave erosion to begin again.

As a cliff is eroded, it seldom wears away evenly. Softer parts weather rapidly and leave more resistant rocks standing for a time. Rocks with a number of joints or cracks are also more vulnerable to wave erosion. Unequal retreat of the cliffs often leaves narrow projections or headlands in their places. If erosion is extended to the sides of these projections, *sea arches* or isolated columns called *stacks* may be created. See Figure 14–3A&B. When erosion penetrates deeply into a cliff along a joint, the formation of a large hole or *sea cave* is likely to result.

Beaches. Along any shoreline the rocks which make up the land are covered with a layer of very small fragments. This layer is produced by wave erosion which always works to grind up the rocks at the boundary between land and sea. A *beach* is made of the layers of rock fragments carried toward the land by wave action. The size and kind of rock fragments found along the beach differ widely. The various sizes of rock fragments found along beaches are given in Table 14–1.

The composition of beach material depends upon its source. For example, granitic rocks yield light colored fragments. These are often found in the form of a coarse sand composed of quartz and feldspar. Beaches made of this material are common along the North American coasts. But other rocks along the shore, such as basalt, yield dark fragments. In many locations the most common sources of sand for beaches are rivers. Thus much of the sand found along beaches may have come from sources far up the channels of nearby rivers.

Although beaches may differ a great deal in the size and color of their rock fragments, their materials always undergo the same type of movements. All sizes of rock fragments which make up beach materials are moved onshore and offshore by wave action. Although the movement of all beach materials is influenced by the same erosional activity, the smaller the particles, the more readily they move about.

In the following discussion only the motion of sand grains will be considered. The individual sand grains are moved slightly forward with each succeeding wave. Several thousand waves each day can move sand grains a good distance, though each advance is very small. The shoreward motion by small waves piles up sand on the shore producing a sloping platform of sand. This platform extends from the farthest reach of the waves at high tide to the beginning of the land itself. This inner portion of the beach is called the *berm*. See Figure 14–4. To most people, the berm is the beach on which they rest and play.

When the waves are large, the movement of sand on the beaches is changed. Larger waves hold a heavy load of suspended sand until the onrushing water carries the sand up over the berm and deposits it on top. Thus large waves can increase the height of the berm. But when the mass of water pours back down, it picks up sand and moves it seaward. This action raises the slope of the berm and great quantities of sand are carried offshore. When the currents

Table 14–1 Beach Materials

Description	Size (millimeters)
Boulders	More than 200 (over 8 in)
Cobbles	76 to 200 (3 to 8 in)
Gravel	
Coarse	19 to 76
Fine	5 to 19
Sand	
Coarse	2 to 5
Medium	0.4 to 2
Fine	0.07 to 0.4
Silt	less than 0.07

FIG. 14–4. A berm developed along this beach; it is seen along the right side of the photograph. (Ramsey)

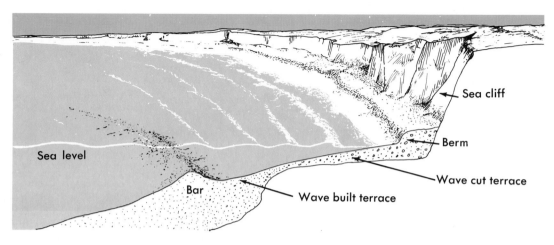

FIG. 14–5. Features that might be found at various times along a beach are shown in this diagram.

Interpret
The presence or absence of a continental shelf can influence the type of shoreline that develops. Describe the shoreline features which could be expected to develop where a continental shelf exists.

FIG. 14–6. Longshore currents (black arrows) are produced when waves strike the shore at an angle. Such currents are the result of local conditions which cause water to flow parallel to the shore.

carrying the sand away from the shore reach deeper water, the sand load is deposited. The load settles to the bottom in long underwater ridges called *bars*. See Figure 14–5.

Most beaches go through a cycle of change from summer to winter as the varying sizes of waves alter the movement of sand. In summer, when waves are smaller, the berm is low and wide. With the larger waves that accompany winter storms, the berm becomes steeper and higher. Sand removed from the berm is redeposited beneath the shallow offshore waters to form an offshore bar.

Movement of materials along the shore. Beach materials actually move in two directions. Besides the in-and-out motion of waves, there are *longshore currents* which carry sand parallel to the shoreline. See Figure 14–6. These currents are created by waves striking the shore at an angle.

At any particular place along the shore, movement of sand depends upon the amount of local wave action. The effect of waves carrying sand in and out combined with longshore currents can only be studied by carefully observing the actual sand movements. It is still likely to be very complicated. However, there is usually a general movement of sand parallel to the shoreline in one direction or another.

When sand is moved by longshore currents, it will keep moving until the shoreline changes direction. A projection of the shore (a headland) or an indentation (a bay) slows down the longshore current, and part of the sand load is deposited. For example, a headland or island may contain a deposit of sand built up on one side, as shown in Figure 14–7. Such a deposit is called a *spit*. It is formed in the relatively still water on the side of the projection opposite the direction of the usual longshore current. Commonly, tidal currents force the end of a spit to curve inward or outward forming a hook or curved spit. Often a chain of spits will develop from one island to another. These connecting ridges of sand, called *tombolos*, may also link an island to the mainland.

HOW A SHORELINE DEVELOPS

Origin of shoreline features. The line where the land and water meet is called the *shoreline*. All shorelines are the result of two factors. First, the land next to any large body of water is constantly exposed to the forces of weathering and erosion. Second, the exposure of land to the erosional activity of wave action gives the shoreline its present form. For many shorelines it is difficult to determine the particular set of conditions responsible for its characteristics.

Shorelines are often described as though the rise and fall of sea level were responsible for their features. A shoreline dotted with many sea cliffs is said to be *emergent*. The presence of the sea cliffs is interpreted as evidence that sea level fell and the exposed land was subjected to wave attack which eventually produced the cliffs. See Figure 14–8. If the land rises or there is an actual decrease of water in the sea, a drop in sea level is a likely event. Where there are many bays and inlets along a

FIG. 14–7. A spit is formed on a headland or island near the mouth of a bay or inlet by offshore currents moving parallel to the shore. An island connected to nearby land by similar deposits becomes a tombolo. (Fairchild Aerial Surveys, Inc.)

FIG. 14–8. The erosional cycle in a shoreline with features formed mainly by emergence of the land or lowering of water level. Top left – initial stage; top right – youth; bottom left – maturity; bottom right – old age.

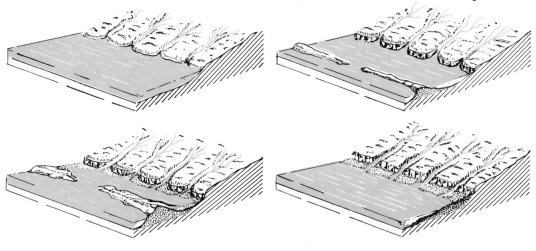

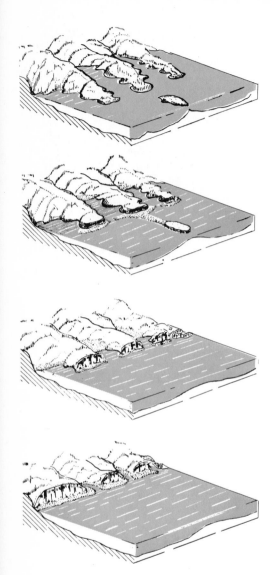

FIG. 14–9A. The erosional cycle in a coastline with features formed mainly by submergence due to rising water level or sinking land. From top to bottom: initial stage, youth, maturity, and old age.

shoreline, it is classified as *submergent*. The bays are believed to be formed by the sinking of the land next to the shoreline, or because sea level has risen. Either method would result in the drowning of existing stream valleys to produce the bays and inlets. See Figure 14–9.

This scheme of shoreline classification has proved difficult to apply. The features of many coasts can be accounted for by either emergence or submergence. However, most shorelines have certain features which are clearly caused by rise or fall of the water level in relation to the land.

There is definite evidence to show that many bays are actually drowned river valleys. These valleys were originally eroded by streams flowing to the sea. Then the land sank or the level of the sea rose and the valleys filled with sea water to become bays. Divides between neighboring valleys became headlands and points jutting into the sea. The highest parts of the submerged land often remained above water as offshore islands. Irregular coastlines in many parts of the world were apparently formed by this process.

When an occupied stream valley with a gentle gradient submerges, it may produce a shallow bay which extends far inland. Such a bay, located at the mouth of a river, is called an *estuary* (*es*-chu-ehr-ee). See Figure 14–9A. Many of the world's best harbors are estuaries. They provide shelter and are easily entered from the sea. San Francisco Bay, an unusually fine harbor, is an example of a system of drowned river valleys or estuaries.

The coasts of Alaska, Norway, Greenland, Chile, and British Columbia show an unusual characteristic of submergence. Extending down to the sea along these coasts are many deep, U-shaped drowned glacial valleys. These valleys give the shoreline many very narrow, deep, steep-walled bays called *fiords*. See Figure 14–9C.

The formation of a regular shoreline with few bays or headlands is entirely different. Here the emergence of land exposes the inner edge of the continental shelf. Since sediments deposited on the shelf can make it smooth, a shoreline formed by emergence can be very regular. Almost the entire Gulf Coast and the east coast of the United States, from Rhode Island south to Florida are examples of a shoreline of emergence.

Such coasts often have a wide shelf of sand deposits built up offshore just beneath the water. The gentle sloping of the sea floor allows the waves to build a sand bar during storms. As often happens, this bar builds up above mean sea level and is then referred to as a *barrier island.* The water between a barrier island and the mainlands is called a *lagoon.* As it fills with materials brought in by streams it usually becomes swampy.

There are many barrier islands off the shorelines of Texas and Florida. This is also true of the Atlantic coast south of Rhode Island. Many of the resort cities of the Atlantic coast, such as Atlantic City and Miami Beach, are located on these narrow sandy islands. The south shore of Long Island, New York, illustrates the difficulty of shoreline classification. Long Island has a relatively regular shore with a string of off-shore barrier islands. However, it was not produced by emergence. Instead, a glacial outwash plain drowned, and the original shore built up to form the barrier islands as submergence took place. The result is very similar to nearby shorelines which were produced by the emerging continental shelf. See Figure 14–10.

FIG. 14–9B. An estuary in the Mississippi River delta, showing a drowned bay at the mouth of several smaller streams. (Litton Industries-Aero Service Division)

FIG. 14–9C. The Geiranger Fiord in Norway illustrates the characteristic high, steep walls of a drowned glacial valley. (Springs)

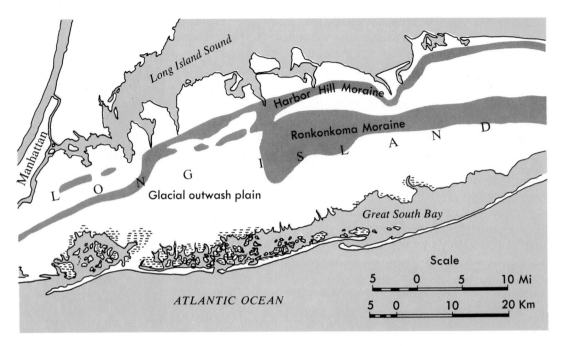

FIG. 14–10. Coastal features, moraines, and outwash plains of western Long Island, New York.

Construct

Study the diagrams in Fig. 14–11B. Use modeling clay or plaster of Paris to produce a class model of each of the four stages in the formation of an atoll.

Coral reefs. In tropical and subtropical waters coral reefs are common shoreline features. But they differ sharply from the reefs or underwater bars formed by wave action. Coral reefs are the work of living organisms. Corals are a simple type of animal which exists in a wide variety of forms. Their body form, known as a *polyp,* is soft and small. It can extract calcium carbonate from sea water to build a hard skeleton. Polyps grow rapidly and usually remain attached to each other in large colonies. A typical coral reef is built of great numbers of the accumulated skeletons of these colonies together with other kinds of animals and plants.

Three forms of coral reefs are found. A *fringing reef* is closely attached to the shore of an island or land mass. A *barrier reef* is a long narrow strip some distance offshore with a lagoon in between. A barrier reef around an island may be nearly circular. An *atoll* is a circular reef enclosing only an open water lagoon. See Figure 14–11A.

Several theories have been proposed to account for the development of the various kinds of coral reefs. One theory suggests that the reefs represent stages in the gradual sinking of an island. At first corals and other organisms growing in the shallow water around the island build a fringing reef. As the island slowly subsides, its shoreline sinks, but upward growth of the reef keeps the coral in shallow water to produce a barrier reef. Finally the island disappears beneath the sea. If the reef organisms can survive, an atoll is finally formed. This theory is illustrated in Figure 14–11B. Stages A, B, and C show the steps by which subsidence of the central island could result in a circular atoll.

Another theory explains barrier reefs and atolls on the basis of changes in sea level during the last ice age. According to this theory, coral reefs formed on the outer edges of rock platforms. These platforms were formed when the water level was low. Wave action cut broad platforms around the islands or completely trimmed off the projecting rock. As the glaciers retreated and sea level rose, coral growth set in at the edges of the existing platforms. Probably no single theory can account for all the different types of coral reef structures.

The changing sea level. At least three times in its history the earth has experienced the advance of continental glaciers that marks the beginning of an ice age. Each time, the accumulation of snow and ice in the glaciers held a large share of the earth's water supply. The level of the ocean dropped several hundred feet. The subsequent melting of the glaciers is accompanied by a rise in sea level. Each change in sea level has had its influence on the shorelines of the continents. For example, one of the most obvious evidences of the former height of the sea is the presence of wave-cut terraces well above the shoreline along many modern coasts. See Figure 14–12.

FIG. 14–11A. An inhabited coral reef in the Pacific Ocean, near the Philippines. (Shostal)

FIG. 14–11B. In the diagrams the possible stages in the formation of an atoll are shown in addition to a cross-section of a coral reef. Note that growth of corals does not proceed below a depth of about 120 feet.

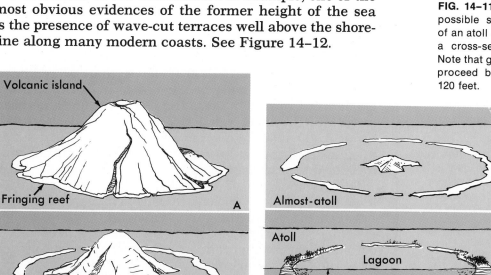

FIG. 14-12. A series of ancient wave-cut terraces exposed above the present shoreline. (John Shelton)

activity

Read the following features of emergent and submergent shorelines. Then classify each of the shorelines in the figures listed as either an emergent or submergent shoreline. Give your reasons for each answer.

Submergent Shorelines (rising sea level) may show:

sea cliffs; coves and stacks; irregular coastline; headlands; absence of wide beaches; drowned stream valley; peninsulas; fiords; bays; harbors, estuaries.

Emergent Shorelines (lowering sea level) may show:

low, smooth, gentle sloping coastal plain; regular coast line with few bays or headlands; wave cut cliffs and terraces above sea level; wide beaches and bars; shallow water barrier islands; lagoons; swamps.

Classify these figures:
Fig. 14-1, p. 297　　Fig. 14-9B, p. 303
Fig. 14-3A, p. 298　　Fig. 14-9C, p. 303
Fig. 14-7, p. 301　　Fig. 14-12, p. 306
Fig. 14-13, p. 307

In some cases, these terraces have been elevated by an uplift of the shoreline. Recently scientists have established a link between the time when terraces were cut by wave action and the time of the Ice Age. The evidence for this relationship lies in the discovery of similar type fossils found buried along the ancient shorelines.

This shows that a higher sea level must have existed at the same time in many different locations. Since uplift of the land could hardly have happened at the same time everywhere, fossil evidence buried in the ancient terraces indicates that sea level must have changed.

Examination of the inner parts of the continental shelves also shows the presence of submerged beaches. These indicate that the level of the sea in the past has at times been much lower than it is now. Radioactive dating methods have been used on fossils dredged up from these submerged shorelines. They tell us that at the height of the last glacial period, about 17,000 years ago, sea level was about 100 meters (300 ft) lower than today. Since then, the sea has risen as the water from the melting glaciers returned to the sea. The rise in sea level has been irregular. The most rapid rise seems to have occurred about 6,000 years ago when the sea level rose at a rate of almost 10 meters per century. Heavier rainfall over most of the earth accompanied the rising level of the sea.

The combination of these factors must have caused violent flooding of coastal regions which were caught between the rising sea and overflowing rivers. Geologic evidence seems to support the legends of a flood as described in the ancient folklore of almost all people, as well as in the Old Testament.

The trapping and release of water is not the only effect the ice ages have had upon the world's shorelines. Each time the ice sheets formed, their great weight pushed the land down. As the land sank, there was a corresponding rise in the sea floor. When the glaciers melted, the continents were released from their ice burden. Land masses rose as the sea floor dropped. This balance between the land and the sea floor is described by the principle of isostasy (see page 212).

At present, the land areas covered by glaciers during the last ice age are rising. Parts of the sea floor and coastal plains are sinking. See Figure 10–13 (page 213).

In some parts of Europe, particularly in Scandinavia, a rise in the land surface has taken place dramatically. Docks at the water's edge have actually been lifted completely out of the water during the past several centuries. But this rising of the land is by no means common all over the earth. Other conditions, such as volcanic activity and depositing of sediments, also influence the up-and-down movements of shorelines.

The changes in sea level produced by release of water from melting glaciers are much more rapid than the accompanying drop in the sea floor. Thus the level of the sea rises during a post-glacial period and reaches a maximum in the long period between ice ages. It is possible that we are in an interglacial period right now. Records indicate that climate is changing. Glaciers continue to recede, parts of the frozen Arctic Ocean are free of ice, and sea level is rising. All of these conditions indicate a warming trend. At some time in the distant future we may witness the second great flood in human history.

Where people and oceans meet. The shorelines are not only a narrow margin of land where the oceans meet the continents. They are also a valuable resource to the people who live on the land. Among the most important uses of the coastal lands are ports, shipbuilding, industrial and residential development, recreation, commercial fishing, and as a dumping ground for wastes. Any shoreline is a mixture of cliffs and beaches, bays and harbors, lagoons and estuaries, islands and spits. All of these are combined in different ways to produce the individual shorelines of the world. But these coastal resources are fragile and can be easily damaged or destroyed by human use.

As world population grows, the shoreline resources will be put to even heavier use. It is estimated that by the year 2000 half of the total population of the United States will live on 5 percent of the country along the coasts of the Atlantic, Pacific and Great Lakes. Dredging and filling

FIG. 14–13. Two scenes showing the same coastal region of Florida taken twenty years apart. (Airflite)

Investigate
Locate a lake or a portion of a
river near where you live. Find
out if any of it is used for drinking
or recreation, and if any part of
it is polluted. See what you can
find out about any actions taken
to correct or control possible
growing problems.

operations as shown in Figure 14–13 will increase. Pollution in all forms, which has already destroyed one-tenth of the shellfish producing waters of the country, will continue to increase at a rapid rate. If the shoreline resources are used in the future as they have been used in the past, much of the existing coastal land of the nation will be destroyed.

Wise use of the coastal zone depends upon understanding the natural forces that created the shoreline features. With such understanding, it should be possible to predict harmful changes in the natural scheme and control them. It may even be possible to correct the past mistakes. But knowledge that some changes in shorelines might be destructive may not solve the problems. Control over development of coastal regions is divided among many separate groups. Some of the shore is privately owned. The remainder is managed by towns, cities, counties, states and the Federal Government. There is no single group with authority to give leadership for an organized development of the coastal resources.

VOCABULARY REVIEW

Match the word or words in the column on the right with the correct phrase in the column on the left. *Do not write in this book.*

1. Separate column of rock found along oceanside cliffs.
2. The raised portion of the beach on which people rest and play.
3. Ridges of sand just off shore.
4. Causes the movement of beach sand parallel to the shoreline.
5. Sand piled up in the quiet water protected from longshore currents.
6. Ridges of sand connecting two land bodies.
7. A bay located at the mouth of a river.
8. The result of submergence of glacial valleys.
9. A bar which is above average sea level.
10. A body of water found between a barrier island and the mainland.
11. Similar to a bar but built of skeletons of once living organisms.
12. A circular reef enclosing a lagoon.

a. fiord
b. lagoon
c. atoll
d. tombolos
e. berm
f. coral reef
g. estuary
h. barrier island
i. sea stacks
j. spit
k. sea cave
l. bar
m. sea arch
n. longshore current
o. wave-built terrace

QUESTIONS

Group A

Select the best term to complete the following statements. *Do not write in this book.*

1. Which one of the following events is *not* considered a part of mechanical wave erosion? (a) air and water pressure breaking the rock (b) pieces of rocks thrown against the shore (c) chemical weathering of rock (d) small fragments of rock rubbing each other in the tumbling water.

2. When waves act directly against the rock of the land, erosion forms a (a) wave-cut cliff (b) bar (c) spit (d) hook.

3. Waves attacking the shoreline of Cape Cod caused some of the shoreline to retreat (a) about 10 meters since the time of the Romans (b) several meters each year (c) several kilometers each year (d) about 1 meter in the past century.

4. In the formation of a wave-cut terrace, some of the rock may be transported seaward to form an extension called a (a) bar (b) spit (c) hook (d) wave-built terrace.

5. Rock terraces along the shore can reduce wave erosion. It may be halted entirely unless (a) the sea level is unchanged (b) the sea level is raised (c) a bar forms (d) a spit forms.

6. Sea caves result when wave erosion (a) occurs along a deep joint (b) forms a headland (c) forms sea arches (d) causes chemical weathering.

7. Analysis of some beach sand shows it is composed of rock fragments which are foreign to the shoreline. This sand most likely came from (a) the ocean floor (b) a volcano (c) sea caves (d) rivers.

8. If a beach is composed of coarse sand containing quartz and feldspar, it most likely is the result of wave action on (a) basalt rock (b) granite rocks (c) limestone (d) clay.

9. Although beaches may be composed of various sized rock fragments, their materials always (a) make the same type of movements (b) have the same composition (c) have a white color (d) have the same shape.

10. Small waves can move sand particles (a) toward the shore (b) away from the shore (c) in circles (d) up and down in the same spot.

11. Large waves can (a) cover the berm with coarse rock (b) cause sand to move seaward (c) smooth out the berm (d) decrease the height of the berm.

12. When currents from the beach carry sand seaward, bars are formed (a) during low tide only (b) during high tide only (c) mostly during winter (d) mostly during the summer.

13. Winter storms produce larger waves which can form a (a) low, wide berm (b) berm with a gentle slope (c) higher, steeper berm (d) beach made up of smaller rock fragments.

14. Longshore currents are usually the direct result of (a) waves striking the shore at an angle (b) deep sea currents (c) surface currents coming close to shore (d) very wide beaches.

15. Longshore currents result in the movement of sand (a) in and out from shore (b) higher on the berm (c) out to deeper water (d) parallel to the shoreline.

16. A curved spit called a "hook" is often formed by the action of (a) large waves (b) small waves (c) deep water currents (d) tidal currents.

17. A shoreline resulting from the fall of sea level is said to be (a) emergent (b) submergent (c) divergent (d) ancient.

18. The sinking of land or a rise in sea level can form (a) sea cliffs (b) arches (c) bays and inlets (d) tombolos.

19. San Francisco Bay is a good example of (a) a drowned river valley (b) emergence (c) tombolo (d) wave-built terrace.

20. A shoreline feature often resulting from emergence is (a) a fiord (b) an estuary (c) a sea cliff (d) a headland.

21. The city of Miami Beach is a good example of a (a) headland (b) barrier island (c) tombolo (d) coral reef.

22. Coral reefs are produced by living organisms which have the ability to (a) accumulate large quantities of sand (b) swim about causing a deflection of ocean currents (c) form very hard skeletons by extracting minerals from sea water (d) attach themselves to sand bars.

23. A circular reef enclosing an open water lagoon is called (a) a barrier reef (b) a fringing reef (c) an atoll (d) an estuary.

24. A type of coral reef which is attached to an island is (a) a barrier reef (b) a fringing reef (c) an atoll (d) an estuary.

25. Which of the following ideas is not a part of present theories explaining the formation of reefs? (a) Reefs are the result of a barrier island forming off the mainland. (b) Reefs represent stages in the gradual sinking of an island. (c) Reefs are formed on the outer edges of rock platforms. (d) Sea level has risen due to the retreat of glaciers.

26. It is known that at least four times in the earth's history continental glaciers have advanced causing the oceans to (a) become very cold (b) lower their level several hundred feet (c) become unusually warm (d) raise their level several hundred feet.

27. Which of the following is considered to be evidence of changes in sea level? (a) wave-cut terraces well above present shorelines (b) higher tides than in the past (c) presence of sand on the deep sea floor (d) many beaches along modern shores.

28. A great increase in continental glaciers appears to have resulted in (a) the sinking of land with a corresponding rise in the ocean floor (b) the raising of land with a corresponding sinking of the ocean floor (c) the sinking of the land and sinking of the ocean floor (d) the raising of the land and raising of the ocean floor.

29. Present evidence indicates that the earth may be entering a warming period in its history which, if it continues, may result in (a) lowering of sea level (b) lowering of the level of the land (c) an unusual dry period (d) a second great flood in the history of man.

30. Which of the following will most likely preserve the fragile coastal zone as a valuable resource? (a) Allow no more than 50% of the continental population to live in the coastal zone. (b) Stop the filling and dredging operations. (c) Have the coastal zone under one authority. (d) Have those who own the coastal zone understand the natural forces that created the shoreline features.

Group B

1. Describe the four ways waves attack the shoreline rock.

2. What process other than wave action causes deterioration of the shoreline?

3. Compare the formation of a wave-cut terrace with that of a wave-built terrace.

4. What effect does the development of a terrace have on the erosion of the shoreline? What condition may alter this effect?

5. Name three prominent shoreline formations that often result from uneven erosion of sea cliffs.

6. How do beaches get their sand?

7. Describe the general movement pattern of small rock fragments near the shoreline.

8. How can you tell by observations of the berm whether the recent waves have been larger than usual?

9. What causes sand to be moved parallel to the shore?

10. What features are typical of an emergent shoreline? Of a submergent shoreline?

11. Classify the following shoreline features as due to emergence or submergence. (a) San Francisco Bay (b) Coast of the Gulf of Mexico (c) the fiords of Alaska.

12. Polyps, the coral's soft body form, living in the warm waters of the ocean, form reefs which are hard and sharp enough to cut through the hull of a steel ship. Explain.

13. Compare the manner in which the following were formed: (a) the island on which Miami Beach is found (b) the island off the south coast of Long Island, New York (c) an atoll.

14. What geological evidence exists to indicate that a great flood occurred on earth about 6,000 years ago as described in the Old Testament of the Bible?

15. Discuss the source of rock fragments which go to make up the following beaches:
(a) coarse light colored sand composed of quartz and feldspar.
(b) coarse dark rock fragments.
(c) fine sand composed of materials not found along the shoreline.

16. What effects are produced when great continental glaciers form and melt that indicate the crust of the earth rests upon a somewhat plastic (soft) base?

17. What geologic evidence is there for believing that the level of the sea has changed?

18. What cause or causes are believed to account for changes in sea level?

19. Describe the geologic results on a shoreline for a drop in sea level and for a rise in sea level.

20. For each of the six uses of coastal lands listed in the text, give one specific example of how the coastline can be harmed if used carelessly.

time and change

Will the Ice Return?

During most of the past two million years the earth has been locked in a series of ice ages. We are now in an interval of time that is warm. In the past, very similar warm intervals occurred between periods when ice covered large parts of the northern hemisphere. Above is a picture of an icefield in Juneau, Alaska. Similar icefields during the last ice age covered Canada and parts of the United States. But, will the ice return again? One of the scientific theories that may provide an answer is based on knowledge of slight changes in the earth's orbit. There is known to be an advance in earth's elliptical orbit that allows the earth to be farther from the sun in June or July. This would make the summer temperatures cooler. Also it is known that the earth has a precession (Chapter 3) that causes a change in the tilt of the axis and this in turn causes the northern hemisphere to receive less heat. Another known change is in the shape of the earth's orbit every 10,000 years, becoming more circular than elliptical. When this happens, the earth is not as near to the sun at perihelion and is cooler in winter. These changes over a period of time could account for the ice ages since each change would cause a greater amount of ice to grow at the north polar ice cap due to the cooler temperatures. At the present time, the earth's orbit is becoming more circular, thus contributing to the general cooling effect.

Some scientists believe that the combined effects of all these changes in the earth's orbit are responsible for the ice ages as they occur in slow cycles. Some opposing theories are discussed in Chapter 13. But if another ice age is coming, at least there is another 6,000 years or so to prepare for its effects. With the possibility of another ice age coming upon us, the activity pictured at the left might be replacing swimming as a summer activity in Connecticut, that is, in a few hundred years or so.

Balancing Rocks

An unusual result of wind erosion are balanced rocks. These strange formations are made up of heavy boulders or capstones delicately balanced on a narrow stem of softer rock. The upper rock remains balanced during the hundreds or thousands of years needed to erode away the base.

This balancing act is actually performed by the capstone that tilts back and forth as erosion proceeds. The shifting of the capstone weight causes the soft rock of the base to be compressed on one side. That side of the base erodes more slowly than the side under less pressure. In time, the more rapid erosion on the side under less strain results in tilting of the capstone toward that side. When this happens, the erosion rate then increases on the opposite side. This tilting back and forth continues for many years while the base wears away first on one side, then the other, until it becomes only a narrow stem. Finally, the base becomes too narrow for the balancing act to continue and the capstone falls. Curious people have been known to greatly hasten the final crash by pushing on balanced rocks, thus destroying a rare and unique natural object.

People Who Find Patterns and Cycles

Deep under an icefield in Juneau, Alaska, a geologist explores an ice cave. The many layers tell of years of accumulation and change. Few places on earth have ice fields extensive enough to provide material for an in-depth study. So, these rugged scientists must travel to the farthest northern and southern reaches of the earth in order to explore the past and make predictions of future conditions.

If the soil on this Georgia farm washes away, it will be a problem for the soil scientist pictured to solve. By making accurate observations and using various testing techniques, she can make a determination of the best way to cultivate this land. Our soil resources can be preserved with soil conservationists on the job. Using soil conservation methods, the forces that act to erode the land can be reduced.

unit

4

the earth's
envelope
of water

15
the oceans

objectives

- ☐ Define the term "ocean-ography."
- ☐ Name the seven seas.
- ☐ Describe several methods used to examine the sea's characteristics.
- ☐ Describe, using the correct terms, the shape of the earth's crust beneath the sea.
- ☐ Describe the different kinds of sea-floor sediments.

A large part of history is the story of the explorers who sailed across the world's oceans. These daring adventurers discovered the boundaries of the sea that cover most of the earth's surface. But the true mystery of the sea lies beneath its surface. Only during the past hundred years has there been any significant progress in the scientific understanding of the ocean depths.

The first scientific exploration of the oceans began with the voyage of the British Navy Ship, H.M.S. *Challenger*, in 1873–76. The team aboard this small ship made thousands of observations as it crossed the Atlantic, Antarctic, and South Pacific oceans. Their findings marked the beginning of the new science of *oceanography,* the scientific study of the oceans. Oceanography is a blend of the four basic sciences of physics, chemistry, geology, and biology.

Since the voyage of the *Challenger*, oceanography has grown rapidly into one of the most important branches of science. Many ships are now equipped to perform oceanographic research. An example of a research ship is the *Glomar Challenger,* shown in Figure 15–1. This ship is able to drill into the rock of the sea floor that lies as much as 4 km beneath the ocean's surface. Samples of the sea floor have been taken by the *Glomar Challenger* from all over the world. These samples have given scientists a foundation for understanding how the oceans formed. Other modern methods of oceanographic research are used to study the basic processes that control ever-changing sea conditions. The once mysterious ocean depths are slowly becoming as familiar as the earth's surface that lies above the sea.

CHARACTERISTICS OF THE SEA

The seven seas. The oceans are only a small part of the earth. The water in the oceans weighs 1560 million billion tons. Yet this enormous mass is only 240 millionths of the total mass of the earth. The volume of the solid earth is about 800 times greater than the volume of the water in the oceans. However, the oceans are a more prominent feature of the earth than the overall quantity of water suggests. Ocean water is spread in a thin layer over about 362 million square kilometers, covering about 70 per cent of the earth's surface. As far as we know, no other planet has such an extensive covering of liquid water. Among the planets, the earth can be called the "water planet."

Since ancient times, the oceans covering most of the earth's surface have been incorrectly referred to as the "Seven Seas." The seven included the North Atlantic, South Atlantic, North Pacific, South Pacific, Indian, Antarctic, and Arctic oceans. See Figure 15-2. Actually, all seven are parts of one continuous body of water.

Nevertheless, the division of the sea into oceans is not totally artificial. Each ocean has its own characteristics. For example, water from the North Atlantic is saltier than water from the other oceans. This would not be unusual,

FIG. 15–1. The *Glomar Challenger* is an oceanographic research ship equipped for deep-sea drilling. (Scripps Institute)

FIG. 15–2. The oceans of the world. It is evident from this map that the various oceans are parts of one continuous sea.

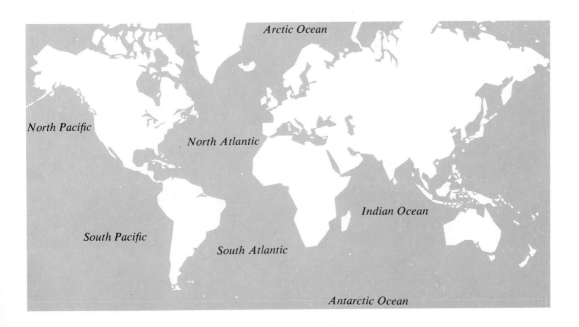

Arctic Ocean

North Pacific

North Atlantic

South Pacific

South Atlantic

Indian Ocean

Antarctic Ocean

except that the entire Atlantic receives more fresh water from rain and run-off than the other oceans do. Thus, the salt content should be more diluted. The Arctic Ocean is completely surrounded by land. The Antarctic Ocean completely surrounds a continent. Both of these polar oceans are less salty than the others. Melting ice constantly dilutes the water, and no rivers empty into them to supply additional salts.

The Pacific is the largest of the three principal oceans. With an average depth of 3.9 km, it is also the deepest. It contains more than half of all the water found in the sea. Next in size is the Atlantic Ocean. It includes the Mediterranean, Caribbean, and Baltic seas, and the Gulf of Mexico. Because of some shallow parts, the Atlantic is the shallowest of the major oceans. Its average depth is 3.3 km. The Indian Ocean ranks third in size and has an average depth of 3.8 km. The oceans are not evenly divided between the earth's two hemispheres. About 39 percent of the Northern Hemisphere is land; the Southern Hemisphere is only about 19 percent land.

Exploring the depths. Scientific study of the sea is based on measurement. For centuries, sailors have measured depth in shallow water with a weighted rope line. From this simple beginning, a great number of special instruments have been developed. These instruments allow

FIG 15–3. Tracing the depth of the sea bottom in the Caribbean, using a sonic depth recorder. (Woods Hole Oceanographic Institution)

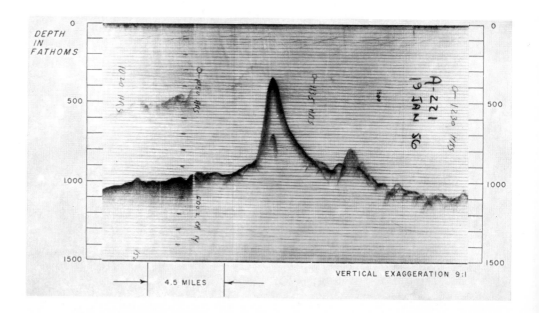

modern oceanographers to measure many properties of the sea. The following is a description of the properties of the sea, and the instruments and techniques commonly used for measuring these properties.

1. **Measurements made from a moving ship.** Sound waves, which pass through water easily, make it possible for a moving ship to draw a profile of the sea floor. A transmitter on the ship sends out a continuous series of sound pulses. The sound waves are reflected from the bottom back up to receiving equipment on the ship. The time it takes for the sound impulse to return is automatically translated into a depth reading. A graph is produced that shows the profile of the sea floor along the ship's course. See Figure 15–3. The kind of sound waves sent out and the technique for recording them can be adjusted to show various features on the sea bottom. For example, one type of sound signal penetrates the sediment layers and shallow rock structures on the bottom. The returning signal shows the thickness of the bottom sediments and the nature of the underlying rocks in that part of the ocean.

One of the most revealing properties of sea water is its temperature. Because the temperature of water changes slowly, it can remain constant in a certain mass of water. Thus, temperature readings are often used to identify and trace moving masses of sea water. Temperature can be measured as a ship moves by towing a cable that has electronic temperature sensing devices attached at certain depths. This gives a continuous record of the temperature at the selected depths. Another type of temperature measuring device is shown in Figure 15–4. As it is lowered through the water, it produces a record of the temperature.

Another kind of electronic device towed behind a ship is able to detect surface currents. This device measures the action of the ocean on the earth's magnetic field. In order to collect samples of different forms of sea life, nets are towed slowly through the water at various depths.

2. **Measurements made at fixed locations in the ocean.** Sea water is a very complex solution that contains many dissolved substances. Chemical analysis of water samples taken from certain locations and depths yields some of the most important information used by oceanographers. Water samples for chemical analysis are usually taken with a special sampling container attached to a line. See Figure 15–6. When lowered to the desired depth, the container is closed by a weight dropped down the line. The

FIG. 15–4. An instrument called a bathythermograph is used to measure water temperature. As it is lowered through the water, it traces a record of the temperature on the inserted glass slide. A bathythermograph is used only to measure temperatures in the upper water layers and cannot be used at great depths. (Ramsey)

FIG. 15–5. A giant buoy containing experimental oceanographic instruments is being towed to its permanent station. (General Dynamics)

FIG. 15–6. Water samples are taken at desired depths with this Nansen bottle. Attached thermometers record and register the water temperature of the various samples. (Ramsey)

trapped water is taken on ship and is either analyzed or preserved for detailed work later. Some chemical properties of sea water can be measured directly by instruments lowered into the water.

Measurements of the movement of sea water are also made from fixed locations. To detect movement of sea water, two basic measurements have to be made. First, the up-and-down movements are measured. These are waves and tidal movements. Another common kind of wave detector uses electronic methods that record water motions. The electronic wave sensor can be aboard a ship, attached to a floating buoy, or placed on the sea bottom near the shore. A continuous record is then available that shows wave activity at a particular location.

Currents, or horizontal movements of the water, also have to be measured. The speed, direction, and location of the current are the most important information needed. The velocity or speed of the current can be measured with several different kinds of current meters. One type of electronic current meter can be installed in the water at a particular depth and left unattended. The meter will produce a record of the current velocity at that location. Floats are often used to track the direction of a current. The floats are followed as the current carries them along. Current direction can also be measured by adding certain "tracer" materials to the water. The movements of the tracer are followed by identifying its presence in water samples taken at different locations.

Valuable information can also be obtained from samples of the sediments that cover the sea floor. These samples are usually obtained through various devices lowered from ships and anchored or held at fixed locations. Some sampling devices have jaws that take a "bite" of exposed sediments on the bottom. Deeper sediments are sampled with long tubes that are driven into the sediment layers. When brought back aboard ship, the *core samples* are removed as long cylinders. These samples show the vertical arrangement of the layers. Core samples of this kind can be as long as 30 meters and can be taken from the deepest parts of the ocean. Other sampling devices drill into rock beneath the sediment layers to obtain samples of solid rock.

Cameras are used to photograph the surface of the sea floor. The photographs are useful in preparing maps and in locating areas that might yield valuable resources.

Probes that are able to measure temperature can be driven into the ocean floor. These temperatures give clues about the nature of the rock layers and the deeper parts of the crust beneath the oceans.

Humans beneath the sea. Instruments are vital to studying the sea. However, nothing will replace the first-hand knowledge that comes from direct observation of conditions beneath the sea. No matter how fascinating, the sea is still a foreign environment for humans. To penetrate this vast world of "inner space" beneath the sea, special equipment and great skill are needed.

A basic requirement for survival beneath the sea is a continuous supply of oxygen. But a more serious problem is pressure increases during descent. This pressure is created by the weight of the water pressing down from above. Naturally, it increases as the diver goes deeper and there is more water above. Pressure nearly doubles at 10 meters of depth and increases the same amount for each additional 10 meters. At a depth of 1000 meters, the pressure is 100 times that of the surface.

The human body can withstand fairly high pressures if breathing air is provided at close to normal pressure. Breathing air at above normal pressures can dissolve gases in the blood and in other body fluids. When returned to normal pressure, these dissolved gases form bubbles in the body. These bubbles then produce the often fatal condition known as "bends." To prevent this from happening, air must be supplied to a person in such a way that its pressure is nearly normal, no matter how great the water pressure. An apparatus usually called *scuba* (self-contained underwater breathing apparatus) is designed to be carried by an individual diver. See Figure 15–7. It supplies air at the correct pressure for periods up to about an hour. But it must be used only by trained divers and is safe only for relatively shallow water. At depths greater than about 50 meters, it is necessary to enclose the diver completely in a protective suit. See Figure 15–8. With this outfit, a diver can work efficiently in depths up to about 200 meters.

Divers are able to dive to greater depths through the use of more complicated methods. Gases such as helium can be mixed with oxygen and breathed by divers under high pressure. Such a mixture of gases will not dissolve in the

FIG. 15–7. Much exploration of the shallower parts of the sea can be done by divers using this kind of self-contained apparatus. (Rapho Guillumette–Ron Church)

FIG. 15–8. In the "hard hat" helmet-suit, air is forced down through a pipe to the helmet. (E. R. Degginger)

body as readily as ordinary air. This allows divers to remain under the higher pressures at greater depths for longer periods. Other techniques have divers slowly exposed to high pressures in special chambers on the mother ship. Since their bodies are already adjusted to the high pressures, they can then quickly descend to working depth. In these chambers, they can be slowly brought back to normal pressure or be kept at the higher pressure for later dives. With the use of these systems, divers have been able to reach depths of over 300 meters without protective suits.

Divers can also live for long periods of time in undersea chambers that are kept at high pressures. They can come and go from these chambers without any problems caused by pressure changes.

To reach very great depths in the sea, a human must be enclosed in some kind of submarine vehicle. It is possible to reach bottom in the deepest parts of the sea in a vehicle that acts as a kind of underwater balloon. Gasoline, which is lighter than water, is used along with ballast to control the craft under water in the same way a balloon rises or sinks in air. Such a vehicle has reached the record depth of about 10,900 meters. But this type

Find Out

In most large cities, there are underwater tunnels that allow cars to pass from one body of land to another. Find out how people worked under the water in air-tight chambers to join the separate units of the tunnel.

FIG. 15–9. A number of research submarines designed for deep-sea diving and living.

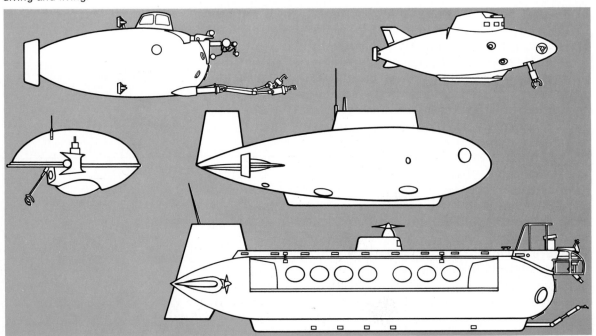

of undersea vessel is not able to move around very much once it has reached the bottom. Other research submarines have been designed that allow deep diving along with the ability to move over considerable distances while underwater. See Figure 15–9.

THE SHAPE OF THE CRUST BENEATH THE SEA

Ocean basins. If the earth's oceans dried up and the sea floor could be seen as part of the solid crust, the earth's landscape would be very different. Mountains taller than any on the continents would become visible. The deepest canyons and longest range of mountains on the earth would be visible for the first time. Oceanographers have discovered that the features of the sea floor are very similar to those found on the land surfaces. See Figure 15–10. Many of these features result from the processes

FIG. 15–10. The ocean floor is covered by a great diversity of features.

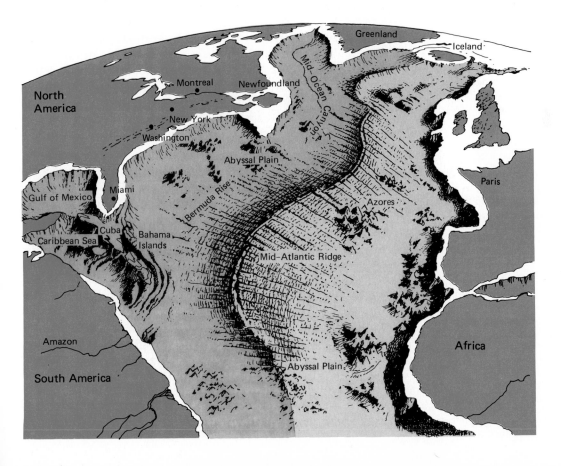

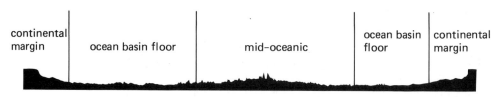

| continental margin | ocean basin floor | mid-oceanic | ocean basin floor | continental margin |

FIG. 15–11. The major divisions of the floor of the North Atlantic on a profile from the coast of North America to Africa. The various features have been exaggerated by forty times to make them more visible. The sea floor is not nearly so rugged as this profile shows.

Explore
Recent deep-sea expeditions of the ocean floor have uncovered many interesting discoveries. From the origin of the continents, a source of food and an untouched wealth of raw materials, the sea will provide us with a new world of knowledge. Look up some recent magazine and newspaper articles that give accounts of deep-sea findings.

FIG. 15–12. This diagram shows the features found along the continental margins. Notice how sediments have accumulated on the continental shelf and moved down to form the continental rise.

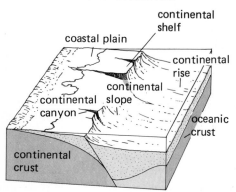

that formed the ocean basins. These basins were created as the continents were broken apart by the action of the moving crustal plates.

The study of the oceans has slowly revealed the true shape of the crust beneath the sea. Thus, it is possible to divide the ocean basins into three major areas. These are the *continental margins,* the *ocean basin floor,* and the *mid-ocean ridges.* Figure 15–11 outlines these three regions as they are found in a typical ocean basin like that of the North Atlantic.

The edges of the continents. The line that divides the land masses from the sea is not as sharp as it seems. Shorelines, where the water washes against the continents, are not always the true boundaries between sea and land. The real boundary usually lies some distance off the shoreline, beneath the water. Every continent is bounded in most places by a zone of shallow water where the sea has flooded the edges of the continental land masses. This zone of shallow water is called the *continental shelf.* It is present beneath the water around the margins of all the continents.

Continental shelves may have come into existence as a result of changes in sea level. During an ice age, when great amounts of water are locked up in glaciers, sea level falls. Later, with passage of the ice age, glaciers melt and sea level rises. Apparently, the earth is now in a warm period when the sea level is relatively high. Thus the sea covers a part of the continental edges.

The continental shelf is part of the continent itself, rather than part of the ocean basin. At one time or another, parts of the shelf's surface were exposed to the forces that carved the land surface into hills and valleys. This means that the surface of the continental shelf is not likely to be perfectly flat. In general, the surface of the continental shelf is a smoother version of the land surface at the shoreline. The shelf usually slopes gradually away from the shoreline until it reaches a steeper slope. This marks the edge of the continent and is called the *conti-*

nental slope. See Figure 15–12.

The width of the continental shelf is not the same at all places. In some places, like off the west coast of South America, it is only a few kilometers wide. Off the west coast of Florida, in the Gulf of Mexico, the shelf reaches out 760 km. The widest continental shelves in the world extend about 1280 km off Siberia and Alaska into the eastern Bering Sea. But a more typical continental shelf is that found along the east coast of the United States. It averages 170 km in width.

Because the continental shelf generally slopes very gently as it extends out to sea, the water is shallow. Average water depth over the world's continental shelves is about 60 meters. However, at the continental slope, there is usually a very sharp increase in depth. Within a distance of a few kilometers, the depth increases to several thousand meters. At the base of the continental slope, there may be an accumulation of sediments that have moved down the slope. This bulge at the base of the slope is known as a *continental rise*. The continental slope itself may be cut by valleys running along its surface. Occasionally, there can be an unusually deep valley called a *submarine canyon*. See Figure 15–12. The origin of these deep canyons is not completely known. They may be caused by turbidity currents that move down the continental slopes. See page 354.

In some places, the continental slopes plunge downward into the deepest parts of the sea floor. These long, narrow depressions at the base of some continental slopes are called *trenches*. Their depth is usually greater than 6000 meters. Most of the trenches of the world are found in the Pacific Ocean. See Table 15–1.

Ocean basin floor. Between the margins of the continents and the mid-ocean ridges, lies the ocean basin floor. Most of this part of the sea floor is made up of *abyssal plains*. These are extremely level regions that cover about half of all the deep ocean floor. The abyssal plains are the flattest regions on the earth. Some parts have changes in elevation of less than 3 m over distances greater than 1300 km. Sound signals used to study the sea floor reveal that the abyssal plains consist of thick layers of sediment deposited on the rougher features of the ocean bottom. The sediments appear to have been deposited by currents that carried them out into the deep sea from the con-

Table 15–1 Major Ocean Trenches

Trench	Depth	
	Meters	Feet
Pacific Ocean		
(1) Aleutian	7672	25,165
(2) Kurile	10,543	34,580
(3) Japan	9800	32,153
(4) Mariana	10,915	35,800
(5) Philippine	10,033	32,907
(6) Tonga	10,853	35,597
(7) Kermadec	10,003	32,809
(8) Peru-Chile	8057	26,427
Atlantic Ocean		
(9) South sandwich	8262	27,100
(10) Puerto Rico	9392	30,184
Indian Ocean		
(11) Sunda (Java) Trench	7252	24,442

tinental margins. Trenches found along the edges of some continents act as traps for the sediments carried down the continental slopes. Thus, abyssal plains are most widespread in parts of the sea floor where there are no trenches along the continental margins. The Atlantic Ocean, with few trenches, has much of its floor covered with abyssal plains. The Pacific, with more trenches, has less extensive abyssal plains.

Scattered across the ocean basin floor are a number of separate volcanic mountains called *seamounts*. Most of them have a height of at least 1000 m. Hundreds of seamounts dot the Pacific Ocean floor but are less common in the other oceans. Many of these underwater volcanic peaks are formed near the spreading plate boundaries marked by mid-ocean ridges. Volcanic activity is very common in this part of the sea floor. Seamounts often grow high enough to raise their tops above the water surface, making an island. Some of these volcanic islands have their tops eroded away by waves. Then they slowly sink back beneath the sea as they are carried farther from the mid-ocean ridge area. These flat-topped seamounts are called *guyots*.

In a few places along the sea floor, a chain of volcanic islands and seamounts were produced by a *hot spot*. A hot spot exists where a continuous stream of magma erupts through the crust. Volcanoes are periodically built where hot spots are located. The crustal plate making up the sea floor moves, while the hot spot remains fixed. As a result, a chain of seamounts or volcanic islands form as the moving plate carries the volcanic peaks away from the hot spot. The Hawaiian Islands are part of a long chain of volcanic peaks. They were built over millions

Interpret
The Hawaiian Islands are really the peaks of shield volcanoes whose bases lie on the sea floor. Find out if these volcanoes were ever seamounts and if they existed before or after the ocean was there.

FIG. 15–13. The Hawaiian Islands are the most modern part of a chain of volcanic peaks that have grown over a permanent hot spot in the crust and upper mantle.

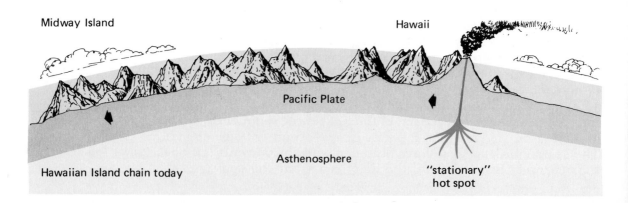

Midway Island

Hawaii

Pacific Plate

Hawaiian Island chain today

Asthenosphere

"stationary" hot spot

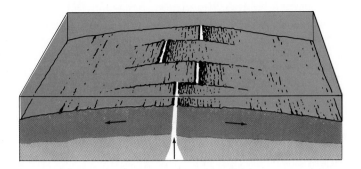

of years as the Pacific plate slowly moved over the hot spot now marked by the island of Hawaii. See Figure 15–13.

Mid-ocean ridges. The most prominent features of the sea floor are the *mid-ocean ridges*. See Figure 15–10. This continuous series of underwater mountain ranges runs through all of the ocean floors. See page 323, mid-ocean ridges exist where new sea floor is formed. This happens when molten rock is forced up from below the crust and spreads out. Because the molten rock is forced up, the central region of the sea floor occupied by the ridges is elevated. See Figure 15–11. Mid-ocean ridges usually have a narrow depression, or *rift*, running along their centers. The rift is evidence that the crust pulled apart as the plates moved away on each side of the ridge. Volcanic activity is very common along the rift zone, since this is where magma reaches the surface. Each side of the mid-ocean ridge system is covered by a series of parallel hills and valleys. A rugged landscape has been produced because the plate movement breaks the surrounding crust into tilted blocks. In many places, the mid-ocean ridges are cut across by large faults. See Figure 15–14. The faults are probably produced by the molten rock that rises in neighboring parts of the ridges. This causes the rate of spreading out to be greater in some parts of the ridges than in others. Due to the unequal rate of spreading, the ridge breaks into sections.

FIG. 15–14. Notice the large faults that appear along the mid-ocean ridge.

SEA FLOOR SEDIMENTS

Origin of marine sediments. The kinds of sea bottom

activity

Fill a large test tube about one fourth full of sandy soil. Add water to the test tube until it is filled to within 1 cm of the top. Cover the top of the test tube and shake the mixture until the soil is mixed throughout the water.

Set the test tube upright and observe the soil particles as they settle to the bottom. Use a magnifier. When the soil has nearly settled to the bottom, answer these questions:

1. How does the particle size at the bottom of the test tube compare with that found farther up?

2. What size particles settled to the bottom first?

3. What size particles settled to the bottom last?

4. Do any pieces of material float to the surface?

5. How does the size of the particles that remain suspended in the water, but not floating on the surface, compare to those that settle out?

Rivers carry soil particles, and even rocks, downstream and to the ocean. When the river water reaches the calm of the ocean, the soil particles settle out.

6. Where, in the ocean, would you expect to find the larger soil particles carried by rivers? Explain.

7. What kind of river-carried particles would you expect to find on the ocean floor, far from shore? Explain.

Investigate

The first Atlantic telephone cable was mysteriously ripped apart by some unknown powerful force on the sea floor. Further investigation led to the discovery of deep-sea turbidity currents. See what you can find out about how these powerful currents could have torn the cable apart.

sediments differ a great deal from one part of the sea to another. Over the continental shelves and slopes, the bottom is covered mainly with fragments of rock. These come mostly from rivers emptying into the sea and from wave action against the shoreline. Some of the sediments also seem to have been produced before the area was covered by the sea. Usually, the sediments are fairly well separated according to size. Since it takes more force to carry larger rock fragments, coarse gravel and sand are usually found closer to the shore. Fine particles (mud) usually cover the bottom farther from shore.

In the deep sea, beyond the continental slopes, the bottom sediments are different from those found in shallower water. Most of the sediments in the deep sea are formed from materials that settle from above. These materials come from a variety of sources. A brief description of the origin of the main types of deep sea bottom sediment follows.

1. **Particles carried from the land.** The total amount of sediment carried into the oceans by streams can be accurately estimated. When this amount is compared to the amount already known to be deposited along the shore and on the continental shelf, a large quantity of sediment is found to be missing. What has happened to it? Careful study of the problem revealed that great quantities of sediments occasionally slide down continental slopes to the deep sea floor. These movements are like great underwater landslides and may be set loose by earthquake shocks. In a single movement, millions of tons of material can be involved. The force of the material sliding down creates powerful currents. These currents probably cause the moving sediments to be spread over large areas of the deep sea bottom.

Wind is another way that land sediments reach the deep sea. Fine particles of rock blown great distances out to sea by the wind finally settle on the water's surface. Eventually, they sink to the bottom and become part of the sediment accumulation. Volcanic dust thrown into the air often becomes part of sea floor sediment after the wind has carried it some distance. Volcanic activity that occurs on the sea floor also forms part of the bottom sediments.

Icebergs also carry material from the land into the sea. When the iceberg melts, any glacial material that it happens to be carrying is dropped and sinks to the bottom.

2. **Organic sediments.** Many of the plants and animals that live in the sea have hard parts, including their skeletons. When these organisms die, their skeletal remains become part of the bottom sediment. In many places on the sea floor, almost all the sediment comes from this source. The two most common substances from organic sources are silica (SiO_2) and calcium carbonate ($CaCO_3$). Silica comes mainly from microscopic

FIG. 15–15. Common microscopic organisms that contribute to ocean floor sediments. (A) radiolarians, (B) diatoms, and (C) foraminifera specimens.

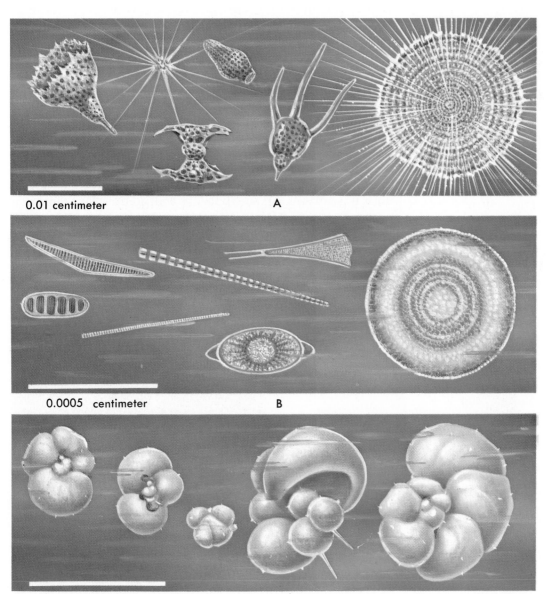

0.01 centimeter   A

0.0005 centimeter   B

0.1 centimeter   C

FIG. 15–16. A concentration of manganese nodules on the floor of the Atlantic Ocean at a depth of 4100 meters. (Lamont Geological Observatory)

FIG. 15–17. A core sample of sea floor sediments is being removed from the hollow coring device at the right. (ESSA)

animals called *radiolaria*. Microscopic plants called *diatoms* (di-a-toms) may also have silica skeletons that form sediments. Calcium carbonate comes mostly from the skeletons of tiny animals called *foraminifera*. See Figure 15–15. Other animals, like corals and clams, can also contribute calcium carbonate to sea floor sediments.

3. **Chemical deposits.** Many chemical changes take place in the sea. Some of these chemical reactions form solid materials that settle to the bottom. An example is the presence of lumps or nodules over large areas in certain regions of the sea floor. These nodules are composed mainly of oxides of manganese, nickel and iron. See Figure 15–16. They seem to have been produced from these substances dissolved in sea water. Other minerals found in sea bottom sediments were formed directly from materials in the sea water. However, these minerals are not abundant and the way they were formed is not completely known.

Results of sediment accumulation. Many devices are available, such as the one shown in Figure 15–17, to obtain samples of sea bottom sediments. Examination of such samples shows that two general types of sediments are common. One kind of sediment is *red clay*. This is composed of at least 40 percent clay particles.

The remainder is made up of silt, sand, and organic material. Some red clays also contain material from meteorites. Actually, red clay is usually brown in color and is sometimes called "brown clay." The first samples obtained happened to have a red color and the name "red clay" became established. Sediments similar to red clay but containing more silt are called *muds*. Various types of muds are common near the continental shelves and slopes.

About 40 percent of the ocean floor seems to be covered with a soft organic sediment called *ooze*. Again, the name is misleading since most ooze looks and feels like sand. However, ooze is not made up of sand. At least 30 percent of it is made up of the remains of radiolarians, foraminifera, and diatoms.

VOCABULARY REVIEW

Match the word or words in the column on the right with the correct phrase in the column on the left. *Do not write in this book.*

1. The scientific study of the oceans.
2. The name given to the water covering most of the earth's surface.
3. Used to determine the depth of the water.
4. The horizontal movements of water.
5. A self-contained underwater breathing apparatus.
6. Samples that show sediment layers.
7. The zone of shallow water around a continent.
8. The true edge of the continent.
9. Name of the bulge caused by an accumulation of sediments at the base of the continental slopes.
10. An unusually deep valley on continental slopes.
11. Long narrow depression at the base of some continental slopes.
12. Extremely level region of half the sea floor.
13. A series of great underwater mountain chains.
14. Sharp volcanic peaks on the sea floor.
15. A very common sea floor sediment.

a. submarine canyon
b. temperature
c. continental slope
d. core samples
e. mid-ocean ridges
f. oceanography
g. continental rise
h. sound waves
i. abyssal plains
j. scuba
k. red clay
l. the sea
m. seamounts
n. currents
o. depth
p. continental shelf
q. trench

QUESTIONS

Group A

Select the best term to complete the following statements. *Do not write in this book.*

1. About how many years ago was the first voyage made for a scientific study of the oceans? (a) 25 (b) 50 (c) 75 (d) 100.

2. Which oceans were included in the first scientific study of the oceans? (a) Arctic, Antarctic, and South Pacific (b) Arctic, Atlantic, and Indian (c) Atlantic, Antarctic, and South Pacific (d) Atlantic, Arctic, and South Pacific.

3. Oceanography is a (a) blend of several sciences (b) blend of only two sciences (c) science independent of other sciences (d) brand new science.

4. All the water in the oceans weighs about (a) 1560 million tons (b) 1560 million million tons (c) 1560 million billion tons (d) 1560 billion billion tons.

5. What part of the earth's mass is water? (a) about 10% (b) about 5% (c) about 1% (d) much less than 1%.

6. What part of the earth's surface is covered with water? (a) 70% (b) 50% (c) 30% (d) 10%.

7. Which of the following is *not* one of the seven seas? (a) Arctic (b) Antarctic (c) Baltic (d) Indian.

8. One of the characteristics of the polar oceans is that they (a) both surround continents (b) are both surrounded by land (c) are more salty than the others (d) are less salty than the others.

9. The temperature tends to remain constant for a certain mass of water because (a) the temperature of water changes slowly (b) the oceans are quite deep (c) currents tend to mix the water (d) thermometers react slowly.

10. Sea water is a solution that (a) is very simple (b) has a few dissolved mineral substances (c) is very complex with many dissolved substances (d) contains mostly gases.

11. An electronic sensing device is used to record (a) temperature (b) dissolved gases (c) depth (d) water motions.

12. Beneath the sea surface, the pressure doubles every (a) 10 meters (b) 32.8 meters (c) 100 meters (d) 1000 meters.

13. A diver enclosed in a protective suit can work in depths up to (a) 36,000 m (b) 11,000 m (c) 200 m (d) 50 m.

14. Bubbles that produce the condition called "bends" can be prevented by (a) breathing air at high pressure (b) breathing air at nearly normal pressure (c) breathing only nitrogen (d) breathing deeply.

15. If the earth's oceans dried up, the sea floor would look (a) similar to earth's landscape (b) exactly the same as earth's landscape (c) different from earth's landscape.

16. Which of the following features is *not* found as part of the ocean floor? (a) barrier islands (b) abyssal plains (c) trenches (d) submarine canyons.

17. The continental shelf is a part of (a) the ocean basin (b) a mountain ridge (c) the continental slope (d) the continent.

18. The continental slope is (a) part of the sea floor (b) the edge of the continent (c) a shallow area near shore (d) the end of a glacier.

19. How wide are the widest continental shelves in the world? (a) 760 km (b) 800 km (c) 1280 km (d) 170 km.

20. The average water depth over the continental shelves is (a) 60 m (b) 105 m (c) 200 m (d) 10,000 m.

21. The depth of trenches is usually greater than (a) 6000 m (b) 60,000 m (c) 600,000 m (d) 6,000,000 m.

22. The level region of the sea floor is called (a) a continental shelf (b) a mid-ocean ridge (c) an abyssal plain (d) a seamount.

23. Separate volcanic mountains on the ocean basin floor are called (a) continental shelves (b) mid-ocean ridges (c) abyssal plains (d) seamounts.

24. Which of the following is *not* a major way that sediment is carried from land to the sea? (a) streams (b) rain (c) wind (d) icebergs.

25. Sea floor sediments containing silica and calcium carbonate come from (a) rocks (b) rain water falling into the sea (c) both plants and animals (d) plant skeletons only.

26. The largest particles of sediment on the sea floor are found (a) near shore (b) far from shore (c) on abyssal plains (d) on top of seamounts.

27. Compared to the total amount of sediment carried by streams, the amount of sediment found on continental shelves is (a) too small (b) about the same (c) slightly greater (d) much too great.

28. Nodules on the sea floor are *not* composed of (a) manganese (b) nickel (c) silver (d) iron.

29. Red clay is usually colored (a) red (b) brown (c) blue (d) green.

30. About 40% of the ocean floor is covered with a soft, organic sediment called (a) sand (b) clay (c) ooze (d) nodules.

Group B

1. Why is the earth known as the "water planet"?
2. Name the "seven seas."
3. Why is the British Navy ship, *H.M.S. Challenger,* famous?
4. What characteristics of the sea are most commonly measured?
5. Why is the science of oceanography said to be a blend of other basic sciences?
6. Describe several instruments used to study the sea.
7. What basic difference is there between the Arctic and Antarctic oceans, other than that they are at opposite poles of the earth?
8. Why is gasoline used in the tanks of some underwater research vehicles?
9. How do the mountains and valleys in the ocean compare to those on land?
10. How are the continental shelf, the continental slope, and the continental rise related?
11. How are abyssal plains formed?
12. Describe a seamount and how it is formed.

13. What is the relationship between seamounts and guyots?

14. Explain how a chain of seamounts can be formed from a single source.

15. Why is a mid-ocean ridge usually broken into sections?

16. Name three types of sediment deposits found on the sea floor.

17. How are most sediments that are found on the continental shelf formed?

18. Describe the nature and origin of the common sea floor sediments.

19. What is meant by the statement, "The total volume of sediments on the sea floor seems to be too small"? How do scientists account for this?

20. What is red clay?

16
sea water

objectives

☐ Describe the physical properties of sea water.

☐ Describe several conditions that change these properties.

☐ Describe the chemical properties of sea water.

☐ Explain how the salt content and dissolved gases in sea water affect sea life.

☐ Explain how the sea can be a valuable resource.

As far as we know, the earth is the only planet that has liquid water on its surface. All other planets in the solar system are either too hot or too cold. Surface temperatures on the earth are temperate and allow water to remain in a liquid state. If the earth were too cold, its water would be another of the solid materials in its crust. If it were too hot, water would be found mostly as a gas. Only in liquid water do we find the properties that make the oceans one of earth's most distinctive features.

All the properties of liquid water in the sea can be divided into two main groups. The characteristics of sea water which permit it to dissolve other substances are generally listed as *chemical properties*. Other characteristics, not connected with dissolved materials, are called *physical properties* of sea water. For example, one of the most important physical properties of sea water is its temperature. Together, the chemical and physical properties of sea water help to give us some idea of what a complicated liquid sea water actually is. A brief study of these properties can also reveal some of the many processes which continuously take place in the sea. A study of the liquid water of the oceans will help us understand the complex relationships which exist between the land, the air and sea.

PHYSICAL PROPERTIES OF SEA WATER

The sea and energy from the sun. The sun is one of the most important single influences on the sea. Energy from the sun reaches the earth in the form of light. This light travels in various wavelengths. Since the sea covers a

FIG. 16–1. Some of the light falling on the sea surface is reflected causing the sparkling effect shown in this photograph. The angle of the sun's rays does much to determine whether the light is reflected or penetrates the water. (Ramsey)

major part of the earth's surface, most of the sun's energy falls upon the oceans. Nearly all of this solar energy penetrates the sea surfaces and is absorbed into the water. Although water appears to be transparent in small amounts, it is actually able to quickly absorb visible light and most other forms of radiant energy as well.

Of the various wavelengths in visible light, only the blue wavelengths are able to travel very far into the water before they are absorbed. At depths greater than 10 meters (about 33 feet), only a blue-green light from the sun can be seen. All other wavelengths or colors of light usually present in sunlight have already been absorbed. It is the ability of blue light to penetrate water that gives the sea and other large bodies of water a blue color. If the water is clear and you look almost straight down into the sea, the water appears blue because it is the last color to be absorbed. The blue color is not so apparent to a viewer looking horizontally across the water surface. Light rays which strike the water at much of an angle are reflected rather than absorbed. See Figure 16–1. Many times the natural blue color of the water is clouded by small particles suspended in the water. No light of any kind can penetrate the sea to depths below a few hundred meters. All but the upper layers of the sea are in total darkness.

Table 16–1 Average Surface Temperature (°C) of the Oceans Between Parallels of Latitude

North latitude	Atlantic Ocean	Indian Ocean	Pacific Ocean	South latitude	Atlantic Ocean	Indian Ocean	Pacific Ocean
70°–60°......	5.60	.....	.....	70°–60°......	− 1.30	− 1.50	− 1.30
60°–50°......	8.66	.....	5.74	60°–50°......	1.76	1.63	5.00
50°–40°......	13.16	.....	9.99	50°–40°......	8.68	8.67	11.16
40°–30°......	20.40	.....	18.62	40°–30°......	16.90	17.00	16.98
30°–20°......	24.16	26.14	23.38	30°–20°......	21.20	22.53	21.53
20°–10°......	25.81	27.23	26.42	20°–10°......	23.16	25.85	25.11
10°– 0°......	26.66	27.88	27.20	10°– 0°......	25.18	27.41	26.01

Surface temperature of the sea. In addition to its ability to absorb the visible wavelengths of solar energy, sea water is capable of also absorbing the longer infrared wavelengths. Energy from these infrared rays reach the earth in the form of heat. Thus they play an important role in determining the surface temperature of the sea. Like visible light, infrared rays are completely absorbed within the upper layers of the sea water. This means that the sun can only heat the upper part of the oceans directly. In the deeper parts of the sea the temperature of the water is always close to freezing (0°C).

The total amount of solar heat falling upon the surface of the sea is much greater at the equator than at the poles. At high latitudes, near the poles, the sun's most slanted rays strike the earth. See page 66. These are weakest in ability to heat the water. Surface temperatures in polar seas usually drop below 0°C. Table 16–1 lists some average surface temperatures for the three major oceans. For sea water to freeze, it must be chilled to about −2°C (28.5°F). The dissolved salts in sea water lower its freezing point below that of water. Vast areas of sea ice exist in both arctic and antarctic waters. If the floating layer of ice completely covers the sea surface it is called *pack ice*. The layer of ice is usually not more than 5 meters (about 17 feet) thick because the bulk of the ice acts as an insulating cover to prevent the water below from freezing. The Arctic Ocean is covered by pack ice during most of the year. The ice pack is brittle and will crack and buckle under pressure when the forces of wind and currents become too great. Broken-off pieces of the ice pack called

FIG. 16–2A. Pressure ridges cause buckling of the pack ice in the sea off the coast of Antarctica. (Black Star-Schulthess)

ice floes are sometimes strewn about stretches of open water. These passages of open water quickly freeze over in winter but usually remain open during the summer. Buckling of the ice pack disturbs its smooth surface and builds up *pressure ridges,* as shown in Figure 16–2A.

In tropical waters near the equator, surface temperatures around 30°C (86°F) are not unusual. One of the most important effects of the high surface temperatures of tropical water is rapid evaporation.

When water absorbs the longer infrared rays from the sun, the heat energy increases the movement of water molecules. It is this increased amount of molecular motion which is recorded by thermometers as an increase in temperature. When the temperature of sea water rises, many water molecules move fast enough to enter the atmosphere. The process of water molecules leaving liquid water and becoming water vapor in the air is called *evaporation.*

Notice that evaporation produces two results. First, the liquid water that is lost by the sea is taken into the atmosphere as water vapor. Second, heat energy is transferred. Water molecules moving from the surface of the sea into the atmosphere removes heat energy represented by their motion. Thus evaporation creates a loss of heat in the sea and a gain for the atmosphere. The energy that is transferred in this process creates a very close relationship between activities in the sea and the atmosphere.

The large amount of water evaporated near the surface also influences the amount of dissolved salts present in sea water. During evaporation only water molecules are removed. Dissolved salts remain behind. This results in an increase in the relative amount of dissolved salts in surface water where evaporation is high. For this reason, tropical waters will have higher concentrations of dissolved salts at the surface than will polar waters.

Temperature and depth. Mainly because the sun does not directly heat sea water below the surface layers, the temperature dips sharply with increasing depth. To understand the reason for this sudden drop in temperature, it is necessary to consider the factors which affect the *density* of sea water. Density expresses how much mass a given volume of sea water will have. It is generally described in grams per cubic centimeter (g/cm³).

Two factors have a major effect on the density of water in the oceans. First, there are the dissolved salts. These add mass to the water in direct proportion to the amount present. The large amount of dissolved substances in sea water make it more dense than fresh water under the same conditions. A second factor affecting sea water density is temperature. When liquid water is heated, it expands slightly; when it is cooled, it shrinks a small degree. This means that if liquid water is warmed, its density generally will be reduced. Cooling will have the opposite effect. Its density will increase. Actually, pure water has a maximum density at about 4°C. Below this temperature liquid water expands as it reaches the freezing point. In addition to dissolved salts and temperature, pressure is considered a minor factor in determining the density of water.

In most places in the sea, measurements indicate a sudden temperature drop not far below the surface. This zone of rapid temperature change, called the *thermocline,* marks the distinct separation between a warm surface layer and colder deep water. See Figure 16–2B. A thermocline exists because heat which enters the water lowers its density as warming takes place. This warm water cannot mix easily with the cold, dense water below. A thermocline is established at the boundary zone between the upper and lower layers of water. The differing densities help to keep them separate. Changing conditions of heat or currents may alter the depth of the thermocline or cause it to disappear completely. Regardless of these possibilities, a thermocline is usually present beneath much of the sea surface. Thermoclines that can definitely be measured are also observed in many lakes.

Describe

In a room with the windows and doors shut, record the temperature near the floor. Take a few readings. Now repeat, taking the temperature near the ceiling. How is this distribution of temperature the same or different from that found in the ocean?

FIG. 16–2B. An idealized graph of changes in temperature and density with increasing depth in the open ocean.

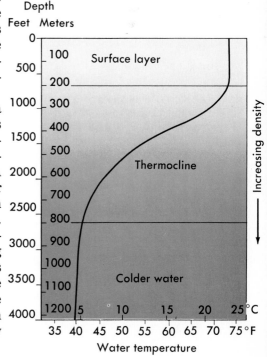

CHEMICAL PROPERTIES OF SEA WATER

activity

Part I

Make a millimeter scale on a piece of masking tape. Make every fifth line extra long. Stick the tape to one end of a plastic straw and run it lengthwise along the straw. Clog up the other end of the straw with clay. Keep adding clay to this end until the straw floats upright in water with about 5 cm of the marked tape still above water. This instrument is a hydrometer.

Make some salt water similar in density to sea water. (Add 3.6 grams of table salt to each 100 ml of distilled water.)

Use a large test tube to hold the water being tested. Record the readings when you float the hydrometer in distilled water, tap water, and sea water.

1. How does the hydrometer level differ in the three liquids? (The hydrometer floats highest in the liquid that is denser.)
2. Which of the liquids is the most dense?

Part II

Add several drops of dark food color to about 100 ml of imitation sea water. Fill a plastic box about half full of tap water. Raise one end and place a book under it. Carefully pour the colored sea water into the raised end of the box and observe what happens.

1. Do the tap water and colored water mix immediately?
2. Where a river enters the ocean, would the water be saltier at the surface or near the bottom?

Dissolved salts in sea water. During the millions of years the oceans have been in existence, sea water has been continuously evaporating. After condensing, this very same water falls as rain. Some of this rain forms the streams and rivers that wash over the land on their way back to the oceans. Large amounts of mineral matter are dissolved in the water as it runs over the land surfaces. Each year the world's rivers carry about 400 million tons of dissolved minerals, including salts, into the sea. These dissolved salts remain trapped in the sea as the water molecules evaporate again in their endless cycle.

Not all of the dissolved matter in the sea originated in minerals found in the rocks. Some of the substances now in sea water probably came from gases produced by volcanoes. Gases that were present in the earth's early atmosphere may also have dissolved in the sea. Other unknown processes have played a role in making the sea a huge reservoir of dissolved substances. If all of these materials could be recovered, they would form a layer 136 meters (about 450 feet) thick over the surfaces of all the continents.

During the great span of time that the sea has been receiving dissolved materials, it has probably reached a state of balance. The processes which are constantly adding new salts to the oceans seem to be balanced by the processes which remove dissolved materials. The formation of sediments on the sea floor removes millions of tons of material from the sea each year. Living things also use up the salts as part of their life processes.

The dissolved salts found in sea water are evidence of the many complicated chemical processes which are at work. At the present time, we have only a very limited knowledge of the chemical reactions which control the composition of sea water. When a sample of sea water is analyzed, a few substances are always found to make up more than 99.9% of the dissolved salts. These substances are listed in Table 16–2. Notice that these materials are represented in the form of ions, a result of the dissolving action of the water molecules. This was described in Chapter 7. Almost all of the known chemical elements have been found in sea water. Most of these are present in very small amounts as ions. All of the known elements are probably present in sea water, but in many cases have not been detected because they exist in such small amounts.

The total amount of dissolved solids present in a sample of sea water is described as its *salinity*. For example, suppose that one kilogram of sea water (1000 grams or about 2.2 pounds) is dried and the total weight of salts which remains weighs 35.0 grams. The salinity of this sample of sea water would then be very near to 35 parts per thousand. This is usually written as: salinity = 35 ‰. Sea water with a salinity of 35 ‰ would have almost 3.5 per cent of its total weight made up of dissolved salts. See Figure 16–3.

The process of evaporation, which removes water, increases the salinity of sea water. On the other hand, heavy rainfall or large amounts of fresh water from rivers bring about a decrease in salinity. Figure 16–4 gives the average surface salinity at different latitudes between the equator and poles. You will notice that because of heavy rainfall near the equator, salinity will be lower in this area than in dry regions. Another factor which produces low salinity near the poles is that the great masses of melting ice at the poles are almost entirely fresh water. Over most of the sea surface, salinity usually ranges between 33 and 36 ‰, with an average value for all the oceans of 34.7 ‰.

Although the salinity or total dissolved salts of sea water may differ from one location to another, the relative amounts of the dissolved salts do not change significantly. This means that the amount of any of the principal ions compared to any other is a nearly constant ratio. This state of equilibrium is brought about by the continuous mixing of all parts of the sea with all other parts. Any dissolved substance emptied into the sea is eventually

Table 16–2 Principal Dissolved Substances in Sea Water.

Ion	Percent of total dissolved solids (By weight)
Chloride (Cl⁻)	55.04
Sodium (Na⁺)	30.61
Sulfate (SO₄⁻²)	7.68
Magnesium (Mg⁺²)	3.69
Calcium (Ca⁺²)	1.16
Potassium (K⁺)	1.10
Bicarbonate (HCO₃⁻)	0.41
Bromide (Br⁻)	0.19
Borate (H₂BO₃⁻)	0.07
Strontium (Sr⁺²)	0.04

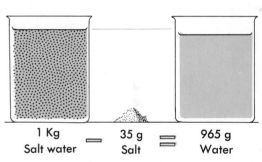

1 Kg Salt water = 35 g Salt = 965 g Water

FIG. 16–3. A kilogram of sea water which contains 35 grams of dissolved salts has a salinity of 35 ‰.

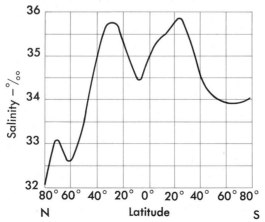

FIG. 16–4. Average salinity for surface water at various latitudes.

FIG. 16–5. A sample of sea water for analysis is being taken from the sampling bottle. (Ramsey)

FIG. 16–6. Potash salt deposits remain after sea water evaporates. (Magnum-Brian Brake)

spread around uniformly through all the water. Thus each dissolved salt will usually be found in the same proportion everywhere in the sea.

In addition to substances found naturally in sea water, some materials are present as a result of human activities. Until recently, people were able to safely use the oceans as a dump. Wastes would be made harmless by being spread through the sea water and become very dilute or be destroyed. But the growth of population and increase in industry has seriously changed the situation. The ability of the oceans to absorb wastes is far less than the ever increasing amounts that are discharged all over the world. Coastal waters are particularly in danger but pollutants can now be found everywhere in the sea.

Traces of the insecticide DDT have been detected in various places in the open ocean. Measurable amounts of lead material in the Pacific have increased by ten times since it first began to be used in gasoline about fifty years ago. Radioactivity from nuclear fallout can be detected in water samples taken from any ocean. While none of these materials has built up to harmful levels in sea water, their presence is a warning that the ocean can not forever be used as a garbage dump.

Dissolved gases in sea water. In addition to the ions which form the solid salts left when sea water evaporates completely, the oceans contain large amounts of dissolved gases, almost all of which enter the sea from the atmosphere. The principal gases of the atmosphere are nitrogen (N_2), oxygen (O_2), argon (Ar), and carbon dioxide (CO_2). All four of these gases are found in the sea, but unlike the dissolved solids, generally not as ions. The dissolved gases usually appear in molecular form just as they do in the atmosphere.

An exception is carbon dioxide which chemically combines with the water to form carbonic acid (H_2CO_3). The reaction itself dissolves large amounts of carbon dioxide. Carbon dioxide is found in sea water in a variety of forms. The total amount of carbon dioxide dissolved in the sea is much greater than the amount which exists in the atmosphere. The other atmospheric gases cannot dissolve as easily in water and are present in the sea in smaller amounts.

Temperature has a strong effect on the amount of gas that dissolves in water. For instance, cold water can dissolve gases better than warm water. In colder regions, water at the surface of the sea will usually dissolve larger amounts of gases from the air than warm tropical waters.

Certain conditions allow dissolved gases to leave the sea and return to the atmosphere. Any time the amount of a dissolved gas becomes greater than the amount which can be dissolved at a particular temperature, excess gas will leave the water. The sea and the atmosphere are continuously exchanging gases as conditions change with time from place to place.

Life and the chemistry of the sea. Once a substance is dissolved in sea water, it becomes part of a very complex, always changing system. Many factors are at work in the sea to change the composition of its water. Evaporation has already been mentioned as a process which concentrates the dissolved substance. Chemical deposits which become part of the bottom sediments remove certain substances and thus also change the composition of sea water. Volcanic action both adds and removes materials from sea water. Many other natural but unknown processes which occur in or near the sea probably act to change the makeup of sea water.

Among the most outstanding of these processes are the changes caused by living things in the sea. These animals and plants must receive from the water, directly or indirectly, all of the materials required for their life processes. And finally, the substances produced in their bodies return a tremendous variety of by-products to the water. Each step in this continuous series involves changes in the chemical composition of the surrounding water.

All life in the sea is regulated by its plant life. Plants remove certain dissolved substances from the water needed for growth and turn them into organic or living matter. To meet these needs, plants absorb large amounts of various substances containing the elements carbon, hydrogen, oxygen, and sulfur. These materials are present in such large amounts in sea water that biological processes do not have any significant effect on their concentrations. More critical to the growth of plants are compounds containing the elements nitrogen, phosphorus, and silicon. Since these substances are not as abundant, heavy plant growth reduces their concentration to nearly zero. In this case the growth of plants in the depleted water will be greatly slowed down or stopped completely. This situation can be compared to the problems of plant growth in poor soil on the land.

Investigate
Put some water in a dish and then cover the dish with clear plastic food wrap. Place the bowl in direct sunlight for at least two hours. How could a method similar to this be used for getting fresh water from the oceans?

In addition to certain dissolved substances from the water, almost all plants in the sea require sunlight. This means that plant growth in the sea is restricted to the upper few meters into which light penetrates. Below about 80 meters (265 feet), there is never enough light to meet all the needs of plants. However, within the depth zone where there is sufficient light, most regions of the sea contain large amounts of free-floating, microscopic living forms called *plankton*. See Figure 16–7. There are two main types. The microscopic plants in plankton are known as *phytoplankton*. These tiny plants remove dissolved materials from the water and use energy from light to carry on photosynthesis. In the cycle of sea life, the phytoplankton then serve as the source of food for microscopic animals or *zooplankton*.

Both forms of plankton are eaten by larger life forms such as small fishes and squid. These, in turn, become the food of adult fish and other large marine animals. However, some large animals, certain whales for example, feed directly on plankton. In any case, phytoplankton are always the first link in the complicated food chain which supports life in the sea. See Figure 16–8. In addition, because phytoplankton require light to carry on photosynthesis, most of the chemical changes associated with life in the sea are confined to the upper layers of the water.

All of the elements needed to support life in the sea are removed from the sea water. In time these elements are returned to the water when the plant or animal remains decay. Bacteria in the water attack the organic remains and again release the trapped elements. The decay processes occur at all depths, but gravity will pull the organic

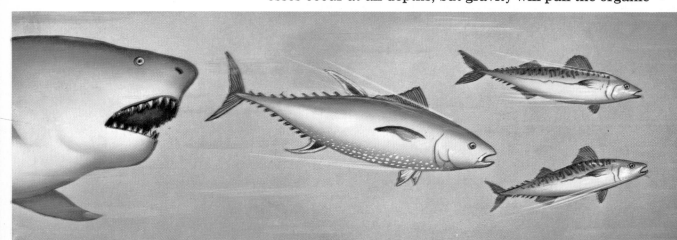

remains slowly downward from the near-surface layers where biological activity is greatest. A tendency exists for the necessary elements in water to be consumed near the surface but released at greater depths. Thus deeper water becomes a storage region for the vital materials needed to support life. Some deep-dwelling animals use the material coming from above as a source of food. Yet most living forms in the sea depend upon the needed substances being returned again to the surface.

Several processes seem to be at work in the oceans causing deep water to move upward. Among the processes which appear to play an important role in returning the needed materials in deep water to the surface are:

1. *Upwelling.* When wind blows steadily away from the shore along a coastline, surface water is moved out to sea. Deep water then moves upward to replace the surface water. See Figure 16–9. This situation is commonly found off the west coast of South America, the California coast, and the northern coast of Florida.

2. *Overturn.* When surface water is chilled it becomes more dense and will sink. Warmer water at a greater depth then moves upward.

3. *Mixing.* In shallow water wave action on the shore may be powerful enough to cause deep water to mix with surface water. Tides can also serve this function.

Other processes are probably also at work to make deep water move upward. The search for these processes is of great interest to oceanographers. The distribution of life in the sea depends, to a large extent, on the way these life-supporting materials return from the depths.

FIG. 16–7. Microscopic photograph of typical plankton organism. (Dr. Roman Vishniac)

FIG. 16–8. A typical food chain in the sea is shown beginning with microscopic plankton and ending with a shark.

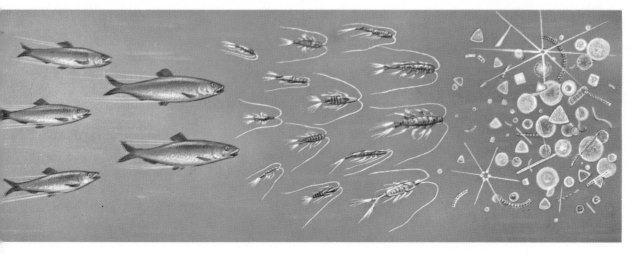

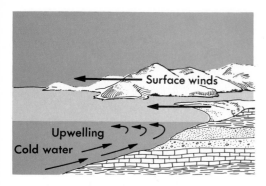

FIG. 16–9. Upwelling may cause cold water currents present in coastal areas. It tends to cool the land nearby and at the same time force to the surface, microscopic marine animals which help supply the food chain.

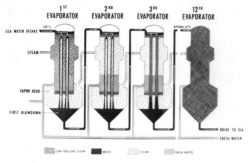

FIG. 16–10. A desalting plant located at Freeport, Texas. This plant can produce one million gallons of fresh water per day. (U.S. Department of the Interior—Office of Saline Waters)

THE SEA AS A SOURCE OF WEALTH

Fresh water from the sea. The sea has always been a source of food and a means of transportation. Recently, however, we have begun to realize its importance as a source of vital resources needed to support our civilization. Increased knowledge of the oceans, combined with the earth's expanding population has caused new interest in the sea as a source of water, minerals, and food.

One of the most important resources of the sea is its water. The world's need for water is increasing at a rapid rate. Developing countries need huge new supplies of water for industry and irrigation. Nations like the United States, that formerly had abundant water supplies are facing shortages. The demand for water in the future can be met in two ways: Most important, the water now available must be used carefully to avoid waste. The long term solution to the problem is to increase the amount of water available. This can be done, if we find a way to convert sea water to fresh water, at reasonable cost.

Several methods can be used to extract fresh water from salt water. The most common method now used is *distillation*, which involves heating sea water. This causes the water to be changed into vapor that is carried off to be condensed into pure fresh water. See Figure 16–10. However, changing liquid water into vapor requires a great deal of costly heat energy. Making fresh water by distillation will be expensive unless a way to use cheap solar or nuclear energy can also be developed.

Another method for desalting water involves freezing it. When sea water freezes, the first ice crystals that form are free of salt. The salt remains in pockets of liquid water in the ice. The ice can then be melted to obtain fresh water, using only about one-sixth as much energy as is needed for distillation. Other methods for desalting sea water use special membranes that allow water to pass through, but block the dissolved salts. Chemicals can also be used to combine temporarily with either the water or salt to separate them.

Minerals from the sea. Since the first oceanographic research voyage made by the *Challenger*, strange black lumps of material called *nodules* have been discovered on the sea floor. Investigation has shown that huge areas of the ocean basin floor, particularly in the Pacific, are covered with these nodules. See Figure 16–11. They contain valuable amounts of manganese that is used in making certain kinds of steel. They also contain a large amount of iron, and small amounts of copper, nickel, and cobalt. Others contain phosphates that are useful as fertilizers.

Recovery of nodules from the sea floor is difficult. They are found mostly in very deep water. Several methods of mining these minerals are shown in Figure 16–12. The way the nodules are formed is not completely understood. However, measurement of their rate of growth suggests they grow fast enough to supply the world's present needs for some of the metals they contain. In the future, the mineral-rich nodules from the sea floor may largely replace the disappearing mines on land.

Some materials of value are now extracted easily from the oceans. Common salt has been obtained by evaporation of sea water for many centuries. The sea is the main source of magnesium metal and bromine. These two materials are manufactured from the dissolved salts in sea water. However, most of the useful minerals dissolved in the oceans are in such small concentrations that it is not practical to extract them. For example, sea water contains about 6 kg of gold in each cubic km. It would require processing about 4 million liters of sea water to obtain a few cents worth of gold.

The most valuable mineral resource taken from the ocean is the petroleum found beneath the sea floor. Huge deposits of oil and natural gas are found along the continental margins in many parts of the world. As the need for energy grows, these fuel resources have become very important. New techniques of drilling have allowed oil and gas to be produced from water depths as great as 100 meters as far as 100 km offshore. In the future, it will be possible to drill wells at depths of several thousand meters along the continental shelves and on the continental rise.

Food from the sea. Of all the resources that the sea is capable of supplying, the one in greatest immediate demand is protein food. At present, a large part of the world's population must rely on a diet of starchy food. While such a diet can maintain life, the lack of protein to build strong tissue allows disease to become a more serious problem. Perhaps half a billion people in the entire world suffer from some form of disease due to a lack of protein in their diets. The only source of high protein food able to meet this mounting need is fish from the sea.

But the present methods for fishing in the sea are not

Fig. 16–11. Nodules found on the floor of the Pacific Ocean. (Kennecott Copper Company)

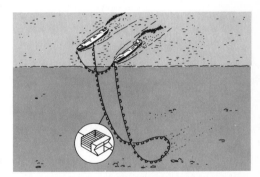

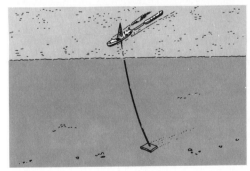

FIG. 16–12. Top. Two ships carry a cable with a continuous line of dredge buckets that scoop up nodules. Bottom. Air pumped into a pipe running from a ship to a dredge creates a pressure that brings nodules to the surface.

satisfactory. While humans have learned to manage the resources of the land to grow animals for food, they still hunt the animals of the sea. The full food resources of the oceans cannot be utilized until humans learn to cultivate the sea the way farmers do the land.

In the future, the practice of *aquaculture* or farming of the sea may become as important as agriculture to provide food for the world's population. Aquaculture involves developing and raising special breeds of water dwelling animals and plants that yield large amounts of food. The principles of aquaculture have already been used to grow fish such as trout in large aquatic farms. See Figure 16–13. Similar methods might be used to breed fish, shellfish, and plants especially suited to grow in ocean farms. These farms would be closed off areas of the sea that provide a suitable environment for aquaculture.

Under the best conditions, water farming could produce more valuable protein food than an equal amount of land. Fish and shellfish are more efficient than most land animals in changing their food supply into protein. Water dwelling animals do not waste energy keeping an even body temperature or in having to support their weight against the force of gravity. Also, agriculture on the land can only use the top layers of soil, while ocean farms will be able to use the entire depth of water to produce crops. The sea may even be "plowed" by pumping the bottom water that is rich in nutrients to the surface. This kind of artificial upwelling would greatly speed up the growth of the "crop." With additional research and experience, it will be possible to develop the food resources of the sea far beyond the limited methods now used.

FIG. 16–13. An aquatic farm where trout are raised for food. (Thousand Springs Trout Farm)

VOCABULARY REVIEW

Match the word or words in the column on the right with the correct phrase in the column on the left. *Do not write in this book.*

1. Properties of the sea connected with its ability to dissolve other substances.
2. Properties of the sea which include temperature and depth.
3. A floating layer of ice completely covering the sea surface.
4. Broken off pieces of an ice pack.
5. Disturbances on the smooth surface of an ice pack due to buckling.
6. Is measured in units of grams per cubic centimeter.
7. Zone of rapid temperature change which separates a warm surface layer of water from colder deep water.
8. Describes the total amount of dissolved solids present in a sample of sea water.
9. The general name of free-floating microscopic plants and animals in the sea.
10. The food for most zooplankton.
11. Free-floating microscopic animals in the sea.
12. Deep water moving upward to replace the surface water moved away from shore.
13. A method for removing dissolved salts from sea water.

a. density
b. pack ice
c. sea water
d. pressure ridges
e. salinity
f. chemical properties
g. upwelling
h. ice floes
i. plankton
j. physical properties
k. distillation
l. thermocline
m. overturn
n. phytoplankton
o. glacier
p. zooplankton
q. mixing

QUESTIONS

Group A

Select the best term to complete the following statements. *Do not write in this book.*

1. The temperature of sea water is an (a) important chemical property (b) unimportant chemical property (c) important physical property (d) unimportant physical property.
2. The most important single influence on the sea is (a) the sun (b) the wind (c) mixing (d) upwelling.
3. Which color of visible light penetrates farthest into the ocean? (a) red (b) orange (c) yellow (d) blue.
4. The sun heats the upper layers of the sea (a) less than the deeper layers (b) about the same as the deeper layers (c) more than the deeper layers (d) by deep penetration.
5. For sea water to freeze it must be chilled to about (a) 0°C (b) −2°C (c) −28°C (d) +2°C.
6. The process of water molecules escaping into the atmosphere after receiving sufficient energy from the sun is called (a) evaporation (b) density variation (c) condensation (d) rain.

7. Evaporation in the sea causes heat to be (a) gained by the sea and gained by the atmosphere (b) lost by the sea and gained by the atmosphere (c) gained by the sea and lost by the atmosphere (d) lost by both the sea and atmosphere.

8. Water reaches its maximum density at (a) 0°C (b) 0°F (c) 4°C (d) 4°F.

9. The amount of dissolved materials in the sea is probably (a) increasing (b) decreasing (c) remaining about the same (d) very small.

10. Analysis of sea water shows that the substances that are found in greatest amounts are ions of (a) sodium and sulfate (b) chloride and sulfate (c) magnesium and sodium (d) sodium and chloride.

11. The total weight of salts in one kilogram of sea water is 33 grams. The salinity of this water is expressed as (a) .33 ‰ (b) 3.3 ‰ (c) 33 ‰ (d) 330 ‰.

12. The gas found dissolved in the sea in an amount greater than its amount in the atmosphere is (a) nitrogen (b) oxygen (c) argon (d) carbon dioxide.

13. Which of the following usually has the greatest effect on the chemical properties of sea water? (a) plants and animals (b) currents (c) temperature (d) depth.

14. Which of the following groups of elements may have their concentration in the sea reduced as a result of heavy plant growth? (a) carbon, hydrogen and oxygen (b) carbon, nitrogen and sulfur (c) sulfur, phosphorus and silicon (d) nitrogen, phosphorus and silicon.

15. Plant life in the sea is restricted to the upper few meters because most plants require (a) low temperatures (b) sunlight (c) rain (d) high pressure.

16. The primary reason that zooplankton are found mostly in the upper layers of the sea is because (a) they need sunlight (b) the temperatures are higher (c) the pressure is less (d) phytoplankton are restricted to this region.

17. Which of the following terms is *not* closely related to the others? (a) upwelling (b) overturn (c) mixing (d) freezing.

18. Desalting of sea water is a solution for the problem of (a) fresh water not always being located where it is needed (b) salt water fish being relocated (c) rain reducing the salt concentration in sea water (d) river water increasing the salt concentration in sea water.

19. Which of the following is *not* a practical method of desalting? (a) evaporation (b) filtering (c) freezing (d) none of these.

20. Desalting of sea water has not been done on a large scale because (a) it is too dangerous (b) no one knows how (c) it costs too much (d) energy cannot be produced to do it.

21. Which of the following is *not* presently obtained from the sea in large amounts? (a) salt (b) bromine (c) gold (d) magnesium.

Group B

1. Explain why the earth is likely to be the only planet in the solar system with oceans of liquid water.

2. Explain what is meant by physical and chemical properties of sea water.

3. Why is desalting of sea water important?

4. Explain the process of evaporation.

5. Why are fish found primarily in the upper 200 meters of the ocean?

6. Why is the ocean usually blue?

7. A sample of sea water is found to contain 34 grams of dissolved materials in one kilogram of the sample. What is the salinity of the sample?

8. A rock has a mass of 350 grams and a volume of 70 cubic centimeters. What is its density?

9. Why do we say that living things have a great influence on sea water?

10. Explain why heating of the ocean by the sun is greater near the equator than near the poles.

11. Give two reasons why the upper part of the sea is warmer than the deeper part.

12. Four hundred million tons of dissolved materials are carried to the oceans each year by rivers. Why is there probably no great increase in the salinity of the sea?

13. A sample of sea water whose mass is 2000 grams is heated until only solids remain. These solids have a mass of 72 grams. What was the salinity of the sample?

14. When water in a pond freezes, the water at the bottom is warmer than at the surface. Therefore the surface freezes first. Explain why this happens.

15. An object will float if its density is less than the density of the fluid it is to float in. The density of sea water is about 1 gram per cubic centimeter while that of steel is about 7.5 grams per cubic centimeter. Under these conditions how can a steel ship float?

16. The volume of a sphere is given by the formula $V = 4/3 \pi r^3$, where r is the radius. A spherical rock has a diameter of 4.0 cm and a mass of 200 grams. Find its density.

17. Vast amounts of minerals are found dissolved in the oceans, including gold, bromine and magnesium. Why are bromine and magnesium obtained from ocean water while gold is not?

18. Why is the ocean considered to be a valuable source of food?

19. Discuss the proportion, types, uses, and origin of mineral matter in sea water.

20. Compare and contrast ice floes and icebergs.

21. Discuss the cause and properties of the thermocline.

22. What kind of chemical change does life in the sea cause?

23. Distinguish between plankton, phytoplankton, and zooplankton.

24. Describe the food chain found in the sea, ending with the large fish.

17
motions of the sea

objectives

☐ Identify the causes of sea water movements.

☐ Describe the patterns of circulation near the sea surface.

☐ Describe the characteristics and effects of the Gulf Stream.

☐ Compare deep currents with surface currents.

☐ Describe the characteristics of ocean waves.

☐ Explain how ocean waves change near shore.

☐ Identify the causes and effects of a tsunami and tides.

Forces acting upon the surface waters of the ocean stir the seas into constant motion. Energy from the sun powers the circulation of air in the atmosphere. Winds are created in the lower atmosphere by the sun's energy and are affected by the earth's rotation. These winds push the surface waters of the ocean into an ever-changing pattern of near-surface currents called "circulation cells" or *gyres* (*ji*-ers). Gravitational forces of the moon and sun add to the swirl of the waters through the rhythm of the tides. Even the rotation of the earth on its axis contributes to changes in the direction of current movement in the sea.

Many of the motions of the sea are quite complex and difficult to trace. It is hard to follow the movements of sea water to find patterns that might lead to the discovery of all the causes. Moving masses of water can be identified by their physical and chemical characteristics. Then their movements must be tracked until it is clear where the water has come from, where it is going, and what causes it to move. Frequently there is not enough information available. Many of the movements of water in the sea are poorly understood, but as oceanographers slowly collect accurate and detailed observations, more pieces begin to fit into the puzzle. Although the picture is far from complete, the general motions of the sea and their possible causes can be outlined.

GENERAL CIRCULATION IN THE SEA

Causes of sea water movements. The liquid water of the sea can be set into motion only if it receives energy. For the oceans, the most important source of energy is the

352

sun. Almost all water movements in the sea can be traced back to the sun as the original source of energy. But most of its energy of motion is not received "directly" from the sun. It comes to the sea by way of the atmosphere.

The atmosphere, like the oceans, is in continuous motion. These atmospheric movements are caused by absorption of solar energy by the gases of the air. This atmospheric motion creates a pattern of winds over the surface of the earth. Global winds will be studied in detail in Chapter 20. Some of the energy possessed by moving air is transferred to the water of the sea. Winds blowing across a body of water will move the water in the direction of the wind. Steady winds that blow constantly in the same direction can move large masses of water near the surface of the sea. Almost all of the surface currents of the oceans are the result of established wind patterns over the entire globe. See Figure 17–1.

Some solar energy is directly absorbed by sea water and is eventually returned to the atmosphere. This heating and cooling of the water as a direct result of the sun's radiant energy also results in sea water movement. This movement may occur in one of two ways. First, when water is heated or cooled its density changes. Cool water becomes denser and sinks. Warm water is lighter and rises. Thus colder regions of the sea have a downward movement of cold water. Warmer water then moves in near the surface to replace the colder sinking water. Temperature differences near the upper layers create a steady cyclical movement of cold water downward close to the ocean floor, as the warmer water moves upward close to the sea surface.

Second, water motion may result from evaporation of water by solar energy. When this happens the relative concentration of dissolved salts increases. Then the higher relative density of the water makes it sink. In regions of the sea where evaporation is high, salty water at the surface will sink and be replaced by less dense water from below.

Winds and differences in density, already mentioned as causes of sea water movement, operate at all times to produce ocean currents. Other factors also produce water motion but only occasionally. For example, a very strong temporary current may be caused by an underwater landslide. These occur when large masses of sediment accumulated on a sloping part of the sea bottom suddenly break loose and slide downward. This seems to be a fairly

FIG. 17–1. The earth from 22,300 miles in space. The swirling cloud formations (white areas) are the result of the continuous motion of the atmosphere, which also affects surface movements of the sea. (NASA)

Demonstrate
Fill a small pyrex beaker with water. Place the beaker on a stand. Use an alcohol or bunsen burner to heat one side of the beaker. Pour a few drops of food coloring on the cold side of the beaker. Describe what you see. How is this similar to convection currents?

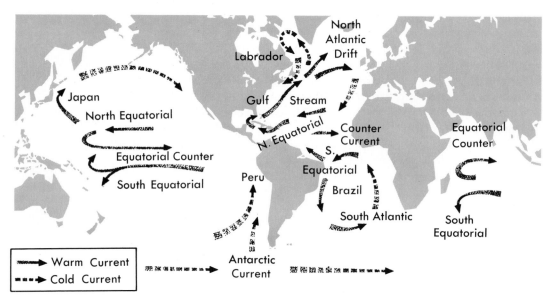

FIG. 17–2. Ocean currents of the world.

activity

When light shines on water, it heats the water. In this activity you will determine if the heating is the same at all depths of water.

Fill a beaker nearly full of water. Use rubber bands to hold three thermometers together. The bulb of one should be at the bottom of the beaker, one in the water near the surface, and one midway between the other two.

Make a data table to record the temperature readings of each thermometer every two minutes.

With the thermometers in place, move the apparatus into the sun (or use a 250 watt infrared lamp). Cover the side of the beaker so that light enters only at the top.

Describe the effects of sunlight on the heating of the ocean at various depths.

common event on the continental slopes. Because a large amount of sediment carried along by the water turns it cloudy or turbid, these movements are called *turbidity currents*. Since these currents are difficult to observe directly, little is known of their effects. They may, however, play a major role in the movement of sediment away from the continental shelves to the deep sea floor.

Earthquakes and volcanic activity on the sea floor may also be a temporary factor in the motions of sea water. Of special interest because of their destructiveness are the poorly named "tidal" waves, or large-scale water movements caused by earthquakes on the sea bottom. These waves will be discussed later in this chapter.

Patterns of circulation near the sea surface. Water within the upper layers of the sea moves mainly in response to winds. But it is not free to move in the direction influenced by the global wind patterns. Land masses of the continents acts as barriers to the motion of wind-driven currents. The rotation of the earth also has an effect on the system of ocean currents. Because of the earth's rotation, the path of a moving object is deflected from a straight line. This shift in direction is to the right in the Northern Hemisphere and to the left in the Southern Hemisphere. This is called the *Coriolis effect*, after the nineteenth-century French mathematician who first described it. Ocean currents are subject to the Coriolis effect and move in a direction that is partly determined by it. However, the effect is more noticeable in the atmosphere. The Coriolis effect in relation to winds will be discussed in Chapter 20.

In all the oceans there is a powerful surface current near the equator moving toward the west. It is driven by the steady winds characteristic of the warm equatorial regions. If there were no continents to deflect this current, it would continuously circle the earth like a great river in the sea. The continents, however, form barriers which turn the current to the north or south. Equatorial currents are found in the Atlantic, Pacific, and Indian Oceans. In each ocean there are two westward flowing parts, the *North Equatorial Current* and the *South Equatorial Current*. Separating these is the east-flowing Equatorial Counter Current. See Figure 17–2.

In the Atlantic and Pacific Oceans, the North and South Equatorial Currents are turned to the north and south along the shores of the continents. At higher latitudes, steady winds push the water from west to east, opposite the direction of current flow near the equator. This action, along with the Coriolis effect, creates great circles of moving water in each ocean. These circular movements, plus many smaller movements form the basis for the principal currents. See Figure 17–3.

In the North Atlantic, the North Equatorial Current piles water against the east coast of North America around the Gulf of Mexico. This water then moves north along the east coast of the United States as a warm, swift current called the *Gulf Stream*. The Coriolis effect then forces the Gulf Stream to move to the right, toward the

FIG. 17–3. Major wind patterns of the earth shown in (A) combine with the effects of the earth's rotation to produce a generally clockwise movement in northern oceans and a counterclockwise one in the Southern Hemisphere (B).

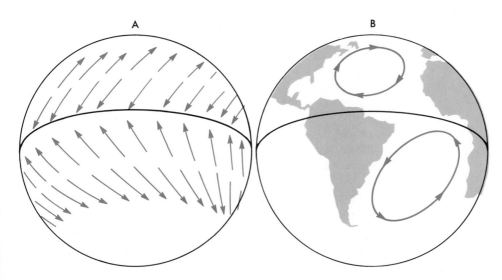

A   B

northeast. This takes it into the North Atlantic. There it branches into three weaker currents. One branch, the *Labrador Current,* doubles back to the south and carries colder water back along the northeast coast of the United States. Another branch turns southward in the direction of the mid-Atlantic, where it eventually disappears. A third branch is the *North Atlantic Drift.* A drift is a weak current. It crosses the North Atlantic and then turns south along the west coast of Europe, finally rejoining the North Equatorial Current.

These currents completely circle the North Atlantic, leaving in the center a vast still area which is relatively free of current action. This is called the *Sargasso Sea.* Great quantities of seaweed are found floating in this quiet region of the Atlantic, named for the *Sargassum* seaweed typical of these waters.

In the Pacific, the pattern of currents is similar to those in the Atlantic. The North Equatorial Current of the Pacific sweeps all the way across the mid-Pacific from Panama to the Philippines. When it meets the island barriers in the east Pacific, the greater part of the current turns to the north and becomes the *Japan Current.* The warm waters of the Japan Current, the Pacific equivalent of the Gulf Stream, pass along the Asian coast. They then swing to the northeast as the *North Pacific Current.* This current carries the cool water from the North Pacific down the west coast of North America to join the North Equatorial Current off the coast of Mexico.

In the southern parts of the Atlantic and Pacific, the current patterns also appear in the form of huge circles. Whereas the currents circle in a clockwise direction in the northern oceans, they move in a counterclockwise direction in the southern oceans. See Figure 17–2.

In the most southerly regions of the Atlantic and Pacific, constant west winds produce the *Antarctic Current.* This powerful current completely circles the Antarctic continent. There are no land masses to interfere with its movement.

Within the Indian Ocean, surface currents follow a two-part pattern. In the southern part of the ocean, the South Equatorial Current is deflected to the south along the coast of Africa. It then flows east toward Australia, where it mingles with the many smaller currents among the islands of the South Pacific. Currents in the northern region of the Indian Ocean are governed by winds that change with the seasons.

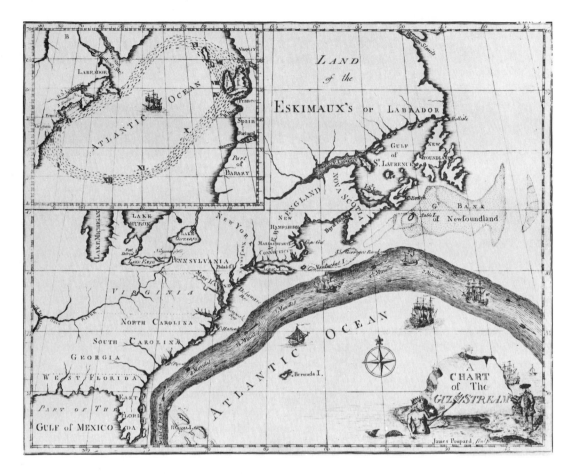

FIG. 17-4. Franklin's chart of the Gulf Stream. (Black Star—Werner Wolff)

The Gulf Stream. Sometime before 1770, Benjamin Franklin became aware that many of the New England whaling ships crossed the Atlantic from England about two weeks faster than the regular mail ships. The whaling captains told Franklin of a swift eastward flowing current in the North Atlantic. They avoided this current by sailing south rather than on a straight line from England. From what he could learn from sailors, Franklin was able to draw and publish in 1779 the first chart of this current. See Figure 17-4. It is still known as the Gulf Stream, the name Franklin gave it, and illustrates the characteristics of many of the sea's surface currents.

The Gulf Stream begins where the North Equatorial Current is deflected by the Panama ridge connecting North and South America. The Equatorial current piles up water in the Gulf of Mexico with such force that sea level there is significantly higher than in the Atlantic.

activity

Density currents like those in the ocean can be produced in the laboratory.

Place an ice cube in a cup of cold water. Then, add some dark food color.

Fill a clear rectangular box (shoe box) about half full of warm (not hot) tap water. Raise one end of the box by placing it on a book. Carefully pour the colored ice water into the raised end of the box, allowing it to flow down the end of the box, so it does not splash. Observe the results.

1. Did the two water masses mix immediately?

2. Where does the colder water tend to go?

3. In the ocean or a deep lake, where would you find the coldest water?

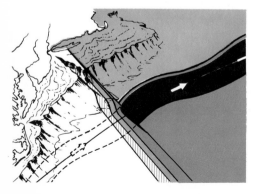

FIG. 17–5. Relative speed of movement of water in the Gulf Stream. Darker colors indicate higher speeds.

The mass of trapped water discharges in a swift flow past the tip of Florida entering the Atlantic as the Gulf Stream.

Once in the open sea, the Gulf Stream flows swiftly in a wavy path along the southeast coast of the United States. Its speed of movement does change, but it usually moves along at the rate of about 10 to 15 kilometers per day. The moving stream of water that makes up the Gulf Stream averages about 100 kilometers in width and at least one kilometer in depth. Because of its size, it carries an enormous amount of water. The Gulf Stream is estimated to transport between 75 to 90 million cubic meters of water per second. By comparison, the Mississippi River averages 20,000 cubic meters per second.

Because it originates in the equatorial regions, the water carried by the Gulf Stream is warmed at its source. The diagram in Figure 17–5 shows how the warm water flows along just off the continental shelf at the latitude of North Carolina. Strangely, this warm water flowing off the Atlantic coastline is often the cause of a drop in temperature in the eastern United States. This climatic change occurs because cooler air moves toward the Gulf Stream to replace the rising air warmed by the current. Thus cold air from inland may move seaward bringing cool weather to the coast. Some of the severest winters along the Atlantic coast have occurred when the water of the Gulf Stream was at its warmest.

Moving north, the stream is deflected to its right by the Coriolis effect. The force that twists the stream off to the right piles water up along the right side of the current. Because of this, water along the Cuban coast is about 50 centimeters (around 20 inches) higher than on the side of the current toward the mainland.

Near Greenland the warm water meets the southward moving cold water of the Labrador Current. Great banks of fog are created when the warm and cold water meet. The sea off Labrador is one of the foggiest regions of all the oceans. Collision with the Labrador Current helps to deflect the Gulf Stream and slow it down. It splits into three parts and sends one branch north to the western shore of Greenland. Another branch passes along the southwest coast of Iceland. But the main branch proceeds eastward as the North Atlantic Drift. This slow movement of warm water across the North Atlantic greatly influences the climate of northern Europe and the British Isles. After warming Europe, the water moves slowly south and eventually becomes part of the westward-flowing Equatorial Current. In this way the water is carried back to its beginning in the Gulf of Mexico.

Deep Currents. In addition to wind driven surface currents, there are slower but equally powerful currents flowing deep beneath the surface. An increase in density due to cooling or an increase in salt content will make water sink. These slower currents seem to be produced as surface water is pulled downward. Very little is known about the way water moves at great depths in the various oceans. However, the deep circulation of the Atlantic has been worked out in sufficient detail for us to understand how water moves at great depths.

In a small region of the North Atlantic, just south of Greenland, the water is very cold and salty. Its salt content is increased by the formation of ice. This cool water with high salinity sinks and moves to the south as a deep current. It flows southward beneath the Gulf Stream. Near the equator, part of this deep water begins a return flow back north again. The above facts tell us that the North Atlantic has a circling deep current which runs in a direction opposite that of the surface currents.

Part of the deep water from the North Atlantic moves into the South Atlantic. As it nears Antarctica, the deep Atlantic water meets similar cold, salty water formed off the coast of Antarctica. The sinking Antarctic water seems to move outward in all directions from the polar region. Deep water from the Atlantic mixes with the Antarctic water and is probably carried into other oceans. Some of the Atlantic water eventually finds its way back to the surface. The complete pattern of deep circulation in the Atlantic is shown in Figure 17–6.

Deep currents are generally more slow moving than those near the surface. An amount of cold, salty water sinking in the polar regions seems to take about 30 years to find its way to the equator. At the surface, the same amount of water would take only a few years to complete the entire circle of an ocean such as the North Atlantic.

WAVES

Wind and waves. Waves on the surface of the water are created when energy is added to a body of water. Energy

FIG. 17–6. Dark blue areas show where cold water is sinking to cause deep currents in the oceans of the two hemispheres.

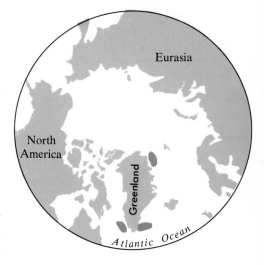

FIG. 17–7. Waves breaking on this beach are the result of swells approaching the shore. (Greene-FPG)

to create waves is added to the sea in many ways. Earthquakes and undersea landslides, gravity of the moon and sun, changes in atmospheric pressure, and the movement of ships are examples of ways that waves are raised on the sea. But one source of energy far exceeds all the others as a force creating ocean waves. That force is the wind.

If the air movements generated by solar heat did not exist, the surface of the sea would be almost as smooth as glass. Even the smallest breeze creates ripples as a result of friction between moving air and the water. The wind is able to push directly against the side of a small wave. As the wave grows larger, more energy can be transferred to it from the moving air. Thus as long as the wind blows the wave increases in size. Actually the wind seldom blows constantly from the same direction. Usually the sea surface is covered with a confused pattern of small waves moving in all directions in response to the shifting winds. However, the larger waves are better able to capture energy from the wind than the small waves. This means that although the wind usually produces waves of all sizes, the larger ones continue to grow while the smaller waves quickly die out.

The usual pattern of waves on the sea surface becomes one of small waves and ripples continually being formed and disappearing on the slopes of larger waves. Groups of large waves, which are alike in size, are called *swells*. See Figure 17–7. They are able to move great distances over the sea surface with very little change. Swells move in groups, one following the other away from the area of their origin. The swells which finally arrive at the shore may have been produced a thousand miles out at sea. From careful analysis of the waves, oceanographers can usually determine where the swells originated. They can also predict the height of waves likely to occur at any point on the water's surface or along the ocean shore.

Characteristics of an ocean wave. A wave on the surface of the water has two basic parts. A ridge which is elevated above the surrounding water surface is called a *crest*. On either side of the crest is a depression or *trough*. Any particular wave can be described by three characteristics. One is its *height:* the distance from the bottom of the trough to the top of the next crest. The second characteristic is its *wavelength;* the distance between crests. The third is its *period:* the time it takes for two consecutive crests to pass a given point. See Figure 17–8. The speed at which a wave moves is determined by a simple relationship between its wavelength and period. This relationship is given by the following formula:

$$\text{wave speed} = \frac{\text{wavelength}}{\text{period}}$$

For example, if a wave has a wavelength of 225 meters and a period of 12 seconds, its speed would be 225/12 = 18.8 meters/second (about 42 miles per hour).

A wave moving through water does not carry water along with it. To illustrate, observe the behavior of a floating object as a wave passes under it. A cork floating in water moves forward slightly as the crest of a wave approaches. Then if falls back an almost equal distance as the wave passes into a trough, ending up almost in its starting position. Analysis of this movement will lead us to conclude that waves are produced when the individual particles of water move in circles.

As the wave passes a given point, the water particles at the surface trace a circle whose diameter is equal to the height of the wave. Each water particle makes one complete circuit as the wave passes by, ending up almost exactly where it started. See Figure 17–8. Only the wave itself moves any distance over the surface of the water. Thus, the surface water remains in place, transmitting only the motion of the wave by the circular movements of individual water particles.

Waves near the shore. When a wave approaches the shore it undergoes change. The reason is found in the circular motion of the water particles which produce the wave. When a wave reaches shallow water, the movement of the water particles is slowed as they rub against the sea floor. If the wave approaches the shore at an angle, the slowdown causes *refraction*. This means the path of the wave

FIG. 17–8. The mechanics of motion in a water wave. As a wave moves forward, the water particles (represented by black dots) move in circles.

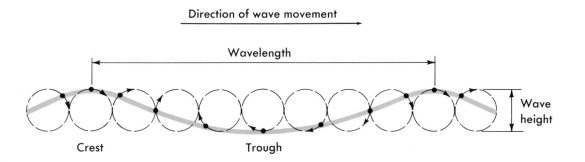

FIG. 17–9. When waves approach the shore they are refracted, as shown in the diagram above. Refraction causes the waves approaching the curved shoreline in the photo to hit all parts directly. (Ramsey)

FIG. 17–10A. Wave stages a to d show the development of a breaker. The wave advances on the shore in the direction indicated by the black arrows above. The white arrows indicate the undertow, and the lower black arrows show the movement of bottom sediments below the surf.

is bent. Bending is produced when one part of the angled wave strikes the shallow bottom first. A part of the wave slowing down before the rest causes a bending or refraction in the entire wave as it moves toward the shore. See Figure 17–9.

Wave refraction causes the incoming waves to line up parallel to the shore. Thus waves tend to approach the shoreline head-on, no matter what direction they had originally. This is an important fact in the development of shorelines.

Directional changes in waves are due to refraction; in addition, changes in height occur as waves meet the shore. The series of height changes in an incoming wave are shown in Figure 17–10A. When a wave reaches water where depth is about twice the height of the wave, the circular paths of the water particles are squeezed upward by the resistance they meet at the bottom. The circular motion is changed to an ellipse, raising the crest of the wave to a greater height (as shown at position a). As the wave moves into more shallow water, the crest rises higher and higher. Finally it tumbles forward into the trough and the wave becomes a *breaker* (at position b and c).

After the wave breaks, the water is usually thrown up on the shore as a foamy sheet. This action uses up the wave's energy and the water runs back into the breakers (as in position d). The size and violence of the breakers is determined by the original wave height and the steepness of the sea floor close to shore. If the bottom is steep, the height of the wave increases rapidly and the wave breaks violently. On the other hand, if the shore slopes gently, the wave rises slowly, spilling forward with a rolling motion that continues for some distance as the wave advances.

Water carried up on the shore by the breaking waves escapes back to deeper water in the form of a usually weak and irregular current that is called the *undertow*. Where the undertow exists, it is a current running seaward along the bottom far below the breakers. An undertow current is seldom very strong and would almost never

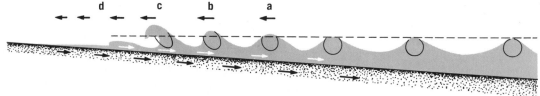

be a problem to swimmers. Some people confuse the relatively weak undertow with a more dangerous but lesser known type. Swift *rip currents* are caused by water returning to the ocean through breaks in underwater sandbars close to the beach. See Figure 17–11. Rip currents are a hazard to swimmers because they carry the person out to deeper water in the midst of the breakers. Generally, rip currents occur in spurts, lasting only a few minutes at any particular spot along the shore. Their presence can be detected by a gap in the line of breakers and by the yellowish color of the water where sand is being stirred up by the current. Anyone caught in a rip current should swim to one side or the other to escape the current and get to a region where the action of the breakers will carry him back to shore.

FIG. 17–10B. The spilling breakers shown in the photo begin rather far from shore and continue landward as the crest gradually collapses. (Photo-Researchers-Winch)

Giant waves. When the wind creates waves on the sea, three factors determine their size. There is (1) the speed at which the wind blows, (2) the length of time the wind blows and (3) the distance the wind blows across the open water (often called the *fetch*). Very large waves are produced by strong, steady winds with a long fetch. Such conditions are most likely to occur during a storm when a few waves are able to gather enough energy from the wind to reach very great size. Waves more than 33.5 meters in height (about 100 feet) have been observed during severe storms. Even such large waves as these are seldom dangerous to ships. A floating object tends to move in the same circular pattern as the water itself. A ship is not usually damaged unless it moves in a different direction from the water. This happens only if the crest of a wave is blown off by the wind. Then the waves break in the open sea, forming what is usually called *whitecaps*. Breaking waves that collide violently with ships can create severe damage.

FIG. 17–11. Rip currents (darker areas in the surf) are clearly visible in this aerial view of the beach near Monterey, California. (Hydraulic Research Laboratory, University of California)

Tsunamis. The most destructive waves in the sea are not created by wind action. By far the most dangerous are the giant seismic sea waves or *tsunamis* ((t) su-*na*-mees) that are produced by earthquake or seismic disturbances on the sea floor. In the past, these waves have been called "tidal" waves. The name is misleading because they have no connection with the tides. In an effort to remove this confusion, the Japanese word "tsunami" has been substituted.

Tsunamis seem to originate close to the deep sea trenches around the edges of the Pacific Ocean. They are probably caused by two kinds of events that accompany an earthquake on the sea floor. Faulting may bring about a sudden drop or rise in a part of the sea bottom. Then a large amount of water drops or rises at the same time. This mass of water circulates up and down as it tries to adjust again to sea level. A series of low waves is sent out by this water movement. Tsunamis may also be triggered by a severe underwater landslide set off by an earthquake. When this happens, a large mass of water above the landslide is thrown into an up and down motion creating a series of tsunamis. Volcanic eruptions may also disturb great amounts of water and produce tsunamis.

Tsunami waves are not very tall but they have a long wavelength. Their height is usually less than a meter in deep water but the wavelength may be as much as 240 kilometers (about 150 miles). A period of about 1000 seconds is common. Because their heights are so small in the open sea, tsunamis cannot be felt aboard ships or seen from the air. But the amount of energy they possess is tremendous. The movement of a tsunami represents motion of the entire mass of water from the surface to the bottom. As the tsunami approaches a coastline, it delivers its destructive energy against the shore, causing great damage to populated regions.

Near the shore the height of the tsunami greatly increases as its speed diminishes. The arrival of a tsunami may be announced by a sudden pull-back of the water along the shore. This occurs when a trough arrives before a crest. If a crest arrives first, there will be a sudden, rapid rise in the water level. The coastline itself seems to affect the height of the tsunami. It may be small enough to cause little harm at one point but large enough to cause great destruction at nearby locations.

More than two hundred tsunamis are known to have caused damage within the span of recorded history. Their effects have been observed mostly along the coasts of Japan, southeast Asia, the Caribbean Sea, Mexico, South America and Alaska. The most destructive tsunami in modern times took place in 1883 when the volcanic island of Krakatoa blew up. More recently, an earthquake on the sea floor near Alaska brought on a series of tsunamis. The waves struck the Hawaiian Islands on April 1, 1946, killing 159 people. See Figure 17–12. Following these disastrous Hawaiian tsunamis, a warning network was set up to observe unusual water movements and give advance notice of any onrushing tsunamis.

However in May of 1960, a tsunami arising from an earthquake off the coast of Chile caused extensive damage and loss of life in Chile, Hawaii, the Philippines, Okinawa, and Japan. In the Hawaiian city of Hilo, 61 people who ignored the warning given six hours before were killed. In Japan and the eastern Pacific, no general alert was issued since it was thought that tsunami waves could not cross the entire Pacific. The waves did strike, resulting in great loss of life and damage to property. These experiences led to expansion and improvement of the Seismic Sea Wave Warning System to function over the entire Pacific area. In the United States the warning network is operated by the National Oceanic and Atmospheric Administration. Tsunami warnings for threatened regions are issued to the general public through the Weather Bureau's links with radio and television stations.

Tides. The basic reason that tides occur is due to the gravity of the moon and sun. However, the tides that come in at any particular place along the shore are determined mainly by conditions here on the surface of the earth. If the moon and sun were the only influences producing tides, there would still always be two high and two low tides each day. Both are linked to the passage of the two tidal bulges as the earth rotates. Along the shores of the Atlantic, tides generally follow the basic pattern of two high and two low tides every 24 hours and 50 minutes. At some locations along the shore of the Gulf of Mexico, there is only one high and one low tide each day. On the Caribbean side of the Panama Canal, there is only one tide change a day and it is very small. On the Pacific side, the average height is fourteen times greater and there are tides twice a day.

There are, then, great differences in the timing and range of tides at different places on the earth. Apparently, varying parts of the sea respond in their own way to the rhythmic gravitational pull of the moon and sun. Remember that the sea is all one body of water. But the unevenness of the sea floor and the position of the continents divide it into several parts. For each part, there seems to be a particular tidal pattern.

Another important factor in determining the tides in any particular part of the sea is the size and depth of its basin. Both of these factors have great influence on a

FIG. 17–12. Tsunami approaching the shore near Hilo, Hawaii. Photos show oncoming waves destroying a pier. Arrow points to a man succumbing to the torrent of water. (World Wide Photo)

Observe

All bodies of water undergo the affect of tides. If there are any small lakes in the area see if you can pick a point at the water level, and after a few hours see if the water level is still at the same point.

FIG. 17–13. These two photographs, taken at different hours on the same day, show the difference between high and low tides in the Bay of Fundy. (Photo Researchers-Russ Kinne)

certain kind of motion in the water. *Tidal oscillations* (*os*-uh-*lay*-shuns) are very slow rocking motions which occur in various parts of the sea in response to the movement of tidal bulges caused by the sun and moon. Similar oscillations can be seen in any container filled with water which is stirred with just the right rhythm. The water in a bathtub, for example, can be kept rocking back and forth easily if it is stirred up with a rhythmic motion suited to its particular size. In parts of the sea, the rhythm of the passing tidal bulges also creates such oscillations. These may add to or cancel out the up and down motion created by the tidal bulges themselves.

A good example of oscillations which greatly add to the flow of tides occurs in the Bay of Fundy. This bay is located on the shore of New Brunswick, Canada, at the end of the Gulf of Maine. The water in the Bay of Fundy is set into rapid oscillations by the twice-daily passage of the tidal bulges. As the water rocks back and forth in the Gulf, it first floods the Bay of Fundy, raising tides 50 feet high. Then, as the water rocks slowly back to the other end of the Gulf, the resulting low tide almost completely drains the bay. See Figure 17–13. Nantucket Island, not far from the Bay of Fundy, does not receive the effect of the tidal oscillations in the Gulf of Maine. Its tidal range is only about 1½ feet. Along straight coastlines and in the open sea, tidal oscillations are not as apparent as they are in smaller bodies of water.

The rise and fall of tides often produce very strong currents. These *tidal currents* are most powerful when there are two regions close together which have large differences in tidal height. A narrow connection between the areas, such as the entrance to a bay, will increase the effect of the tidal current. Ships may have great difficulty in navigating some narrow passages when the tidal current is running.

When a river enters the ocean through a long bay, a *tidal bore* may be produced. This is a wave of water which passes up the river from the sea as the tide rises. See Figure 17–14. In a few rivers the tidal bore moves rapidly upstream in the form of a large wave which eventually exhausts itself. The Amazon River has a tidal bore that is said to resemble a small waterfall, moving rapidly upstream for a great distance.

The power of tides in moving large amounts of water represents a source of energy that could be useful to man. To make use of tidal power, dams must be constructed to trap the water at high tide. A location must be chosen where such a dam can be constructed without too much difficulty. Then water can be trapped behind the dam at high tide and released at low tide. As the water flows through the dam it could be used to generate electricity.

An experimental tidal power plant has been constructed at the mouth of the Rance River in France. This is the first in a series of tidal power projects which the French hope will help supply their increasing need for electricity. A similar project has been considered as a joint Canadian-American venture in the region of the Bay of Fundy. Perhaps the tides will be called upon to supply a part of the increasingly great world need for electricity in the years to come.

FIG. 17-14. The tidal bore of the River Severn near Gloucester, England. The force of the high tide is strong enough to push the wave of water far upstream. (Weston Kemp)

VOCABULARY REVIEW

Match the word or words in the column on the right with the correct phrase in the column on the left. *Do not write in this book.*

1. Ocean current caused by underwater landslides.
2. A change in direction of current flow due to the earth's rotation.
3. An eastward flowing ocean current found near the equator.
4. A swift warm ocean current found off the Eastern coast of the U.S.A.
5. A weak ocean current flowing eastward across the North Atlantic.
6. A vast, almost motionless area in the center of the Atlantic.
7. Groups of large waves which are alike in size.
8. The elevated portion of a wave.
9. The depressed portion of a wave.
10. The distance between crests of adjacent waves.
11. The change in direction of motion of waves on reaching shallow water.
12. Swift currents of water returning to the sea from the beach area.
13. How far the wind blows across open water.
14. Seismic sea waves incorrectly called "tidal waves."
15. A wave of water passing up a river as the tide rises.

a. Gulf stream
b. Sargasso Sea
c. swells
d. wavelength
e. crest
f. turbidity current
g. period
h. trough
i. Coriolis effect
j. Japan Current
k. North Atlantic Drift
l. Equatorial Counter Current
m. tidal bore
n. tsunami
o. fetch
p. refraction
q. rip current

QUESTIONS

Group A

Select the best term to complete the following statements. *Do not write in this book.*

1. Most of the energy which sets the seas in motion comes from the (a) sun (b) moon (c) rotation of the earth (d) many earthquakes.
2. Almost all of the ocean's surface currents are caused directly by the (a) rotation of the earth (b) differences in water temperature (c) global wind patterns (d) gravity effects from the moon.
3. As the ocean water cools it will (a) migrate toward the poles (b) move downward (c) move upward (d) spread out equally in all directions.
4. Where evaporation is occurring rapidly, the saltiest sea water is found (a) near the ocean floor (b) near the ocean surface (c) at moderate depths (d) at all depths.
5. Which one of the following causes of ocean currents is *not* considered an occasional or temporary cause? (a) density differences (b) underwater landslides (c) earthquakes (d) volcanic activity.
6. The rotation of the earth on its axis causes ocean currents in the northern hemisphere to be deflected (a) northward (b) southward (c) to the left (d) to the right.

7. The eastward moving Equatorial Current is bounded by the North Equatorial Current and the South Equatorial Current which flow (a) north and south respectively (b) eastward (c) south and north respectively (d) westward.

8. The Gulf Stream is caused by the piling up of water in the Gulf of Mexico from the (a) Labrador Current (b) Mid-Atlantic Current (c) North Equatorial Current (d) South Equatorial Current.

9. A current which is considered a branch of the Gulf Stream is the (a) Japan Current (b) Labrador Current (c) North Atlantic Drift (d) North Pacific Current.

10. The current in the Pacific Ocean which is equivalent to the Gulf Stream is the (a) Pacific North Equatorial Current (b) Japan Current (c) North Pacific Current (d) Sargasso Current.

11. The large current circles found north of the equator move (a) clockwise (b) counter-clockwise.

12. There is no Arctic Current similar to the Antarctic Current due to the (a) warmer climate in the Arctic (b) land barriers found in the Arctic (c) slower speed of winds in the Arctic (d) fact that there is no land near the north pole.

13. Deep currents are powerful ocean currents thought to be caused by (a) typhoons (b) volcanoes (c) sinking of heavier water (d) the earth's rotation.

14. Although the Gulf Stream moves much slower than the Mississippi River, it transports (a) twice as much water (b) one hundred times as much water (c) one thousand times as much water (d) four thousand times as much water.

15. The Eastern Seaboard frequently has colder weather than expected due directly to the (a) North Atlantic Drift (b) Gulf Stream (c) Sargasso Sea (d) North Equatorial Current.

16. Great fog banks form off Labrador as a result of (a) the meeting of the Gulf Stream and the Labrador Current (b) the unusually warm climate of Labrador (c) the unusually cold climate of Labrador (d) the vast amount of ocean surrounding Labrador.

17. The North Atlantic Drift influences the climate of Europe and the British Isles causing (a) much cooler weather than usual (b) very stormy weather (c) unusually clear skies (d) much warmer weather than usual.

18. The source of almost all waves on the seas is (a) gravity of the moon (b) wind (c) earthquakes (d) undersea landslides.

19. Large waves naturally grow larger due to (a) more surface area being exposed to the push of the wind (b) the elastic nature of water (c) the absorption of the smaller waves (d) the formation of whitecaps.

20. The period of a wave can be defined as (a) the distance between two adjacent wave crests (b) the distance between the crest and trough (c) the time it takes for two consecutive wave crests to pass a given point (d) the distance from the bottom of the trough to the top of the next crest.

21. A wave whose wavelength is 100 meters and whose period is 5 seconds would have a speed of (a) 100 meters/sec (b) 500 meters/sec (c) .05 meters/sec (d) 20 meters/sec.

22. Waves cause the individual particles near the surface of deep water to move primarily (a) in the direction of the wave (b) back and forth (c) up and down (d) in circles.

23. Waves approaching the shore are refracted (a) when they come straight in (b) when they approach at an angle (c) only when the shoreline curves inward (d) only when the shoreline curves outward.

24. Waves usually approach the shore head-on no matter what direction they originally had. This is due to (a) refraction (b) reflection (c) rip currents (d) the action of breakers.

25. Breakers form when waves (a) approach the beach from shallower water than that near the beach (b) tend to speed up in the more shallow water (c) enter shallow water, causing crests to build so high they tumble into the trough ahead (d) cause rip currents.

Group B

1. Sea water is set in motion only if it receives energy. (a) What is the main source of this energy? (b) Describe the ways water receives this energy.

2. Describe the ways temporary ocean currents are produced.

3. How does the rotation of the earth influence the current patterns of the oceans in (a) the Northern Hemisphere (b) the Southern Hemisphere?

4. It is said that water normally spirals clockwise as it goes down a sink drain. Do you think this might be due to the Coriolis effect? Explain your answer.

5. Using drawings, show the ocean current pattern in the Atlantic Ocean for both hemispheres.

6. How is the ocean current pattern of the Pacific Ocean similar to that of the Atlantic Ocean? How do they differ?

7. What causes deep currents? How does their direction of flow compare to the surface currents?

8. How does the Gulf Stream form from the Atlantic North Equatorial Current?

9. Describe the results of the Coriolis effect on the Gulf Stream.

10. Large surface waves on the ocean grow larger while the small waves usually die out rather quickly. Explain why this is so.

11. What are the three characteristics which describe any ocean wave?

12. You notice a group of swells passing by a stationary object. Measurements which you make indicate their wavelength is 200 meters. You notice that six pass the stationary object in one minute. What is the speed of the swells.

13. Analysis of the movement of a free floating object shows it to have a very egg-shaped motion. Describe the condition which would cause the wave to have this motion.

14. Why do waves approaching a beach at an angle tend to line up parallel to the beach?

15. What is necessary for an undertow to become a rip current?

16. During World War II many concrete cargo vessels were built due to the shortage of steel. These vessels were soon found to be unsatisfactory because they were not flexible enough to bend in the waves. It was also found that the engineers chose the wrong length to make them. Why would their length be a critical factor?

17. Why is the name "tsunami" being used more often today in place of the term "tidal wave"?

18. In the Bay of Fundy on the East Coast of North America, 50-foot-high tides are common. Explain why this is so.

19. Some scientists in Russia have recently suggested that the Bering Straights should be partially dammed with an underwater rock ridge from the Alaskan coast to coast of Siberia. The idea is to shut off the deep current of cold water that enters the Pacific Ocean from the Arctic Ocean but allow the warm water to enter the Arctic. What would be the purpose behind this project?

earth, the water planet

A Super Flood

Approximately 100,000 years ago, a tongue of ice that reached out from the continental ice sheet made a dam across a large river. The water collected behind the ice dam and formed a giant lake. Eventually the water reached the top of the ice dam. As the water flowed over the ice, rapid melting caused the dam to suddenly collapse. During a short time, a few days at most, all of the water was released from the lake. This sudden flood as the water made its way to the sea had a flow ten times greater than all the rivers of the world. It was probably the greatest flood the earth has ever experienced.

Today the evidence of that flood can be found in a region of eastern Washington called the "scablands." It is an area of about 40,000 km² made up mostly of bare rock cut with large channels, great potholes, dry waterfalls, giant ripple marks, and other erosion features created by running water. At top is a picture of the eastern Washington scablands. Compare its appearance to features seen on the surface of Mars (bottom). There is no water on that part of the Martian surface at the present time. However, the presence of scabland features on the surface of Mars indicates that sudden great floods have occurred during its past.

Sending Icebergs to Solve the Water Supply Problem

The water supply for coastal regions of the world could in part be supplemented by icy resources from Antarctica. These regions could be supplied with fresh water from one of the numerous icebergs that constantly breaks off the Antarctic continental glacier. Although approximately 70% of the earth's surface is covered with water, most of it is found in the sea. No inexpensive method for obtaining fresh water from sea water is available. The iceberg method would compare favorably with other methods, including distillation (Chapter 16).

Icebergs as large as 16 km high and 0.8 km wide could be towed from Antarctica to almost any coastal location in the world. A route would be chosen that would take advantage of ocean currents. The trip would take about one year. Even if the iceberg melted to half its original size during the trip, it would still contain one billion cubic meters (about 980 billion liters) of fresh water. When it reached its destination, the iceberg would be surrounded by a floating dam. The less dense melted water from the iceberg would float on the surrounding sea water behind the dam and could be drawn off as needed.

Explorers of the Sea

Below, a research team out of Woods Hole Oceanographic Institution looks over dredgings from the ocean floor. From the first voyages of H.M.S. Challenger to modern undersea vessels, oceanographers such as these have been exploring the ocean with instruments specially equipped for the watery realm of the ocean.

When more direct contact with the ocean is required, a diver such as the one at the right performs the task. Diving with SCUBA gear has also become a popular sport in areas such as the tropics where colorful fish and unusual life forms populate the sea.

unit

5

*the record of
earth history*

18
the rock record

objectives

- [ ] Explain the principle of superposition.

- [ ] Describe how this principle relates to conformity, unconformity, and disconformity.

- [ ] Explain how the half-life of a radioactive element is used to learn the age of a rock.

- [ ] Describe three means other than radioactivity used to measure geologic time.

- [ ] Describe several ways fossils are formed.

- [ ] Explain how fossils can be used to relate rock layers to one another.

Within every piece of rock there is a story. Part of that story involves minerals and the way in which these minerals combined to form the rock. Actually, rocks begin to change the moment they come into existence. Through careful study, geologists can often reveal the history of change that a particular rock has undergone. To unravel this mystery very patient detective work is needed. In addition, a knowledge of the way various processes act upon the rocks of the earth's crust is needed. The researcher finds that his reward is a better understanding of the history of the earth.

However, the clues that lead to a complete history of the rocks are not easily found. The search is only successful when certain principles which control the patterns of rock structures are understood. When these principles are applied, the arrangement of rocks in the crust becomes a record of their history.

THE PATTERN OF THE PAST

Rock layers. Throughout most of the earth's history there has been a steady weathering and erosion of rocks at the earth's surface. These powerful erosive forces have steadily built up an accumulation of rock fragments deposited in layers or strata of sedimentary rocks. It is in these rock strata that clues to much of the earth's past can be found. An estimate of the total thickness of sedimentary rocks deposited since the earth was formed is more than 100 kilometers. This does not mean that the total thickness of sedimentary strata was laid down in any one place. Conditions in some locations and at certain times have not been favorable for the deposition of sediments. At other times however, thick deposits have been laid down. For example, a series of strata up to 30 kilometers in depth were built up in geosynclines. These geosynclines were formed as the floors of ancient shallow seas slowly sank. If the greatest known thickness of each layer in all these different places were added together, we could get some idea of the total record available in these rocks.

The most basic principle used in understanding the rock record is a simple one. It is generally believed that when the layers are horizontal, or nearly so, each overlying bed is younger than the one beneath it. This is called the *principle of superposition*. When strata of sedimentary rock are known to have been deposited by a body of water or as wind-carried material, there can be no doubt that the build-up was according to the principle of superposition.

In addition, the earth's crust is twisted and deformed by internal movements. The evidence that is found in these rocks indicates that periods of violent disturbances brought changed conditions over a wide region. These periods of disturbance, called *revolutions*, have caused large scale folding and faulting of the rock layers. We must keep this in mind when applying the principle of superposition to any rock layers being studied. Violent movements, such as those which take place during a revolution, might push older layers up over younger ones. If overturning does occur, then the upper strata would actually be older than the layers below.

Interpret
An exposed section of sedimentary rock layers are visible along a road cut. On identification of the rock types, conglomerate and sandstone were the first two layers. These were followed by limestone and shale. What may have occurred in the early history of the rock layers?

FIG. 18–1. Sedimentary rock layers are clearly visible in this photograph of Badlands National Monument. (Ramsey)

FIG. 18–2A. Cross-bedding in sandstone near Zion National Park. (Ramsey)

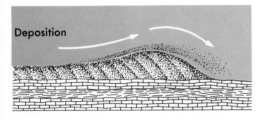

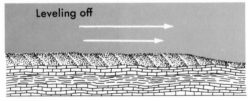

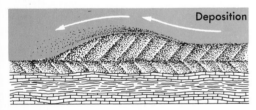

FIG. 18–2B. Cross-bedding generally originates as tilted layers of sand on a migrating dune. As the wind changes its direction and speed, the layers are tilted in different directions.

FIG. 18–2C. Ripple marks in an exposed layer of shale. (Ramsey)

To prevent errors in the use of the principle of superposition, it will be necessary to observe very carefully certain details of the rock structure. In some sedimentary layers, for example, *cross bedding* has occurred, as shown in Figure 18–2A. The angled layers in cross-bedded strata were produced by the movement of wind or water in a certain direction. See Figure 18–2B. When we understand fully the nature of this process, the appearance of cross-bedding will be very useful in determining whether or not a rock layer has been overturned. Ripple marks and the position of fossils can also be used to determine the original position of many sedimentary layers. See Figure 18–2C.

The principle of superposition does not give us any clue to the actual age of the various rock layers. However, if the rock layers were deposited in an uninterrupted sequence, it would accurately show the order of deposition. Then the relative age of the rocks could be determined. The boundary between any two of these layers is called a *conformity*. On the other hand, if the rock layers were folded, faulted and tilted, they would eventually have been eroded. This would leave the disturbed layers exposed. If the area underwent a second period of deposition, new horizontal rock layers would form on the eroded surface. When the beds on either side of the eroded surface are no longer parallel, the boundary between two such layers is called an *unconformity*. See Figure 18–3A.

Sometimes after a period of erosion the rock layers are left in a relatively horizontal position. Then any new sediments will be deposited on top of the almost parallel but

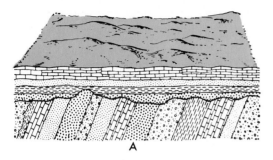
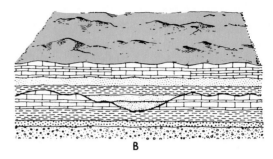

eroded surface. Where the boundary between an eroded surface and the younger overlying layers is nearly horizontal, it is called a *disconformity*. See Figure 18–3B.

Thus, when either an unconformity or disconformity is found in the rock layers, it indicates that a gap exists in the rock record between two adjacent layers.

Evidence from the rocks. In addition to their layered arrangement, the make-up of the rocks themselves reveals some of their early history. Again, it is the sedimentary rocks that provide the most clues to their origins. By examing the texture and composition of sedimentary rocks, we can learn about the conditions which produced them. Interpretation of ancient sediments is made possible by a knowledge of the conditions under which similar sediments are being deposited today.

The last traces of ancient shallow seas, extinct lakes, desert regions, glaciers and other features of the distant past lie buried in these sediments. Rocks that are sedimentary in origin contain information about the original rocks, the relief of the land, the climate of the region, and the kind of life that once existed. These layers can be compared to the pages in a book of geologic history. However, reading this book requires skill and knowledge.

Igneous rocks are also a useful tool in understanding earth history. The rocks which form dikes and sills in sedimentary rocks for example, are younger than the rocks they intrude. Likewise, extrusive formations such as lava flows are obviously younger than the rocks they cover. It is not always easy to tell which rocks are older. However, it is fairly simple to determine whether a formation of igneous rock found horizontally between layers of sedimentary rock, is intrusive or extrusive. If it is intrusive, the igneous formation must have invaded the existing structure, and should of course be younger than the rock layers above and below. In the case of an extrusive

FIG. 18–3. A at the left is an unconformity where tilted, folded, or faulted layers have been eroded, then covered again. B at the right is a diagram of a disconformity where parallel layers are separated by an old erosional surface.

Identify

Refer to Figure 18–3. Identify at least two geologic changes that took place before the sedimentary layers were deposited on the unconformity. How many changes have occurred below the disconformity?

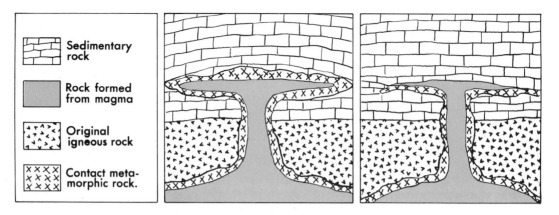

□	Sedimentary rock
▨	Rock formed from magma
▨	Original igneous rock
▨	Contact meta-morphic rock.

FIG. 18-4. Which diagram illustrates an igneous extrusion; which illustrates an igneous intrusion? Upon what evidence do you base your conclusions?

Construct

Refer to the diagram on page 395. Construct your own cross-section and have someone try to interpret the proper order in which the processes you indicated took place.

FIG. 18-5. From the evidence shown here it is easy to see why the rock in intrusion B is younger than that in intrusion A.

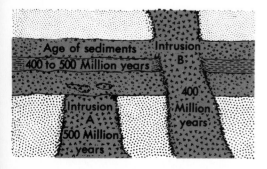

lava flow, the underlying rock structure is older and the younger rock layers are found above. The situation becomes more complicated when the hot magma changes the existing rocks with which it comes in contact. This process is called *contact metamorphism*. See Figure 18-4.

In some situations, it is difficult to determine the order in which these processes have taken place. But in other cases, certain time relationships can usually be worked out if the types of rocks present are carefully examined. For example, a conglomerate may contain granite pebbles identical to those in a granite batholith located close by. It is almost certain that the granite in the batholith would be older than the granite in the conglomerate. Suppose a second mass of igneous rock, such as a dike, cut through the layers of this conglomerate and the granite. This line of thought is illustrated in Figure 18-5.

The above examples are intended as simple and very brief illustrations of how it is possible to judge the age relationships of various rock formations. With actual rocks, it is often very difficult to observe these relationships. Many times, even trained scientists disagree on the relative age of rock formations. For the most part, however, geologists have obtained a fairly clear picture of the rock relationships over the earth's surface. Their experience has led them to make some conclusions about the age and history of the rocks.

Gaps in the record. The general history of the earth as revealed in the rock record is clear up to a point. Beyond that, scientists must search painstakingly for clues to the earth's early history. For this reason, the sedimentary rocks that were laid down in the earliest geological periods are more difficult to identify and study. Our knowledge of the earlier geologic history of the earth is an incomplete one. Nearly all of the rocks formed during the first one and a half billion years cannot be studied directly. These rocks may have been deeply buried, lost through erosion, or else greatly metamorphosed. The early geologists made no attempt to work out the history of the most ancient rocks. Instead, they merely classified the oldest rocks as the *basement complex.*

Modern geologists, however, are gradually developing a more complete picture. In recent years they have gathered additional information from all parts of the earth. This has enabled them to compare individual types of rocks with related strata in other regions. Recently developed methods for drilling deep into the crust on both land and on the sea floor have helped close these gaps in the rock record. However, a part of the earth's most ancient history will probably remain permanently sealed as a record left in the rocks.

MEASURING GEOLOGIC TIME

Methods based on radioactivity. Our knowledge of the ways that rocks are laid down makes it possible to compare the relative age of one group of rocks to another. The actual age of the rocks in terms of the number of years before the present must be found by other means. One of the most accurate ways used to determine the absolute age of rock material is based on the rate of radioactive decay of certain kinds of atoms.

Atoms of the various chemical elements differ in the stability of their nuclei. Some atoms have an unstable nucleus, which will, of its own accord, undergo change by emitting some electrically charged particles along with electromagnetic energy. For example, the nuclei of the uranium isotope with a mass of 238 are radioactive (symbol, U, atomic number = 92). One of these nuclei may throw off an *alpha particle*. The make up of this electrically charged particle includes two protons and two neutrons. When the alpha particle is ejected the loss of protons and neutrons in the uranium nucleus brings about a decrease in both atomic number and mass. A new nucleus is then formed with atomic number 90 and mass of 234. It is called an atom of the element thorium.

activity

Determining whether a given radioactive atom will or will not decay is similar to predicting whether a coin, thrown into the air, will come to rest "heads up" or "tails up." Therefore, you can simulate the rate of radioactive decay by using coins or discs marked so that one side can be identified from the other.

Obtain at least 100 identical coins or discs. Place them in a covered box, like a shoebox. Shake to mix them well. Then remove all coins that come to rest in the box "heads up." Coins that come up "heads" will represent atoms that have decayed radioactively to become stable atoms. Record this as shake 1, together with the number of coins remaining. See the table below. Continue shaking and removing coins until only 2 or 3 coins remain. Graph your results showing the shake number on the horizontal axis and the coins remaining on the vertical axis. Compare the shape of your curve with that in Figure 18–7, page 383.

Number of Shake	Coins Remaining
0.	
1.	
2.	
etc.	

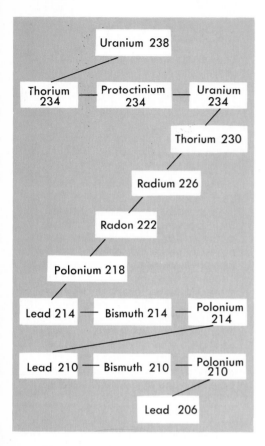

FIG. 18–6. Series of elements formed when uranium-238 undergoes radioactive decay forming lead-206.

The thorium nucleus is also radioactive and it gives off a *beta particle* consisting of one electron. When the beta particle is lost, a neutron in the thorium (Th) nucleus becomes a proton. The atomic number is increased by one and thorium is changed to protactinium (symbol, Pa, atomic number = 91). Protactinium is also radioactive, and the processes of radioactive change continue until a stable nucleus of the element, lead (Pb), is produced. This isotope of lead has a mass of 206 and is the final product of this radioactive chain. Thus an atom of U-238 eventually will become an atom of Pb-206 by radioactive decay. See Figure 18–6.

The conversion of uranium into lead goes on at a very slow but constant rate. Measurements show that for half of a given amount of uranium it takes 4.5 billion years for it to change into lead. This constant rate of atomic decay is not affected by changes in temperature, pressure, or other environmental conditions. At the end of 4.5 billion years half of the original uranium remains. By giving off electrified particles and radiant energy, the other half is eventually changed to lead. The remaining uranium continues to decay at the same rate as before. At the end of another 4.5 billion years just half of that (one-fourth of the original uranium) is left. We can say the *half-life* of uranium is 4.5 billion years. Its half-life, regardless of the amount, is the length of time it takes for half of its atoms to decay. A graph showing the rate of radioactive decay of uranium is given in Figure 18–7.

Knowing that the half-life for uranium is 4.5 billion years makes it possible to determine the age of any rock which contains uranium. Each year that passes $1/7,700,000,000$ of the amount of uranium is changed into lead. If the amount of uranium and lead in a mineral is measured, comparison of their amounts will show how many years the mineral crystal has existed. To determine the uranium-lead ratio in rocks requires complicated equipment and very accurate measurement.

The U/Pb ratio can only be used with rocks where there is reason to believe that all lead in the rocks has come from decay of uranium. Since uranium is most common in igneous rocks, the method would be less useful in finding the age of sedimentary rocks. Other radioactive atoms with long-lives which are more abundant than uranium can also be used in this way. Two such elements are rubidium-87, which changes into strontium-87 with half-life of 50 billion years, and potassium-40/argon-40 with a half-life of 12.5 billion years.

For solving age problems involving shorter periods of time (1000 to about 60,000 years), a form of radioactive carbon is often used in place of uranium. The value of this method lies in the fact that radioactive carbon-14 is constantly being produced in the upper atmosphere. Carbon-14 is formed when cosmic rays from outer space occasionally strike nitrogen atoms and change them into this radioactive form of carbon. Living things absorb small amounts of carbon-14 along with ordinary carbon atoms all during their lives. After death, the carbon-14 in the organic matter decreases as the radioactive carbon atoms decay and are not replaced. To establish the age of a small amount of organic material, the amount of carbon-14 must be measured and compared with the normal amount in the atmosphere. The half-life of carbon-14 is known to be about 6,000 years. We are able to calculate the age of the organism because it stopped taking in carbon-14 at the time of its death. Carbon-14 dating methods are a valuable tool in establishing the age of such things as wood, bones, shells, and the remains left by early people.

Other methods for measuring geologic time. Radioactivity measurements for dating of rocks and other materials have become a vital means for finding the age of the earth. They do not, however, accurately measure the age of the rock itself. Amounts of both the original radioactive substance and its final produce may have undergone change after the rock was formed. In that case, the radioactivity measurements would show only when the rock was changed. For example, measuring the age of a mineral crystal in a sedimentary rock does not tell us when the rock itself was created. Nevertheless, radioactive methods are still the most accurate available for establishing the age of the earth's older rocks.

Before the discovery of "radioactive clocks" several decades ago, other approaches were used to estimate geologic time. Some of these methods include the following:

1. *Rate of erosion.* If we can determine the rate at which a stream cuts its channel or erodes other features in its bed, we can determine its age. For example, the rate at which Niagara Falls is retreating by erosion of its rocky

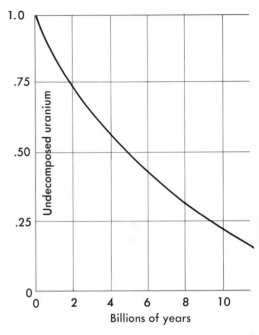

FIG. 18-7. The rate of decay of uranium is shown on this graph.

Discuss

Explain why radioactive dating is not as useful for determining the age of sedimentary or metamorphic rocks as it is for igneous rocks.

FIG. 18–8. Niagara Falls was formed by weak rocks that rapidly wore back, cutting a long gorge. (Photo Researchers-Gianni Tortoli)

FIG. 18–9. The age of some rock strata can be estimated by counting the varves found in some sedimentary rocks. (U.S. Geological Survey)

ledge has been studied for about two hundred years. Geologists have found that the falls move back at an average rate of four feet a year. See Figure 18-8. A gorge about seven and a half miles long runs just below Niagara Falls. A simple calculation shows that it has been 9900 years since the Niagara River started cutting its gorge back from Lake Ontario. The cut-back probably began when ice from the glaciers of the last ice age started to melt.

Estimating geologic age by rates of erosion is most practical with relatively recent geologic features such as Niagara Falls. For older features, like the Grand Canyon, the method is much less dependable. The Grand Canyon developed over millions of years and conditions change during such long intervals of time. There is no way to be sure that the rate of erosion remains constant over many millions of years.

2. *Rate of deposition of sediments.* Before the use of radioactive methods, the most accurate way to date rocks was to study the time it took for them to form sedimentary layers. Attempts were made to determine the average rates of deposition for the different sediments which form the common sedimentary rocks, such as limestone, shale, and sandstone. On the average, it takes between 4000 and 10,000 years for a one-foot layer of sedimentary rock to be deposited. With these figures, it would be possible to discover how long it took to form a layer of a particular thickness. However, there is no way to be certain that any given sedimentary layer was deposited at an average rate. A flood can deposit many feet of mud in just one day. On the other hand, the same kind of mud could accumulate slowly on a lake bottom.

Some sedimentary deposits show definite annual layers, or *varves*, as seen in Figure 18–9. Since each layer represents one year's deposition, a series of varves indicate the number of years it took for the deposits to form. This is much the same way that the annual growth rings in the trunks of trees show the age of the tree when it was cut.

Varves are often found in lake deposits. Some were formed from sediments that were deposited on the bottom of a lake at the edge of a melting ice sheet during the last ice age. During the summer the ice probably melted rapidly, resulting in a rush of water that carried large amounts of sediment into the lake. Most of the coarser particles settled quickly to form a thick layer on the bottom. With the coming of winter, the melting stopped and the lake froze over. The finer clay particles, still suspended in the water, settled slowly to form a thin layer above the coarser sediments. A summer layer and the overlying winter layer make up one varve. Thus, each varve represents one year's deposition.

3. *Amount of salt in the sea.* The third method for estimating the age of the earth also includes finding the age of the oceans. Here the geologist calculates the amount of salt that is now present in all the oceans. Next the yearly average amount of salt carried into the oceans by all of the rivers on earth is figured. From these calculations it is then possible to estimate how many years the oceans have been receiving salt. However, this method contains two serious errors. First it assumes that the rate at which salt has been added to the oceans has remained the same. This is probably not true. In the past salt was very likely added at a different rate than at the present. A second problem is the rate at which salt is removed from the oceans by formation of bottom sediments and various life processes of sea creatures. It is not known how much allowance should be made for the salt lost from the oceans over millions of years. Thus we cannot use the saltiness of the sea as an accurate guide in dealing with the problem of geologic time.

Investigate
The rate of salt accumulation in the oceans throughout geologic time has not remained constant. Discuss at least two factors that may be responsible for varying the rate of salt accumulation.

THE FOSSIL RECORD

What can be learned from fossils? As we have seen, piecing together the record of the rocks is largely a matter of discovering their relation to each other. The study of fossils of ancient plants and animals, called *paleontology* (*pay*-lee-on-*tol*-uh-jee), is an important source of evidence for the age of events in the geologic past. Fossils are the remains of ancient plants and animals or indications of their presence, preserved in the rocks. The paleontologist compares the appearance, composition, and fossil content of rocks in the various layers.

One reason that fossils are so useful is that they can be used to establish the age of similar rocks from different places. For instance, we could assume that a sandstone of similar composition and appearance, in a particular region was formed at the same time. But similar sandstone coming from another continent may have an entirely different age. This is because sandstones were formed in nearly all periods of earth history. Thus we cannot assume that similar sandstones from different places have the same age.

But suppose it is discovered that rocks from many different locations contain the same general kinds of fossils. This would prove that the rocks were formed under similar conditions and would therefore be nearly the same age. Rocks of the same age need not necessarily contain identical fossils. Allowance must be made for differences between related plants and animals which lived in different parts of the world at the same time. In general, there is a similarity but no exact resemblance between fossils formed on different continents during the same geologic period.

However, certain fossils are found in rock layers of only one geologic age. These fossils are called *index* or *guide fossils*. They serve to identify specific rock layers no matter where they are found. These index fossils must meet certain requirements. First, they must be found in rocks scattered over a wide area of the earth's surface. Second, they must have features which clearly distinguish them from other forms. Third, the organisms that formed the index fossils must have lived during only a relatively short span of geologic time. Fourth, they must occur in fairly large numbers within the rock layers. A fossil that meets most of these requirements gives important clues to the history of a particular series of rocks. The presence of index fossils is especially important because they quickly establish the relative ages of rock layers. You will find this illustrated in Figure 18–10.

An economically valuable use of index fossils is in the location of oil and gas formations. Petroleum and natural

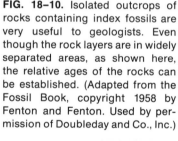

FIG. 18–10. Isolated outcrops of rocks containing index fossils are very useful to geologists. Even though the rock layers are in widely separated areas, as shown here, the relative ages of the rocks can be established. (Adapted from the Fossil Book, copyright 1958 by Fenton and Fenton. Used by permission of Doubleday and Co., Inc.)

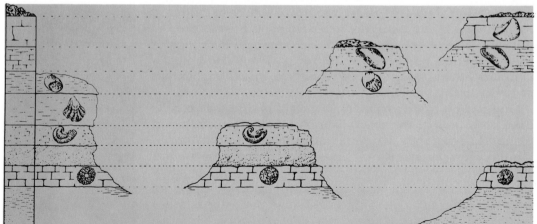

gas are apparently produced from ancient plants and animals which have been changed by complex chemical processes over millions of years. Index fossils of certain types are commonly found in the layers of sedimentary rock above the gas and oil formations. They often aid geologists in locating the oil-bearing rocks. Scientifically trained field and laboratory workers are needed to locate and identify these fossils.

Fossils also furnish us with important clues to the environments that have existed in the past. For example, fossils of tropical plants have been found in the polar regions. This indicates that the climates over the earth's surface have changed greatly from one geologic period to another. The boundaries of ancient seas have been traced through the remains of marine animals and plants. Analysis of oxygen isotopes in certain fossil shells can even tell the relative temperature of the water the animals lived in.

Formation of fossils. Many different conditions that exist when certain animals or plants die allow fossils to be formed. In rare cases, an entire animal may be completely preserved. The frozen bodies of an extinct elephant and several woolly mammoths have been found in Siberia and Alaska. Frequently however, only the harder parts of organisms are preserved in a relatively unchanged condition. Some fossils are so altered that little of the original material remains. Only shapes and textures can be recognized. Other fossils are not remains at all, but evidence that life has existed in some form. Even so, such evidence is often remarkably clear and detailed. In general, fossils can be classified in one of the following groups, according to the way they were formed:

1. *Unaltered remains of organisms.* A dead plant or animal must be protected for it to remain intact so long. Usually it must be buried to prevent other organisms from eating it. It must also be protected from decay caused by bacteria. Low temperatures found in frozen soil and solid ice will effectively protect and preserve organisms. Unaltered remains of animals have also been found in soil saturated with oil, deposits of volcanic ash, and in caves. Some remains, such as those often found in caves, are preserved by drying out and are said to be mummified. See Figure 18–11. Tar beds, formed by petroleum coming to the surface, are also effective in trapping animals and preserving their bones. See Figure 18–12.

Discuss

Explain how fossils of different ages have helped provide evidence for a theory of evolution. Find out about fossils found in the layers of the Grand Canyon. How do they change from bottom to top?

FIG. 18–11. Above, the mummified remains of Trachodon, the duck-billed dinosaur; left, closer detail of the skin. (The American Museum of Natural History)

2. *Altered remains of organisms.* Under certain conditions, fossil specimens may be completely altered. In many cases the original parts have been replaced by mineral substances. This process is said to *petrify* the organism. This does not mean that the original organic matter has been changed to stone. Actually, what happened was that minerals replaced the original cell materials which had long since disappeared. Some of the common petrifying minerals are silica, calcite, and pyrite. Usually the replacement of the original materials is a very slow process. It probably takes place molecule by molecule. The substitution of mineral for organic material is often nearly

FIG. 18–12. An artist's conception of a scene at the Rancho La Brea tar pits in prehistoric times. (The American Museum of Natural History)

perfect, such as the fossil shown in Figure 18–13. Seen under a microscope, the detailed cell structure of the original tissues is clearly visible in thin sections of some petrified fossils. In other fossils, only the general outline of the organism remains.

The vast coal deposits of today are fossil remains of plants which have been changed by the process of *carbonization*. This takes place when the remains of trees and other plants decompose in swamp water and then are buried. Some of the plant material changes to marsh gas or methane (CH_4), water (H_2O), and carbon dioxide (CO_2). In the formation of these coal deposits much carbon remained because such products carried away more oxygen and hydrogen than carbon. The original, complex chemical compounds present in the plants were gradually changed by a concentration of this carbon. After a long period under pressure, the final product we know as coal was produced.

3. *Indirect fossil evidence of life.* Some fossils have left definite imprints, including such things as the footprints of an animal or the outline of a leaf. Let us trace the history of such an imprint. Suppose a giant dinosaur left deep footprints in the soft mud. What happened then? Later sand or silt may have blown or washed into the footprints so gently that they were not destroyed. Still later, more sediment may have been deposited above the prints. Then, as the ages passed, the mud containing the footprints hardened into solid rock. Thus the footprints were preserved. An imprint like this one is shown in Figure 18–14. The imprints of fossil leaves, stems, flowers, and fish were often laid down in soft mud or clay and then preserved in a similar way. Footprints of amphibians, birds, and mammals, and even early human-like creatures have been found.

FIG. 18–13. Fossil crinoids, perfectly preserved by the process of petrifaction. Crinoids are marine animals often resembling plants. (Allan Roberts)

FIG. 18–14. Fossilized dinosaur tracks made in mud more than a hundred million years ago. (The American Museum of Natural History)

FIG. 18–15. An ant, perfectly preserved in amber. Amber is a fossil gum derived from the sap of prehistoric plants. (The American Museum of Natural History)

FIG. 18–16. Prehistoric worm trails, believed to be the fossil burrow of worms. (E. R. Degginger)

Shells of snails, parts of trees and plants, and similar organic remains were often buried in sediments which later formed rock. Eventually these remains decayed or dissolved. In some of them an empty cavity, called a *mold*, remained. The mold reveals many characteristics of the original organism. Natural molds may be studied just as they are. Or they may be filled with plaster or some other substance to make duplicates of the original organism. These replicas are called artificial casts. Sometimes natural casts were formed when sand or mud hardened in a natural mold. An unusually perfect type of mold was formed when insects millions of years ago were trapped in sticky resin flowing from trees. Later the resin was changed to hard amber, with the forms of the insects perfectly preserved as cavities in the transparent amber. See Figure 18–15.

Trails and burrows may also be preserved as fossils. See Figure 18–16. Some ancient snails, sponges, and crabs are known as much by their borings or burrows as they are by their remains. For instance, ancient sea worms may have swallowed sand or mud. Then they digested the small organisms contained in the sand. After the food was extracted, the sand or mud was left in the form of castings. These are found as fossils in some marine sediments. *Coprolites* are fossilized masses of waste materials from animals. They can be cut into thin sections. Analysis of these sections reveals the feeding habits of animals long since extinct.

Most people are familiar with the gizzard of a bird such as a chicken. The gizzard contains small stones which help grind the food. Some dinosaurs also had such gizzard stones or *gastroliths* to grind their food. They can often be recognized by their smooth, rounded, and polished surfaces. However, their identification is certain only if they are found within the remains of a dinosaur.

Finding fossils. Anyone with sufficient interest can successfully hunt fossils. The first requirement is to look in places where fossils are likely to be found. Maps of good fossil-collecting areas are available for most regions. Information about the best collecting areas can be obtained from museums, colleges, and other collectors. Many students have found fossils in the exposed rocks of gorges and canyons, quarries, and rock cuts along railroads and highways. However, permission must be obtained from owners before collecting from private property such as railroad rights-of-way.

Fossils may be found in almost any kind of sedimentary rock. Shales and limestones are especially likely sources. Dolomite and sandstone are also often good. Other sources of unusual or well-preserved fossils include chalk, diatomaceous earth (remains of microscopic water plants) and coal. Fossils are not found in igneous rock. Those originally present in metamorphic rocks were almost always destroyed or changed into unrecognizable forms. Among sedimentary rocks, conglomerates and chert generally give poor results. Any fossils that exist in these rocks are usually difficult to remove.

The collector should note the locality and rock layer from which his fossils come. A label with this information should be made out for each specimen as soon as possible. Collecting tools might include a hammer with one pointed end, such as a geologist's hammer. This is used to loosen rock-containing fossils from their surroundings. An ordinary hammer with a cold chisel can also be used for this purpose. A stout knife blade is useful for separating small rock layers such as those found in soft shale. A large iron bar is also useful for moving large pieces of rock. Other valuable equipment includes a notebook, knapsack and a magnifier. Small boxes and sacks are convenient.

Specimens should be cleaned at home so that they can be handled carefully. The cleaning can best be done with plain water. The rock surrounding them can be trimmed with a knife, chisel or hack saw. Needles are often used to clean small specimens.

Some rock structures resemble fossils in appearance, but they were not made by animals or plants. These are called *pseudofossils* (false fossils). One type of pseudofossil is a rounded accumulation of mineral matter known as a *concretion*. They often look like petrified eggs or potatoes. Others have cracks filled with minerals and may be mistaken for fossilized turtle shells. Although concretions are not fossils, some do contain fossils.

If you decide to make a collection of fossils, you will want to classify them into their proper groups. Each group is composed of organisms that have some common characteristics. The scheme of classification can be found in most paleontology reference books.

VOCABULARY REVIEW

Match the word or words in the column on the right with the correct phrase in the column on the left. *Do not write in this book.*

1. Period of violent earth crust disturbance.
2. The boundary between layers of rock deposited at greatly differing times.
3. An early classification of the oldest rocks.
4. An electrically charged particle consisting of two protons and two neutrons.
5. An electron given off by a nucleus.
6. Alternating dark and light layers found in sedimentary rock.
7. The study of fossils.
8. Fossils used to give the age of a rock layer.
9. Changing of an organism to stone.
10. The method by which coal is produced.
11. An empty cavity left in rock by an organism.
12. Undigested sand or mud fossils left by small organisms.
13. Gizzard stones left by dinosaurs.
14. Rock structures resembling fossils.

a. petrification
b. coprolites
c. pseudofossils
d. carbonization
e. revolution
f. basement complex
g. gastroliths
h. unconformity
i. concretions
j. alpha particle
k. varves
l. index fossils
m. paleontology
n. beta particle
o. castings
p. mold

QUESTIONS

Group A

Select the best term to complete the following statements. *Do not write in this book.*

1. The type of rock which gives the most distinct clues to the earth's past is (a) igneous (b) sedimentary (c) metamorphic (d) contact metamorphic.

2. A basic principle used in interpreting the earth's history through rock layers is the principle of (a) superposition (b) sediment deposition (c) contact metamorphosis (d) unconformity.

3. Violent disturbances evidenced by the twisted and deformed layers of the earth's crust are called (a) castings (b) concretions (c) revolutions (d) varves.

4. An unconformity in rock layers results when (a) a revolution occurs (b) layers of different thicknesses are formed (c) magma intrudes between the layers (d) a long period of time passes before the next layer of rock is formed.

5. What else besides a layered arrangement tells us most about the history of the rock? (a) the composition of the rocks (b) the thickness of the rock layers (c) the location of the rocks on the earth's surface (d) the present condition of the rock's surroundings.

6. When hot magma changes the make-up of nearby rocks, the process is known as (a) revolution (b) unconformity (c) contact metamorphosis (d) radioactivity.

7. Of the following, which type of igneous rock formation is most useful in interpreting the rock record? (a) extrusive formations (b) intrusive formations (c) lava flows (d) laccolith.

8. The greatest gap in the earth's rock history occurs (a) in the first 1.5 billion years (b) about the time of the great flood (c) during the time of the first ice age (d) about the time of the last ice age.

9. An atom whose nucleus is unstable undergoes change by throwing off electromagnetic energy and (a) heat (b) electrical particles (c) magnetism (d) light.

10. The particle given off by an atom, which changes both the atomic mass and atomic number is the (a) alpha particle (b) beta particle (c) gamma particle (d) electromagnetic particle.

11. The particle given off by an atom, which increases the atomic number by one is the (a) alpha particle (b) beta particle (c) gamma particle (d) electromagnetic particle.

12. When an atom of U-238 undergoes radioactive decay, eventual stability is reached when it becomes an atom of (a) thorium (b) protactinium (c) lead (d) strontium.

13. The half-life of uranium-238 is (a) 50 billion years (b) 12.5 billion years (c) 4.5 billion years (d) 100,000 years.

14. Another element having a long half-life which can be used for radioactive dating of rocks is (a) thorium-234 (b) strontium-87 (c) carbon-14 (d) protactinium-234.

15. Carbon-14 is most useful in radioactive dating to determine the age of (a) igneous rocks (b) sedimentary rocks (c) mineral crystals (d) remains of plants and animals.

16. Carbon-14 is produced by (a) lightning striking carbon dioxide (b) cosmic rays striking oxygen (c) cosmic rays striking nitrogen (d) lightning striking nitrogen.

17. The age of Niagara Falls has been estimated by a method using (a) rate of erosion (b) rate of deposition of sediments (c) U-238 half-life (d) C-14 half-life.

18. Determining geologic age by counting varves is an example of the method using (a) rate of erosion (b) seasonal change in the deposition of sediments (c) strontium-87 (d) carbon-14.

19. The dating method which is believed to be most accurate is (a) rate of erosion (b) rate of deposition of sediments (c) amount of salt in the sea (d) radioactivity.

20. A paleontologist is concerned mostly with the (a) layers of the earth (b) fossil content of rocks (c) relationship of earthquakes to the structure of the earth (d) age of igneous rocks.

21. Index fossils serve to identify (a) specific rock layers no matter where they are found (b) the composition of different rock layers (c) the location of ancient oceans (d) the origin of rock layers.

22. A practical use of index fossils is (a) to use them as fuel (b) to help locate oil and gas deposits (c) to determine the composition of a given rock layer (d) in understanding the origin of the continents.

23. One reason paleontologists believe the polar regions once had a tropical climate is that (a) fossils of tropical plants have been found there (b) evidence shows that the polar continents were once at the equator (c) the temperatures of polar regions sometimes are as high as in the tropics (d) evidence indicates that snow was absent at one time in the polar regions.

24. Analysis of oxygen isotopes in certain fossil shells has enabled scientists to determine (a) the age of the fossil (b) the relative temperature of the water where the shells grew (c) the kind of water where the shells grew (d) the growth rate of the shells.

25. In rare cases entire plant or animal bodies preserved as fossils have been found, such as (a) trees in the Arizona petrified forest (b) wooly mammoths in Alaska (c) dinosaur imprints in mud flats (d) shell molds in limestone.

26. When an organism's body has been completely replaced by minerals, it is (a) petrified (b) carbonized (c) a casting (d) a coprolite.

27. Carbonization of plants results in the formation of (a) oil and gas (b) petrified forests (c) coal (d) castings.

28. An example of an imprint fossil would be (a) the wooly mammoths found in Alaska (b) the castings left by worms (c) coprolites left by large numbers of animals (d) the outline of a fern leaf.

29. Preserved waste materials left by ancient animals are called (a) gastroliths (b) imprints (c) coprolites (d) molds.

30. Concretions are a good example of (a) pseudofossils (b) castings (c) coprolites (d) gastroliths.

Group B

1. Why is it that sedimentary rock layers offer the most clues to the earth's history?

2. What is the principle of superposition?

3. What is the principal reason that errors occur in the use of the principle of superposition?

4. Why is an unconformity a problem in determining the age of two neighboring layers of rock?

5. List two general ways cited in the text of how igneous rock formations can be useful in understanding earth history.

6. Why is the rock record for the earth's first 1.5 billion years not available?

7. Describe how an unstable atomic nucleus becomes stable.

8. What is meant by the term "half-life?"

9. How can the half-life of a radioactive substance be used to determine the age of rock in which it is found?

10. Explain why carbon-14 can be used in dating a piece of wood but is not generally used for dating rock.

11. Briefly describe three methods not based on radioactivity which have been used for measuring geologic time.

12. In what ways does the work of the paleontologist differ from that of the geologist?

13. What are the four requirements a fossil must meet to be used as an index fossil?

14. What explanation can you give for the fact that gas, oil, and coal are called "fossil fuels?"

15. In what ways besides determining the relative ages of rock layers, are fossils useful in understanding the earth's history?

16. Why are fossils consisting of an entire animal or plant rarely found?

17. How do organisms become petrified?

18. Give several examples showing how indirect fossil evidence proves the existence of life.

19-20. The following questions refer to the cross section below. Formations are indicated by letters A-H.

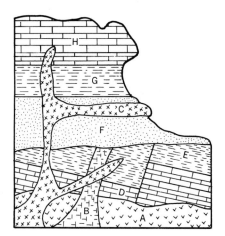

19. Between which two lettered formations does an unconformity exist? Explain.

20. Using the letters for each formation, list the correct order in which the formations were deposited. (From oldest to youngest)

19
earth history

objectives

☐ Explain how the divisions of a geologic calendar are determined.

☐ List the names of the divisions of the geologic calendar.

☐ Describe the events that occurred during the Precambrian era.

☐ List the periods of the Paleozoic era.

☐ Describe the conditions and life that existed during each period of the Paleozoic era.

☐ Describe the tectonic activity that occurred during the Mesozoic era.

☐ List the two periods of the Cenozoic era.

☐ Describe the conditions and life that existed during each period of the Cenozoic era.

The story of the earth's past can be compared to a dramatic play. The earth serves as the stage. Just as scenes shift and one act passes to another, the earth's surface shifts and continues to change. All living things that appear and disappear are the cast of characters. A continuous plot unfolds as life struggles to exist on the changing earth. A portion of this drama is a mystery. We have witnessed only the latest part of the performance. The earlier parts are unknown and can only be discovered by studying the present.

In this chapter, we will study what is currently known about the past development of the earth's history. This chapter will deal only with the processes involved in the building of the continent of North America. It is important to realize that the history of geologic changes and changes in life forms that appear in this chapter are based on available scientific evidence and not on proven facts.

THE GEOLOGIC CALENDAR

Earth history. Two kinds of events make up the earth's history. Physical changes that altered the earth and affected the conditions on its surface are one of the events. The other event is found in the record of the many kinds of living things that have existed on the earth.

The physical history of the earth began when it became a separate body orbiting around the sun. In time, the young planet Earth developed a solid surface. The earth's surface has been a scene of constant change. Continents formed and shifted over the globe as the crust was broken into separate plates. The oceans grew to fill the basins between the continents and finally covered more than three-fourths of the surface. Mountains were thrust up on the continents and the sea floor. On the land, mountains were worn away by erosion and slowly replaced by new ones. As the continents shifted, the sea partly covered them at times. When the sea later retreated, it left sediment behind as evidence of the temporary conquest of the land. Great ice sheets advanced to cover much of the continents. As the climate warmed the glaciers melted and their water returned to the oceans.

While physical changes continuously modified the earth's surface, a variety of living things made their appearances. Most flourished for a time, then disappeared, leaving only traces of their existence in the rocks. For the most part, the development of living things has been closely linked to the physical events that took place throughout the earth's history.

An earth calendar can be constructed by tracing the record that physical changes left in the earth's rock. This calendar can serve as a guide to the events involved in forming the earth's history. The end of one major time period and the beginning of another on this calendar marks a *geologic revolution*. One of the most important effects of a geologic revolution is the disappearance of many kinds of living things. New forms of life, better suited to the new conditions, replace the old. Proof of such an occurrence can be found in the fossil remains left in rocks. If there is no fossil record, it can be very difficult to establish the divisions of the geologic calendar. About five-sixths of the earth's history took place before fossils became common. This means that only about one-sixth of the earth's history is known well enough to be included on a calendar.

The largest unit of time in the geologic calendar is called an *era*. One era ends and another begins when there is a geologic revolution that causes a widespread change in existing conditions. Because geologic revolutions do not occur at definite times, the different eras are not all the same length. For example, the oldest and longest era is the *Precambrian*. It began when the earth came into existence. It ended about six hundred million years ago, when the fossil record shows that life became abundant. It is believed that many events must have occurred during the Precambrian era. However, because it was so long ago, evidence of these changes is very difficult to find. The Precambrian Era can be divided into two

Geologic Calendar

ERAS	PERIODS	EPOCHS	DURATION	Years before present
CENOZOIC	QUATERNARY	RECENT	25,000	
		PLEISTOCENE	975,000	
	TERTIARY	PLIOCENE	11,000,000	1,000,000
		MIOCENE	13,000,000	12,000,000
		OLIGOCENE	10,000,000	25,000,000
		EOCENE	25,000,000	35,000,000
		PALEOCENE	10,000,000	60,000,000
MESOZOIC	CRETACEOUS		65,000,000	70,000,000
	JURASSIC		45,000,000	135,000,000
	TRIASSIC		45,000,000	180,000,000
PALEOZOIC	PERMIAN		45,000,000	225,000,000
	PENNSYLVANIAN		60,000,000	270,000,000
	MISSISSIPPIAN		20,000,000	330,000,000
	DEVONIAN		50,000,000	350,000,000
	SILURIAN		40,000,000	400,000,000
	ORDOVICIAN		60,000,000	440,000,000
	CAMBRIAN		100,000,000	500,000,000
PRECAMBRIAN	LATE PRECAMBRIAN		1,100,000,000	600,000,000
		KEWEENAWAN		1,700,000,000
		HURONIAN		
	EARLY PRECAMBRIAN	TIMISKAMING	2,800,000,000	
		ONTARIAN		4,500,000,000

Geologic Calendar

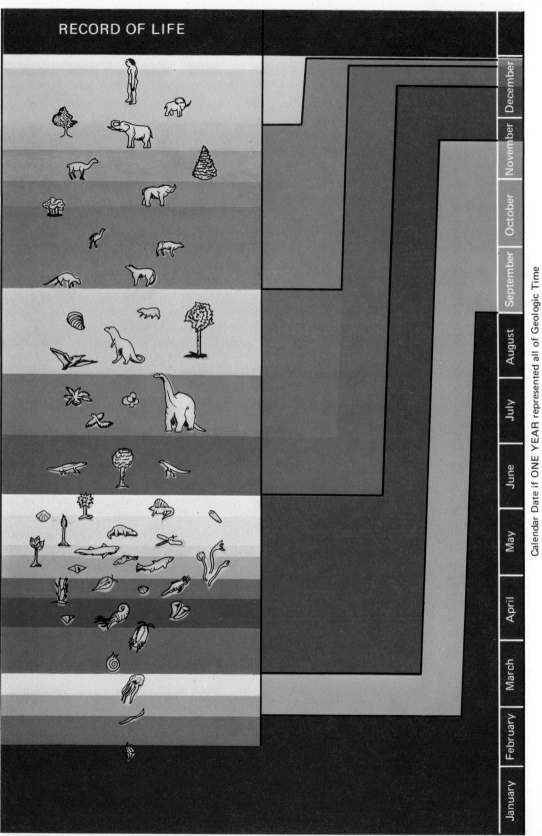

RECORD OF LIFE

Calendar Date if ONE YEAR represented all of Geologic Time

January | February | March | April | May | June | July | August | September | October | November | December

Investigate
Refer to Figure 19–1. Using the bar graph that depicts the relative time period for each era, set up a similar scale based on portions of a 24-hour day. Approximately what part of a day would be represented by the Precambrian Era?

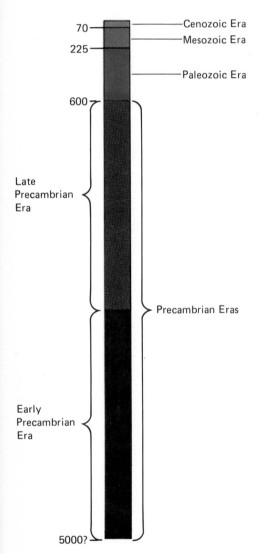

FIG. **19–1.** Precambrian time, of which little is known, makes up ⁵⁄₆ of the earth's history.

major parts. They are the *Early Precambrian era,* which has no record of life, and the *Late Precambrian era* during which the first faint record of life appears.

The time from the Precambrian era to the present can be divided into three more eras. Following the Precambrian era was the *Paleozoic era.* It lasted about 375 million years. Then came the *Mesozoic era,* which lasted approximately 155 million years. The *Cenozoic era* covers the remaining 70 million years up to the present. The relative length of time covered by each of these geologic eras is illustrated in Figure 19–1.

These three most recent eras are known well enough to be broken down into smaller divisions called *periods.* Within an era, one period is separated from another by evidence of crustal changes and the presence of characteristic fossils. A period is usually named for the location where these changes were best observed. For example, there is a Devonian period (from Devon, England) in the Paleozoic era. The term "Devonian" applied to rock layers means that they were deposited in the same time span as those rocks in England that identify the period.

Since the Cenozoic era is the most recent, there is enough evidence to divide its two periods into even shorter *epochs.*

The dividing of geologic time into eras, periods, and epochs was originally done based on evidence found of changes in the earth's history. Fossil records showed that many animals and plants suddenly died out and were then replaced by new forms. Many of the original divisions of geologic time do not have as clear boundaries as was once thought. However, the original plan of the geologic calendar is still used by geologists to describe the earth's history. A complete summary of the geologic calendar is given in Table 19–1, pages 398–99. It shows the eras, periods, and epochs with their approximate time durations. You should refer to this chart frequently while studying the following sections.

DIVISION OF GEOLOGIC TIME

The Precambrian era. Almost 90 percent of geologic time is shown through rocks formed during the Precambrian era. But the record left in these rocks is very difficult to read. It is believed that the earth's crust was broken into moving plates very early in its history. Thus, the continents were probably arranged differently and surrounded by different oceans. How those continents and oceans were arranged may never be known with complete certainty. However, study of the world's oldest rocks indicates that the ancestors of today's continents did exist at least 3 billion years ago.

In North America, for example, Precambrian rocks are exposed over a wide area of eastern Canada and parts of the United States. This large region of ancient rocks is called the *Canadian Shield*. See Figure 19–2. It has been found to contain large deposits of valuable minerals. The Canadian Shield represents the nucleus around which the continent of North America grew. Nearly all continents have a center of Precambrian rocks from which they formed. These centers are called *cratons*. Most of these cratons of very old rock are about 1,000 km across. A continent's craton can be exposed as the Canadian Shield is, or it can be buried under younger rocks. In some areas, these very old rocks are exposed by erosion. The Grand Canyon has been cut through younger rock layers to reveal the Precambrian rocks below.

The rocks of the Canadian Shield, as well as the cratons of other continents, are mostly metamorphic rock. In some places, these rocks are almost unchanged. This shows that the geologic forces that originally formed the Precambrian rocks were probably the same as the forces that operate today. There is evidence of many forms of volcanic activity in these rocks. Sedimentary rocks made from the deposits of streams and larger bodies of water are also present in the Canadian Shield. Evidence of glaciers found in late Precambrian rock shows that there were at least two ice ages.

Precambrian rocks almost completely lack fossils. However, it is likely that simple forms of life did gradually appear. The absence of fossils is mainly due to the kind of living things that existed during that era. Plants and animals with hard parts that leave well-preserved fossils did not live during Precambrian times. There is evidence of the existence of simple plants called algae, sponges, and worms. Beds of limestone and graphite, usually formed by the remains of living things, are also common in Precambrian rocks.

Precambrian rocks also show evidence of the mountain-building processes that took place along crustal plate boundaries. Mountain ranges probably formed when the lithospheric plates collided. These mountains then eroded

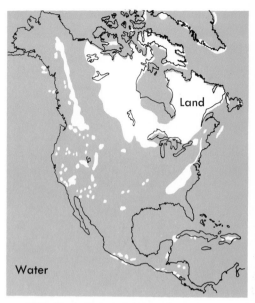

FIG. 19–2. Precambrian rocks exposed in North America. The largest area is the Canadian Shield. Smaller areas of ancient rock have been exposed by mountain building or erosion.

and left characteristic metamorphic rock as evidence of their presence. Precambrian rocks changed so much over the centuries that the outlines of the very ancient continents, or their movements, are difficult to trace. However, there is evidence that when the Precambrian era ended, there was only a single land mass. This was the ancient super-continent called Pangaea.

The Paleozoic era. The Paleozoic era lasted about 375 million years. Throughout this long era, the ancient super-continent of Pangaea remained a single land mass. But it is believed that this giant continent did not remain in the same position. The slow shifting of Pangaea produced the events that allow the era to be divided into periods.

1. **The Cambrian period.** This is the earliest part of the Paleozoic era. During this period, much of the super-continent was covered by sheets of ice. It is believed that at that time, Pangaea lay in the southern hemisphere near the South Pole. This meant that most of the land surface of the earth was cold. As a result, conditions were not suitable for the development of a variety of living things. This accounts for the fact that almost all fossils from the Cambrian period are from sea-dwelling forms of life. Since there were no continents in the Northern Hemisphere to interfere with water circulation, the water was probably warm and provided a good environment for marine life.

There is little fossil evidence of Cambrian life on land surfaces. Only very simple forms of plant life could have

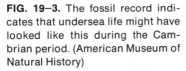

FIG. 19–3. The fossil record indicates that undersea life might have looked like this during the Cambrian period. (American Museum of Natural History)

FIG. **19–4.** A giant trilobite of the Paleozoic Era.

existed on the exposed rocks. The seas, however, flourished with many kinds of living forms. See Figure 19–3. In the sedimentary rocks of this period, the remains of many kinds of animals without backbones are buried. These animals are called *invertebrates.* Typical Cambrian invertebrates include jellyfish, snails, sponges, and a clamlike animal called a brachiopod. The ancestors of the modern squid and octopus, called cephalopods, are also invertebrates.

The most common invertebrate found during the Cambrian period was the *trilobite* (*TRY*-LOH-BYT). See Figure 19–4. They were the dominant life form during Cambrian times. They became extinct as conditions changed near the close of the Permian period. The large number of trilobites remaining in rocks makes them excellent index fossils for dating Cambrian formations in different parts of the world.

2. **The Ordovician** (Or-do-*vish*-an) **period.** During the beginning of this period, there were widespread changes on land surfaces. Parts of the continental area were flooded by the sea and much of what is now North America was covered by shallow seas. See Figure 19–5. A giant mountain range grew along the region that has become the east coast of the U.S.

Sea life continued to develop during this period. Among invertebrates, the trilobites were replaced as the chief form of life by the mollusk group. Mollusks include clams

Explore
In 1938 a type of fish believed to be extinct for over 70 million years was caught alive in the Indian Ocean. The fish was called coelacanth. Find out more about this species and its special characteristics.

FIG. **19–5.** The North American continent during the Ordovician period.

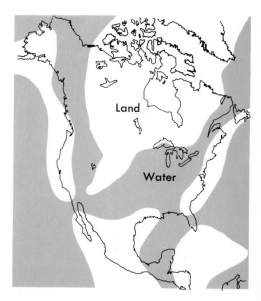

FIG. 19–6. Late Ordovician life as it may have appeared near an inland sea. (Allan Roberts)

and similar animals with hard shells, as well as cephalopods. See Figure 19–6. The first *vertebrate*, or animal with a backbone, was the ancient fish. It initially appeared in the Ordovician period and was the ancestor of the vertebrate animals that followed.

3. **The Silurian period.** During this period, there was a slow uplift of what was to be the eastern coast of North America. This uplift changed the interior land mass into a large inland sea. Most of this relatively short period (20 million years) was fairly quiet with little movement of the

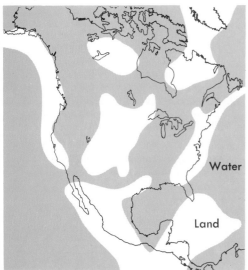

FIG. 19–7. The North American continent during the Silurian period.

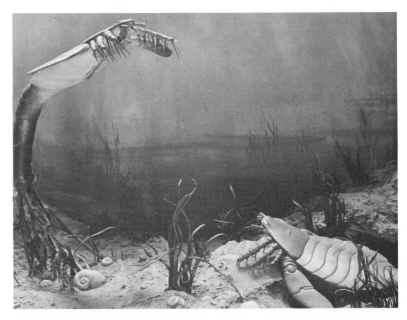

FIG. 19–8. An undersea scene as it might have appeared during Silurian times. (New York State Museum)

land. See Figure 19-7. The development of sea life continued through the Silurian period. See Figure 19-8. The sea gradually evaporated and by the end of the period, a large salt desert was formed. These Silurian salt deposits are not deeply buried beneath the earth's surface. Some of these deposits have been reached and mined in Pennsylvania, Michigan, and New York.

The earliest known land animals made their appearance during the Silurian period. They were small scorpions and millipedes that had the ability to breathe air. Fossil evidence does not show whether these animals spent all their time out of the water or only occasionally came onto dry land.

4. **The Devonian period.** Development of sea-dwelling vertebrates began during the Ordovician period and advanced further during the Devonian period. See Figure 19-9. Many kinds of fossil fish are found in rocks from this period. During the Devonian period, some fish developed the ability to breathe air. They were also apparently able to come onto land for short periods. The fish that could come on land were the beginnings of the amphibious group of animals. Development of amphibious animals was probably encouraged by slow movements of the earth's crust. These movements caused shallow seas to rise and fall. The appearance of North America during this time is shown in Figure 19-10. A large variety of land plants also began to appear during the Devonian period. This period ended with the formation of mountains in what is now northern Maine and eastern Canada.

5. **The Mississippian and Pennsylvanian periods.** About 350 million years ago, it is believed that Pangaea drifted northward. This caused the climate of the earth to become warmer. Great swampy forests grew over much of

FIG. 19-9. A reproduction of some of the earliest fish that existed during the Devonian period. (American Museum of Natural History)

FIG. 19-10. The region that was to become the North American continent as it probably appeared during the Devonian period.

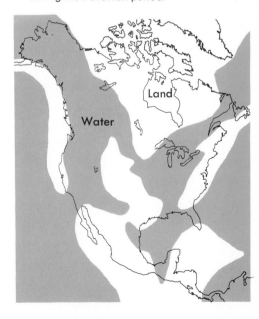

FIG. 19–11. An artist's conception of the types of plant and animal life that were found during the Carboniferous period. (Yale Peabody Museum)

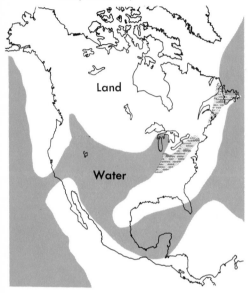

FIG. 19–12. North America during the Pennsylvanian period. Swampy areas near the east coast later became coal deposits.

FIG. 19–13. The North American continent during the Permian period. Dotted area represents the location of an ancient inland sea.

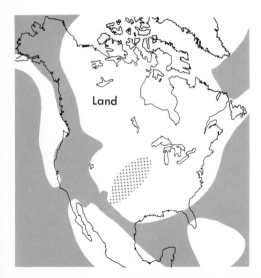

the land surface. See Figure 19–11. These forests had large ferns, horsetails, and distant ancestors of modern trees. See Figure 19–12. The land in some areas where these forests grew sank very gradually. Thus, huge deposits of the remains of plants accumulated and were eventually buried under sediments. Coal deposits were created in this way. These two periods together are often called the *Carboniferous period.* Carboniferous means "carbon bearing" and refers to the vast coal deposits formed during these two periods.

Land animals continued to develop during the Carboniferous period. Many kinds of insects first appeared. A new group, known as reptiles, became the first vertebrates to dwell entirely on land. The development of reptiles marked the first time any group of backboned animals became air breathers. They no longer needed to return to water to lay their eggs. Life continued to develop in the seas, and fish and sharks became particularly abundant.

6. **The Permian period.** This was one of the most violent periods of the earth's history. Great mountain ranges were built, like the ancient Appalachian Mountains in what is now eastern North America. The Appalachian Mountains may have once been as tall and rugged as the modern Rockies. The remains of the Appalachians, in the form of folds and faults, are still visible after more than 200 million years. In addition to disturbances of the land surface, the Permian period marked severe changes in climate. The large inland sea of North America dried up. What remained were a desert and giant salt deposits that reached from Kansas to New Mexico. See Figure 19–13. Life forms, including the trilobites, could not adapt to

FIG. 19–14. An artist's version of the types of life that existed during the Permian period. Notice how animal life differs from earlier times. (Chicago Museum of Natural History — Charles R. Knight)

these changes and became extinct. Reptiles proved themselves to be especially adaptable to these new conditions. They survived and became one of the dominant life forms.

The Mesozoic era. During this era, the face of the earth was greatly changed. It is believed that Pangaea began to break up. First, it separated into two parts called Laurasia and Gondwanaland. See Figure 19–15A. Then each of these fragments of the super-continent in turn broke up again and formed the general outlines of the modern continents. See Figure 19–15B. By the end of the Mesozoic era, the earth's surface had taken on its familiar modern shape. See Figure 19–15C.

The sudden burst of tectonic activity during the Mesozoic era greatly influenced life on earth. The level of the sea rose as much as 500 meters. This caused a large part of the land to be covered with shallow seas. The rise in sea level may have been caused by the lifting of the sea floor along the mid-ocean ridges. The land flooded

FIG. 19–15. A) Pangaea began to break up at the end of the Paleozoic era. B and C) During the Mesozoic era, the continents took on their present form.

A

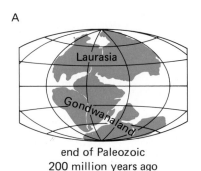

end of Paleozoic
200 million years ago

B

Mesozoic—65 million years ago

C

Cenozoic—present

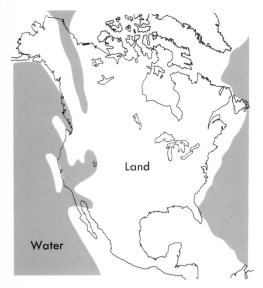

FIG. 19-16. The North American continent during the Triassic period.

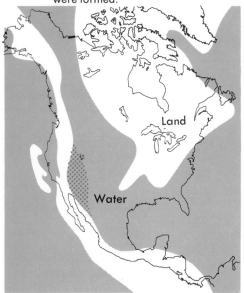

FIG. 19-17. The North American continent during the Cretaceous period. The dotted region indicates large swampy areas in some parts of the West, where coal deposits were formed.

and the climate became warm. Little change occurred between the seasons. These new conditions caused the disappearance of many of the life forms characteristic of the Paleozoic Era. Reptiles, well-suited to the warm and wet conditions, became the dominant life form.

1. **The Triassic period.** During this period, the North American continent began to take on its present shape. See Figure 19-16. The land was generally high and dry, but the Atlantic and Gulf coastal plains had not yet formed. Erosion started to attack the rugged, ancient Appalachians. Some regions of the continent dried up completely. Descendants of the Permian reptiles were the first of the dinosaurs and the ancestors of modern reptiles. For the first time the vertebrates, mostly reptiles, became the chief form of life on the land.

2. **The Jurassic period.** This period is generally known as the "Age of Reptiles." It was dominated almost entirely by the great dinosaurs. They were enormous, awkward, and slow-moving creatures. Some were so large that their legs could not support their bodies out of the shallow water in which they lived. Other dinosaurs were small and were probably able to move around rapidly. Some reptiles returned to the sea while others developed the ability to fly. During the Jurassic period, reptiles were masters of land, sea, and air.

The Jurassic period ended with a disturbance that raised a chain of mountains in western North America. This mountain chain stretched from southern California to Alaska. Today we call it the Rocky Mountains. The Sierra Nevada mountains were also formed at this time.

3. **The Cretaceous period.** During this period, the present Rocky Mountain area was flooded by a shallow sea. The sea extended from the Gulf of Mexico to the Arctic Ocean. See Figure 19-17. Dinosaurs continued to dominate as animal life until near the end of the Cretaceous period when they became extinct. Warm-blooded animals, mammals and birds, which first appeared during the Jurassic period, increased rapidly during Cretaceous times. Land plants, similar to modern trees and flowering plants, also appeared during this period. The Cretaceous period came to an end during a geologic revolution that created the Rocky Mountains. This revolution also re-elevated the ancient Appalachian Mountains, which by this time had been worn almost level. The end of the Cretaceous period also marked the end of the Mesozoic era.

The Cenozoic era. Toward the end of the Mesozoic era, the activity of crustal plates slowed a great deal. The mid-ocean ridges shrank and the oceans returned to their lower level. The climate grew cooler with sharp temperature differences between the seasons. The existing life forms, particularly the great reptiles, could not adapt to these new conditions. They were replaced by the ancestors of the animals and plants that live on the earth today.

By the start of the Cenozoic era, North America had probably reached the familiar shape we know today. See Figure 19–18. Only the Atlantic and Gulf coasts, and parts of what is now California were still submerged. Disturbances continued during the Cenozoic era that caused the land surface of much of North America to rise and fall. Violent movements along many of the faults in the western part of the continent created block mountains. The Sierra Nevadas, which first appeared in the Jurassic period, were lifted closer to their present height. Smaller mountain ranges were created throughout the west as fault blocks were tilted and raised. In the east, the newer Appalachian Mountains continued the lifting process that began during the Cretaceous period.

Many volcanic eruptions also took place during the Cenozoic era. Large areas of Washington, Oregon, Idaho, and northern California were covered by thick layers of lava. Volcanic cones created by these eruptions are still seen in those regions. In addition to changes in the surface of the crust caused by mountain building and volcanism, the Cenozoic also saw the development of the Ice Age. These events made the entire Cenozoic era a time of great change. Because of these factors, it is difficult to separate this time into smaller divisions. The Cenozoic era is often divided into two periods, the *Tertiary* (tersh-i-ary) and *Quaternary* (kwah-*ter*-na-ree). The periods are separated by the Ice Age that occurred late in the era. These periods can be classified according to the kinds of plants and animals living at that time. The Tertiary period is divided into the following epochs:

1. **The Paleocene epoch.** This epoch might be called the "early morning" of the development of modern mammals. Dinosaurs had vanished, although ancestors of present-day reptiles were fairly abundant. The mammals that did exist were generally small in size. Most of them bore only a slight resemblance to modern forms. The first of the primates are thought to have developed during this epoch. This ancient primate was a small, tree-dwelling

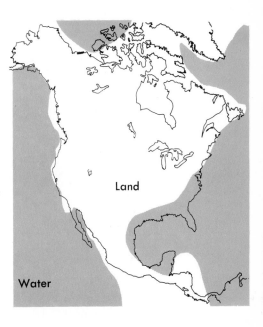

FIG. 19–18. North America during the early part of the Cenozoic Era.

Construct
Prepare a diorama of the Cenozoic era, depicting the major epochs and the accompanying mammal types that dominated this era.

FIG. 19–19. Eocene mammals. Early ancestors of the modern horse and the rhinoceros. (Chicago Museum of Natural History)

activity

You will be given a hypothetical cross section of rock that contains fossils. The cross section probably shows the history of a particular part of the North American continent. To discover what part of the continent it is, do the following:

1. Use Table 19–1 and Figures 19–3, 19–6, 19–8, 19–9, 19–11, 19–14, 19–19 and 19–20, to help identify and label the periods of geologic history that the fossils represent.

2. Decide in which periods the surface was land or submerged underwater. Write "land" or "water" in the blanks.

3. Use the above information and Figures 19–5, 19–7, 19–10, 19–12, 19–13, 19–16, 19–17 and 19–18, to identify what state in the U.S. the cross section best represents, according to geologic history.

animal with bushy hair and a long tail. The plants of this epoch resembled present-day plants that grow in warmer areas of the world.

2. **The Eocene epoch.** In this epoch, the climate remained much as it was during the Paleocene. The mammals in general became larger, though most of them still did not look much like the familiar forms of today. See Figure 19–19. Many kinds of rodents appeared, and hoofed animals began to develop into the more specialized animals that exist today. The ancestor of the horse, which was about the size of a small dog, made its first appearance. It had slender legs for running through heavy forests and teeth shaped for eating the plant food. Lemurs, monkey-like primates with dog-like faces, were also very abundant. Lemurs, in fact, still exist. Birds were also numerous during this time. And in the sea, the first whales had developed.

3. **The Oligocene (ol-i-go-seen) epoch.** During this epoch, volcanic activity was widespread in North America. The climate grew cooler, favoring the growth of grasses, cone-bearing, and hardwood trees. These plants replaced the older tropical forests of earlier epochs. The first mammals disappeared and were replaced by animals more like modern forms. The dog and cat families became distinct groups. Primitive camels (similar to the modern

llama), primitive apes, and the first elephants could be counted among ever-increasing forms of life.

4. **The Miocene epoch.** This is often called the "Golden Age of Mammals." Great herds of early horses and camels roamed the plains of North America feeding on the abundant grasses that grew in the cool, dry climate. Elephants migrated into North America over a land bridge that joined Siberia and Alaska. Members of the deer family, the rhinoceros family, the pig family, and most of the familiar animals of today left abundant fossils during the Miocene epoch. Monkeys similar to the ones we know today also existed at this time.

5. **The Pliocene epoch.** This epoch brought many changes in living forms and closed the Tertiary period. A series of mountain-building and crustal disturbances, especially in western North America, began during this epoch. The activity that took place during this time in the geologic past seems to still be going on. The development of mammals continued, but large numbers of those present during the Miocene and earlier times disappeared. Evidence indicates that many hunting animals of the bear, dog, and cat families probably appeared during this epoch. These flesh eaters hunted among the herds of grazing animals that roamed the plains. Careful study of fossils shows that early forms of human-like creatures probably appeared during the Pliocene epoch.

The close of the Pliocene epoch was accompanied by great changes in the climate all over the world. Continental glaciers from the polar regions began to spread huge ice sheets toward the equator. A large amount of the earth's water was locked in these glaciers. With a portion of its water supply cut off, the level of the sea fell and land bridges linking North and South America in the region of Panama were formed. Existing bridges between Asia and Alaska were widened. Groups of animals crossed these bridges in both directions and mingled with the existing life. In the resulting competition, some forms of life became extinct and many new groups developed.

The Quaternary period was then established. However, there is only one complete epoch in this period.

1. **The Pleistocene epoch.** The climate of this epoch was very different from earlier times. The ice sheets advanced and retreated at least four times during this epoch. With each advance, the climate grew colder and harsher. Mammals and other animals were required to develop protection against the cold or else move to

Interpret
Of all the mammals, man shows the greatest development from as early as the Pleistocene epoch. List some of the factors that allowed man to survive against the forces of nature and the wilderness.

FIG. 19–20. Pleistocene life. In the foreground, American mastodon; to the back, a northern mammoth. Notice the heavy protective covering. (American Museum of Natural History)

FIG. 19–21. A cross section of North America focusing on the United States, which shows the major rock structures controlling surface features.

Cenozoic

Mesozoic

Paleozoic

Mostly Precambrian

warmer climates. Many animals did develop special characteristics that allowed them to survive. See Figure 19–20. Others moved to new regions. But many less adaptable species became extinct. Those that survived became the familiar animals we know today. Evidence indicates that modern humans probably appeared during the Pleistocene epoch.

MODERN NORTH AMERICA

The building of a continent. Like all parts of the earth's crust, the continent of North America has had a long history. Its modern features are a product of hundreds of millions of years of development. The land surface on all the continents is only another chapter in a long and dramatic

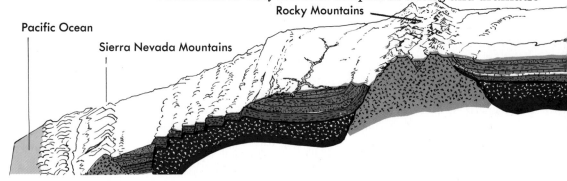

Pacific Ocean

Sierra Nevada Mountains

Rocky Mountains

VOCABULARY REVIEW

Match the word or words in the column on the right with the correct phrase in the column on the left. *Do not write in this book.*

1. The largest unit of geologic time.
2. Begins when the earth came into existence.
3. Era that begins with the Cambrian period.
4. Subdivisions of a period.
5. The most recent era of earth history.
6. Separated from one another by evidence of crustal changes that take place during an era.
7. Wide areas where Precambrian rocks are exposed at the surface.
8. The period that saw the beginning of the amphibian group of animals.
9. Original unit of rock from which continents were formed.
10. Ending of this epoch accompanied by great changes in climate all over the world.

a. periods
b. Paleozoic era
c. shields
d. Precambrian era
e. invertebrate
f. epoch
g. era
h. craton
i. Rocky Mountains
j. Cenozoic era
k. Permian period
l. Devonian period
m. Pleistocene epoch
n. Pliocene epoch
o. physiographic provinces

story. To understand the development of a continent such as our own, it will be necessary to study the hidden rock structure built up over the course of its history. It is the basic rock structure that controls the way the land surface is changed by the many forces that affect it.

If we study the part of the North American continent that is now part of the United States, we will find that it can be divided into nine natural subdivisions. These subdivisions or regions are further subdivided into *physiographic provinces*. These are large land areas of mountains, plains, or plateaus. They are separated from adjacent areas by uniform physical features that come from the rock structures beneath the surface. Figure 19–21 shows this basic rock structure and its age along an east-west cross section of the United States.

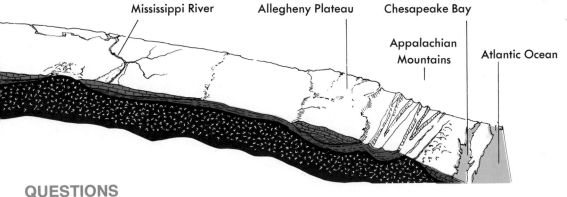

QUESTIONS

Group A

Select the best term to complete the following statements. *Do not write in this book.*

1. Fossils in rock formations have helped us to determine the kind of environment that existed for what fraction of earth's history? (a) 5/6 (b) 2/3 (c) 1/6 (d) 1/3.

2. Which one of the following would not be classified as a geologic revolution? (a) the dying off of an entire species of animals or plants (b) the changes marking the beginning of the Cenozoic era (c) the changes that occurred at the beginning of the Devonian period (d) the changes occurring at the end of the Precambrian era.

3. The oldest and longest era in the earth's history is the (a) Precambrian (b) Cenozoic (c) Mesozoic (d) Paleozoic.

4. The era whose name means "recent life" is (a) Precambrian (b) Cenozoic (c) Mesozoic (d) Paleozoic.

5. Which of the following did not accompany the end of any of the geologic eras? (a) many unconformities were found in the rock structure (b) great mountain belts were formed (c) shallow inland seas were drained or greatly changed (d) all living things had disappeared from the earth.

6. Which of the eras listed shows sufficient evidence for its periods to be divided into epochs? (a) Precambrian (b) Mesozoic (c) Cenozoic (d) Paleozoic.

7. The record left by rocks of the Precambrian era is (a) very difficult to read (b) moderately difficult to read (c) easy to read (d) impossible to read.

8. During the Precambrian era (a) no life forms existed (b) vertebrates probably existed (c) simple forms of life were probably present (d) mollusks were present as shown in the fossils found.

9. The largest of all shields in the world is in (a) Africa (b) Australia (c) South America (d) North America.

10. The shields of the world (a) are mostly sedimentary rocks (b) contain more fossils than any other rocks (c) have the world's largest mountains on their surface (d) have always remained as a foundation while other land masses underwent change.

11. The super-continent that existed during the Cambrian period (a) began to break into two large continents (b) was covered by sheets of ice (c) was warm but without life (d) was located near the North Pole.

12. Sedimentary rocks show that animal life of the Cambrian period included many (a) snakes (b) land animals (c) fish (d) invertebrates.

13. The earliest known land animals appeared during the (a) Cambrian period (b) Ordovician period (c) Silurian period (d) Devonian period.

14. Reptiles developed during the (a) Devonian period (b) Permian period (c) Silurian period (d) Carboniferous period.

15. The Mesozoic era was a time of great changes that are attributed mostly to (a) the great tectonic activity (b) the ice ages (c) meteor bombardment (d) earth changing its orbit.

16. The Rocky Mountains were created in a revolution that ended the (a) Paleozoic era (b) Mesozoic era (c) Cenozoic era (d) Precambrian era.

17. The North American continent first began to take on its present familiar shape during the (a) Jurassic period (b) Triassic period. (c) Permian period (d) Cretaceous period.

18. The period in which the great dinosaurs dominated the earth was the (a) Jurassic (b) Triassic (c) Permian (d) Cretaceous.

19. A fossil commonly found in Paleozoic rock formation is (a) eohippus (b) trilobite (c) mastodon (d) dinosaur.

20. The era called the "Age of Mammals" is the (a) Cambrian (b) Paleozoic (c) Mesozoic (d) Cenozoic.

21. The two periods of the Cenozoic era are separated by the (a) Ice Age (b) Tropic Age (c) Age of the Reptiles (d) formation of the Rocky Mountains.

22. The first three epochs of the Tertiary period were the (a) Pliocene, Eocene, Oligocene (b) Paleocene, Oligocene, Miocene (c) Paleocene, Eocene, Oligocene (d) Pleistocene, Eocene, Oligocene.

23. Many flesh-eating mammals appeared during the (a) Pliocene epoch (b) Eocene epoch (c) Paleocene epoch (d) Pleistocene epoch.

24. The Miocene epoch is often called the (a) Age of Mammals (b) Golden Age of Mammals (c) Age of Reptiles (d) Age of Dinosaurs.

25. How many complete epochs are there in the Quaternary period? (a) 5 (b) 4 (c) 3 (d) 1.

26. Which of the following life forms reached their peak during the Devonian period? (a) fish (b) trilobites (c) amphibians (d) reptiles.

27. The most important development of the Pleistocene epoch was probably the appearance of (a) trilobites (b) human beings (c) reptiles (d) mammals.

28. The way the land surface is changed by the many forces at work is controlled mainly by (a) humans (b) the underlying basic rock stucture (c) the climate (d) action of plants and animals.

Group B

1. What is an era in geologic time?

2. List the eras of the earth's history in the order they occurred.

3. What is a period in geologic time?

4. List the periods of the Mesozoic era in the order they occurred.

5. List the periods of the Paleozoic era in the order they occurred.

6. Describe the two kinds of events that make up the earth's history.

7. What relationship exists between cratons and continents?

8. Why is it difficult to find Precambrian fossils?

9. Describe the major revolution that marked the end of two geologic eras. Name these eras.

10. What is a trilobite? Why are their fossils important?

11. What evidence do geologists use to reconstruct the earth's history?

12. Humans were one of the last organisms to appear on the earth. How do you account for the fact that they now dominate the earth and, to some degree, control their environment?

13. Why do you suppose the earliest forms of life developed in the sea rather than on land?

14. Briefly describe the tectonic activity that occurred from the beginning of the Mesozoic era to its end.

15. What period is known as the "Age of Reptiles?" Describe the conditions that existed during this time.

16. Why are the Pennsylvanian and Mississippian periods commonly combined under the title of "Carboniferous?"

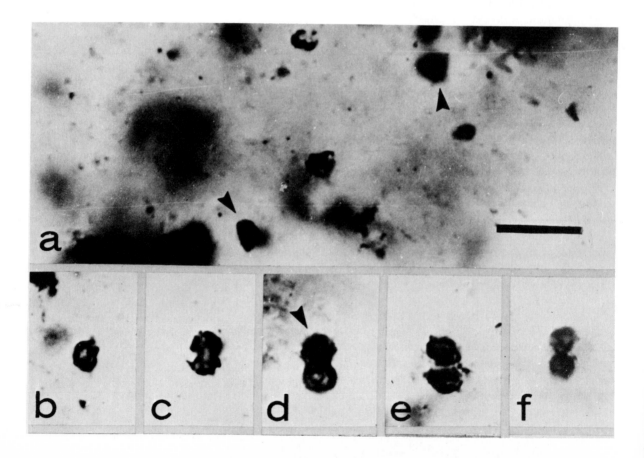

digging up the past

Oldest Record of Life

Sedimentary rocks, 3.1 to 3.4 billion years of age, contain the oldest direct evidence of life. Most sedimentary rock containing traces of the earliest forms of life has been changed into metamorphic rock. This change has destroyed almost all of the fossil record of the beginnings of life. Below is a picture of the oldest fossil known. These cells are similar to the blue-green algae of today.

There is some evidence that life existed more than 3 billion years ago. One type of indirect evidence is based on the knowledge that living things are made up of molecules that contain more of the lighter isotopes of carbon than the carbon found in non-living things. Measurement of the ratios of carbon isotopes from Precambrian rocks suggests that organisms incorporating carbon must have been present when rocks with ages greater than three billion years were formed.

Other direct evidence of ancient life is found in certain iron deposits that appear to have been formed in bodies of water where oxygen-producing plants were present. Although difficult to interpret, scientists believe that the evidence shows that life on the earth has a history almost as old as the planet itself.

A Frozen Fossil

A gold prospector at work in a remote region of Siberia recently uncovered a remarkable fossil. The picture at the right shows this perfectly preserved frozen baby mammoth. Although mammoths were ancestors of modern elephants, they had a heavy covering of fur and long hair. They also had very long curving tusks. Frozen remains of these Pleistocene animals have been discovered previously in Siberia and Alaska, preserved by the frozen soil in these northern regions. However, this most recent discovery in Siberia was the first time scientists were able to examine the remains before they decomposed or were eaten by predators.

The baby mammoth had died at the age of six months, apparently by falling into a lake or swamp, between 9 and 12 thousand years ago. It became frozen in such a perfect state that even its blood could be analyzed when it was found. Fossil remains in such a complete state of preservation can provide information about an earth environment that disappeared thousands of years ago. This unusual discovery of a complete animal gives hope that other such preserved fossils may still lie buried underneath the frozen wastes waiting to be revealed.

People Who Find a Key to the Past

Above and to the left are people who find meaning in the fossils left in rock. They are Paleontologists. A mere trace of bone or piece of shell will enable these scientists to construct the past of a layer of rock. Using a bit of inference, the history of an entire area may be revealed from as little evidence as an imprint of an ancient trilobite.

unit

6

the earth's
atmosphere

20
air and its movements

objectives

☐ List the materials that make up the earth's atmosphere.

☐ Describe air pressure and the instruments used to measure it.

☐ Describe the chemical balance of the atmosphere.

☐ List several types of air pollution and their effects on earth.

☐ Describe the effect of the atmosphere on the solar energy received by the earth.

☐ List the layers of the atmosphere.

☐ Explain how movement of the air is caused by heating the atmosphere.

☐ Describe how the rotation of the earth affects the movement of air.

We are accustomed to believing that we live on the surface of the earth. Seldom do we think of ourselves as creatures who must live beneath a blanket of gases, as a fish must live in water. But man is designed to live only within the narrow region where two of the earth's layers meet. Above this area is a layer of gases, the *atmosphere*, which surrounds the entire planet. Below are the earth's oceans or *hydrosphere*, covering $7/10$ of the earth's solid crust or *lithosphere*. At the boundary between the atmosphere and the lithosphere is the place where we spend most of our lives.

Since we are always submerged in the atmosphere, it is important that we know which gases and other materials are present. It is equally important that we understand the processes which affect their composition. The study of the gaseous region above the solid earth is called *meteorology*. One of the greatest challenges in meteorology is to understand the effects of heat energy in the atmosphere. The atmosphere is actually a giant heat engine. An engine is a device arranged for changing one form of energy into another. The atmosphere is an engine because it changes radiant energy from the sun into heat energy. The workings of the atmospheric heat engine fuel the winds and cause the ever changing patterns of weather.

THE AIR

Composition of the atmosphere. If a sample of air were dried and all foreign matter removed, chemical analysis would uncover two main gases. The gas nitrogen (N_2)

would account for 78 percent of the total volume of the sample; oxygen (O_2) would make up 21 percent. The remaining 1 percent would be mostly argon (Ar) and carbon dioxide (CO_2). Table 20–1 shows the normal composition of air near sea level.

Some of the atmospheric gases, including those listed in Table 20–1, are always present in approximately the same amounts in air near the earth's surface. Other gases are also likely to be present, but not in definite amounts. Water vapor is by far the most important of these variable gases. In very moist air, water vapor makes up as much as three percent of the total volume. Another important variable gas in the atmosphere is ozone (O_3), which is a form of oxygen. The ordinary oxygen that we breath (O_2) has only two atoms per molecule while poisonous ozone has three. There is very little ozone found in the lowest parts of the atmosphere. It is found in large amounts at altitudes of 20 to 50 km.

In addition to the true gases which are composed of molecules, air usually contains larger solid particles which include dust, smoke, and salt crystals. A large number of dust particles are small enough to mix with the gaseous molecules and remain suspended almost indefinitely. Much of the dust in the atmosphere is composed of mineral particles carried by the force of the wind. Smoke is a common source of dust. Volcanic eruptions and meteors which have vaporized in the air also contribute to the amount of dust in the atmosphere. Tiny drops of sea water thrown into the air evaporate, leaving behind minute particles of salt crystals which are important in the formation of clouds.

Air pressure. The gases which make up the atmosphere are held in place by the earth's gravity. Altogether, the atmosphere presses down with a weight of about 6×10^{15} tons. For each square centimeter of surface near sea level, air pressure exerts a force of 1.03 kg. (14.7 pounds/sq. in). An average-size person at sea level must support a force of 10 to 20 tons due to air pressure. Usually you can't feel any pressure because it is the same both inside and outside your body. At higher altitudes the air is thinner and the pressure is reduced. You would become conscious of the change because of the effect of reduced pressure on the outside of your ear drum. The "popping" sensation in your ears as you rise quickly in an elevator or in an automobile is also due to a sudden drop in pressure.

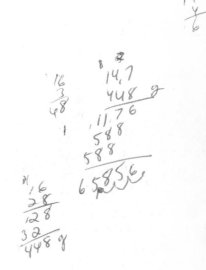

Table 20–1 Gases in Pure Dry Air

Gas	Symbol or Formula	Percent by Volume
Nitrogen	N_2	78.084
Oxygen	O_2	20.946
Argon	Ar	0.934
Carbon dioxide	CO_2	0.033
Neon	Ne	0.00182
Helium	He	0.00053
Krypton	Kr	0.00012
Xenon	Xe	0.00009
Hydrogen	H_2	0.00005
Nitrous oxide	N_2O	0.00005
Methane	CH_4	0.00002

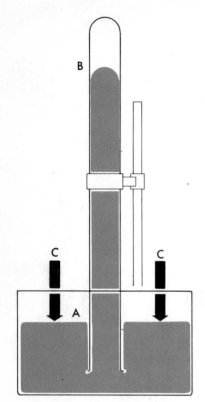

FIG. 20–1. This diagram shows the principle of the Torricellian barometer. The tube containing the mercury has been shortened for convenience.

An instrument which measures atmospheric pressure is called a *barometer*. The type of barometer most commonly used was invented in 1643 by an Italian named Torricelli (tohr-ree-*chel*-ee). The principle of the Torricellian barometer is illustrated in Figure 20–1. Air pressure is indicated by arrow C, pressing down on the surface of mercury in the dish. The height of the mercury column in the tube, A to B, is determined by the atmospheric pressure. The space in the tube above the level of the mercury is nearly a perfect vacuum. Mercury is used because it is very heavy, about 14 times more so than water. Thus a relatively small amount of mercury in the tube is sufficient to balance the pressure exerted by a column of air extending from sea level to the top of the atmosphere. Torricelli found that air pressure at sea level would balance a column of mercury about 760 mm (30 in) high.

The modern version of Torricelli's barometer is shown in Figure 20–2. There is a scale to measure the height of the mercury and a thermometer to indicate corrections for temperature changes. This type of barometer is called a *mercurial barometer.*

Air pressure measured by a mercurial barometer is expressed by how high the mercury rises. For example, the atmospheric pressure at a certain time might be given as "758 mm of mercury" which is the same as "29.87 inches of mercury." A pressure equal to 1 mm of mercury may be called 1 Torr after Torricelli. Thus the pressure mentioned above might also be given as "758 Torr."

Official weather maps may use still another measurement of air pressure called millibars (mb). This is a way of expressing air pressure which is understood all over the world. Millibars are complicated to define and it is usually only necessary to know that normal air pressure at sea level is equal to 1013.25 millibars. Average sea level pressure may also be expressed as 760 mm of mercury or 29.92 inches of mercury.

Another type of barometer is also commonly used. The *aneroid* (*an*-uh-royd) *barometer,* also shown in Figure 20–2, is more portable and rugged than the mercurial barometer. Unlike the mercurial type, an aneroid barometer contains no liquid. The heart of this instrument is a sealed metal container from which most of the air has been removed. When the air pressure increases, the sides of the container bend inward. When the pressure decreases, the sides bulge out again. These changes are indicated by a moving pointer on a dial. The movement of the pointer is regulated by a system of gears and levers. The dial can be marked off to show the pressure in inches of mercury or any other units desired. Aneroid barometers can be constructed so that a continuous record of air pressure is made on a revolving drum. Such an instrument, shown in Figure 20–2, is called a *barograph*.

Aneroid barometers can also be used to measure approximate altitude above sea level. When it is used for this purpose it is called an *altimeter*. At higher altitudes the atmosphere is thinner and exerts less pressure. Thus a lowered pressure reading can be interpreted as an increased altitude reading. To accurately measure altitude,

FIG. 20–2. Types of barometers. Left, a *modern mercurial barometer*. Top, an *aneroid barometer*. Compression and expanison of the sealed metal box provide a measurement of air pressure. Below, a *barograph*. The readings of the aneroid barometer are recorded on the drum. (Bendix Aviation Corporation; U.S. Weather Bureau)

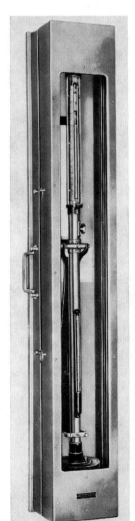

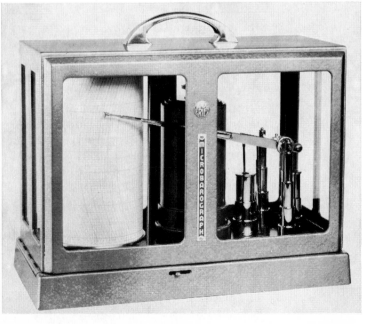

FIG. 20–3. Animals, such as these cattle, play a part in the nitrogen cycle. First nitrogen is removed from the air by nitrogen-fixing bacteria. Secondly nitrogen is removed from the soil by plants, which are in turn eaten by animals. Decomposing plant and animal remains return the nitrogen and its compounds to the atmosphere. (Ramsey)

an altimeter based on air pressure must be corrected to local conditions. Changes in pressure due to weather changes require continual correction adjustments. Aircraft receive information for these corrections from ground stations.

The chemical balance of the atmosphere. The gases of the atmosphere seem to be in a state of chemical balance. Although the amount of each main gas in the air remains almost constant, these gases are always being added and removed from the atmosphere. Nevertheless, processes which remove and produce the principal gases, nitrogen, oxygen and carbon dioxide, seem to be in balance with each other.

The balance of oxygen in the atmosphere is an example of the chemical balance that keeps the composition of air the same. Many of the processes involved in rock weathering and mineral formation remove oxygen from the air. Animals also remove oxygen from the air as part of their life processes. If there were no means of replacement, the supply of oxygen in the atmosphere would probably be exhausted in a few thousand years. Green plants create a counterbalance by giving off oxygen in the manufacture of their food. In the process of photosynthesis, plants use sunlight, water, and carbon dioxide to produce their food. Oxygen is released as a by-product of photosynthesis.

The balance of nitrogen in the atmosphere is maintained by the *nitrogen cycle*. Nitrogen is removed from the air mainly by the action of nitrogen-fixing bacteria. These microscopic plants live in the soil and in the roots of plants of the legume family (beans, peas, clover, alfalfa, and others). They take nitrogen from the air and chemically change it into nitrogen compounds. These nitrogen compounds are vital to the growth of all plants. Through plants, nitrogen compounds enter the bodies of animals. These compounds are then returned to the soil through animal excretions or by decay of their bodies after death.

In the soil, decay processes release nitrogen and return it to the atmosphere as a gas. See Figure 20–3. A similar cycle takes place among water-dwelling plants and animals. To a lesser extent, lightning also removes nitrogen from the air. The weathering of certain rocks results in the formation of nitrogen compounds which are carried out of the atmosphere by rain and snow. Volcanic activity is yet another way nitrogen is released into the air.

The carbon dioxide cycle in the atmosphere is regulated by three great reservoirs: the sea, rocks, and living things. Of these, the sea is the largest storage area for carbon dioxide. The oceans contain about 50 times more carbon dioxide than the entire atmosphere. Most carbon dioxide is in the form of carbonate compounds. However, quantities of this gas are only dissolved in water. Each year the sea and the atmosphere exchange 200 billion tons of carbon dioxide. This exchange seems to keep the amount of carbon dioxide in the atmosphere fairly stable. If the concentration of carbon dioxide in the air is increased, the sea will probably absorb more of this gas. If the concentration of carbon dioxide in the atmosphere falls, the sea will probably release more gas. Figure 20–4 shows all of the processes which maintain the carbon dioxide concentration in the atmosphere.

FIG. 20–4. The process which maintains the carbon dioxide concentration in the atmosphere is shown in this diagram. The numbers indicate how many tons of carbon dioxide are involved in the process.

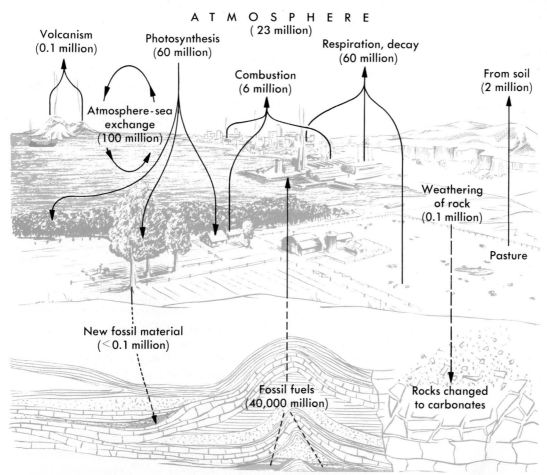

Air pollution. It has taken billions of years for the earth's atmosphere to reach its present form. Sensitive natural cycles now tend to maintain its normal composition. However, like many of the earth's delicate natural cycles, this balance can be upset by man's activities. Consider the case of the atmosphere. Thousands of tons of gases and solids are dumped into the atmosphere each day. These troublesome and often dangerous foreign substances in the atmosphere are called *air pollutants*.

There are two main types of air pollutants. First are the gases which irritate eyes, throats, and lungs, harm plants, and even attack stone and metal. The chief air pollutant gases are produced mostly by burning of fuels. The sulfur which is present in a fuel such as coal or oil will produce irritating and destructive sulfur dioxide gas when either fuel is burned. The operation of automobile engines produces two kinds of pollutant gases. Unburned gasoline containing hydrocarbon-type compounds is usually given off in automobile exhausts. In the air, hydrocarbons mix with other gases to produce many additional substances known to be irritating when breathed. Besides the non-lethal hydrocarbons, automobile exhausts also contain quantities of poisonous nitrogen oxides. Also present is the very deadly gas carbon monoxide.

Another source of unwanted gases in the air are some manufacturing processes. Compounds of fluorine and chlorine may be discharged into the air around certain kinds of industrial plants. Generally speaking, however, this is a very limited problem. The burning of fuels, particularly in automobiles, remains the most serious problem.

A second type of air pollutant is composed of solid particles. Here too, the automobile is a serious problem. The gasoline engine changes about 0.08 percent of the fuel it burns into solid material. About half of this solid material given off in automobile exhausts is in the form of lead and lead compounds. The friction of automobile tires wearing against pavement also throws tiny particles of rubber into the air. Other than motor vehicles, the chief source of solid air pollutants is smoke. If all the solid pollutants in the air over a large city during one month's time settled out at once, it would amount to an average of more than 10 tons per square mile.

Air pollution becomes more serious when weather conditions and the topography of a region prevent the dispersal of pollutants. Under normal daytime conditions,

Investigate

Call your town government and find out if they are keeping a check on local air pollution and how it is done. Report your findings to the class.

the temperature of the air decreases as the altitude above the earth's surface increases. However, there are occasions when, instead of decreasing, the temperature increases with altitude. This is often what occurs when a layer of warm air exists above a layer of cool air next to the ground. Meteorologists refer to this condition as a *temperature inversion.* The inverted layer of warm air traps the pollutants and holds them close to the ground because cool air has a tendency to sink. Temperature inversions or other weather conditions which produce stagnant air commonly occur in a valley topography. The hills forming the walls of a valley trap the inversion layer and prevent the polluted air from drifting away. Some cities, such as Los Angeles, have frequent temperature inversions resulting from the cooler Pacific air being held near the ground by the arrangement of mountains. In locations where these factors are present air pollution is an ever increasing problem. See Figure 20–5.

The total effects of air pollution are difficult to determine. It is obvious that polluted air is much less clear than clean air. Solid particles produce a hazy effect which has led to the term "smog" to describe the appearance of polluted air around many cities. Of more importance is the effect of polluted air on the people who must breathe it. This problem receives much attention on the rare occasions when unusually heavy air pollution causes deaths. In London, England, in 1952, at least 4000 deaths occurred because of exposure to five days of a heavily polluted fog. In Donora, Pennsylvania, in 1948, twenty deaths were apparently caused by an attack of severe air pollution. Less dramatic but perhaps more serious are the still unknown effects on the health of city-dwellers who breathe polluted air for their entire lives. But the problem of air pollution will not be solved quickly or easily. It is deeply rooted in industrial processes and the increasing numbers of automobiles and their exhausts. Only the combined efforts of government and interested citizens can help solve the problem of increasing pollution of the atmosphere.

FIG. 20–5. This view of Los Angeles shows the definite layer of polluted air, or smog, which is caused by a pronounced temperature inversion. (Los Angeles Air Pollution District)

Explore
Find out exactly what caused the air to become deadly in Donora, Pennsylvania in 1948.

THE ATMOSPHERE AND THE SUN

Solar energy and the atmosphere. Remember that almost all of the energy reaching the earth from the sun is

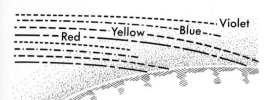

FIG. 20-6. This diagram shows how sunlight which contains all colors can appear red or yellow when the sun is near the horizon.

Table 20-2 Reflection of Solar Energy

Surface	Percent of incoming solar radiant energy which is reflected
Water	
Incoming rays straight down	2
Incoming rays 40° from overhead	2½
Incoming rays at low angle	Very large
Clouds	40–80
Snow, Ice	46–86
Grass	14–37
Fields	3–25
Bare ground	7–20
Forest	3–10

in the form of electromagnetic waves. The earth's atmosphere is the first part of the earth to receive this radiant energy from the sun. As the electromagnetic waves from the sun pass downward they are affected by the atmosphere. First, the very short wavelength X-rays, gamma and ultraviolet rays are absorbed by molecules of nitrogen and oxygen. This happens in the upper parts of the atmosphere above 50 km (30 miles). As solar energy is absorbed, molecules and atoms of nitrogen and oxygen lose electrons and become positively charged ions. At a lower level, from about 20 km up to 35 km (12 to 21 miles) above the earth's surface, ultraviolet rays act upon oxygen molecules (O_2) to form the ozone layer, (O_3).

In their passage through the upper atmosphere, almost all of the shorter wavelengths of solar radiation are lost through absorption. The small number of ultraviolet rays which penetrate to the earth's surface are responsible for sunburn of skin exposed to sunlight for too long a period. Only the longer wavelengths of visible light and the infrared rays reach the lower atmosphere. In this region the air is denser and the gas molecules crowd closer together. These gas molecules are responsible for scattering light. Scattering means that the light rays are bent in all directions. Some of the rays are turned back into space and are lost. Since blue light is most affected by scattering of molecules, a clear sky always has a blue color. When the sun or moon is seen near the horizon, the light from its surface appears yellow or red. This is because dust and water droplets in the air near the earth's surface allow yellow and red light to come through while the blue light is scattered. See Figure 20–6.

The longer infrared rays are not scattered by the atmosphere as is visible light. However, infrared rays may be partly absorbed by water vapor in the air. Clouds also absorb and reflect back into space a part of the solar energy which reaches the lower atmosphere.

Of the total amount of solar energy which enters the atmosphere, an average of about 19 percent is absorbed at some level. Another 35 percent is turned back into space either by reflection or by scattering. See Table 20–2. This leaves 46 percent which actually reaches the land and water surfaces of the earth. The percents given are average figures for the entire earth's surface. The amount of solar energy reaching the ground is affected by clouds, dust in the air and, of course, by the latitude and the season of the year.

Trapping the sun's energy. When the rays carrying solar energy finally reach the earth's surface, two effects are possible. The radiations are either absorbed or reflected. The kind of surface on which the radiation falls determines to a large extent which of the above will take place. The amount of incoming solar radiation that is reflected by various surfaces is shown in Table 20–2. Of the total solar energy arriving from the sun, 35 percent is reflected back into space. Thus the earth is said to have an average reflectivity or *albedo* of 0.35. By comparison, the albedo of the moon is only 0.07, indicating that it reflects 7 percent of the total solar energy it receives.

The solar radiation that is not reflected by the earth's surface is absorbed. Part of the radiation absorbed by the earth's surface is composed of the infrared rays that have penetrated the atmosphere. The rocks, soil, water, and other earth materials absorbing these infrared rays are heated. Then, the heated materials produce their own infrared rays from the heat energy. But the infrared rays produced by the materials of the earth's surface have a much longer wavelength than the infrared rays in sunlight. The gases in the atmosphere allow the shorter infrared rays from the sun to pass through. The longer wavelength infrared rays sent out from the warmed materials of the earth's surface are mostly absorbed by the atmosphere. Water vapor and carbon dioxide in the atmosphere are chiefly responsible for absorbing the longer infrared rays.

The final effect of this action is to trap heat energy from the sun and prevent it from escaping back through the atmosphere into space. This process acts very much like a greenhouse. The principle is the same. In a greenhouse the glass allows the short wavelengths of infrared rays coming from the sun to pass through. But the glass prevents the longer infrared rays leaving the warmed surfaces in the greenhouse from escaping. Since the two effects are so similar, the process of the atmosphere trapping the sun's heat over the earth's surfaces is called the *greenhouse effect*. This is illustrated in Figure 20–7.

By the greenhouse effect, the energy from the sun warms the air after having first been absorbed at the earth's surface. Thus the atmosphere is heated mostly in its lower parts.

Layers of the atmosphere. Gravity pulls the gases of the atmosphere toward the earth's surface. The molecules of

Investigate
Ask your teacher how you can find out how the albedo is effected by the construction of cities, highways, airports, and large housing developments.

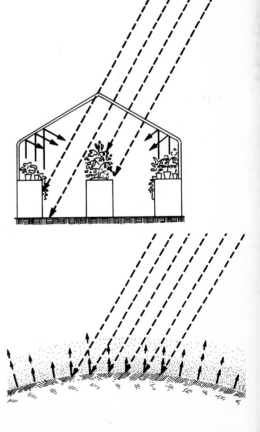

FIG. 20–7. The greenhouse effect. The atmosphere, like the glass in a greenhouse, allows infrared rays to enter, but prevents the escape of longer wavelength infrared rays.

gases move apart and take up as much space as possible. If gravity did not attract the gaseous molecules in the atmosphere, they would eventually move out into space and be lost. Half of the total weight of atmospheric gases is found within 5.6 km (3.5 miles) of the earth's surface. The remaining half of the atmosphere extends upward for hundreds of miles but gets increasingly thinner. Thus air pressure falls rapidly but smoothly with increasing altitude. There are no sharp pressure changes which separate the atmosphere into layers.

However, the atmosphere does show distinct changes in temperature and composition with increasing altitude. Basically, this is a result of the greenhouse effect. The air closest to the earth's surface becomes heated, which in turn warms the rest of the atmospheric zone or layer closest to the earth's surface, called the *troposphere*. Temperature within this layer decreases with height at the average rate of 6.5° C/km. The word "troposphere" comes from a Greek root meaning "change" because almost all weather changes occur in the troposphere. The upper boundary of the troposphere, called the *tropopause*, extends to an average height of 10 km. However, this layer changes with latitude and the season of the year. The tropopause is located higher above the equator than it is above the poles.

Above the tropopause the temperature remains almost constant, even as height increases. This steady temperature zone marks the beginning of the second layer called the *stratosphere*. The only important warming effect upon the air in the stratosphere is the direct absorption of solar rays by the ozone layer. Here the temperature rises slowly with increasing height. The temperature continues to rise steadily above the ozone layer to a height of about 48 km (30 miles). The elevation marking the upper boundary of the stratosphere, is the *stratopause*.

Above the stratopause is the atmospheric layer called the *mesosphere*. In this layer temperature decreases rapidly with height until its upper boundary, the *mesopause*, is reached. Above the mesopause, the temperature increases steadily. Information gathered about temperature changes in this layer, known as the *thermosphere*, has been too incomplete to determine its upper boundary. In this region of very thin air the molecules are so far apart that a thermometer could not accurately indicate the energy of the moving molecules. The air is much too thin to warm any object that might pass through. At high

Construct
Make a chart showing the relationship between the height of our atmosphere and (1) the density of air and (2) the amount of air.

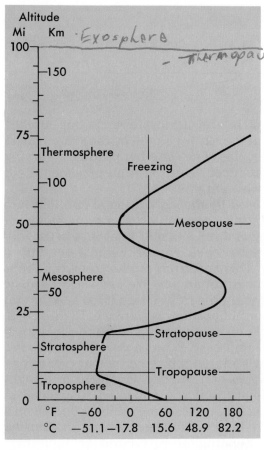

Altitude

Mi   Km   · Exosphere
100 -
         - Thermopause -
        —150

 75 -
        Thermosphere   Freezing
        —100
 50 -   —Mesopause—

        Mesosphere
        —50
 25 -
        —Stratopause—
        Stratosphere
                             —Tropopause—
        Troposphere
  0 -
        °F   —60   0   60   120   180
        °C —51.1 —17.8 15.6 48.9 82.2

FIG. 20–8. Models illustrating the layers of the atmosphere have been made based on temperature differences at various altitudes.

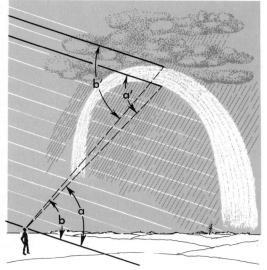

FIG. 20–9. For a rainbow to form, only those raindrops that refract and reflect the sun's rays at a certain angle (a = a′, b = b′) can produce a spectrum that will reach the eye of the observer in the form of an arch. The angle of the reflected rays determines the band of color produced.

altitudes, objects are warmed almost entirely by the solar rays they absorb directly. A thermometer is placed in this layer and shaded from the sun, would register a temperature far below zero in spite of the high speeds at which the few surrounding gas molecules move.

The region of the thermosphere and upper mesosphere is often described as the *ionosphere*. In this zone the effects of absorption of solar rays by the atmospheric gases produce the ions responsible for the auroras (northern and southern lights). Above the thermosphere is the region where the earth's atmosphere blends into the almost complete vacuum of deep space. This indefinite zone is called the *exosphere* because individual atoms and molecules escape the earth's gravity and drift away into space. A summary of the characteristics of the various layers of the atmosphere is given in Figure 20–8.

Visible light from the sun passing through the atmosphere is affected by the gases in the air itself and other

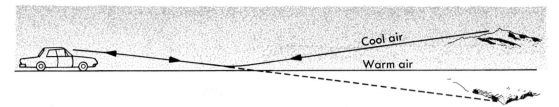

FIG. 20-10. In an inferior mirage light rays are bent in such a way that the ground seems to be a mirror. Distant objects may appear upside down below the horizon, as if reflected from the surface of water. When this type of mirage reflects the sky, the familiar illusion of shimmering water is seen.

FIG. 20-11. The effect of a superior mirage is that of a mirror suspended in the sky, reflecting. the image of objects that are out of sight over the horizon.

particles suspended in the atmosphere. It has already been mentioned that the sky is colored by the scattering of visible light by molecules of gas or other particles. Varying color effects are produced by separation of visible light into its individual wavelengths.

Rainbows are caused by the separation of sunlight into a range of colors. In this case, raindrops are the means of separation. Bending and reflection of light rays in the falling raindrops separate each ray of white light into a spectrum of colors. This spectrum produced by raindrops is visible only from certain angles. We can see a rainbow only when we are facing the falling raindrops and the sun is behind us. The correct position is shown in Figure 20-9.

Among the effects of light passing through the atmosphere, none is more impressive than a *mirage* (mi-*rahzh*). You may remember seeing a mirage on a summer day as water seemed to appear on the dry pavement of a highway. Perhaps a distant building appeared suspended in the sky, or a distant landscape seemed to be reflected on a lake. All of these are forms of mirages. An optical illusion is created by the bending or refraction of light rays passing through layers of air at different temperatures. The density of air varies according to temperature. The path of a light ray is refracted when it passes through several such layers.

The most common type of mirage is an *inferior mirage*. The mirage appears to the observer to be inferior or below his own eye level. Such a mirage is caused by the bending

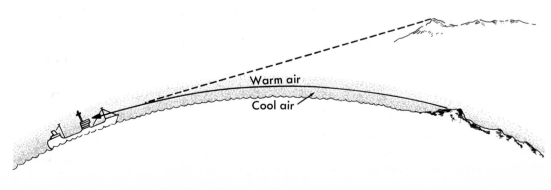

upward of light rays as they enter warmer air near the ground. See Figure 20–10.

In another type of mirage, the image seems to be above the observer. This is called the superior mirage, or *looming mirage*. It appears most commonly over the sea. Ships, icebergs, shorelines, cities, and other objects which would normally be out of sight over the horizon appear suspended in the sky. Sometimes double mirages are seen, one upright and the other upside down. Superior mirages occur when light rays entering cool surface air are bent downward. See Figure 20–11.

THE WINDS

Movement of air caused by heating. There are three ways that heat can be transferred to the atmosphere. Heating by radiation has already been described earlier in this chapter. The atmosphere is warmed when radiant energy is directly absorbed by its gases. Much less important is heating of the atmosphere by _conduction_. In this process heat flows directly to the atmosphere when the air comes in contact with anything that contains more heat. A third way that heat is transferred is by _convection_. This term is used to describe the movement of gases or liquids when they are heated unevenly. It is a very important atmospheric process and is the basic factor controlling the movement of air over the entire earth. It is the main cause of the _planetary circulation system_.

Movement of air due to convection takes place when air is heated by radiation or conduction. Air which has been heated, being less dense and lighter, presses down on the earth with less force. Thus the atmospheric pressure is generally lower beneath a body of warm air than under cooler air. As denser, cooler air, moves into a low pressure region, the lighter, warmer air is forced to rise. The general movement of air is always toward regions of lower pressure. These pressure differences, caused by unequal heating and resulting convection, create winds.

At the equator the earth receives more solar energy than at the poles. Consequently, atmospheric pressure at the equator is lower. The heated air in the region of the equator is constantly rising. At the poles the colder air is heavier and tends to settle, thus creating regions of high atmospheric pressure.

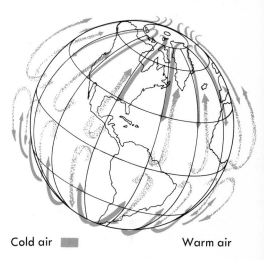

Cold air ▮▮▮   Warm air

FIG. 20–12. If the earth did not rotate, the circulation of the atmosphere would be as shown here.

activity

In this activity, you will feel heat transferred by three methods: conduction, convection, and radiation. Be careful, you should just feel the heat, not get burned.

Light a candle and drip some wax on the center of a index card. Blow out the candle and stand it upright in the soft wax. Light the candle.

Hold a finger about 40 cm above the flame. Bring it closer until you can feel the heat from the rising hot air.

1. At what distance did you feel the heat?

2. Was the heat transferred by conduction, convection, or radiation?

Hold your fingers 40 cm to the side of the lighted candle. Bring your fingers closer until you can feel heat.

3. At what distance did you feel the heat?

4. How was the heat transferred?

Hold a nail (16d finishing nail) near its head. You are going to determine the time it takes for heat to travel the length of the nail. Put the point in the flame for 15 seconds. Then move the nail and your hand until they are more than 40 cm from the candle. Determine how long it takes heat to travel along the nail until you can feel it near your fingers. Be careful!

5. How long did it take?

6. How was the heat transferred?

Pressure differences between air at the equator and the poles create a general movement of the atmosphere. Since air moves from high toward low pressure, there is a general flow of air near the earth's surface from the poles toward the equator. At higher levels there is a general return flow of warm air from the equator toward the poles. See Figure 20–12.

The atmosphere of the earth would probably move as just described if the earth did not rotate and if it were always heated equally on all sides. But the earth does rotate and its shape causes it to be heated unequally. These factors have a strong influence on the circulation of the atmosphere. The general movement of air from the poles toward the equator and its return because of convection results in the planetary circulation of winds.

Effects of the earth's rotation on winds. Everything that moves over the surface of the earth is affected by the rotation of the earth on its axis. To illustrate this, consider what would happen to a rocket fired from the North Pole and aimed toward New York. As the rocket moved south, the earth beneath it would be turning to the east. Since the earth's rotation carries points nearer the equator around faster than points near the poles, the earth would turn faster under the rocket as it moved south. If the rocket took about one hour for its flight, the earth's rotation would cause it to finally land near Chicago. The rocket would appear to have been pushed off to its right during the journey. See Figure 20–13.

All objects moving over the earth's surface will veer off to the right in the Northern Hemisphere and to the left in the Southern Hemisphere. This motion is called the *Coriolis effect*, after the nineteenth-century French mathematician who first described it. A baseball when thrown will curve only slightly due to the Coriolis effect. The ball is usually in the air much too short a time for it to be affected. The world's ocean currents and winds, however, are strongly affected. Winds, instead of blowing only in a direction from high toward lower pressure are turned by the Coriolis effect. This tends to divide the general north-south movement of air over the entire earth into several smaller parts. Within each division the winds generally move in a set direction, determined mainly by pressure differences and the Coriolis effect.

FIG. 20–13. A rocket fired from the North Pole and aimed directly at New York would land near Chicago because of deflection due to the Coriolis effect.

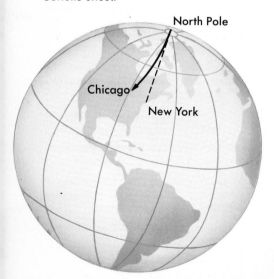

STATION MODEL MAP
STATION MODEL SAMP
HIGH
LOW
AIR MASS SYMBOLS
WIND ARROWS
WIND SPEED FEATHERS
WIND DIRECTION
AIR PRESSURE
DEW POINT
AIR TEMPERATURE
COLD FRONT
WARM FRONT
STATIONARY FRONT
OCCLUDED FRONT
ISOBARS
ISOTHERMS
MILLIBARS
MAXIMUM TEMP.
MINIMUM TEMP.
AREAS OF PRECIPITATION
AMOUNT OF PRECIPITATION
TYPE OF PRECIPITATION
DEPTH OF SNOW ON GROUND
HURRICANE
AUTOMATIC WEATHER STATION
CYCLONE
ANTICYCLONE
MISSING DATA

The earth's wind patterns. Near the equator, winds are controlled mainly by the lifting action of heated air. This is a region of calms, and weak undependable winds near the surface. The motion of the air in this region of low pressure is mostly upward. Because of the general lack of surface winds, the equatorial regions are known as the *doldrums*. The name originated in the days when low pressure and little wind often becalmed sailing ships for long periods.

As warm equatorial air rises and begins to move toward the poles, the Coriolis effect deflects it (to the east in the Northern Hemisphere). Around 30° latitude, the air begins to sink again toward the earth's surface. Part of it descends at these latitudes, forming a belt of high pressure. At the surface the settling air turns and flows both to the north and south. The portion moving back toward the equator creates surface winds called *trade winds*. In the Northern Hemisphere, the trade winds are deflected to the right and become the *northeast trades* because they flow from the northeast. In the Southern Hemisphere they are turned to the left and become the *southeast trades*.

The belt of high pressure mentioned above, which is created by descending air in the vicinity of 30° latitude, is called the *subtropical high*. The surface winds here are weak and changeable. The name "horse latitudes" was given to this region in the days when sailing ships carried horses from Europe to the New World. Horses often had to be thrown overboard to save food and water.

Some of the descending air in the subtropical highs that moves toward the poles is also deflected. It forms a belt of surface winds known as the *westerlies*. In the Northern Hemisphere the westerlies are southwest winds; in the Southern Hemisphere they are northwest winds. The westerlies are located in a belt between 40° and 60° latitude. By comparison they are much less steady than the trade winds.

North and south of the belt of westerlies around latitude 65°, there is another belt of low pressure. The *subpolar lows* are caused by the lifting of warmer air by cold polar air moving toward the equator. Over the polar regions themselves cold sinking air creates areas of high pressure. Within the cold polar air masses the general surface movement is toward the equator. Winds created by this movement are deflected by the Coriolis effect, becoming the *polar easterlies*. Because of the uniformly low temperatures in the polar regions and the resulting lack of

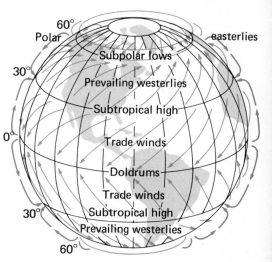

FIG. 20–14. The wind and pressure belts of the earth. Note how wind direction is deflected from straight north and south by the Coriolis effect.

pressure differences, these are usually very weak winds. Wind and pressure belts for the entire earth are shown in Figure 20–14.

As the sun's vertical rays shift north and south of the equator during the year, the positions of the pressure belts and wind belts also shift. The amount of change in their latitude is much less than the $23\frac{1}{2}°$ movement of the sun's vertical rays. The average shift for the pressure and wind belts during the year is about 6° of latitude. However, even this relatively small change is sufficient to cause some places to be in different wind belts during the year. This change may greatly affect the climate of an area, as is the case in southern California, where westerly winds prevail in the winter, and trade winds are dominant in the summer.

VOCABULARY REVIEW

Match the word or words in the column on the right with the correct phrase in the column on the left. *Do not write in this book.*

f 1. The study of the atmosphere.

d 2. An instrument which measures atmospheric pressure.

a 3. The process by which the atmosphere traps the sun's heat.

h 4. The atmospheric zone nearest the earth's surface.

b 5. The atmospheric zone in which the temperature increases slowly with height.

g 6. The atmospheric layer just above the stratosphere.

c 7. The atmospheric layer named for its high temperatures.

e 8. The region of the earth's atmosphere which blends into the vacuum of space.

k 9. The transfer of heat by movement due to uneven heating.

o 10. The apparent push to the right as objects move in the Northern Hemisphere.

j 11. A name often given to the equatorial regions because of their general lack of surface winds.

m 12. Winds moving toward the equator from the subtropical high.

p 13. Winds moving toward the poles from the subtropical high.

q 14. A belt in which warmer air is lifted by cold polar air, around latitude 65°.

l 15. The cold polar air masses moving toward the subpolar low.

a. greenhouse effect
b. stratosphere
c. thermosphere
d. barometer
e. exosphere
f. meteorology
g. mesosphere
h. troposphere
i. albedo
j. doldrums
k. convection
l. polar easterlies
m. trade winds
n. conduction
o. Coriolis effect
p. westerlies
q. subpolar lows

QUESTIONS

Group A

Select the best term to complete the following statements. *Do not write in this book.*

1. Which of the following gases occurs in the atmosphere in the greatest amount? (a) oxygen (b) nitrogen (c) argon (d) carbon dioxide.

2. The most important and usually most abundant of the variable gases in the lower atmosphere is (a) ozone (b) oxygen (c) water vapor (d) argon.

3. Which of the following is closest to the number of pounds of air which press down upon an average person at sea level? (a) 10 (b) 10,000 (c) 20,000 (d) 100,000.

4. The height of the mercury column in a mercury barometer is determined by the (a) atmospheric pressure (b) surface area of the mercury (c) density of water (d) amount of mercury used.

5. The height of a column of mercury which will be balanced by average air pressure at sea level is (a) 76 mm (b) 30 cm (c) 760 mm (d) 760 cm.

6. Aneroid barometers can be made to read any of the following except (a) millimeters of mercury (b) inches of mercury (c) feet of altitude (d) water vapor pressure.

7. An aneroid barometer can be used as an altimeter because at higher altitudes (a) air pressure is greater (b) there is less water vapor (c) the temperature is less (d) air pressure is less.

8. The amount of oxygen in the atmosphere remains relatively constant because oxygen is given off by (a) rocks (b) animals (c) plants (d) the ocean.

9. Plants take nitrogen from the air. It is returned by (a) decaying plants and dead animals (b) breath of animals (c) plants giving it off (d) water evaporating.

10. Compared to the amount of carbon dioxide in the atmosphere, the amount in the oceans is (a) greater (b) less (c) about the same (d) unknown.

11. The danger of air pollution becomes more serious when there is (a) a storm (b) low pressure (c) a clear sky (d) a temperature inversion.

12. In passing through the upper atmosphere, the electromagnetic waves from the sun lose most of their (a) long wavelengths (b) visible wavelengths (c) short wavelengths (d) infrared waves.

13. A clear sky is blue because (a) red light is scattered more than blue (b) blue light is scattered more than red (c) ultra-violet light has been absorbed (d) infrared light has been absorbed.

14. Of the total solar energy which reaches the earth's atmosphere about what percent actually reaches the earth's surface? (a) 19 (b) 35 (c) 54 (d) 46

15. The albedo of the earth is about how many times that of the moon? (a) .07 (b) .2 (c) .35 (d) 5

16. After being heated by solar infrared rays, the materials at the earth's surface produce infrared waves whose wavelengths are (a) longer (b) shorter (c) about the same (d) near ultraviolet.

17. The infrared waves produced by the warmed materials of the earth's surface are absorbed in the atmosphere chiefly by (a) water vapor and oxygen (b) carbon dioxide and oxygen (c) water vapor and carbon dioxide (d) oxygen and nitrogen.

18. Atmospheric pressure changes with altitude. This change in pressure is (a) uniform without sudden change (b) uniform but with several sudden changes (c) uniform but with one sudden change (d) not uniform and not predictable.

19. About half of the total weight of the atmosphere is within what distance of the earth's surface? (a) 5600 mm (b) 5600 mi (c) 5600 m (d) 5600 km.

20. The boundary between the troposphere and the stratosphere is called the (a) ionosphere (b) mesopause (c) stratopause (d) tropopause.

21. The upper mesosphere and the thermosphere are often called the (a) ionosphere (b) mesosphere (c) thermopause (d) exosphere.

Group B

1. List the layers of the atmosphere and their boundaries in proper order.

2. Why is the sky blue?

3. Explain the "greenhouse effect."

4. One inch equals 2.54 centimeters. A barometer reading of 30 inches would equal how many centimeters?

5. Describe the nitrogen cycle.

6. Describe the carbon dioxide cycle.

7. Describe how a mercury barometer measures air pressure.

8. How does a temperature inversion increase the problem of air pollution?

9. What are the principal gases in the atmosphere?

10. Explain the three ways that heat is transferred to the atmosphere.

11. Of what value is a barograph over a barometer?

12. Name and give the approximate locations for the earth's wind belts.

13. Describe the way solar energy is distributed after it enters the earth's atmosphere.

14. Why do surface winds generally blow toward the equator and away from the polar regions?

15. Explain the Coriolis effect.

16. How does the tilt of the earth's axis affect the planetary winds?

objectives

- [ ] Explain how water vapor enters the atmosphere and how it is measured.

- [ ] Explain how water leaves the atmosphere.

- [ ] Describe cloud formation.

- [ ] List and describe the various types of clouds.

- [ ] Describe several causes of precipitation.

- [ ] Describe the conditions needed for the various types of precipitation to occur.

- [ ] Describe several methods used to measure precipitation.

If all the water vapor contained in the earth's atmosphere were to fall suddenly as a worldwide rain, it would add a layer of water only about one inch deep to the earth's surface. It may be somewhat surprising, but water vapor is not one of the most abundant materials in the atmosphere. It is never present in large amounts, yet it is one of the most important components in the atmosphere.

If the earth's atmosphere did not contain water vapor, the greenhouse effect would be much less efficient and the earth would then be a much colder planet. Weather changes that carry life-giving moisture over the land depend upon water vapor in the air. With an atmosphere void of water, this planet's surface would probably be a barren desert swept by clouds of dust and exposed each night to bitter cold. Thus even the small amount of water vapor present is vital to our environment.

Driven by the energy of the sun, billions of tons of water each year pass into the atmosphere as vapor. Tracing the action of this water vapor in the air is the key to understanding the ever-changing weather.

ATMOSPHERIC MOISTURE

How water vapor enters the air. Water can exist as a solid, liquid, or gas (vapor). The difference between water in these three phases is in the amount of energy contained by the water molecules. For example, to change water from liquid to gas, enough energy must be added to overcome the forces holding them together. See Figure 21-1. Then the molecules are able to separate and move around independently as a gas. For 1 gram of liquid water (at

439

FIG. 21-1. Solar energy which causes water to evaporate from the sea surface supplies most of the moisture found in the air. (Photo Researchers—Granitas)

Investigate
Find out more about the process of transpiration and design an experiment that will show that a potted plant carries on transpiration. Report your findings.

40°C) to evaporate, about 600 calories of heat energy must be added. A *calorie* (or gram-calorie) is the amount of heat required to raise the temperature of 1 gram of water through 1°C at near freezing temperatures. It is important to keep in mind that when water vapor enters the atmosphere heat energy is carried along with it. This is called *latent energy* and can be released to the air if the water vapor changes back to liquid water.

Millions of tons of water evaporate each day from the surface of the sea. Because of energy absorbed from the sun, separate molecules of pure water pass into the air. Most of this evaporation takes place in the regions around the equator where the largest amounts of solar energy are received. While the sea is the principal source of atmospheric moisture, evaporation from lakes, ponds, streams and soil also supplies some water vapor to the atmosphere. Plants also give off water vapor as they carry on a part of their life process called *transpiration*. A small amount of moisture enters the air from volcanoes and from burning of fuels.

Measurement of atmospheric moisture. The amount of water vapor in the air is referred to as its *humidity*. If a sample of air is passed over a certain chemical, its moisture will be absorbed and can then be weighed. This kind of investigation would show that the total weight of water

32

vapor in a certain volume of air is relatively small. For
example, a cubic meter of saturated air (35.3 cubic feet)
at 10°C will contain only about 9 grams of water vapor.
This direct measurement of the amount of water vapor
in a certain volume of air is said to express its *absolute
humidity.*

Further investigation would show that the capacity of
air to hold water vapor changes with its temperature.
A given volume of warm air holds more moisture than the
same volume of cool air. This means that the maximum
absolute humidity of a certain volume of air increases
with temperature. The graph in Figure 21–2 shows the
maximum absolute humidity of air in relation to temperature changes. The amount of water vapor an amount of
air can hold at any particular temperature is called its
saturation value.

The humidity of air is often expressed by comparing its
absolute humidity to its saturation value. Suppose a cubic
meter of air at a temperature of 21°C is found to have an
absolute humidity of 13.9 grams per cubic meter (13.9
g/m³). The saturation value for air at 21°C is about 18.5
g/m³. Then,

$$\frac{13.9 \text{ g/m}^3}{18.5 \text{ g/m}^3} = 0.75 \text{ or } 75\%$$

The formula states that air must contain 75% of the moisture it can possibly hold at that temperature. This method
of stating the moisture content of the air is called the
relative humidity. Relative humidity is always given as a
percent. It shows the amount of moisture in the air (absolute humidity) compared to the amount which could be
contained at the given temperature (saturation value).
Notice that relative humidity must change if the air
temperature changes and no additional water vapor
enters the air. As the temperature increases, relative
humidity decreases, if the moisture content remains the
same. This means that on most days relative humidity is
highest in the cool hours of night and morning and lowest
in the afternoon.

The actual amount of water vapor in the air is difficult
to measure accurately. The only way absolute humidity
can be measured directly is by trapping the water with a
chemical drying agent, then weighing the water collected
from a known volume of air. Generally this is not practical
because it takes too much time and complicated equipment. A less accurate but more convenient method for

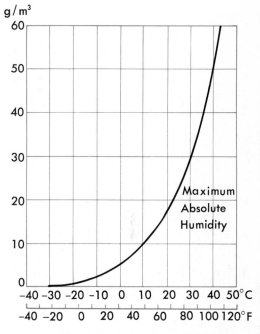

FIG. 21–2. Graph showing the
change in absolute humidity as a
result of changes in air temperature.

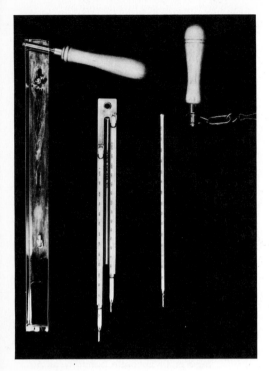

FIG. 21–3. A psychrometer is an instrument used to measure humidity. The wick of the thermometer shown at the left is wet. Evaporation of moisture from the wick cools the thermometer bulb. (U.S. Weather Bureau)

activity

Water vapor condenses on the outside of a glass of ice water because the temperature of the air that comes in contact with the glass is lowered below its dew point. The following technique can be used to measure dew point:

a. Fill a shiny tin can or plastic glass half-full of water.

b. Place an ice cube in the water and stir carefully with a thermometer.

c. When dew first appears, record the temperature. Immediately remove all the ice from the water. (See the photo, top of page 443.)

measuring humidity makes use of two thermometers in an instrument called a *psychrometer*. See Figure 21–3. The bulb of one thermometer is covered with a damp wick while the other remains dry. The wet-bulb thermometer usually gives a lower temperature reading because evaporating water has a cooling effect. When the air is dry, rapid evaporation cools the wet-bulb thermometer faster. Its temperature reading, therefore, differs most from the dry-bulb thermometer when the relative humidity is low. When the air is very moist, there is little evaporation and the two thermometers indicate about the same temperature. Using tables or a special slide rule, the temperature difference between the two thermometers can be translated into approximate relative humidity.

Another type of humidity measuring instrument is based on the fact that human hair stretches when the moisture of air increases. A piece of human hair will stretch about $2\frac{1}{2}$ percent when the relative humidity increases from 0 to 100 percent. The *hair hygrometer* is an instrument which records the changing length of a bundle of hairs when humidity changes. The variation in length moves a pointer on a scale or a pen on a graph to show changes in relative humidity.

To measure humidity at high altitudes an electrical instrument connected to a balloon is used. This measuring device is triggered by pressing an electrical current through a moisture-collecting chemical substance. The amount of moisture collected determines the electrical conductivity. This can then be measured and expressed as the humidity of the surrounding air.

CONDENSATION OF WATER VAPOR

How water leaves the air. A water molecule will remain a part of the water vapor in the air as long as it maintains its supply of latent energy. However, if the air is cooled, its water vapor molecules may release this energy, forcing them to join together. This means that when the temperature of the air is lowered, the capacity of the air for holding water is reduced. If the temperature continues to drop, a point is reached at which the amount of water vapor the air holds is equal to its capacity. At this point the air is said to be saturated and the relative humidity is 100 percent. The temperature to which air must be cooled to reach this level of relative humidity is called its *dew point*. At any temperature lower than the dew point water vapor may begin to change to a liquid or solid in the process called *condensation*.

Dew and frost. On almost any clear autumn night, radiation of heat from grass and leaves close to the ground is relatively rapid, and the temperature of their surfaces soon reaches the dew point of the surrounding air. The resulting form of condensation called *dew* is seen as tiny water droplets on many surfaces which have become cooled by radiation. See Figure 21–4.

If the dew point falls below the freezing temperature of water (0°C), water vapor will condense directly into solid ice crystals to make *frost*. Since frost forms directly from water vapor without first becoming liquid, it is not frozen dew. Frozen dew, which is relatively uncommon, forms as clear beads of ice. Orchards and valuable crops must be protected from frost, since it will kill growing plants. One method of frost protection uses heaters or blowers that circulate the cold air.

Adiabatic cooling. Air may be cooled by two other processes besides radiation. If a quantity of air mixes with a mass of colder air, a drop in temperature will result. A second cooling process is caused by the upward movement of air. When air rises, it expands because pressure is decreased. A gas such as air is cooled when it expands causing collisions between its molecules to take place less often. Less frequent collisions mean less energy and lower temperatures for the individual molecules. Thus, a force that causes air to rise will also cause it to expand and become cooler. Downward movement of air has the opposite effect.

This cooling and heating effect of air can be seen with the kind of hand pump often used to fill bicycle tires. Pushing the pump handle down compresses the air which releases enough heat to make the barrel of the pump quite hot. Yet the escaping compressed air feels cool. This is because it expands rapidly as it escapes. Temperature changes that take place in air with no addition or withdrawal of heat from the outside are called *adiabatic changes*.

Dry air registers an adiabatic temperature change of about 1°C for every 100 meters (5.5°F for each 1000 ft) of altitude change. This means that *rising* air will show a decrease in temperature of about 1°C for every 100 meters

d. Continue to stir until the dew disappears. Record the temperature of the water at this time.

The two temperatures you have recorded should not differ by more than a few degrees. If they do, it would be advisable to repeat the above procedure until they differ by only a few degrees. The average of these two temperatures is the dew point.

FIG. 21–4. Dew forms on objects that are cold enough to lower the temperature of the surrounding air below the dew point. (Allan Roberts)

FIG. 21–5. Cumulus clouds form when rising air is cooled. (Ramsey)

FIG. 21–6. Banner clouds often form around mountain tops. (Ramsey)

it rises. The temperature of *sinking* air will increase by an equal amount for every 100 meters it descends. But the adiabatic rate of 1°C/100 meters applies only to dry air. If there is water vapor in the air, there will also be absorption or release of heat due to evaporation or condensation. For air in which water is changing its phase, the temperature change for either rising or sinking air is about 0.6°C per 100 meters.

Cloud formation. Adiabatic changes are an important factor in cloud formation. These changes provide a simple method for the necessary cooling of air to its dew point. Air rises mainly because of temperature differences. A body of air that becomes warmer than the surrounding air also becomes lighter and rises. For example, on sunny days a small region of air near the ground may receive more heat than the surrounding air. This will happen if the heated air is spread over a part of the ground that reflects more of the sun's energy. The heated parcel of air will rise and undergo adiabatic cooling until the dew point is reached. Then its moisture will condense to form a cloud. See Figure 21–5.

Air is also cooled as it is lifted in its passage over mountains. Entire mountaintops are often covered with clouds formed in this way. See Figure 21–6. Large areas of clouds connected with storms are formed when a mass of warm air is pushed up over a heavier body of cooler air. This type of action will be taken up in Chapter 22.

A cloud can only be formed when a body of air is cooled below its dew point. However, cooling of moisture-laden air below its dew point is not the only requirement a forming cloud must meet. The individual water molecules which make up the water vapor in the air must clump together. Only in this way can the gaseous water in the air form the droplets of liquid water that usually make up a cloud. This means that the rapidly moving gaseous water molecules must collide with other water molecules and collect until a droplet of liquid water has accumulated. This process cannot take place until the temperature has dropped low enough so that the water molecules will stay clumped together.

After the air is swept clean of all foreign particles such as dust, it can be cooled until the relative humidity is much greater than 100 percent. Only then do cloud droplets begin to appear. Air with a relative humidity above 100 percent is said to be *supersaturated*. Only clean air

Cirrus (UNESCO)

Altocumulus (UNESCO)

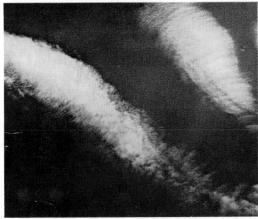

Cirrocumulus (UNESCO)

Stratus (UNESCO)

Cirrostratus

Nimbostratus

can be supersaturated with water vapor and still not contain cloud droplets.

Usually the air is filled with many small particles which prevent supersaturation. These particles include dust, salt particles from the sea, smoke, and perhaps particles showered on the earth by disturbances on the sun. Each particle can serve as a nucleus for condensation of water vapor. These particles are called *condensation nuclei*. As soon as the dew point is reached, water molecules begin to collect on the various condensation nuclei present in air and cloud droplets are produced. If the dew point is below the freezing temperature of water, an ice crystal will form. Each separate cloud droplet or ice crystal usually contains the particular condensation nucleus around which it grew. Because air normally contains large numbers of condensation nuclei, clouds are usually formed when the air is cooled to its dew point. For this reason supersaturation of air with water vapor does not often occur naturally.

Types of clouds. The most convenient method of classifying clouds is based on altitude. The major cloud types and their principal subdivisions are as follows:

1. *High clouds.* The base of these clouds is at an altitude well above 7 km. They are generally thin and composed of ice crystals. Through them the outline of the sun or moon can be seen. Their principal forms are:

   a. *Cirrus*. Thin, featherlike with a delicate appearance, frequently arranged in bands across the sky; sometimes called "mare's tails."

   b. *Cirrocumulus*. Like patches of fluffy cotton or a mass of small white flakes, frequently in groups or lines; sometimes called "mackerel sky."

   c. *Cirrostratus*. Whitish layers, like a sheet or veil, giving the sky a milky appearance. They often produce a halo or ring around the sun or moon. This is a result of the bending of light by ice crystals in the cloud.

2. *Middle clouds.* These clouds range from 2 to 7 km. Their principal forms are:

   a. *Altocumulus*. White or gray patches, or layers of clouds having a rounded appearance.

   b. *Altostratus*. Gray to bluish colored layers, often with a streaked appearance.

3. *Low clouds.* The bases of these clouds range from near the surface to about 2 km.

Investigate
Call your local weather station every day at the same time for a week, to find out the cloud types and heights. Try to relate this information to the kind of precipitation that may be occurring.

Cumulus (UNESCO)

Cumulonimbus (UNESCO)

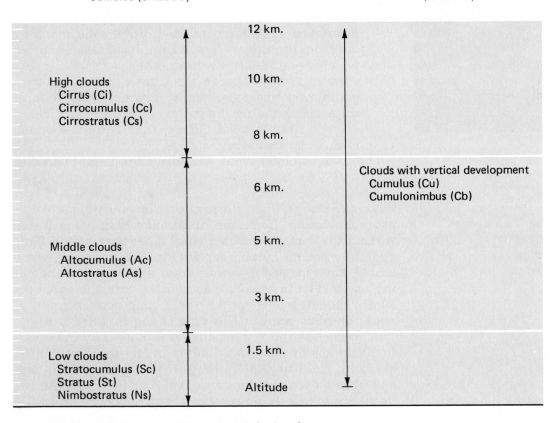

High clouds	12 km.	
Cirrus (Ci)	10 km.	
Cirrocumulus (Cc)		
Cirrostratus (Cs)	8 km.	
	6 km.	Clouds with vertical development
Middle clouds		Cumulus (Cu)
Altocumulus (Ac)	5 km.	Cumulonimbus (Cb)
Altostratus (As)		
	3 km.	
Low clouds	1.5 km.	
Stratocumulus (Sc)		
Stratus (St)	Altitude	
Nimbostratus (Ns)		

FIG. 21–7A. Classification of cloud types and their elevations.

Altostratus (Keith Gunnar/Bruce Coleman)

Stratocumulus (UNESCO)

FIG. 21–7B. A typical upslope fog caused by adiabatic cooling in a glacial valley. (Photo Researchers—Mappes)

a. *Stratus.* Low, uniform, sheetlike clouds similar to fog but not resting on the ground.

b. *Stratocumulus.* Large rounded clouds with a soft appearance, usually arranged in some pattern with spaces in between.

c. *Nimbostratus.* Low, shapeless thick layers, dark gray in color. They usually bring rain or snow.

4. *Clouds with vertical development.* These clouds extend from a lower level of about 2 km to a maximum of over 7 km.

a. *Cumulus.* Thick, dome-shaped clouds, usually with flat bases and many rounded projections reaching out from the upper parts. Cumulus clouds are often widely separated from one another.

b. *Cumulonimbus.* Large, thick, towering clouds with cauliflower-like tops; often crowned with veils of thick cirrus giving the entire cloud a flat top. These are the thunderhead clouds, frequently associated with violent weather.

The various cloud types and forms are summarized in Figure 21–7A.

Fog. Like clouds, fog is the result of condensation of water vapor in the air. The chief difference is that fog forms when air is cooled by some means other than lifting. This generally occurs by contact with a cool surface. For example, one type of fog results from the nightly cooling of the earth. The layer of air in contact with the earth becomes chilled below its dew point and condensation of water droplets occurs. This type of fog is called a *radiation fog* or ground fog because it is caused by the radiation heat lost by the earth. Radiation fogs form most often on calm, clear nights. The fog is thickest in valleys and low places because the dense, cold air in which the fog forms drains to the lower elevations. Fogs are often unusually thick around cities because of the greater amount of smoke and dust particles which act as condensation nuclei.

Another common condition which produces fog is the movement of warm, moist air over cold surfaces. A fog produced in this way is called an *advection fog*, referring to the horizontal air movements. Advection fog is very common along seacoasts as the warm moist air from the water moves in over the cooler land surface. Dense fogs may form on the sea when warm, moist air is carried over cold ocean currents.

A third type of fog, called an *upslope fog*, is formed by the adiabatic cooling of air as it sweeps up rising land slopes. This is really a kind of cloud formation at ground level. See Figure 21–7B.

PRECIPITATION OF MOISTURE

Causes of precipitation. The process of condensation cannot go beyond a certain point without causing precipitation. A cloud produces precipitation when its droplets or ice crystals become large enough to fall as rain or snow. Most cloud droplets have a radius of about 10 microns (1 micron = 0.00004 in). Droplets of this size easily remain suspended in the air. Even slight air movements prevent them from settling out. Before it will fall as a raindrop, a cloud droplet must grow about 100 times larger in radius. See Figure 21–8.

Cloud droplets seem to reach the precipitation stage in two ways. The first involves differences in size between separate cloud droplets. To a large extent the original size of a cloud droplet depends on the kind and size of its condensation nucleus. Large nuclei tend to form large cloud droplets. The size of various cloud droplets is generally matched to those of the condensation nuclei. The larger droplets do not remain suspended in the cloud as well as the smaller ones. Instead they drift downward, forcing the larger droplets to collide and combine with the smaller ones. This process is called *coalescence* (koh-uh-*les*-ens). See Figure 21–9. The larger droplets continue to grow by coalescence until they contain several million times as

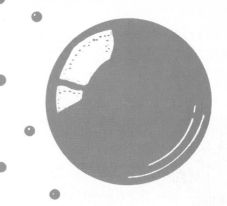

FIG. 21–8. Cloud droplets, compared with a growing raindrop.

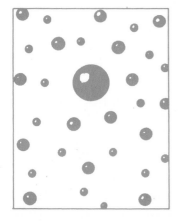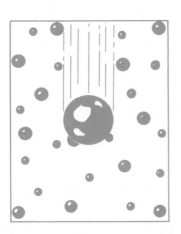

FIG. 21–9. By the process of coalescence, larger raindrops pull the smaller droplets together as they drift downward, forming even larger ones.

FIG. 21–10. The breaks in these clouds were produced by seeding with pulverized dry ice. (U.S. Weather Bureau)

FIG. 21–11. Types of ice crystals. (a) flat plate, (b) six-sided crystal, (c) needle-like. (Courtesy of Carl Zeiss, Inc., N.Y.)

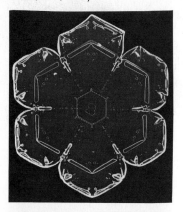

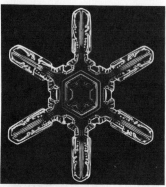

much water as a single cloud droplet. By this time, their weight is such that they fall as raindrops.

A second way that precipitation develops is related to the presence of ice crystals in a cloud. Each microscopic ice crystal grows by capturing and freezing water which evaporates from its neighboring water droplets. In this way the ice crystal grows large enough to fall. The falling ice crystal may melt in passage and reach the earth as a raindrop, or it may fall as a snowflake.

Knowledge of the way in which ice crystals grow in clouds has made it possible to take the first steps toward controlling the weather. This is done by scattering tiny crystals in a cloud with the correct temperature and moisture conditions. The artificial crystals most used are the compound silver iodide. Crystals of this compound resemble ice crystals and are able to act as "seeds" for the formation of ice. Silver iodide can be released into clouds by special rockets fired into the cloud. Flares dropped from aircraft which release silver iodide can also be used. Powdered dry ice has also been used successfully as seeds. These techniques for causing clouds to form precipitation are usually called "cloud seeding." See Figure 21–10.

The seeding of clouds under certain temperature conditions may cause them to release more water than they would naturally without being seeded. Other temperature conditions in seeded clouds can produce less precipitation. Thus cloud seeding seems to be a way of getting some control over the amount of precipitation coming from clouds under the right conditions. Cloud seeding may be a way to make droughts less of a problem as well as to help control floods by reducing snowfall when heavy runoff might cause flooding.

Seeding of clouds may also offer the possibility of reducing lightning which starts many forest fires. How clouds produce lightning is not understood very well but one theory requires the presence of ice crystals in the correct amounts. Cloud seeding might be able to change the production of these crystals and thus reduce lightning. It also may be possible to change hurricanes by cloud seeding. This could be done by causing heavier precipitation in the storm's early stages. Loss of this moisture would remove some of the cause of the strong winds and reduce the strength of the storm.

Increasing ability to control the weather will bring new problems as well as the solutions to some old ones. It is difficult to get everyone at a particular place to agree about the amount of precipitation most helpful to all. Generous amounts of rain or snow, or the lack of it, might please farmers while causing trouble for people with other interests. Weather control also raises serious questions about the future effects on the natural environment. It will be necessary to move very carefully in changing weather patterns in order to avoid upsetting the natural plant growth and stream flow of the land.

Other types of precipitation. The only other type of liquid precipitation is *drizzle*. This occurs when cloud or fog droplets fall to earth because the air is very still. All other forms of precipitation are in solid form.

The most common solid precipitation is, of course, snow. Snow will fall if the ice crystals fail to melt before they reach the ground. These ice crystals come in many interesting shapes. There are three main types: needles, plates, and branching crystals. See Figure 21–11. Temperature and humidity conditions control the type of ice crystal which forms. Lowest temperatures favor the growth of needles and plates, while branching crystals grow at higher temperatures. Naturally, temperatures must be below freezing in the cloud for snow to be produced.

However, if temperatures near the ground are above freezing, some of the falling ice crystals are likely to melt. Under the resulting film of water, clumps of them will stick together, forming large snowflakes. Colder temperatures near the ground produce a hard, fine snow, since the individual ice crystals remain separate.

Extreme temperatures near the ground may sometimes produce solid precipitation from raindrops. The ice pellets which are formed when rain falls through a layer of freezing air are called *sleet*. Occasionally rain does not

FIG. 21–12A. Glaze ice resulting from ice storms may do tremendous damage in woods and forests. (E. R. Degginger)

FIG. 21–12B. Hailstones covering a cornfield. (J. C. Allen)

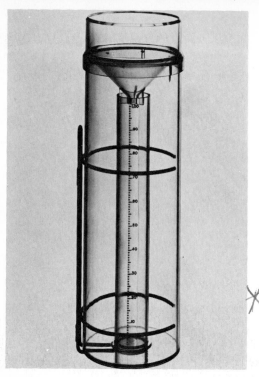

FIG. 21–13. A simple but accurate rain gauge. (Taylor Instrument Co.)

FIG. 21–14. A tipping bucket rain gauge with the tipping buckets visible through the open door. (Bendix Aviation Corporation)

freeze until it actually strikes the ground. In this event it forms a thick and very destructive layer of *glaze ice* over everything. Conditions which produce glaze ice occur in what is generally called an ice storm. See Figure 21–12.

Hail is also produced when raindrops are frozen into ice pellets. Unlike sleet however, hail is formed by layers of ice. Rain freezes, then falls through the warm air again, accumulating a layer of water. Then the hail is carried up again by strong updrafts into freezing air, or falls back through another layer of cold air. The water coating freezes and another layer of ice is formed. If the process is repeated a number of times, the hailstones may accumulate many layers of ice and grow quite large. These large hailstones cause great damage.

Measurement of precipitation. A *rain gauge* consists of a container which measures the depth of the rain water it collects. In one type of rain gauge a wide-mouthed funnel catches the rain and empties it into a cylindrical container below, as shown in Figure 21–13. The mouth of the funnel is much larger than the measuring container beneath it. This magnification of the actual amount of rain makes it possible to accurately measure the rainfall.

In another type of rain gauge, a small divided bucket catches the water from the funnel, fills on one side, then tips, dumping the water and allowing the other half of the small bucket to fill. Each time one side of the tipping bucket fills with exactly $1/100$ inch of rain, it tips and sets off an electrical device that records the amount. The rainwater dumped from the tipping bucket is collected and weighed as a means of checking the accuracy of the record. Another type of rain gauge catches the water in a large bucket which is weighed continuously. The weight is recorded directly on a graph as inches of rain. See Figure 21–14.

Rain gauges can also be used to measure snowfall. The collected snow is melted and the water is weighed to determine its rain equivalent. A light fluffy snow may yield one inch of rain for every 15 inches of snow; a dense well-packed snowfall may produce one inch of rain for every six inches of snow.

Because rain gauges measure only the precipitation falling in one spot, it is difficult to establish the total amount of rain or snow which falls over a large area. The amount of precipitation may differ by large amounts for even short distances. This is especially true in mountainous regions.

VOCABULARY REVIEW

Match the word or words in the column on the right with the correct phrase in the column on the left. *Do not write in this book.*

1. The conversion of a vapor into a liquid or a solid.
A 2. Ice pellets formed when rain falls through a layer of freezing air.
3. A device used to measure relative humidity.
4. The way large water droplets become larger by adding smaller droplets.
5. Air with a relative humidity above 100%.
L 6. The measurement of the weight of water vapor in a certain amount of air.
Q 7. Comparison of the absolute humidity of air to its saturation value and expressed as a percentage.
O 8. Temperature to which air must be cooled to raise its relative humidity to 100%.
C 9. The only form of liquid precipitation other than rain.
G 10. Ice pellets having a number of layers of ice.
D 11. A very high cloud composed of ice crystals.
I 12. Low clouds which usually bring rain or snow.
H 13. A cloud belonging to the "middle" group.
E 14. Clouds which have great vertical development.
K 15. The weight of water vapor an amount of air can hold at any given temperature.

a. psychrometer
b. supersaturated
c. drizzle
d. cirrostratus
e. cumulus
f. sleet
g. hail
h. altostratus
i. nimbostratus
j. calorie
k. saturation value
l. absolute humidity
m. ground fog
n. condensation
o. dew point
p. coalescence
q. relative humidity

QUESTIONS

Group A
Select the best term to complete the following statements. *Do not write in this book.*

1. To change liquid water to solid water (a) energy must be added (b) the temperature must increase (c) energy must be taken away (d) the dew point must be reached.

2. When water vapor enters the air (a) energy is released (b) energy is carried along (c) the relative humidity decreases (d) the saturation value changes.

3. The principal source of moisture in the air is (a) evaporation from lakes, ponds, streams, and soil (b) living plants and animals (c) volcanoes and burning fuel (d) evaporation from oceans.

4. The major source of energy causing moisture to enter the air is (a) the sun (b) the earth (c) volcanoes (d) living substances.

5. A humidity reading of 75% was obtained from an instrument. This is a measurement of (a) relative humidity (b) absolute humidity (c) saturation value (d) dew point.

6. The graph in Figure 21–2 shows the relationship to be (a) an increase in temperature decreases the amount of moisture a given volume of air will hold (b) an in-

crease in volume increases the amount of moisture a given volume of air will hold (c) a decrease in volume increases the amount of moisture a given volume of air will hold (d) an increase in temperature increases the amount of moisture a given volume of air will hold.

7. The most accurate way to measure humidity is to (a) use a psychrometer (b) trap the water in a given volume of air with a chemical drying agent (c) use a human hair hygrometer (d) measure the electrical current which flows through a chemical substance which collects moisture.

8. The temperature to which air must be cooled to give a relative humidity of 100% is called its (a) saturation value (b) condensation point (c) dew point (d) absolute humidity.

9. If the temperature of the air is below the freezing temperature of water (a) the dew point cannot be reached (b) when the dew point is reached, the moisture will condense into a liquid (c) the moisture in the air will condense into ice (d) frost will form when the dew point is reached.

10. Frost (a) does not actually cause plants to die (b) kills growing plants (c) forms only above the dew point (d) is the result of dew freezing.

11. Which one of the following would be *least* likely to cause moisture to condense from the air? (a) radiation of heat at night (b) mixing of warm air with cold (c) downward motion of air (d) upward motion of air.

12. An adiabatic change is one in which air temperature changes due to (a) radiation (b) condensation (c) mixing of warm and cool air (d) rising or sinking air.

13. A cloud may form when a body of air is (a) cooled below its freezing point (b) cooled below its dew point (c) mixed with warmer air (d) sinks 100 meters.

14. For each 100 meters a parcel of dry air rises, it (a) warms 1°C (b) cools 1°C (c) warms 5½°F (d) cools 0.6°C.

15. For clouds to form out of moist air, the air must (a) cool, with condensation nuclei present (b) rise and cool adiabatically until the dew point is reached (c) be cooled below its dew point (d) sink adiabatically until its dew point is reached.

16. The reason water droplets usually form in air when the relative humidity reaches 100% is that (a) the air contains dust, smoke, salt particles, etc. (b) the air can hold more moisture (c) the dew point has not been reached (d) heating is taking place.

17. The type of cloud which does not belong with the rest is the (a) cirrocumulus (b) cirrostratus (c) cirrus (d) stratocumulus.

18. The type of cloud which would be found at the lowest altitude is the (a) altocumulus (b) altostratus (c) cirrus (d) stratus.

19. Clouds composed of ice crystals share a common part to their name, which is (a) alto (b) cirro (c) strato (d) nimbo.

20. Low, shapeless, thick-layered, dark gray clouds usually bringing rain or snow are called (a) stratus (b) altostratus (c) nimbostratus (d) stratocumulus.

21. The "thunderhead" cloud belongs to which of the following cloud types? (a) cumulus (b) altocumulus (c) stratocumulus (d) cumulonimbus.

22. The kind of fog produced from the nightly cooling of the earth is called (a) sea fog (b) advection fog (c) radiation fog (d) night fog.

23. The kind of fog produced by horizontal motion of air is called (a) sea fog (b) advection fog (c) radiation fog (d) wind fog.

24. Before water droplets in clouds can fall as rain or snow, they must grow to about (a) 100 times their normal size (b) 10 microns (c) 0.000004 inches (d) 100 microns.

25. An example of coalescence in precipitation is (a) the formation of large droplets from one large condensation nucleus (b) ice crystal capture of moisture from air (c) seeding with silver iodide (d) the collision of two droplets to form one.

26. Methods of artificially producing rain by "cloud seeding" are based on our knowledge of how (a) clouds form (b) ice crystals grow (c) cloud droplets coalesce (d) fog is formed.

27. Which one of the following forms of precipitation is not like the others? (a) drizzle (b) snow (c) hail (d) sleet.

28. A type of precipitation which does not start in raindrop form is (a) sleet (b) glaze ice (c) hail (d) snow.

29. A tipping bucket measures liquid precipitation to the nearest (a) inch (b) centimeter (c) 0.1 centimeter (d) 0.01 inch.

30. The amount of rainfall which would be equal to 2.5 feet of fluffy snow would most likely be (a) 15 inches (b) 2 inches (c) 6 inches (d) 30 inches.

Group B

1. Water vapor is not very abundant in the atmosphere. If it were completely missing, what effect would this have on the earth's surface?

2. If you spilled 454 grams (approximately 1 pint) of water and it all evaporated into the air, how many calories of energy would it carry with it due to the evaporation?

3. Compare the role of plants, lakes and rivers, burning fuels, and seas as sources of the moisture in air.

4. One liter (one thousandth of a cubic meter) of air is passed over the chemical calcium chloride which absorbs all the moisture. It is found to be .01 gram of water. What is the absolute humidity of the sample of air?

5. The saturation value of a given sample of air is known to be 10 grams at room temperature. The absolute humidity is found to be 8.0 grams. What is the relative humidity of the air sample?

6. What effect does increasing the temperature of a sample of air have on each of the following? (a) absolute humidity (b) saturation value (c) relative humidity.

7. Describe three different methods for measuring relative humidity.

8. What are three ways that air is generally cooled?

9. A parcel of air decreased in temperature by 5°C (9°F). If this change was adiabatic, how high did the parcel rise?

10. What two conditions must be met before clouds will form?

11. Name the form(s) of clouds which would go with each of the following: (a) thunder-head (b) rain clouds (c) high fog (d) ring around the moon (d) mare's tails (f) mackerel sky.

12. Very early in the morning, a coastal city is enveloped in a dense fog. At the same time 100 miles inland, a farm valley is also packed with a dense fog. Explain how these two fogs most likely formed.

13. Describe two ways cloud particles may grow large enough to precipitate as rain or snow.

14. Describe a method of artificially producing precipitation from clouds and discuss the possible benefits and problems which might be encountered.

15. The tipping bucket rain gauge cannot usually be used to measure snowfall. Why?

16. How would you explain the fact that source regions for many violent storms originate over the oceans near the tropics?

17. The normal flow of air is sometimes reversed over certain areas of the continents, causing an unusual hot wind to blow. Such a wind is the "Santa Ana" in the south-western region of the United States. The Santa Ana winds originate over the high desert areas and blow toward the coast. Using your knowledge of the geography of southern California and the characteristics of sinking air, explain why these winds sometimes result in temperatures soaring as high as 110°–115°F in such cities as Los Angeles and San Diego.

18. At present many cities are having problems with a condition in which smoke and a very light fog combine to cause "smog." Smog is most likely to occur when there is a layer of warm air at about 1,000 to 2,000 feet, with cooler air extending down to the ground. (a) Why would this contribute to the production of smog? (b) Why is this layer called an "inversion" layer?

22
weather

Weather is the sum of all the properties of the atmosphere—temperature, pressure, humidity, and winds—at any particular time. Of all the forces of nature, weather influences people's environment most. Its properties may have far reaching effects. Thunderstorms, blizzards, rain or snow, all affect our daily lives in one way or another.

Almost all weather changes are the result of the unequal heating of the earth's atmosphere. These temperature differences in the atmosphere play a key role in the earth's ever-changing weather patterns.

Many of the earth's changes in weather originate at high altitudes, about halfway between the equator and the poles. In this zone cold air moving down from the poles collides with warm air from the equatorial regions. Cold air clings close to the earth. It pushes in under the warm air, thrusting cold fingers toward the warm tropics. At the same time, warm air slides up over the cold and moves toward the poles. The collision of these bodies of warm and cold air produces storm centers. These are responsible for most of the weather experienced in the middle latitudes.

Scientists are only beginning to understand the origins of weather changes. Much remains a mystery. The shifting patterns of the weather are often hard to analyze. However, one important principle is clear as a result of careful study of weather patterns. This principle is the basis for understanding most of the weather of North America. It states that weather changes are largely brought about by movement of giant bodies or masses of air moving in response to the earth's general wind patterns.

objectives

☐ Describe the origin of air masses.

☐ Name the air masses that influence North American weather.

☐ Explain how weather fronts form.

☐ Describe the methods and instruments used to observe weather.

☐ Describe how to make a weather map and forecast weather.

☐ Explain how local weather phenomena develop.

AIR MASSES

Origin of air masses. A body of air covering millions of square miles of the earth's surface, with nearly the same temperature and humidity throughout at any particular altitude, is called an *air mass.* See Figure 22-1. An air mass is created when a large body of air remains relatively stationary over a body of land or water for some time. It then takes on the temperature and humidity characteristic of that region. For example, air remaining over the cold arctic plains of North America becomes very cold and dry. Over a tropical ocean a quiet body of air becomes warm and moist. A region that serves as a source of air masses must be fairly uniform, such as the arctic plains or the sea. Mixed land and water areas are not suitable. The source region must be free of continuous strong winds.

Air masses are classified according to their source region. A system of letters is used to designate the source and characteristics of the various air masses. The principal source regions are the cold polar (P) areas and the warm tropical (T) areas. Air masses are described as coming from the sea, maritime (m), and continental (c), if the source is over land. Naturally, maritime air masses tend to be moist, and continental air masses generally dry. It follows that there are four main types of air masses, described by the following letter combinations: *mP, mT, cP* and *cT.*

Once formed, an air mass moves away from its source region because of the general movements of the atmosphere. An air mass moving away from its source region is said to be *cold* if it is cooler than the surface over which it is moving. The temperature of a *warm* air mass is higher than the surface over which it passes. Several days are usually required for an air mass to move past an area. During this time, the region will have a period of *air mass weather,* meaning the weather usually remains the same for several days. When the air mass moves on, there is a change in the weather. The winds, clouds, and precipitation accompanying these weather changes are the result of the collision of the different air masses.

North American air masses. Air masses that strongly influence North American weather come from seven main source regions. The origins of these air masses and their general direction of movement are shown in Figure 22-2.

FIG. 22-1. A polar air mass is shown moving from Canada into the United States. A cross-section along the line AB shown above indicates the uniform conditions that prevail within the air mass.

Polar Canadian (cP) air masses are formed in northern Canada. They generally move in a southeasterly direction across Canada and into the northern United States. In the winter they create intense cold waves. Occasionally these cold air masses penetrate as far south as the coast of the Gulf of Mexico. In the summer, they usually bring cool, dry weather.

Polar Pacific (mP) air masses originate in the northern Pacific Ocean region. They are very moist and not extremely cold. To the Pacific Coast, the area most affected, they bring rain and snow. In summer, the polar Pacific air brings cool, often foggy, weather.

Polar Atlantic (mP) air masses are formed over the northern Atlantic Ocean. The general direction of their movement is eastward toward Europe. But they may pass over the northeastern portion of North America. In winter they usually bring cold, cloudy weather and light precipitation. In summer these air masses often produce cool weather with low clouds and fog.

Tropical Continental (cT) air masses affect North America only in the summer. They form over Mexico and the southwest United States and generally move northeast. They usually bring clear, dry, and very hot weather.

Tropical Gulf and Tropical Atlantic (mT) air masses form over the warm water of the Gulf of Mexico and the South Atlantic. They move north across the eastern United States. In the winter they bring mild and often cloudy weather. In the summer, they produce hot, very humid weather and frequent thunderstorms.

Tropical Pacific (mT) air masses collect over the warm parts of the Pacific Ocean and reach the Pacific Coast only rarely in the winter. On these infrequent visits they bring very heavy precipitation.

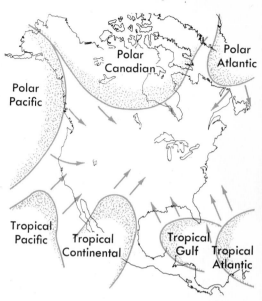

FIG. 22–2. The types of air masses that influence weather in North America. Arrows indicate the general direction in which these air masses usually move.

WEATHER FRONTS

Formation of a front. When two air masses meet, no general mixing of the air contained within the two masses takes place. Temperature differences keep the two air masses separate. The cooler air of one mass is denser and does not mix with the warmer air. Thus a definite boundary is usually formed between air masses. It is called a *front*. The colder of the two air masses is denser, and thus it will push under the warmer air and lift it. Therefore the front always slopes up over the cooler air. See Figure 22–3.

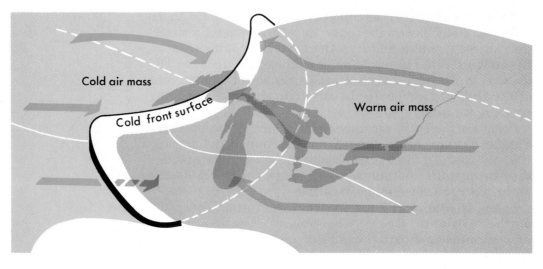

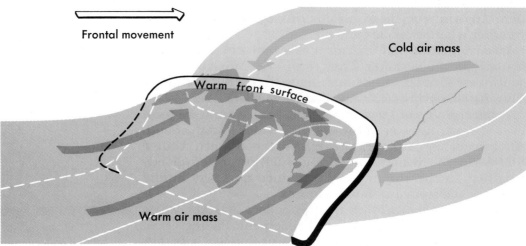

FIG. 22-3. A side-view of air masses along a cold and a warm front. Note that the frontal surface separating the two air masses exists not only on the ground but continues upward along the boundary of the air masses.

Air masses are usually in motion. The nature of a front depends upon how the air masses are moving. When a cold air mass overtakes warmer air, a *cold front* is formed. The advancing cold air is held back by friction against the ground so that cold air tends to pile up in a steep slope. The moving cold air lifts the warmer air and will create a heavy cloud formation if the warm air is moist. Large cumulus and cumulonimbus clouds are typical of a rapidly moving cold front. See Figure 22–4. Storms created along a cold front are usually short and violent. A long line of heavy thunderstorms, called a *squall line,* may advance just ahead of a fast moving cold front. A slowly moving cold front lifts the warm air ahead less rapidly, producing a less concentrated area of cloudiness and precipitation.

When warm air advances over the edge of a mass of colder air, a *warm front* is produced. The slope of a warm front is very gradual, as shown in Figure 22–5. Because of this gentle slope, the clouds may extend far beyond the base of the front. Stratus clouds and heavy, though not violent precipitation over a large area, are conditions generally associated with a warm front. Occasionally a warm front will produce violent storms if the warm air advancing over the cooler air is very moist.

Both a cold and a warm front may come to a halt for several days. When this happens, they form a *stationary front*. The weather around a stationary front is not very different from that produced by a warm front.

Storm centers formed by fronts. Bulges often develop along the edges of slowly moving cold or stationary fronts. A bulge is created when the air on one side moves slightly ahead of the front itself. These bulges or waves in the frontal boundary can signal the beginning of a storm center called a *wave cyclone*. A frontal wave cyclone is not the same kind of storm as a tropical cyclone or hurricane. Nor should it be confused with a tornado that is a small but very violent local storm. A wave cyclone consists of a very large body of air. Its winds blow in circular paths toward a low pressure region at the center. Such storms may cover a large part of an entire continent.

The stages in the development of a typical wave cyclone are illustrated in Figure 22–6. At the beginning (A), there

FIG. 22–4. Weather conditions along a cold front. Clouds shown along a cross-section of the frontal surface are typical of those that accompany an approaching cold front.

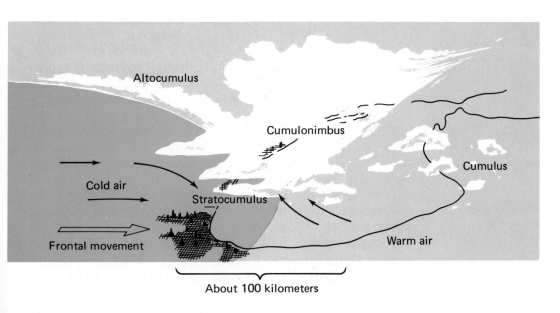

Altocumulus

Cumulonimbus

Cumulus

Cold air

Stratocumulus

Frontal movement

Warm air

About 100 kilometers

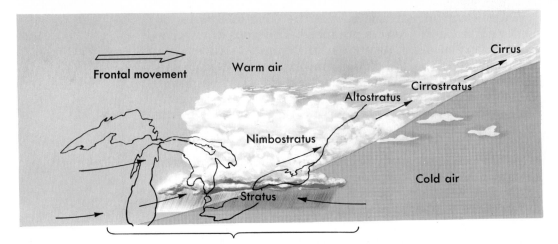

About 500 miles

FIG. 22-5. Weather conditions along a warm front. To an observer on the ground the approach of a warm front can be detected by the appearance of cirrus clouds.

is only the frontal boundary between the two air masses. At this stage, their winds usually move parallel to the front. This is due to the fact that air in the two masses does not mix. The winds on each side of the front are blowing opposite each other at different speeds. Thus they tend to develop a spinning motion. Try the following activity and you will see why this is so. Place a pencil across your two hands, then slide them in opposite directions.

This tendency toward a spinning movement causes slight waves or bulges to form along the front. This action also causes cold air to push into the warm. At the same time, warm air is pushed forward into the colder air. See Part (B) of Figure 22-6. The result is a warm front moving slowly ahead of a rapidly moving cold front.

A low pressure area develops at the point where the two fronts come together (C). The lighter warm air is lifted as it presses into the cold air along the warm front. It is also pushed up by the advancing cold air along the cold front. As a result of these lifting actions, clouds and precipitation spread along both fronts (D).

Soon the swiftly moving cold front overtakes the warm front and the warm air trapped between them is completely lifted off the ground (E). The front at that place is said to be *occluded*. This means that the front then exists only at upper levels and is completely closed off from the ground. At this point, the effects from the strong lifting action on the warm air usually bring the storm to its highest intensity. Now, the winds are moving in a circle toward the low pressure region at the center of the storm (F). In this area pressure is low because air is rising.

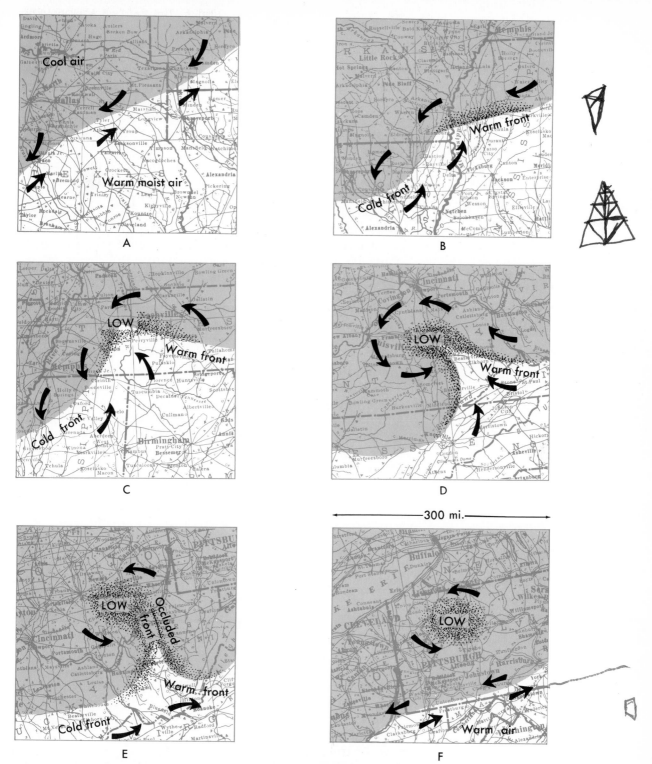

FIG. 22–6. The development of a typical wave cyclone or low pressure area as seen from above. The underlying maps indicate that this storm area moved northeastward.

Investigate
Find out how you can tell the
directions to high and low
pressure regions by just knowing
the wind direction. Then report
your discovery to the class.

It usually takes 12 to 24 hours for a wave cyclone to develop. During this time the air masses are in motion, and the disturbance moves with them. In the Northern Hemisphere wave cyclones generally move in an easterly direction at speeds of around 32 to 64 km/hr (20-40 mi/hr). Most cyclones follow well-known paths established through careful observation.

Figure 22–7 shows some of the most common paths that wave cyclones take. Notice that there are two well established paths in the Northern Hemisphere. One begins in the North Pacific and moves toward Alaska and south along the western coast of North America. The other common cyclone tract originates in eastern North America and crosses the North Atlantic to Europe. Wave cyclones usually develop one after another in a common path along a front.

Without its supply of moist air a cyclonic storm would soon die out. Winds circling into the low pressure center provide the amount of moist air needed. As this moist air rises and cools, the condensation of water vapor releases the heat energy which originally converted it into vapor. This supply of heat energy keeps the cyclone going. Later in its development the cyclone becomes cut off from its supply of moist air and begins to gradually die out. Then it disappears altogether as a separate weather disturbance.

As a separate storm center, the cyclone is a region of low pressure. The general wind direction near the ground

FIG. 22–7. Arrows indicate the most common paths for cyclonic storms.

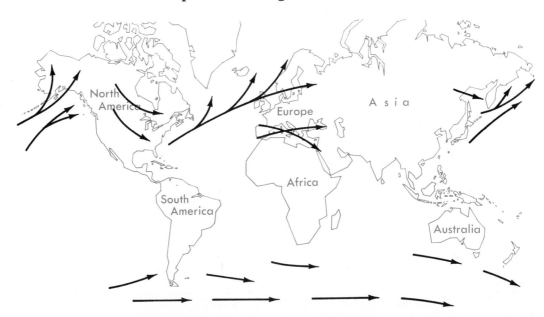

within such a storm is inward toward the low pressure. The Coriolis effect and friction of the winds against the earth's surface create a whirling motion. This twisting motion of winds inward toward a low pressure center moves *counter-clockwise* in the Northern Hemisphere.

In high pressure regions the direction of wind flow is outward from the high pressure center. The air spreads out away from the center. The Coriolis effect and surface friction again deflect the winds. In the Northern Hemisphere the outward moving winds of a high pressure region move in a *clockwise* whirl. The high pressure centers are usually called *highs* or *anticyclones*. See Figure 22–8. Since the air in an anticyclone is sinking, there is little tendency for precipitation or for clouds to form. Thus anticyclones or highs usually bring clear weather.

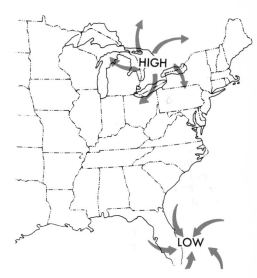

FIG. 22–8. The clockwise movement of winds from a high, contrasted with the counter-clockwise flow of winds into a low.

The polar front. Over the entire earth there is a general movement of warm air rising from the equator and moving toward the poles. Due to the Coriolis effect, air moving in the direction of the poles is turned to the right. In the Northern Hemisphere this means that it is deflected to the east. Then, about halfway between the equator and the North Pole, the eastward moving air meets cold polar air flowing south toward the equator. This zone where warm tropical air meets cold polar air is a more or less continuous weather front circling the entire earth. It is often called the polar front. The eastward motion of the air in the region of the polar front results in a wide band of high altitude winds. This is known as the *circumpolar whirl*. See Figure 22–9. These winds are generally strongest in the region of latitude 30° and weaken toward the equator and poles.

In the Northern Hemisphere, near the southern boundary of the circumpolar whirl, a band of very strong winds often exist. These winds are called the *jet streams*. They are found at an altitude of 10 to 15 kilometers (about 6 to 9 mi), and measure about 100 kilometers wide and two to three kilometers deep. When temperature differences between the two air masses separated by the polar front are greatest, the jet streams may have speeds up to 480 km/hr (300 mi/hr). The jet streams do not blow steadily, but change in speed, direction, and position. In summer when the circumpolar whirl contracts, the jet streams are found closer to the North Pole at 35° to 45° N latitude. In

FIG. 22–9. The circumpolar whirl. The length of the arrows indicates the relative speed of the winds.

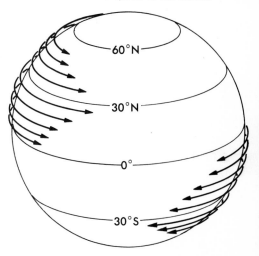

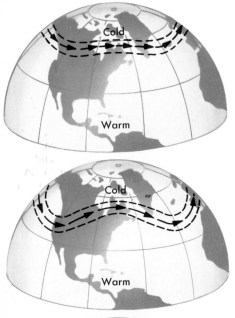

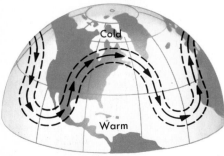

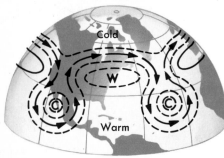

FIG. 22–10. A wave of jet streams. The waves tend to isolate pockets of warm and cold air far north and south of their average locations.

winter as the circumpolar whirl expands, they shift to the south and are found then at 20° to 25° N latitude.

The mixing of polar air and tropical air does not apparently take place continuously at the polar front. Mixing occurs only when waves develop in the circumpolar whirl. These waves carry cold polar air to the south and warm equatorial air to the north. The development of a wave in the circumpolar whirl can be traced by shifts in the position of the jet streams. A wave cycle of the jet streams is shown in Figure 22–10.

The behavior of the jet streams is closely connected with the development of wave cyclones. The cyclonic storms exist at any time in the air layers nearest the earth's surface. They seem to be formed and move under the general control of the jet stream cycles. But the details of this relationship are not completely understood.

WEATHER INSTRUMENTS

Observing weather conditions. One of the most practical applications of meteorology, the science of the atmosphere and the study of changing weather conditions, is the ability to make fairly accurate weather observations. To do this, meteorologists rely on various weather instruments and on techniques that have been developed largely within the last hundred years. The use of new instruments and improved techniques has increased the accuracy of observing the following weather conditions:

1. *Temperature.* Air temperature at ground level is measured on either the *Fahrenheit* (F) or *Celsius* (C) (also called Centigrade) scale. In the United States both scales are now commonly being used. See the appendix for a description of these temperature scales. Special thermometers are constructed to record the highest or lowest reading since the previous observation. See Figure 22–11.

2. *Pressure.* The instruments used to measure atmospheric pressure were described in Chapter 20. In weather reporting and forecasting, pressure readings are usually expressed in *millibars* (mb). A pressure of 2.54 cm of mercury equals 33.86 millibars.

3. *Precipitation.* The amount of precipitation since the last observation is recorded in hundredths of inches or millimeters. Solid forms of precipitation are melted and the depth of the liquid is recorded.

FIG. 22–11. Maximum and minimum thermometers. The maximum thermometer is filled with mercury, whereas the minimum thermometer uses alcohol. The special holder allows the maximum thermometer to be whirled and the minimum thermometer to be tilted in resetting. (Bendix Aviation Corporation)

4. *Wind speed and direction.* To measure wind speed, an instrument called an *anemometer* (an-uh-*mom*-uh-ter) is used. A typical anemometer consists of several small cups attached by spokes to a shaft that is free to rotate. See Figure 22–12. This rotation is then converted into an electrical signal which registers the wind speed on a conveniently located dial. An arrangement which produces a permanent record of wind speed on a graph is especially useful.

To determine wind speed and direction at higher altitudes, meteorologists send up a hydrogen or helium filled balloon. The balloon rises at a constant speed. Its movement in response to prevailing winds is followed with a small telescope, which also measures the angle of the line to the balloon. With this information the speed of the balloon can be calculated as it is carried along by the wind.

Wind speed is expressed in meters per second, miles per hour, or in knots (1.15 mi per hour). The direction of a wind is described according to the direction from which it comes. Thus a wind from the west, blowing toward the east, is called a west wind. In some weather reports, exact directions for winds may be given as one of 32 directions (points) on the compass. Wind direction is also recorded in degrees. They start with 0° at north and move clockwise around to 360° at north again. These two commonly used systems for describing wind direction are shown in Figure 22–13.

Electronic weather instruments. Conditions of the atmosphere near ground level are only a part of the complete picture of weather changes. Conditions at upper levels must also be known. An instrument commonly used to investigate weather conditions in the upper atmosphere is the *radiosonde.* See Figure 22–14. Measurements of temperature, pressure, and humidity are sent out by a radio

FIG. 22–12. An anemometer. (Bendix Aviation Corporation)

FIG. 22–13. The two systems by which wind directions are given.

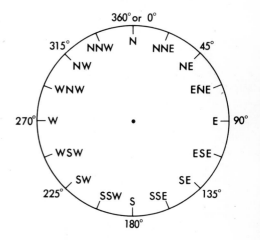

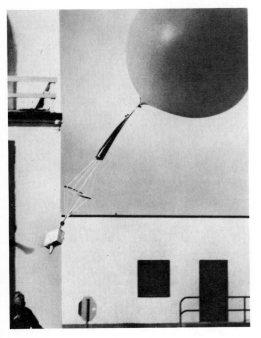

FIG. 22–14. The radiosonde being lifted by a balloon with a parachute which will return the instrument to earth. (Environmental Sciences Services Administration)

FIG. 22–15. Spacecraft launching of a TIROS weather satellite. Close-up of the satellite shows some of its principal parts. (ESSA)

signal as a helium-filled balloon lifts the instrument. A special radio is set up to receive and record the information. The path of the balloon is also followed by radar so that direction and speed of high altitude winds may be determined. When it reaches a very high altitude, the balloon finally bursts. The radiosonde instrument is then parachuted back to earth.

Another valuable electronic weather instrument is *radar*. Particles of water in the form of cloud droplets or precipitation reflect radar waves. Thus weather disturbances are visible on a radar screen. Radar information can also be used to give the precise location and extent of a storm. On a radarscope, one can watch the origin and growth of a storm system and track its movements across the land.

Recent advances in space science have given meteorologists an important instrument for future study of the weather. The weather satellite carries equipment for making the kind of detailed observations that are needed. The first weather satellite was known as TIROS (Television and Infrared Observation Satellite). A number of these satellites were launched in 1960 and orbited the earth from west to east. Later models were launched into a polar orbit allowing television pictures to be made of the entire earth. See Figure 22–15.

NIMBUS, the first of a more advanced series of weather satellites, was launched in 1964. From the results of the TIROS and NIMBUS experiments, a plan for continuous satellite observation of the earth's weather was developed. As time goes on, information obtained from satellites will greatly increase our knowledge of the earth's atmosphere and result in improved weather forecasting.

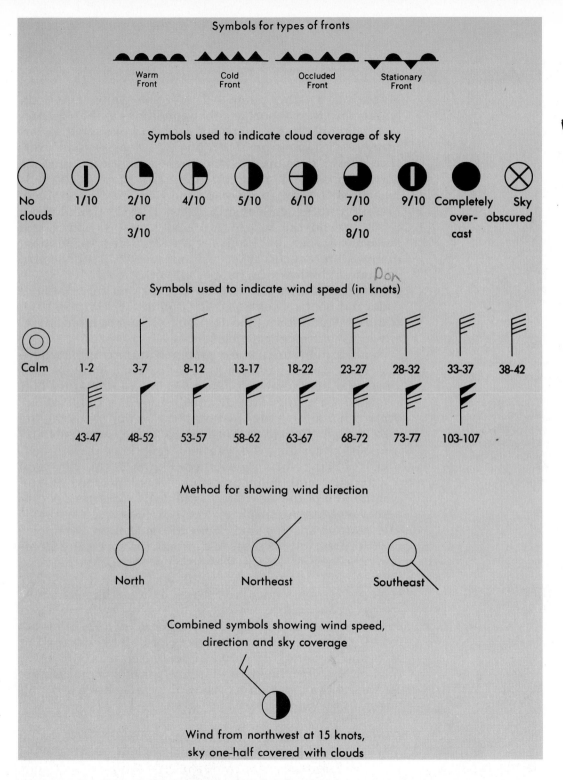

FIG. 22-16. Symbols used in the construction of station models.

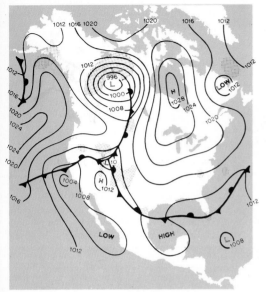

FIG. 22–17. A simplified weather map usually shows isobars, highs and lows, and fronts. It may also indicate other conditions, such as wind direction, temperature, and precipitation. Front symbols: triangle—cold front; half-circle—warm front; both on opposite sides—stationary front; both on same side—occluded front.

Construct

Make a form that could be used to keep a record of local weather observations.

THE WEATHER MAP

Making a weather map. All over the world, every six hours, observers report weather conditions at their respective locations. Barometers are read and corrected to correspond with pressure at sea level. All reports show pressure to be on the same basis regardless of altitude. Surface winds speed and direction are noted. Precipitation is measured, temperature read, and humidity determined. The information collected includes a description of cloud covering and the height to cloud bases. Visibility and general weather conditions are also recorded. In addition, many larger observation stations send up radiosonde equipment to determine upper air conditions.

All this information is then put into an international code and sent to central collection centers within each country. The information is then exchanged internationally by government agencies.

Weather maps are usually prepared at the centers where the coded weather information is received. The reported weather observations are first translated into figures and symbols. Then they are grouped around a small circle, drawn on a map at the position of the station reporting the information. The circle on the map, with its symbols and numbers describing the weather conditions at that location, is called a *station model*. See Figure 22–16.

When the station models have been recorded on the weather map, the next step is to draw lines connecting points of equal atmospheric pressure. See Figure 22–17. The relative spacing and shape of the isobars, when correctly drawn, tells us something about the speed and direction of observed winds. Closely spaced isobars indicate rapid change in pressure and higher wind speeds. Widely separated isobars generally mean light winds. Isobars in rough circles enclose centers of high or low pressure. Such centers are usually marked with a large H or L. Since air moves toward regions of low pressure, the general wind direction is usually toward low pressure areas. However, the Coriolis effect makes winds flow parallel to the isobars. Surface features, such as mountains, also have a strong effect on the wind pattern.

Principles of weather forecasting. The basic tool used by meteorologists in forecasting weather is the weather map. Weather forecasters try to predict the intensity and path of weather systems plotted on the map. They also try to forecast the formation of new weather disturbances. The forecaster relies on the fact that in the middle latitudes the upper air moves in a general easterly direction. The low level weather is usually carried along with this upper air movement. Most cyclonic storms that affect North America enter from the west. They move eastward across the middle of the continent and pass out to sea off the North Atlantic coast.

In determining the speed and direction of a storm center, forecasters make use of charts showing pressures and winds, particularly at upper levels. For their estimates of the future path of the disturbance, they note upper movements, pressures, previous rates, and direction of movement in the system. To learn about other features of the coming weather, such as temperature, humidity, and cloudiness, they must have more information. From their knowledge of how steep the surfaces of other weather fronts have been, they can draw some conclusions about future conditions. The steepness of a front's slope is a very important factor. If the slope is steep enough the lifting of air may be sufficient to cause condensation of moisture producing cloudiness and precipitation. Information obtained from satellites provide pictures of general cloud patterns. These are very useful in helping discover and follow weather systems.

At the present time it is possible to accurately predict general weather conditions up to about five days in advance. Detailed forecasts cannot be made beyond 48 hours. Extended-range forecasts up to 30 days are made by computer analysis of the average movements of the atmosphere. In this procedure slowly changing large-scale movements of the air, such as the jet stream cycles, are revealed. These changes help to predict the general weather pattern, but they cannot forecast the exact weather at a particular place. Accurate and detailed long-range forecasting will not be possible until all of the factors which control the weather are better understood.

LOCAL WEATHER

The scale of weather changes. A weather map shows only large-scale weather systems that affect large areas of the earth's surface. The reason for this is the way that weather observations are made. Weather instruments usually

activity

On a weather map, a line that connects points with the same temperature is called an **isotherm**. Isotherms can be used to help identify air masses. You are to draw isotherms in this activity.

Get a map on which the temperatures at stations in the United States are written. The number above the circle is the temperature. The number below is the number of the station.

The 15° isotherm starts at station 14 in the northeast United States. It goes southwest into Texas, across to Arizona, then north to Washington, and ends between stations 1 and 2.

Notice that the 15° isotherm goes through station 13 because the temperature is 15°C there. It also goes between stations 20 and 21. Since the temperature must be 15°C, the line must pass somewhere between 13.9°C and 15.6°C, the temperatures at stations 20 and 21. The line also goes between stations 40 and 22 as well as between 36 and 37. These are examples of places you can use to draw the 15° isotherm.

Draw the 15° isotherm on the map lightly, in pencil, then go over it and smooth out any bumps.

The 15° isotherm is a single line. Another isotherm can be in two parts, or it may come back to where it started, forming a closed loop.

On your map, draw isotherms for 9°C, 11°C, 13°C, 15°C, 17°C, 19°C, 21°C, 23°C, and 25°C. Label each isotherm with the correct temperature.

1. What is the lowest temperature for which you have drawn an isotherm? what is the highest?

2. Which forms a closed loop, the low-temperature isotherm or the high-temperature isotherm?

3. Is the air mass identified by the closed isotherm a cold air mass or a warm air mass?

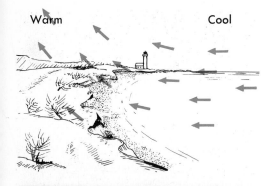

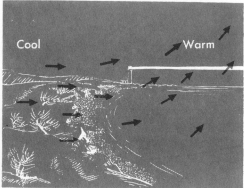

FIG. 22–18. Unequal heating of land and water results in land and sea breezes.

register only the slower changes in atmospheric conditions. For example, most anemometers are designed to record only general wind speed. Short gusts of wind are not registered. Weather observations are also made at widely spaced locations and must be averaged to give a picture of the general weather pattern. This averaging process eliminates small-weather changes from a common weather map. Only the weather disturbances that last for days are described by weather maps.

At any particular place, the movements of air are influenced by local conditions. A breeze that suddenly springs up is not usually part of a large-scale weather pattern. Small-scale winds that extend over distances of less than 100 kilometers are more often caused by some local feature that produces temperature differences. Some examples of this type of air movement are:

1. *Land and sea breezes.* Equal areas of land and water may receive the same total amount of energy from the sun. But the land surface is soon heated to a higher temperature than the water. During daylight hours this causes a sharp temperature difference to develop between the water and land along a shore. This temperature difference also appears in the air above the land and water. The warmer air above the land rises and the cool air from above the water moves in to replace it. A cool sea breeze will begin generally in the late morning hours. In late afternoon or early evening the sea breeze is replaced by a land breeze as the temperature situation is reversed. See Figure 22–18.

2. *Mountain and valley breezes.* In the daytime, mountains heat faster than surrounding valleys. This is mainly because the exposed slopes of mountains easily absorb the sun's energy. The valleys, because they are generally covered with forests and other vegetation, absorb energy more slowly. In the daylight hours, a gentle valley breeze blows up the slopes. It is caused by cooler air from the valleys moving up to replace the warmer mountain air. At night the mountains cool faster than the valleys. Then cooler air descends on the mountain slopes creating a mountain breeze.

Hurricanes. For reasons that are not yet clearly understood, hurricanes frequently develop over warm, tropical sea water near the equator. These storms have some resemblance to the wave cyclones of the middle latitudes and may develop in a similar way. However, hurricanes are seldom more than 700 kilometers (450 mi) in diameter, compared with the diameter of a wave cyclone which is about 2,500 kilometers. A full-scale hurricane carries an unusually large amount of energy. This energy comes almost entirely from the heat that is released as moisture condenses from warm tropical air.

A hurricane begins when very warm moist air over the sea rises rapidly. When moisture in the rising warm air condenses, large amounts of heat are released. This forces the air to rise even faster. More moist tropical air is drawn into the column of rising air at lower levels. A continuous supply of water vapor is needed to keep the process going. The entire system develops a spin which is thought to be caused by the Coriolis effect. See Figure 22–19A.

A fully developed hurricane consists of a series of thick cloud bands spiraling upward into the storm's center. See Figure 22–19B. Precipitation is very heavy in these clouds. Winds increase in velocity toward the center or eye of the storm, reaching maximum speeds of over 100 miles/hour. Their lifetime is around nine to twelve days. During this time the hurricane moves in the direction of the existing large-scale wind pattern.

The principal paths of hurricanes over the entire earth are shown in Figure 22–20.

FIG. 22–19A. A cyclonic storm developing over the Pacific as seen from space. (Shostal)

FIG. 22–19B. Diagram of a hurricane.

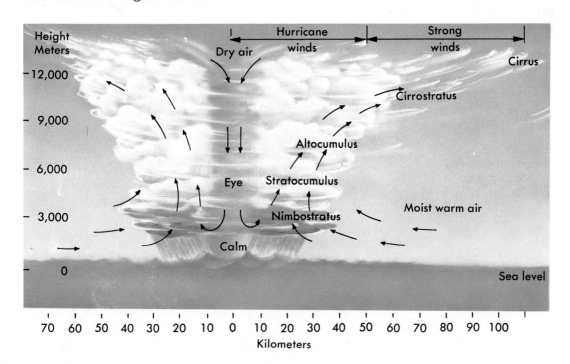

FIG. 22–20. The principal hurricane and typhoon areas of the world are shown in color. Arrows indicate the normal paths of the storms. Note that these storms all originate over water in tropical regions.

FIG. 22–21. The first stage (A) in the development of a thunderstorm is the cumulus stage. A swiftly moving current of warm air rises to high altitudes. The second, (B) or mature, stage of a thunderstorm begins as precipitation falls from the upper levels of the cloud. The precipitation causes violent up-and-down movements of air within the cloud. Final stage of a thunderstorm (C). Upward movement air has stopped and precipitation has slowed or stopped. Ice crystals at the top of the cloud spread out into the anvil-shaped top typical of a well developed thunderhead cloud.

Thunderstorms. When a small-scale upward movement of warm moist air takes place, a thunderstorm may result. Such an event is likely when the air in a moist, warm (mT) air mass is heated in one location. The air may also be lifted by mountains or by contact with an occluded front.

A local thunderstorm develops in three distinct stages. These stages are illustrated in Figure 22–21. How some clouds develop the electrical features necessary to produce lightning is not completely understood. The electrical

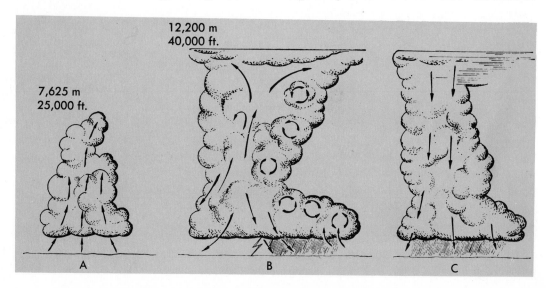

charges separate, causing the upper part of the cloud to carry positive electricity while the lower part carries both positive and negative charges. What seems to cause this separation is that large water droplets carrying negative charges fall to the lower parts of the cloud. Small droplets carrying positive charges are lifted up by strong updrafts to the top of the cloud. See Figure 22–22.

Each year lightning kills a large number of people. A few simple precautions would lessen the risk of being struck by lightning. The first rule is to avoid any prominent feature or high location in open land. Lightning always follows the shortest path between cloud and ground. Therefore, it is most likely to strike the highest part of any particular location. Trees, tall metal objects, and bodies of water are all likely targets and should be avoided during a thunderstorm. The interior of buildings, particularly those with a metal frame, are relatively safe, as are the interiors of automobiles.

Tornadoes. The smallest, most violent and short-lived of all storms is the tornado. A tornado is most likely to occur on a warm, humid day when the sky is filled with heavy thunderclouds. One of the cloud bases may suddenly develop a funnel-shaped, rapidly turning extension that reaches down toward the ground. The tip of the funnel may reach the ground or it may not. If the tip of the funnel does touch the earth, it generally moves in a wandering path at a speed faster than a man can run. Frequently the funnel rises, then touches down again a short distance away. The tornado generally sweeps a path 100 meters or less in width. But within that path there is usually complete destruction of everything on the ground surface. See Figure 22–23.

The destructive power of a tornado is a result of the speed of the winds whirling within the funnel. These winds have never been measured, but their great power indicates speeds as high as several hundred miles per hour. Buildings are also damaged because the pressure within the funnel is very low. The sudden pressure drop often causes buildings to burst open as the air at normal pressure trapped inside expands. Most of the injuries caused by tornadoes occur when people are struck by objects flung around by the winds. It is usually safest to lie down as low as possible with the head covered. However, only a hole in the ground can offer much real protection.

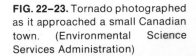

FIG. 22–22. Lightning results from the accumulation of electric charges within a cloud.

FIG. 22–23. Tornado photographed as it approached a small Canadian town. (Environmental Science Services Administration)

How tornadoes develop their tremendous energy is not understood. To form, they seem to require moist warm air at low levels and cool dry air at upper levels. Some sudden lifting action, such as that caused by an advancing cold front, is frequently associated with tornadoes. Electrical charges formed within the funnel may also be a source of energy for these storms.

Tornadoes are most common in the Midwest region of the United States during the late spring or early summer. They have been known to occur in many locations. Tornadoes over the sea are called *waterspouts*.

FIG. 22-24. A waterspout at sea. (Pix—Hoflinger)

VOCABULARY REVIEW

Match the word or words in the column on the right with the correct phrase in the column on the left. *Do not write in this book.*

1. A large body of air of about the same temperature and humidity throughout. e
2. A definite boundary between different air masses. c
3. A long line of heavy thunderstorms which may advance just ahead of a fast moving cold front. a
4. A large body of air with winds tending to blow in circular paths toward a low pressure region at the center. g
5. A front which exists only at upper levels. d.
6. High pressure center. b.
7. A band of very strong winds near the southern boundary of the circumpolar whirl. i
8. Centigrade scale. f
9. An instrument used to measure wind speeds. m
10. Measures temperature, pressure, and humidity at upper levels. k
11. Symbols and numbers grouped around a circle on a map to describe weather conditions. o
12. Some resemblance to the wave cyclone of the middle latitudes but seldom more than 700 km in diameter. q
13. Electrically charged clouds resulting from a small-scale upward movement of warm moist air. p
14. The smallest, most violent, and short lived of all storms. n
15. Tornadoes over the sea. l

a. squall line
b. anticyclone
c. front
d. occluded front
e. air mass
f. Celsius
g. wave cyclone
h. radar
i. jet stream
j. Fahrenheit
k. radiosonde
l. waterspouts
m. anemometer
n. tornadoes
o. station model
p. thunderstorm
q. hurricane

QUESTIONS

Group A

Select the best term to complete the following statements. *Do not write in this book.*

1. The sum of all the properties of the atmosphere at any particular time is called (a) the weather (b) a storm (c) the temperature (d) energy.

2. The main cause of the earth's changing weather patterns is (a) the oceans (b) temperature differences in the atmosphere (c) pressure differences (d) humidity.

3. An air mass is created when a large body of air (a) moves across land (b) moves across the ocean (c) remains stationary over a uniform region for a time (d) remains stationary over land only.

4. Which of the following air masses would most likely be dry and warm? (a) mP (b) mT (c) cP (d) cT.

5. Which of the following air masses would most likely be moist and cool? (a) mP (b) mT (c) cP (d) cT.

6. Which of the following is most likely a polar Pacific air mass? (a) mP (b) mT (c) cP (d) cT.

7. Which of the following is most likely a tropical gulf air mass? (a) mP (b) mT (c) cP (d) cT.

8. The weather at a stationary front is generally the same as that produced by a (a) squall line (b) warm front (c) cold front (d) anticyclone.

9. Wave cyclones result in an area of (a) low pressure (b) anticyclones (c) high pressure (d) clear weather.

10. The twisting motion of the winds toward a low pressure center in the Southern Hemisphere is (a) southward (b) northward (c) clockwise (d) counterclockwise.

11. Which of the following is *not* related to the others? (a) temperature (b) Celsius (c) Fahrenheit (d) millibar.

12. An anemometer would most likely be read in (a) degrees Celsius (b) knots (c) millibars (d) centimeters.

13. Which of the following does *not* belong with the others? (a) temperature (b) pressure (c) precipitation (d) knots.

14. A northwest wind would be from a compass direction of (a) 315° (b) 225° (c) 135° (d) 45°.

15. On a weather map, lines which connect points of equal atmospheric pressure are called (a) isotherms (b) thermoclines (c) isobars (d) pressure barriers.

16. On a weather map, closely spaced isobars mean (a) rapid temperature change (b) higher wind speeds (c) lower wind speeds (d) slow change in pressure.

17. The Coriolis effect tends to make winds blow (a) perpendicular to isobars (b) clockwise around a low pressure region in the Northern Hemisphere (c) clockwise around a high pressure region in the Northern Hemisphere (d) toward low pressure regions.

18. Most cyclonic storms that affect North America move (a) eastward (b) westward (c) northward (d) southward.

19. International weather maps are drawn once every (a) hour (b) 2 hours (c) 6 hours (d) 24 hours.

20. Detailed weather forecasts can be made accurately up to (a) 5 days (b) 2 days (c) 30 days (d) 60 days.

21. How many weather details are shown in a complete station model? (a) exactly 5 (b) less than 5 (c) more than 5 but less than 10 (d) more than 15.

22. Sea breezes are most likely to begin in (a) late morning (b) mid-afternoon (c) late afternoon (d) late evening.

23. Which of the following is likely to occur at about the same time of day and for the same reason as a mountain breeze? (a) valley breeze (b) sea breeze (c) land breeze (d) early morning breeze.

Group B

1. What role do satellites have in weather forecasting?

2. Why is a polar orbit an advantage for a weather satellite?

3. Name the seven source regions for air masses which strongly influence North American weather. Name the type of air mass from each region.

4. What effect does the earth's rotation have on winds?

5. What are the steps in constructing a weather map?

6. Explain the formation of a warm front.

7. Explain the formation of a cold front.

8. Change a wind speed of 16 knots into miles per hour.

9. Change a pressure of 29.75 inches of mercury into millibars.

10. Draw a station model with the following information: (a) wind: NW at 16 knots (b) temperature: 64°F (c) sky: overcast (d) clouds: stratus (e) pressure: 1002.4 millibars.

11. Temperature in degrees Fahrenheit (F) can be changed to degrees Celsius (C) by using the formula $C = \frac{5}{9}(F - 32)$. How many degrees Celsius is normal body temperature (98.6°F)?

12. How many degrees Fahrenheit is −40° Celsius?

13. Change 1015.8 millibars pressure into inches of mercury.

14. What advantage does the Celsius temperature scale have over the Fahrenheit scale?

23
elements of climate

The day-to-day changes in the weather are part of a greater pattern of changes in solar energy, wind, and moisture. This overall pattern is called *climate*. The word climate comes from a Greek word meaning "slope." The ancient Greeks believed that the earth sloped toward the North Pole and that differences in climate depended upon this factor.

Climate is the average weather in a locality over a number of years. It is the result of many influences. Places on the same latitude around the globe may have widely different climates. Local conditions alter the temperature or modify moisture-bearing winds. Thus the surface of the earth is subject to a countless variety of climates. These blend and form every possible environment between the extremes of hot, humid tropics and cold, dry polar regions.

Basic to all climates are the forces that govern the weather. The nature of climate lies in the actions of heat, moisture, and air movements. These all make up the daily conditions of weather. In this chapter the influences of the sun, the winds, and the moisture of the air will be studied in relation to their effect on climate as well as their role in the classification of climates.

CONTROLS OF CLIMATE

Temperature. The total solar energy received by any location on the earth's surface is the major influence affecting its average temperature. At any particular place the amount of solar energy received is determined by two factors: (1) the angle at which the sun's rays strike, and (2) the length of time the sun shines during a day. Both of

objectives

☐ Define the term climate.

☐ Explain how temperature, moisture, and air movement control climate.

☐ Name and briefly describe the earth's three main climatic zones.

☐ Describe several different climates found in each of the three main climatic zones.

☐ List the three major climate controls.

☐ Describe how these climate controls influence North American weather.

479

Table 23–1

Latitude	Longest Night
0°	12 hours
17°	13 hours
41°	15 hours
49°	16 hours
63°	20 hours
66°30′	24 hours
67° 21′	1 month
69° 51′	2 months
78° 11′	4 months
90°	6 months

activity

Using an almanac, list the dates of the year on which your area has 14 or more hours of daylight. Also list the dates of the year on which your area has 10 or less hours of daylight. Obtain from your local weather station or from newspaper files at the library the high and low temperatures recorded for each day. Make a separate graph for each set of data. Use the vertical axis to show temperature and the horizontal axis for the date. Plot the high temperatures in red and the low temperatures in blue. Join the red points with a red line; the blue points with a blue line. Write an explanation of the results shown on your two graphs.

these factors depend on the latitude of the place in question.

Because of the inclination or tilt of the earth's axis, the sun's rays are always nearly perpendicular at the equator. For the same reason the length of day and night near the equator does not change very much throughout the year. Thus the equatorial regions have steady high temperatures all year long. In the higher latitudes, closer to the poles, the sun's rays are more slanted and less effective in heating. Thus temperatures are lower. Winter nights are long and the days short. Table 23–1 shows how the length of the longest night increases as the latitude increases. Very long nights in regions near the poles allow very low temperatures to be reached. This reduces the yearly average temperature in the polar regions to a low level.

It should be remembered that the length of the longest night in winter is the same as the length of the longest daylight period in summer. During the summer the high latitudes have very long daylight periods. The sun's rays then are more direct, giving these regions summer temperature maximums which are a great deal higher than the winter readings.

Latitude is not the only factor which determines the average temperature at a particular place. If it were, all locations at the same latitude would have the same average temperature conditions. When the average temperatures over the earth's surface are examined by means of *isotherms* drawn on a map, we find that this is not the case. An isotherm is a line which connects all locations having the same average temperature. See Figure 23–1. Notice that the isotherms do not follow the parallels of latitude. This indicates that factors other than position relative to the equator and poles influence the average temperature.

The following are some of the other principal influences on the average yearly temperature at a specific place:

1. *Land and water differences.* A quantity of solar energy heats land faster and to a higher temperature than it does a water surface. One reason for this difference is that water is mixed by waves, currents, and other motions. The warm surface water is thus continuously replaced by cooler water from below. This prevents the surface temperature from increasing rapidly.

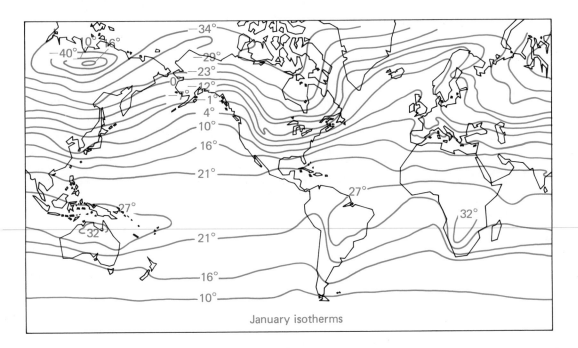

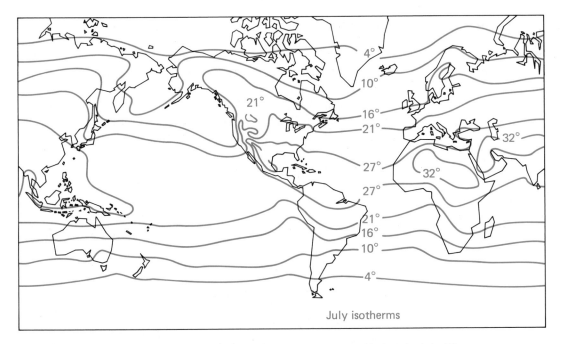

FIG. 23–1. Isotherms of average temperature: above in January: and below, in July. Why are differences much more pronounced in the Northern Hemisphere than in the Southern Hemisphere? Temperatures are in degrees Celsius.

Descending air warms at
the dry adiabatic rate:
1°C per 100 m

FIG. 23-2. Foehn or chinook winds are warm and drying because of the difference between the moist adiabatic and dry adiabatic changes.

Land and water also have different capacities for absorbing and releasing heat. The *specific heat* of water is higher than that of land materials. A quantity of water requires more heat than does the same weight of rock or other land materials to increase its temperature the same number of degrees. Even if it is not in motion, water will warm up more slowly than land. Water also cools off more slowly than land.

There is yet another reason for differing temperatures of land and water at the same latitude. Some of the solar energy received by water is released in the process of evaporation. This heat cannot be used to raise the temperature of the water. The land is not affected in this way because there is less evaporation of water from land surfaces. In addition, some of the heat used in vaporizing water is transferred to the air masses over the land when water vapor condenses to form clouds.

2. *Altitude.* Because the lower layers of the troposphere are warmed as a result of the Greenhouse effect, average temperature decreases with altitude. In mountainous regions average temperatures are always lower than in surrounding lowlands. Even along the equator the tops of some high mountains are cool enough to be covered with snow.

Mountains also affect the temperature of passing air masses. When a moving air mass enters a mountainous region, it is lifted and cooled adiabatically. As the air rises, it loses most of its water vapor by condensation. Then as the air descends on the other side, it is warmed at the dry adiabatic rate of about 1°C per 100 meters. This means that air flowing down mountain slopes is usually warm and dry. See Figure 23-2. Such a wind occurs in the Alps Mountains and is known as a *foehn* (fain). On the eastern slopes of the Rocky Mountains, these warm winds are called *chinook winds*.

3. *Ocean currents.* Nearness to the sea has a strong influence on temperature. This is particularly true if ocean currents passing close to land masses have a different temperature than the land. Such currents have a stronger effect on the coastal air masses as winds consistently blow toward the shore. For example, the average temperature of northwestern Europe is unusually high for its latitude. Here we see the combined effects of a warm current, the North Atlantic Drift, and the steady westerly winds. On the other hand, the warm Gulf Stream has less of an effect along the east coast of the United States. Here winds come more often from the west, blowing away from the coast.

Wind and moisture. Other than average temperature, the climate of a region is determined by its usual winds and precipitation. For the most part, the general direction of the wind in a particular place is determined by its location in one of the global wind belts described in Chapter 20. Storms and local weather disturb and mask this effect. But in most places winds are able to overcome these factors and blow in the direction set by the global wind pattern.

Some locations experience two distinct and directly opposite wind directions during a year. They are caused by differences in heating between large land masses and oceans. During the summer, the land is heated more than the sea. If a low pressure center develops over the land the air will move away from the water toward the land. During the winter, the land receives less heat than the water. Then the wind direction is reversed, as air flows out to the oceans. Such seasonal winds are called *monsoons*. They are strongest over the larger continental land regions near the equator. For example, monsoons occur in southern Asia as a result of the heating and cooling of a large land area in the northern Indian peninsula. See Figure 23–3. Southerly winds in the summer usually bring heavy rainfall as they carry moisture from the sea. In winter the northeast winds bring dry weather. Monsoon conditions also occur in eastern Asia and in other places near the equator where a long stretch of coastline faces toward the equator.

While precipitation may be controlled by factors such as a monsoon, it usually remains under the main influence of the global pattern of air movements. Within the world's belts of average low pressure, the doldrums and the subpolar lows, there is a continuous lifting of air. As a result,

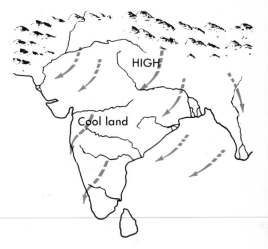

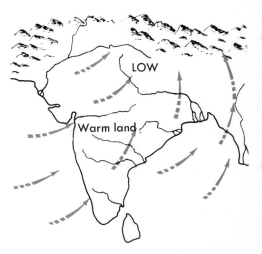

FIG. 23–3. Monsoons are seasonal winds. Why are they more pronounced over large land masses near the equator?

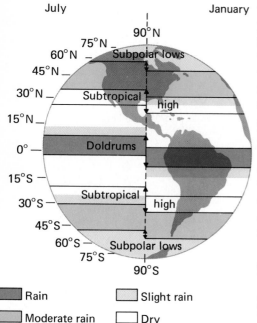

July   January

FIG. 23–4. The general rainfall pattern over the earth at two different seasons. Note that this pattern shifts with seasonal shifts in the planetary wind belts and pressure areas.

the regions of the earth covered by these areas of low pressure generally experience heavy precipitation. The most abundant rainfall of the earth occurs in a belt around the equator. Moving from the equator toward the poles, the amount of rainfall steadily decreases. It reaches a minimum at around 20° to 30° latitude, in the region of the sinking air of the subtropical high. Here there is hardly any rainfall since the air is mostly sinking and there is little condensation.

Closer to the poles, at the subpolar lows around 45° to 55° latitude, there is another belt of high average precipitation. Here warm air meets cold polar air and wave cyclones frequently develop. Above latitudes 50° to 55°, average precipitation decreases in the cold, dry polar air masses.

With the changing seasons, the global wind pattern shifts in a north-south direction. As the wind and pressure belts shift, the belts of precipitation associated with them also change position. The positions of the precipitation belts and their seasonal shifts are shown in Figure 23–4.

CLIMATIC REGIONS

Classification of climates. The warm zone immediately around the equator is the zone of *tropical climates*. To be classed as tropical, a region must have an average temperature of at least 18°C (64°F) during the coldest month of the year. Tropical climates fall under the influence of the continental tropical (cT) and maritime tropical (mT) air masses formed in the source regions close to the equator.

At the other extreme are the *polar climates*. In these regions average temperature never climbs higher than 10°C (50°F) during the warmest month. Continental polar (cP) and maritime polar (mP) air masses are produced in the source regions of these areas.

Between the tropical and polar climate zones is the belt of *middle latitude climates*. In these climates the average temperature of the coldest month is below 18°C (64°F) and the temperature of the warmest month is above 10°C (50°F). Weather is changeable in the middle latitude climates since both the tropical and polar air masses invade this zone. They are also exposed to frequent cyclonic storms which are produced along the polar front. The general boundaries of the three major climate zones are shown in Figure 23–5.

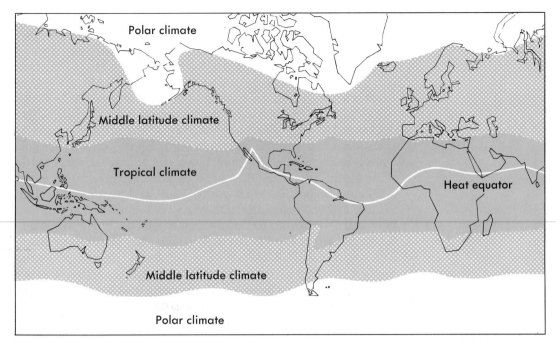

Within each of the principal climate zones there are many types of climates. They are mainly the result of differences in precipitation. On this basis, each of the large climate zones can be divided into smaller parts as follows:

Tropical climates

1. *Tropical rain forest climates.* In a region extending about 5° or 10° on each side of the equator, rainfall is very heavy. Moist maritime tropical air masses are heated and experience the upward movements that bring heavy precipitation. Annual rainfall is usually greater than 250 cm (100 in). Rain is abundant all year and there is little temperature difference between the various seasons of the year. The constant, warm humid weather produces the heavy plant growth called the "rain forest." See Figure 23-6.

Regions of the world where the tropical rain forest type of climate is found include the Congo area of Africa, the Amazon Basin of South America, parts of Central America, Indonesia, and the Malay peninsula. Conditions typical of the tropical rain forest climate may reach far north and south of the equator along the eastern coasts of continents. Mountains along the eastern edge of the land cause the moist trade winds to be lifted. Such wet east coast

FIG. 23–5. The general boundaries of the major zones of climate. The heat equator connects those places with the highest average yearly temperatures.

FIG. 23–6. Rain forest in El Yunque Mt., Puerto Rico. (Shostal)

climates are found in Central America, the West Indies, Southeast Asia, the Philippines, Africa, Brazil, Madagascar and Australia.

2. *Tropical deserts.* In the regions lying fairly close to the edge of the Tropics of Cancer and Capricorn are found the sinking dry air masses of the subtropical high. Land areas within this zone contain some of the world's driest deserts. Usually the deserts are in the western half of continents or on the western side of mountain ranges. These regions are shielded from the moist easterly trade winds and serve as the sources for continental tropical (cT) air masses. Rainfall in the tropical deserts is less than 25 cm (10 in) annually. Desert vegetation is limited to a few hardy plants. In many areas plants are completely lacking, and there is only a bare surface of rocks or sand. See Figure 23–7. The lack of moisture in the air allows the sun's harsh rays to penetrate the ground farther during daylight but also speeds cooling at night. Thus differences between day and night temperatures are likely to be wide on the tropical deserts. Summer maximum temperatures are often very high.

A belt of tropical deserts extends across North Africa, the Near East and southwest Asia. Smaller tropical deserts are also found in southwest Africa, the interior of Australia, northern Mexico, the west coast of Peru and northern Chile, and the extreme southwestern United States. Along the margin of the tropical deserts are regions where the rainfall is between 25 to 50 cm (10 to 20 in). These regions are known as the *steppes.*

3. *Savanna climates.* Between the rain forests and tropical deserts are regions which have a wet and dry climate cycle each year. This cycle effect is a result of the polward shift of the precipitation belts with the seasons. For savanna climates the summer brings weather controlled by the moist (mT) air masses with their heavy precipitation. During this wet season, which becomes shorter as distance from the equator increases, rainfall averages

FIG. 23–7. This sand covered region of the Sahara desert is typical of the very dry tropical deserts. (Photo Researchers – Thomas Hollyman)

about 13 to 25 cm (5 to 10 in) each month. The temperature is warm both day and night. Winter brings the cT air masses with their dry weather, warm days and cool nights. The savanna climate favors the growth of tall grasses with scattered trees and shrubs. See Figure 23–8.

The main areas of savanna climates include African and South American regions bordering the rain forests. Northern Australia also has a large savanna climate. India and parts of southeast Asia have savanna climates produced by the seasonal shift of the monsoon winds.

Polar climates. 1. *Icecap climate.* In the immediate neighborhood of the poles the temperature never rises above freezing. Precipitation is low, less than 25 cm (10 in) per year, since air begins to sink as temperature drops. Almost all precipitation is in the form of snow which remains without melting. The snow covered surface reflects back into space almost all the rays received from the sun during the long summer period of daylight. Icecap climates are found on the continental glaciers of Antarctica and Greenland.

2. *Tundra climates.* In the most northern parts of the continents of North America, Europe and Asia are lands which border the north polar icecaps. These zones have a tundra climate. It is a cold region though the temperatures never fall as low as they do on the icecap. Because of the low temperatures the relative humidity is high with cloudy skies much of the time. However, yearly precipitation, mostly in the form of snow is less than 25 cm (10 in). Plant life on the tundra consists of a carpet of mosses and related plants, along with a few scattered small shrubs. See Figure 23–9. During the summer, the surface of the ground thaws to form a mud. The deeper soil remains permanently frozen. It is the *permafrost* layer. The tundra is easily damaged by any human activity that causes deep melting of the permafrost because the existence of plant cover depends on constantly frozen ground.

3. *Subarctic climates.* Between latitude 50° to 65° in the Northern Hemisphere across North America, Europe and Asia lies a wide belt of land with a cold climate. The winters are very cold and the summer is short. However, the summer may be surprisingly warm due to the very long periods of daylight at these high latitudes. Annual precipitation is between 25 to 50 cm (10 to 20 in). The lands with a subarctic climate are the principal source regions for the cold, dry continental polar (cP) air masses. Plants

FIG. 23–8. Vegetation of the savanna region in Southwestern Brazil in South America (Jacques Jangous)

FIG. 23–9. Tundra of northwest Canada. (Photo Researchers— John Lewis Stage)

FIG. 23–10. The taiga forest in northwest Canada. (E. R. Degginger)

found in subarctic regions make up a kind of thin forest of stunted pines, spruce, fir and smaller cone-bearing trees. This light forest is called *taiga* (*ty*-ga), the Russian word used to describe it in Siberia. See Figure 23–10. Subarctic climates are found in Europe and Asia from Sweden, Norway and Finland across northern Siberia to the Pacific. In North America the subarctic climates reach across Alaska and Canada to Labrador. All of these areas are spotted by many lakes and bogs. These formed in depressions created by the glaciers which formerly covered the northern parts of the continents.

Middle latitude climates. The middle latitudes lie between the warm tropical climates and the cold polar regions. Thus both tropical and polar air masses effect middle latitude climates. The middle latitudes are also located in the belt of westerly winds close to the sources of the cyclonic storms along the polar fronts. Thus the middle latitudes are exposed to many varieties of weather. Large fluctuations in the weather from day to day and between the seasons are the most common feature of these climates. A strong factor in producing the variety of climates in the middle latitudes is large land masses. The continents of North America, Europe and Asia lie mostly in the middle latitudes about halfway between the equator and poles. Since the westerlies are the prevailing winds, the climates of the western edges of these continents fall under their influence. The westerlies bringing in moisture from the sea give the western margins of the continents a moist climate. The westerlies, however, lose much of their moisture as they blow toward the middle of the continents. Thus the interiors of the continents are usually dry. The eastern margins of the continents often have distinct wet and dry seasons produced by monsoon winds. The principal climates found in the middle latitudes are:

1. *Marine west coast climate.* This climate is found along the western margins of continents in the middle latitudes, from about 40° latitude to the edges of the polar regions. It is produced by the flow of cool moist air as the westerlies blow inland from the sea. This constant flow of moist air produces abundant precipitation throughout the year. Average yearly precipitation in marine west coast climates is 50 to 75 cm (20 to 30 in). Unusually heavy precipitation (250 to 400 cm annually) may occur where mountains near the coast block the eastward movement of the moist westerly winds. Most regions with marine west coast

FIG. 23-11. A marine west coast forest in Oregon. Coastal fog is approaching through the valley. (Ramsey)

climates are covered with heavy forests of cone-bearing trees. See Figure 23-11. Areas of the world that have this type of climate are the northwest coast of North America, southern Chile, northwestern Europe and the British Isles, western Australia and New Zealand.

2. *Middle latitude deserts and steppes.* Toward the interior of continents in the middle latitudes, the climates become much drier. Within the interior of the largest land areas, North America and Asia, are the dry middle latitude deserts and their surrounding steppes. Precipitation is very low in these regions, usually less than 25 cm per year (10 in). Unlike the tropical deserts, the middle latitude deserts have a definite winter season which may be cold. Summers range from warm to hot when they serve as sources for hot, dry continental tropical (cT) air masses. As might be expected in regions of so little moisture, plant life is not abundant. It generally consists of only hardy plants and cacti. See Figure 23-12.

Middle latitude deserts occupy large parts of the interiors of North America and Asia. Like the tropical deserts, the middle latitude deserts merge into the less dry steppes. Here the rainfall is sufficient to support a heavy growth of grasses. Middle latitude steppes make up much of the midwestern United States, the pampas region of Argentina, and southeastern Russia.

3. *Mediterranean climates.* The southwest coasts of continents in the middle latitudes lie along the edges of the belt of descending air of the subtropical highs. These regions do not extend very far into the belt of westerly

winds. With the seasonal shift of the subtropical high and westerlies, climate changes. During the summer, the dry air of the subtropical high affects the Mediterranean climates. Rainfall is very low and temperatures are warm. Winter brings the southward shift of the westerlies and their cyclonic storms. The yearly average rainfall of about 25 cm (10 in) falls almost entirely during this time. Winters are usually mild and freezing temperatures are not common. However, plants growing naturally in regions with a Mediterranean climate must be able to survive the long summer dry period. See Figure 23–13. Some of the important land areas having Mediterranean climates are the shores of the Mediterranean Sea, the southern-most part of South Africa, central Chile, central and southern California, and southern Australia.

4. *Humid subtropical climates.* These climates are found in the same general latitudes (around 30°) as the Mediterranean climates. However, the humid subtropical climates are located on the opposite or eastern edges of the continents. Here the climates are controlled in summer by humid, tropical (mT) air masses moving north. They usually bring warm humid weather, often with heavy rains. In winter, continental polar (cP) air masses moving from inland may bring brief but sharp cold waves. Annual precipitation is between 75 to 165 cm (about 30 to 65 in). The land is usually covered with heavy plant growth and forests. The regions with humid subtropical climates include eastern and southern Asia and Japan, northern Argentina, Uruguay, southern Brazil, and the southeastern United States.

5. *Humid continental climates.* From the heart of the continents of North America, Europe and Asia, eastward

FIG. 23–12. Sparse plant growth is typical of desert regions. (Ramsey)

FIG. 23–13. Scattered trees and low shrubs typical of Mediterranean climate vegetation. (Photo Researchers—John Lewis Stage)

to their coasts, climates are controlled by cold, dry, continental polar (cP) and moist, tropical (mT) air masses. One of the main features of these climates is the west-to-east movement of wave cyclones. These storms are formed out of friction between the cold polar and warm tropical air masses. They are responsible for most of the yearly average precipitation of at least 75 cm (about 30 in). In many locations summer thunderstorms bring a sizeable part of the total yearly precipitation. A monsoon wind shift can also be present bringing heavy summer rains. This is particularly true along the east coast of Asia. One outstanding feature of humid continental climates is the wide difference between summer and winter temperatures. Summers are usually warm and humid as tropical air masses move north. But winters are usually very cold as polar air masses move south.

In general, inland regions have the coldest winters because they are not helped by the warming influence of the sea. Inland areas also have less moisture than the coastal areas. Heavy forests of both hardwood and softwood trees are common in the more moist regions. The less humid areas are usually prairies covered with heavy growth of grasses. See Figure 23–14A and 23–14B. Eastern Europe, eastern Asia, and much of the eastern half of North America are the principal regions of humid continental climates. There are no large land areas in the

Interpret

Suppose you lived on a tiny island halfway between the coast of northeastern United States and Spain. The island is rather flat and circular. Describe the climate on your island, using all the information presented so far.

similar latitudes in the Southern Hemisphere to support this type of climate.

NORTH AMERICAN CLIMATES

Continental climate controls. The most important influence on the climate of North America is the position of the continent. It stretches from Central America near the equator to northern Canada and Alaska in the north polar regions. Thus the North American continent is exposed to both tropical heat and polar cold. But the greatest part of the continent, which includes the United States and Canada, lies in the middle latitudes. This means that the largest part of North America is influenced by air masses which come from both tropical and polar regions. Meeting of the warm and cool air masses over much of North America produces the cyclonic storms which furnish most of the precipitation falling on the continent. However, this precipitation is not evenly spread over the entire face of North America.

Since most of its land area is in the belt of westerly winds, air masses and weather disturbances usually move

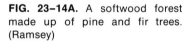

FIG. 23–14A. A softwood forest made up of pine and fir trees. (Ramsey)

FIG. 23–14B. Cattle grazing is quite common in prairie regions. (E. R. Degginger)

across North America from west to east. The west coast receives a large amount of precipitation as moist air moves toward the east from the Pacific. But the air becomes drier as it moves inland across the continent. Precipitation again increases toward the east coast as a result of moist tropical air moving northward. Contact between this warm, moist air and the cool air masses releases much of this moisture, particularly in summer thunderstorms. These precipitation patterns are illustrated in Figure 23–15.

A second important control of North American climate is due to the arrangement of mountains on the continent. Large mountain ranges run in a north-south direction along the western side of North America. Smaller mountain systems are located in the eastern part of the continent. Between them is the great central lowland region.

The arrangement of highlands and lowlands in North America has a strong influence both on precipitation and temperature. Moist air coming from the west is forced to drop much of its moisture as it is lifted over the western highlands. This produces heavy precipitation along much of the west coast. Loss of moisture from the eastward moving air also produces the dry desert and steppe regions covering large areas of the land in the western part of the central lowlands.

The absence of mountains in an eastward direction across the continent allows cold polar air to sweep south

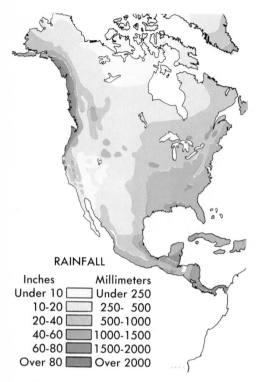

RAINFALL

Inches	Millimeters
Under 10	Under 250
10-20	250- 500
20-40	500-1000
40-60	1000-1500
60-80	1500-2000
Over 80	Over 2000

FIG. 23–15. Pattern of average annual rainfall over North America.

Interpret

Suppose all the mountain ranges in the United States ran east and west. What kind of climates would the United States have.

as far as the Gulf of Mexico. The central lowlands are frequently invaded by these very cold polar air masses moving south in the winter. Warm tropical air moves freely toward the north across the central lowlands. The meeting of these different air masses is responsible for a number of the weather systems which frequently appear in the central part of North America.

A third major influence on the climate of North America comes from the ocean currents along its shores. Naturally, these currents have their main influence on coastal regions. Two principal currents flow along both coasts of North America. Along the west coast, the North Pacific Drift or Alaska Current brings relatively warm water to the coasts of southern Alaska and western Canada. Farther south, the cool California Current influences the west coast of the United States. Along the eastern slope of North America, the northward moving Gulf Stream flows along the southeastern coast of the United States. The Labrador Current carries cool water toward the south along the east coast of Canada and the northeastern shores of the United States as far south as North Carolina. All of these currents tend to make the coastal climates less extreme in temperature than would be true if air masses alone were the chief influence.

Figure 23–16 includes all of the climatic regions of the United States except Hawaii and Alaska. Alaska has predominantly subarctic and tundra climates. Hawaii shows both tropical rain forest and savanna climates. Because of their location in the belt of Trade Winds, the Hawaiian Islands receive heavy precipitation on the sides facing these moist winds. But their mountainous nature also shields many parts of the larger islands from these wind effects. The areas receiving lesser rainfall generally have savanna climates. The rest of the United States has a very representative sampling of the world's climates.

Local climates. A description of the general climate influences on a large land mass such as North America cannot be complete without a discussion of local climate controls. At any location on the continent certain major climate influences are present. But, at the same time, there are minor climate influences at that spot which also help to determine its climate. As a result, there exists within all major climate areas a great number of smaller local climates.

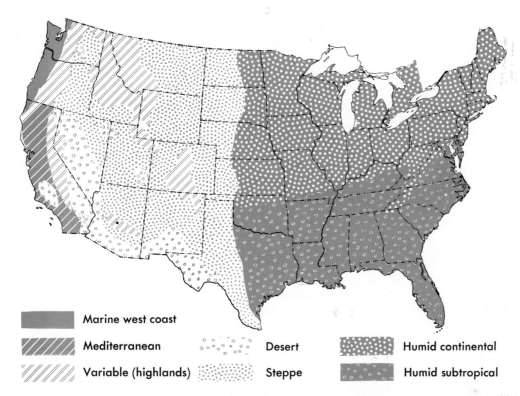

Marine west coast

Mediterranean

Variable (highlands)

Desert

Steppe

Humid continental

Humid subtropical

FIG. 23–16. The major climatic regions of the United States.

The most common example of local influence on climate is altitude. Temperature goes down rapidly with increasing elevation, giving highlands generally lower average temperatures than surrounding lowlands. In addition, the thin air at higher altitudes retains less heat at night. Thus the difference between day and night temperatures is greater. Mountains create local climate effects by their various exposures to winds. Mountain slopes facing the wind receive more precipitation than the opposite slopes or the surrounding lowlands. However, mountains do not create a special type of highland climate. A mountainous region is likely to have a variety of local climates as altitude and direction of slope change. As a result, highlands do not fit into any scheme of general climate classification.

Local climates may also be produced by lakes. They can keep temperatures from rising very high or falling very low, and often cause higher precipitation. This is especially true for places on the shore of a lake away from the common wind direction. For example, places on the eastern shore of Lake Michigan generally have more moderate temperatures and higher precipitation than locations on the western shore.

Interpret

Select one of your local climate controls. Pretend that it would disappear. How would your climate change?

Forests also affect local climates, by reducing wind speed. They also increase humidity as compared with nearby open land. Even cities create their own climates as large buildings reduce wind speed and produce haze and smog. Haze and smog reduce the loss of heat from the ground, resulting in locally higher temperatures.

VOCABULARY REVIEW

Match the word or words in the column on the right with the correct phrase in the column on the left. *Do not write in this book.*

1. Warm dry winds caused by sinking air which is heated adiabatically.

2. The average weather in a particular place.

3. Found along the margin of tropical deserts, having rainfall between 25–50 cm.

4. Average temperature no higher than 10°C during the warmest month.

5. A line connecting locations having the same temperature.

6. A measure of how a substance changes its temperature compared to water.

7. Coldest month average temperature at least 18°C.

8. Seasonal winds.

9. Coldest month average temperatures below 18°C; warmest month average temperatures above 10°C.

10. A thin forest of stunted pines, spruce, fir, and similar cone-bearing trees.

a. steppes
b. polar climates
c. isotherm
d. chinook
e. monsoon
f. taiga
g. climate
h. temperature
i. specific heat
j. middle latitude climates
k. altitude
l. tropical climates

QUESTIONS

Group A

Select the best term to complete the following statements. *Do not write in this book.*

1. The word "climate" comes from a Greek word meaning (a) weather (b) wind (c) slope (d) atmosphere.

2. Probably the most important single influence on climate is (a) the general movement of air over the entire planet (b) altitude (c) longitude (d) local conditions.

3. The amount of solar energy received at a particular location is governed mainly by its (a) latitude (b) longitude (c) altitude (d) distance from the prime meridian.

4. The polar regions of the earth are cold due to very long nights and (a) their altitude (b) the fact that they receive very slanted rays of the sun (c) absence of moisture (d) very long days.

5. Examination of the isotherms of the earth as in Figure 23–1 shows that temperature (a) depends entirely on the latitude (b) depends entirely on the longitude (c) increases to a high near the poles (d) is determined by other factors as well as latitude.

6. Bodies of water change temperature more slowly than land mainly because water (a) reflects most of the heat (b) has a smooth surface (c) has a high specific heat (d) has a low specific heat.

7. Because of the Greenhouse effect, the average temperature of the atmosphere decreases with (a) latitude (b) distance from the equator (c) longitude (d) altitude.

8. Air flowing down mountain slopes, as in chinook winds, is usually (a) cold and dry (b) warm and dry (c) cold and moist (d) warm and moist.

9. The Gulf Stream usually does not increase the temperature along the east coast of the United States because (a) it is nearly the same temperature as the land (b) it flows too far out at sea (c) it flows too fast (d) winds generally blow off shore.

10. In southern Asia the monsoon winds during the winter bring (a) heavy rainfall (b) dry weather (c) changeable weather (d) heavy snowfall.

11. The most abundant rainfall of the earth occurs in (a) the polar regions (b) mountainous regions (c) the middle latitudes (d) the equatorial regions.

12. At the horse latitudes of 20°–30° rainfall is rare because (a) air is sinking and thus warming up (b) air is rising and thus cooling down (c) there are no mountains at these latitudes (d) there are no land masses at these latitudes.

13. The climatic zone that has climates characteristic of changeable weather is the (a) polar (b) middle latitude (c) tropical (d) subtropical.

14. Tropical rain forest climates can occur farther than 10° away from the equator in places such as (a) the Congo (b) the Amazon basin (c) Indonesia (d) Brazil.

15. The great difference in temperature between day and night on the desert is caused by (a) lack of moisture in the air (b) lack of moisture in the ground (c) the high altitude of most deserts (d) the longer days on the desert.

16. Savanna climate is best for the growth of (a) tall pine trees (b) dense forests (c) tall grasses (d) scrub brush.

17. The icecap and tropical desert climates both have low precipitation due to (a) the sinking of air (b) a wide temperature range (c) similar temperatures (d) presence of tall mountains.

18. The polar climate which is found only in the Northern Hemisphere is the (a) icecap (b) tundra (c) subarctic (d) steppe.

19. Middle latitude climates have weather which is best described as (a) consistently warm (b) consistently cool (c) changeable day to day (d) quite dry.

20. Marine west coast climates generally have considerable precipitation caused by (a) northwest trade winds (b) sinking of air masses (c) westerlies blowing inland (d) easterlies blowing to sea.

21. Middle latitude deserts are unlike the tropical deserts since they (a) are dry only part of the year (b) have higher temperatures (c) have less precipitation (d) have a definite winter season.

22. Steppes are found at the edge of middle latitude deserts. They are also found around (a) subarctic climates (b) marine west coast climates (c) tropical deserts (d) tropical rain forests.

23. In the United States the Mediterranean climates would be found in (a) California (b) Texas (c) Louisiana (d) Florida.

24. Most of the eastern half of the United States has a climate which is (a) Mediterranean (b) humid subtropical (c) humid continental (d) steppe.

25. Most of the precipitation falling on the North American continent comes from (a) sinking air masses (b) cyclonic storms (c) prevailing easterlies (d) hurricanes.

26. If the mountains of the North American continent ran east and west, the cold polar air would (a) not sweep so far south (b) be warmed much more readily (c) sweep all the way to South America (d) mix more readily with the warm tropical air.

27. Ocean currents flowing near continental shores change the weather by making the temperature (a) warmer (b) colder (c) less extreme (d) more extreme.

28. The two ocean currents which influence the weather of the west coast of North America are the California Current and the (a) Labrador Current (b) Gulf Stream (c) North Atlantic Drift (d) North Pacific Drift.

29. The most common example of local influence on climate is (a) altitude (b) latitude (c) nearness of large bodies of water (d) presence of forests.

30. The climates of highlands (a) are readily classified into the general climate scheme (b) do not fit into any scheme of climate classification (c) are always cold and dry (d) depend entirely on altitude.

Group B

1. What factors other than latitude determine the main effect of global wind patterns on climate?

2. What is the chief influence effecting the average temperature of a given place?

3. Give three reasons why the surface of water heats or cools more slowly than the surface of land.

4. What causes the temperature to decrease as altitude increases?

5. Why is the average temperature of northwestern Europe unusually high for its latitude?

6. Describe the kind of weather which is typical of monsoons.

7. What effect do the changing seasons have on the global belts of precipitation?

8. Name the three general kinds of climate, the type of air mass associated with each, and the temperature range of each climate.

9. What is the principal cause of the tropical deserts?

10. Soils of the tundra climate regions have an abundant supply of water, yet plant life is restricted to mosses and a few scattered small shrubs. Why is this?

11. Why do the middle latitude climates have such changeable weather?

12. Why does the west coast of the United States receive most of its precipitation in the winter months, while the middle and east coast portions receive most of their precipitation in the summer?

13. Classification of climates often fails when you consider a specific location. Why is this?

14. Prepare a table listing the various middle latitude climates giving (a) general position (b) type of air involved (c) annual precipitation (d) typical plant growth (e) areas having this climate.

15. Erecting any sizeable building such as radar stations, in regions having tundra climate is quite an engineering challenge. Can you suggest why this is so?

weather woes

Weather and Health

Air masses that control much of the earth's weather also influence people. Research indicates that almost every change in the weather has an effect on how we feel and behave. Clouds, winds, heat, cold, humidity, air pressure, and other conditions of the atmosphere have a measurable effect on the human body. These effects show up in body temperature, blood pressure, pulse rate, and in the chemistry of various body functions.

Individuals respond differently to changes in the weather. The effects of a particular kind of weather or climate may be helpful to some persons but harmful to others. Some people are never bothered by sudden changes in temperatures, humidity, or air pressure. Others find that even the slightest change causes a physical reaction. In general, people seem to feel best when the sky is clear, but humidity is not too low, and air pressure is rising. Marked changes in weather produce the greatest response from people who suffer from arthritis, headaches, asthma, heart disease, or mental illness. Much of the way that weather and climate affect people's feelings and behavior is not yet explained.

Sunspots and Climate

In the 1930's, a drought made much of the midwestern United States a dust bowl. More recently in the early 1970's, a drought has caused widespread famine in Africa and Asia. The parched land as shown in Kenya is strewn with carcasses and skulls of drought-starved animals. Could these periods of low rainfall, which are a serious threat to the world's food supply, be caused by sunspots? Records of the pattern of climate changes suggest that there is a connection between the appearance of sunspots and extreme changes in the earth's temperature and rainfall.

It has been well established that sunspot activity occurs in cycles. In each cycle, the number of sunspots reaches a maximum then falls to a minimum. There are three of these cycles: An approximate 11-year cycle, a double cycle of about 22 years in which the sun's magnetic field reverses, and a long cycle of about 100 years. The two shorter cycles represent shorter term changes within the 100-year cycle. The 11-year cycle, which actually ranges from 9 to 14 years, shows little connection with the earth's climate. However, some scientists believe that there is a relationship to the longer 22-year and 100-year cycles. Sunspots might cause a change in the amount of heat produced by the sun. Sunspots might also directly affect the earth's upper atmosphere by producing a shower of atomic particles. As yet, there is no proof that sunspots are responsible. If a connection can be found between sunspot activity and climate changes, scientists may be able to predict droughts and other climatic problems that trouble the earth's population.

Weather 'Casters

A network of weather stations covers the globe, providing valuable information on the movement of large air masses and intensities of storms. Weather in the frozen Arctic is monitored just as it is anywhere else in the world. The meteorologist pictured above works under the most adverse conditions to complete the important link in the weather station network.

A meteorologist, at the right, checks the quality of air in San Diego, California, while the computer keeps an accurate record of air quality readings. Weather information correlated with information from the computer will give her a good indication of future conditions.

unit

7

people
and their
planet

24
AND
earth's resources and energy needs

objectives

- ☐ Define the term "ore."
- ☐ Describe several ways an ore can be formed.
- ☐ List some nonmetallic mineral resources and explain how they are used.
- ☐ Explain why metals like iron and aluminum should be recycled.
- ☐ Describe how the fossil fuels were formed.
- ☐ Describe the methods used to obtain fossil fuels from the earth.
- ☐ Explain how energy is obtained from the nucleus of atoms.
- ☐ Describe several ways geothermal and tidal energy can be used.

The great Gold Rush in California began in 1849 when a few pieces of gold were discovered in a stream. Thousands of people joined the wild rush to find gold. Simple methods such as panning streams were used to recover the precious metal. As gold became harder to find, the miners attempted to trace the gold to its source. The source was the "mother lode" of rocks from which the gold had originally come. These gold-bearing rocks had probably formed hundreds of millions of years before during the Mesozoic era. The gold deposits became part of the Sierra Nevadas, where they were discovered by the "forty-niners." Great fortunes were taken from the mother lode in California. However, after many years of mining, the gold became much harder to find. The Gold Rush was over.

Gold is only one of many resources provided by the earth. A *resource* is a natural material found in the earth that we can use in some way. Some resources such as air, water, soil, plants, animal life, and solar energy are never actually used up. These resources are always being renewed or replaced. Some of the renewable resources, like air and water, must be protected from pollution even though they are never actually used up.

Some resources cannot be replaced or take a very long time to replace. These are nonrenewable resources. Gold is nonrenewable because it takes millions of years to replace in the earth. Other nonrenewable resources are metals, oil and coal. These nonrenewable earth resources are being used up very rapidly. What problems will arise in the future when these resources become even more scarce? How will we solve these problems? This chapter will cover some of the basic information

504

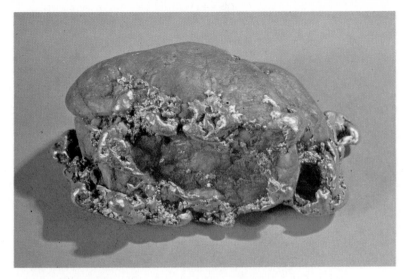

needed to understand and solve these problems.

FIG. 24–1. Gold ore is an example of a non-renewable resource. (Smithsonian Institution)

THE CRUST AS A SOURCE OF MATERIALS

Metals from the earth's crust. A few metals, such as gold, silver, platinum and copper, can also be found in the earth's crust as *native metals*. A native metal is a metal that exists in the pure form. However, most metals exist as compounds or as parts of different metals. These metal-bearing minerals are scattered throughout the crust. To obtain the metals at reasonable cost, we must find places where the metal-bearing minerals are concentrated. An accumulation of minerals containing a substance that can be recovered for profit is called an *ore.* There are several ways that ores are formed within the earth's crust.

Once common type of ore is formed within a body of magma as it cools. As the magma becomes solid, dense crystals form. These crystals sink through the remaining liquid magma. Layers or bodies of certain minerals accumulate within the mass of cooled magma and form deposits of ore. The ores of chromium, nickel, and iron are commonly formed by separation from magma.

The high temperature of magma can also cause ores to be formed when the magma comes into contact with surrounding rock. Hot water carrying dissolved minerals comes from the magma and soaks into the surrounding rock. The original minerals in the surrounding rock are replaced by those carried in the hot magma fluids. This process often forms a band of ore minerals around a body

FIG. 24–2. Large deposits of copper ore are being uncovered in this open-pit copper mine near Salt Lake City, Utah. (Kennecott Copper Co.)

of cooled magma. The ores of lead, copper, and zinc are formed in this way.

Often a body of magma will, in later stages of cooling, produce large amounts of hot solutions containing dissolved minerals. These hot solutions can enter cracks and deposit their minerals in narrow bands called *veins*. Veins often contain valuable minerals like gold and silver. A large number of thick veins filled with mineral deposits is called a *lode*. Sometimes the mineral solution can spread through many small cracks in a large mass of rock. Some of the largest deposits of copper ore were formed this way. See Figure 24–2.

Ores can also be deposited by the action of streams or by waves along the shore. Small fragments of native metals such as gold or platinum are released from their parent rock by weathering. These metal fragments are carried away by streams. Along the stream bed, or sometimes on beaches, the density of the metal causes it to become concentrated in layers of gravel. These layers are called *placer deposits*. Ores of tin or diamonds and other precious gems are sometimes found in placer deposits.

FIG. 24–3. Along stream beds, minerals become concentrated in layers of gravel. Buckets bite into a stream bed in order to obtain the minerals gold and platinum. (Joe Monroe/Photo Researchers)

Water moving downward through the ground can also produce ore deposits. As rainwater moves down from the surface through the cracks in the rocks, it dissolves minerals. These dissolved minerals can accumulate in some places and become ore deposits. Ores of iron, lead, copper, zinc, and aluminum are often formed this way.

Nonmetallic mineral resources. In addition to metals, large amounts of other substances are taken from the earth. Two of the most important of these nonmetallic earth resources are coal and petroleum. These important energy sources will be discussed later in this chapter. The nonmetals taken from the earth's crust include many different substances. Many of these materials are used in building. Some of the building materials include sand, gravel, crushed rock, and blocks of stone. Cement is made from limestone and clay. Gypsum is used to make plaster and wallboard. Almost all the nonmetallic materials used in building come directly from the earth.

Other minerals are used to manufacture the many products we use in everyday life. One of these minerals is sulfur. Huge amounts of sulfur are used in the production of iron and steel, chemicals, fertilizers, rubber, and other things that we use. Some minerals like phosphate rock, are used directly as fertilizers. Without fertilizers like phosphates, it would be impossible to raise enough crops to feed the world's population. Natural rock salt, or halite, is also a valuable mineral. It is used in the manufacture

FIG. 24–4. Sulfur is a non-metallic mineral that is a valuable resource. Large amounts must be mined in order to meet manufacturing needs. (Freeport Sulfur Co.)

of chemicals. The list of minerals used in industry and agriculture is very long. The minerals mentioned earlier provide an example of these important nonmetallic earth resources.

Mineral resources in the future. The natural processes that concentrate minerals into usable ores take millions of years. Once gone, these deposits cannot be replaced for several hundred million years. The rate at which iron and aluminum ore are being used increases every year. If this continues for many years, the known supply of these high-grade ores will soon be exhausted. The current supply of aluminum ore, for example, is estimated to be enough to last for about 30 years. Iron ore is more abundant and will probably last for about 90 years. The amount of ore is different for each metal, but it is believed all known reserves of high-grade ores will probably be gone within the next 100 years. However, this does not mean that the world will run out of metals within the next century.

Metals can be *recycled*. Aluminum used to make cans can be recycled rather than being thrown away. Discarded metals usually end up buried or dumped into the sea. If this happens, they are lost to further use. But if the metals that we use can be recycled, less will have to be taken from the available reserves of ore. The supplies of ore can then last many more years.

As our reserves of high-grade ores are used up, low-

FIG. 24–5. In this photograph, aluminum cans are being collected for recycling. (Alcoa)

grade ores containing a lower concentration of the metals can be used. These low-grade ores can only be used profitably when the prices of metals have risen to a certain point. Recycling metals and turning to less rich ores makes the future of our metal supply seem more encouraging.

But the same encouraging picture does not exist for some other essential substances that come from the earth. These are the materials which, unlike metals, are destroyed by being totally depleted.

ENERGY FROM THE EARTH'S CRUST

Fossil fuels. In many parts of the earth's crust, the remains of living things have become part of the sediments that form rocks. These organic remains consist mostly of carbon and hydrogen, along with some oxygen. They are generally called *hydrocarbons.* Coal, petroleum, and natural gas are made up of hydrocarbon compounds and are known as the *fossil fuels.* The fossil fuels are now the main source of energy for the world.

About 200 to 300 million years ago, during the late Paleozoic era, huge deposits of coal were formed within the ancient super-continent of Pangaea. These coal deposits are now found scattered throughout the continents that separated from Pangaea. Coal was also formed during the late Mesozoic era, about 100 million years ago, when the continents were separating. Today, the largest amounts of coal are found in North America, Asia, and Europe.

Among the fossil fuels, only coal can be considered rock-like. Coal is found where swamps existed millions of years ago. In these ancient swamps, the plant remains accumulated but did not completely decay because there was a lack of oxygen in the swamp water. In time, the plant material formed a soft brown or black material called *peat*. Layers of sediment were deposited on top of the peat. The weight of the sediment squeezed the water out of the peat and changed it into a more dense material called *lignite* or "brown coal." As more sediments accumulated, they caused higher pressure. The lignite then became *bituminous* or "soft coal." Higher pressures were created when rocks folded during mountain building and produced a very hard form of coal called *anthracite*. Millions of years are needed to change peat into bituminous or anthracite coal. These forms of coal contain about 80–90 per cent pure carbon and produce a large amount of heat when they burn.

FIG. 24–6. Soil and rock covering surface deposits of coal are removed by strip-mining. (Camilla Smith/Rainbow)

Coal is also found in layers between beds of shale, limestone, and sandstone. The "seams" of coal range from a few centimeters up to several meters in thickness. Much of the coal is found in seams that are buried deep beneath the surface. Deep mines are needed to reach these deposits and horizontal tunnels are build that follow the coal seams. Where layers of coal are exposed along a mountain slope, the mine can be driven directly into the coal seam. When coal deposits are close to the surface, *strip-mining* is used. In this method, soil and rock lying over the coal seam are removed with heavy machinery. See Figure 24–6. The coal can then be removed. In the past, the soil and rock removed during strip-mining was left in large piles. This destroyed the natural environment of the area. Now, when the strip-mining method is used, the soil and rock are replaced.

The United States has about 250 billion tons of coal that are still to be recovered with ordinary mining methods. In addition, there is another 1200 billion tons that could only be recovered by more expensive methods of mining.

It is difficult to predict how long the supply of coal will last. At the present rate of use, there is enough to supply the energy needs of the world for many centuries. However, the future will certainly place an increased demand on the reserves of coal. It is only possible to predict that coal will probably be the most abundant fossil fuel for the next few centuries.

The situation with the other fossil fuels, petroleum (oil), and natural gas, is very different. Oil and gas are believed to have been formed by processes similar to those that made coal. But petroleum seems to come from the remains of both plants and animals that lived in the seas. As the organic remains accumulated on the sea floor, tiny drops of oil formed and became part of the sediments. As the sediments were slowly changed into rock, the petroleum became trapped. Natural gas is produced along with the petroleum. Both move through porous rocks and accumulate to form an *oil pool.* An oil pool is usually formed where a layer of solid, nonporous rock traps the oil and gas. Two common types of traps are shown in Figure 24–7. Petroleum accumulates only in special locations among the sedimentary rocks that were once ocean basins. This makes it much more difficult to find than coal.

Petroleum is made up of a number of different kinds of hydrocarbon molecules. These molecules range in size from the largest, which make up the tars and asphalts, to

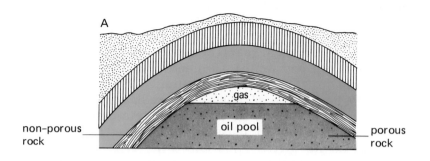

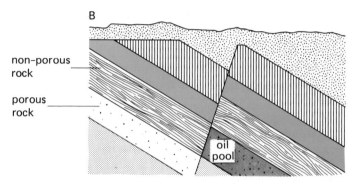

FIG 24–7. Two common types of oil pool traps. In both A and B, oil and gas are trapped in a porous rock area beneath a layer of non-porous rock.

smaller molecules composing oils, gasoline, and kerosene. The smallest hydrocarbon molecule is found in natural gas. When petroleum is refined, these different substances are separated. The heavier molecules are changed by heating to make gasoline and other similar fuels, such as kerosene.

A mixture of hydrocarbons similar to petroleum is also found as a heavy waxy substance in a sedimentary rock called *oil shale*. Large deposits of oil shale are found in the Rocky Mountains and in parts of Canada. When oil shale is crushed and heated, its hydrocarbons are changed into petroleum. The oil shale can be mined and brought to the surface and then heated to separate out the oil. Another mining technique is for the shale to be heated underground and then pump out the oil that is produced.

No one knows how much petroleum and natural gas still remain within the earth's crust. For this reason, it is not possible to predict when the supplies of oil and natural gas will be depleted. However, it is certain that the reserves of these vital fuels are limited. Scientists generally believe that if petroleum and natural gas continue to be used at the present rate, it is likely that they will be used up within the lifetime of today's students.

THE NEED FOR ALTERNATE ENERGY SOURCES

For many years, people thought that the earth's fuel resources would last forever. However, today people are beginning to realize that these resources are quickly diminishing. In the future, the world's energy sources may not be large enough to meet our increasing demands. Where will we turn to for our energy supply? The following alternate sources of energy are being considered today.

Nuclear Energy. At the center of every atom is a tiny *nucleus*. Compared to the size of the whole atom, the nucleus takes up about the same space as a gumdrop in a football stadium. However, all of the atom's protons and neutrons are contained in this small nucleus. The protons and neutrons are held in the nucleus by a very strong force. If something happens to the atom that causes its nucleus to change, some of the nuclear force may be given off as a large amount of energy. An example of such a nuclear change is *fission*. Nuclear fission happens when the nucleus of a large atom splits into two or more smaller nuceli. When the fission or splitting occurs, a large amount of energy is produced. Most of this energy is in the form of heat. Nuclear fission can then be used as an energy source in the same way as fossil fuels.

Only one kind of atom found in nature can be used as a source of nuclear energy. This kind of atom makes up a very rare form of uranium called Uranium 235 or U – 235. To use U – 235 as a nuclear fuel, it is first necessary to extract this scarce metal from deposits of natural uranium in the earth's crust. The U – 235 is then made into rods. These fuel rods are put together in a bundle. When one U – 235 nucleus fissions in these fuel rods, it releases energy along with neutrons. These neutrons strike neighboring U – 235 atoms and cause them to fission. This process is called a *chain reaction*. It continues causing the bundle of fuel rods to become very hot. A liquid, usually water, is pumped around the fuel rods to carry away the heat. The hot liquid or steam is used as an energy source for running electrical generators. This chain reaction can be controlled. Rods are pushed into the nuclear fuel assembly

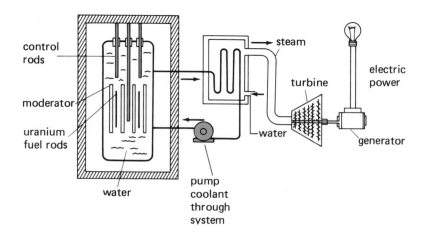

control
rods

moderator

uranium
fuel rods

water

pump
coolant
through
system

steam

turbine

electric
power

water

generator

FIG. 24–8. In a nuclear reactor, the coolant carries heat produced by nuclear fission to a heat exchanger. Here the heat from the reactor converts water into steam. The steam drives a steam turbine and generates electricity.

and absorb the neutrons needed to cause the fissions to continue. The equipment needed to carry out controlled nuclear fission is called a *nuclear reactor*. See Figure 24–8.

One gram of U–235 is a piece about the size of a grain of rice. A piece this size yields an amount of energy equal to the burning of 3 metric tons of coal or about 14 barrels of petroleum. No one knows how much uranium is available in the earth's crust. Very few parts of the world have been thoroughly explored for the presence of this valuable

FIG. 24–9. Photograph of a nuclear power plant in Connecticut. (AEC)

nuclear fuel. Some areas, like the United States, have been thoroughly investigated. These findings seem to show that the amount of uranium reserves will yield much more energy than the petroleum and natural gas that remain. However, the coal reserves of the world are a much larger future source of energy than the natural nuclear fuel. It may be possible to build a certain kind of nuclear reactor that produces more nuclear fuel than it consumes. This is called a *breeder reactor*. If breeder reactors can be successfully built and operated, the amount of nuclear fuel available will be a far greater energy source than any fossil fuel.

At some time in the future, it may also be possible to obtain large amounts of energy from a nuclear process called *fusion*. Nuclear fusion is the joining of the nuclei of atoms to form other nuclei. When this occurs, large amounts of heat energy are released. This heat energy can be used to run electric generators. The sun and all other stars get their energy from nuclear fusion. If in the future fusion reactors were used as a source of energy, the fuel would be hydrogen taken from the water of the oceans. The amount of energy available from this source would be almost limitless.

Geothermal energy. The earth's interior is very hot. This heat represents a huge supply of energy called *geothermal energy*. Some of this energy appears on the earth's surface naturally in the form of volcanoes and hot springs. Earthquakes may also get their energy from this source. In some places on the earth, it is possible to use geothermal energy to supply a part of our energy needs. Wells drilled in a few special locations yield a steady flow of steam or very hot water. See Figure 24–10. The steam can be used directly, or the hot water can be used to make steam to run electric power generators. The city of San Francisco gets part of its electricity from geothermal steam produced by wells in nearby mountains. Other geothermal steam fields now used to produce electricity are located in Italy, Mexico, Japan, Iceland, New Zealand, and the Soviet Union. Many other locations able to supply geothermal energy have been discovered and can be developed as additional sources of energy.

Geothermal energy sources that have been developed up to the present time use steam or hot water that comes from the earth's interior. Another method for making energy would be to drill deep wells into the body of hot rock within the crust. Water pumped down these wells

activity

Heating wood in the absence of air causes it to decompose into a number of useful materials. To identify them, do the following steps:

A. Partially fill a 10 cm test tube with wood splints or dry sawdust. Connect the apparatus as shown below and heat with a burner.

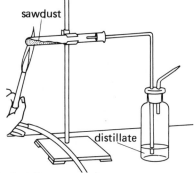

B. Test the gas at the jet with moist neutral litmus paper.
   1. What happens to the litmus paper?
C. The gas being produced is a mixture of CO, CO_2, CH_4, and some very volatile liquids.
   2. Does the gas burn?
D. Continue heating until no more liquid distillate collects in the bottle. Remove the stopper from the bottle and then stop heating.
   3. Describe the color and odor of the liquid distillate.

The distillate usually forms two liquid layers. The top layer is called pyroligenous acid, the bottom is wood tar.
   4. What is the solid residue left in the tube called?
   5. List the products you have made from wood and use reference books to help you describe the uses of each.

FIG. 24–10. Steam wells in Guerneyville, California. (J.R. Eyerman/Uniphoto)

would be changed into steam and returned to the surface. The steam would then be used to operate power stations. However, development of all geothermal energy sources will probably not supply more than a small fraction of the energy requirements for the world.

Tidal power. Harnessing the power of the tides is another possible source of energy. A dam could be built across the mouth of a bay. Water could be allowed to enter the bay at high tide. After high tide the dam would be closed, trapping the water behind it. At low tide, the water could then be allowed to run out through the dam. The moving water would be used to turn water turbines connected to electrical generators. Tidal power plants could only be built at certain locations along the coast. So energy gained from tides would never be a major source of energy.

Geothermal energy and tidal power, along with power provided by the wind, will be small but important sources of energy in the future. In Chapter 2, page 27, solar energy is discussed as a possible energy source that might fill our future needs.

VOCABULARY REVIEW

Match the words in the column on the right with the correct phrase in the column on the left. *Do not write in this book.*

1. A natural material provided by the earth.
2. An accumulation of minerals containing a substance that can be obtained for a profit.
3. A large number of thick mineral veins.
4. Reusing certain metals such as iron and aluminum.
5. Organic remains consisting mostly of carbon and hydrogen.
6. Commonly called brown coal.
7. Known as soft coal.
8. A very hard form of coal.
9. Method of mining coal deposits close to the surface.
10. A sedimentary rock containing a heavy, waxy form of hydrocarbons.
11. Splitting of a large nucleus into two or more smaller nuclei.
12. Process by which energy is released during fission.
13. Produces more nuclear fuel than it uses.
14. Nuclear process by which the sun releases energy.
15. Energy found in the hot interior of the earth.

a. anthracite
b. bituminous
c. breeder reactor
d. chain reaction
e. fission
f. fusion
g. geothermal energy
h. hydrocarbons
i. lignite
j. lode
k. oil shale
l. ore
m. recycle
n. resource
o. strip-mining

QUESTIONS

Group A

Select the best term to complete the following statements. *Do not write in this book.*

1. A nonrenewable resource is (a) air (b) water (c) soil (d) gold.
2. When natural processes concentrate metal-bearing minerals, the deposit is called (a) native metals (b) an ore (c) magma (d) hydrocarbons.
3. Ores of lead, copper, and zinc have been made by (a) original minerals in surrounding rock being replaced by those carried in the hot fluids from (a) magma (b) action of streams or waves along the shore (c) decaying organic remains (d) fusion.
4. The most important nonmetallic earth resources are (a) sand and gravel (b) crushed rock and blocks of stone (c) coal and petroleum (d) limestone and clay.
5. A nonmetallic mineral resource that is used in almost every product available is (a) phosphate rock (b) halite (c) sulfur (d) gypsum.

6. High-grade mineral ores of some metals may be in short supply in the near future since (a) the natural processes that concentrate minerals into useful ores take millions of years (b) once an ore deposit is used up, it will not be replaced for another several million years (c) some metal ores are being used up at a rate that increases every year (d) all of these choices are correct.

7. One way that is presently used to extend the length of time high grade ores will last is (a) rationing (b) recycling (c) limiting exports (d) making more profit on the metal.

8. The fuel not considered a fossil fuel is (a) coal (b) petroleum (c) Uranium-235 (d) natural gas.

9. Fossil fuels are organic remains consisting mostly of (a) carbon and hydrogen (b) carbon and oxygen (c) hydrogen and oxygen (d) carbon and uranium.

10. Of the following, which is the hardest and most pure form of coal? (a) peat (b) anthracite (c) bituminous (d) lignite.

11. Coal is now found as a sedimentary rock (a) only on a few continents (b) only deep within the crust (c) scattered equally throughout the continents (d) in layers between beds of shale, limestone, and sandstone.

12. At the present rate of use, coal could supply the energy needs of the world for (a) several decades (b) several centuries (c) many centuries (d) several years.

13. A method of extracting coal that is close to the surface is (a) strip-mining (b) deep-mining (c) scam-mining (d) pit-mining.

14. Coal and petroleum are limited on earth since (a) they are not renewable resources (b) they are destroyed when they are used (c) earth conditions are such that new coal and petroleum are not being produced in nature at a rapid rate (d) all of these ideas are involved in limiting coal and petroleum.

15. The largest hydrocarbon molecules found in petroleum make up (a) gasoline (b) tars and aspalts (c) natural gas (d) kerosene.

16. Nuclear fission happens when (a) a nucleus of a large atom splits into two or more smaller nuclei (b) an atom is heated (c) two or more nuclei form a larger nucleus (d) the sun gives off energy.

17. During the fission of an atom, the useable energy is given off as (a) light (b) heat (c) electricity (d) radioactivity.

18. Using nuclear fission as an energy resource is presently limited since (a) only one kind of a scarce natural atom can be used (b) it requires a lot of uranium to make energy (c) it is hard to control the reaction (d) plants must be located far from where the energy is needed.

19. The chain reaction in a nuclear generator is controlled by (a) releasing excess steam to the atmosphere (b) limiting the amount of fuel in the rods (c) pushing rods into the fuel bundles to absorb neutrons (d) pumping more liquid around the fuel rods.

20. It is possible that the supply of nuclear energy may be far greater than any fossil fuel if (a) more atoms can be made to fission (b) an artificial nuclear source is found (c) nuclear reactions are sped up (d) breeder reactors can be used.

21. The nuclear reaction that occurs on the sun is (a) fission (b) fusion (c) radioactivity (d) burning.

22. The source of geothermal energy is (a) earthquakes (b) volcanoes (c) the earth's hot interior (d) hot springs.

23. The hot water or steam produced from geothermal energy is generally used to (a) run electric generators (b) heat homes (c) power machines (d) produce fresh water.

24. Only a small fraction of the energy requirements of the world is expected to come from (a) geothermal energy (b) wind (c) tides (d) all these are expected to represent only small fractions of the energy requirements.

Group B

1. Describe three ways that magma is involved in forming ores.
2. What is meant by a placer deposit? List two native metals found in placer deposits.
3. Describe two ways that water is involved in forming ores.
4. List one or more nonmetallic mineral resource used for each of the following:
   a. building materials
   b. production of iron and steel
   c. production of fertilizers
   d. production of chemicals
5. Explain why the picture of the future supply of metals seems to be an encouraging one.
6. Describe the processes involved in the natural formation of coal.
7. Explain how petroleum and natural gas are thought to have been formed.
8. How is petroleum produced from oil shale?
9. Describe what happens in the nuclear fission process.
10. Give a brief description of how Uranium-235 is obtained and processed to produce fuel.
11. Describe how energy is produced and used in a nuclear reactor.
12. Why would the fusion reaction give us an almost limitless energy source?
13. List several natural ways geothermal energy is released at the earth's surface.
14. Most geothermal energy comes from deep wells that contain hot water or steam. Describe another future method for obtaining geothermal energy.
15. Explain how tidal power could be used as a source of energy.

polluting our planet

Dispersal and Photochemical Smog

Pollution is a by-product of a technological society. In keeping our present standard of living, we use energy and the earth's resources. From that use there is waste. A past practice to dispose of all types of waste all over the world was the method of dispersion. In some areas, this is still the case. The theory behind the dispersion method was that if the waste could be distributed over a wide enough area, it would hardly be noticed. But, as years of dispersion passed, there were fewer and fewer places where the pollution could go. Lake Erie was one of the first to rebel. It became so overladen with waste that it was known to actually burst into flames now and then. Then came the advent of "photochemical smog" in Los Angeles—too many cars in one place belching out fumes and the sky turned yellowish brown with deadly gases. The dispersion method wasn't holding up under the strain.

Previous to the Los Angeles air pollution problem, sulfur dioxide was the main culprit causing serious threats to people's health. A 1952 air pollution "episode" in London was attributed as the cause of 4,000 more deaths for a 5-day period than was normal in December. Specifically, sulfur dioxide, a by-product of burning coal, was suggested as the problem. Several other such "episodes" in other parts of the world were also traced to sulfur dioxide as the causative agent. It was natural then to assume that Los Angeles' problem was also sulfur dioxide. But this wasn't the answer. Something different was happening in Los Angeles and it involved a damaging and irritating haze.

Los Angeles' location in a "bowl" didn't allow for dispersal of exhaust fumes. This was the problem. Also, one of Los Angeles' biggest attributes, lots of sunshine, was in this case an enemy. Ultraviolet light from the sun, along with exhaust fumes from cars and industry, resulted in a series of photochemical reactions. The final end-products were entirely new contaminants, some of which haven't even been identified yet.

Above is a picture of what Los Angeles looks like when atmospheric conditions are right for photochemical smog.

Spray Cans and Ozone

Along with hairspray, paint, and other useful materials that come out of spray cans, there is also an invisible gas. This gas is used to propel the product out of the can. One type of gas that has been commonly used is made up of various compounds of carbon, chlorine, and fluorine, called fluorocarbons. These fluorocarbon gases have been used in spray cans because they are odorless, colorless, non-poisonous, and do not react with the contents of the can or anything else. These properties also make fluorocarbons useful for many other purposes. But their ability to remain unchanged when released into the air has made these gases a possible danger.

Eventually the fluorocarbons build up in the atmosphere and finally reach the upper stratosphere. At this high altitude, ultraviolet rays from the sun are powerful enough to break down the fluorocarbon molecules. Chlorine released by this process then attacks the ozone layer found in the stratosphere and changes it into ordinary oxygen. The ozone layer normally acts as a shield protecting the earth's surface from the sun's intense ultraviolet rays. The most dangerous effect of increased ultraviolet radiation would be an increase in skin cancer. There is no evidence that the fluorocarbons so far released into the atmosphere have damaged the ozone layer. However, if this gas continues to accumulate in the atmosphere, it may become an uncontrollable threat to the earth's ozone shield.

Working Against Pollution

Below, an hydrologist monitors the environment by checking water quality. Various tests can be made and conditions observed that give an idea of pollution levels. Mud and soil can be termed "natural" pollutants. Their presence, as in this case, may cause the river to be unuseable for many purposes.

Citizens groups, scout groups, and other community organizations band together to clean up a local eyesore, above. A clean environment has an esthetic appeal as well as promoting better health.

appendix

A. How to Use This Book

B. Geology in the National Parks

C. Key for Identification of Minerals

D. International System of Measurement and Temperature Units

A. How to Use This Book

objectives

☐ Describe the theory of continental drift.

☐ Describe the theory of plate
tectonics.

☐ Identify the main layers of
the earth and state the evidence of their existence.

☐ List and describe three kinds
of crustal plate boundaries.

In 1912 a theory wa:
thought was contrary to
that the continents were
along the earth's surfa
continental drift and w
German scientist. Most
not believe that his the
sible. They did not und
could drift and plow t
rounds them. Others be
tinents showed evidenc
joined together and the
lines of western Africa a
ample, seem to fit toget}

Objectives
A list of objectives appears on the
first page of each chapter. The
objectives state the important
concepts to be learned in that
chapter.

Italic Words or Phrases
Words or phrases associated with
the development of new subject
material in a chapter are printed
in *italics*.

PLATE TECTONICS: A THEORY

The earth's interior. According to the modern theory of
plate tectonics, the earth's crust is made up of a number
of separate rigid plates. The plates are exposed to forces
that cause them to move. This movement creates the familiar landscape of the earth's surface. "Tectonics" refers

Major Topic Headings
Major topic headings introduce
the general theme of that section
of the text.

Boldface
Boldface phrases at the beginning
of each new section point out the
main idea of that section.

substances are the source
this heat remains trapped
adioactive materials. Very
Most of the energy on the
the sun. The lighter rocks
ontain the largest amount
s means that the earth's

Interpret
From the graph in Figure 9—12,
determine the temperature and
pressure at the boundary
between the mantle and the core.

Student Marginal Notes
These notes are intended to extend the learning process beyond
the realm of the text. They are designed to offer the student the opportunity to examine, discover,
construct, experiment, discuss,
observe, and interpret many interesting ideas in earth science.

Activity
A marginal activity appears in every chapter. The purpose of the activity is to further develop a concept or several concepts covered in the chapter.

activity

Most volcanic activity occurs in certain regions of the earth's surface. In this activity you will locate several volcanoes on a map of the world.

Obtain a map of the world that has the crustal plates outlined. Your teacher may provide you with one.

Volcano	Name	Longitude	Latitude
A	Aconcagua	70W	35S
B	Tungurahua	80W	0N
C	Pelee	61W	15N

Tables
Tables present information in a concise way for easy reference.

Table 14–1 Beach Materials

Description	Size (millimeters)
Boulders	More than 200 (over 8 in)
Cobbles	76 to 200 (3 to 8 in)
Gravel	
Coarse	19 to 76
Fine	5 to 19
Sand	
Coarse	2 to 5
Medium	0.4 to 2
Fine	0.07 to 0.4
Silt	less than 0.07

Vocabulary Review
The vocabulary review at the end of each chapter covers key words used in the chapter.

VOCABULARY REVIEW

Match the word or words from the column on the right with the correct phrase in the column on the left. *Do not write in this book.*

1. The super continent as proposed by the continental drift theory.
2. The modern theory stating that the earth's crust is made up of several rigid parts.
3. The thin outer layer of the earth.
4. The layer of the earth between the crust and the core.
5. The Mohorovicic discontinuity.
6. The hard shell of the earth.

a. mid-ocean ridges
b. core
c. mantle
d. batholiths
e. trench
f. Pangaea
g. island arc
h. crust
i. sills

QUESTIONS

Group A

Select the best term to complete the following statements. *Do not write in this book.*

1. The theory of continental drift is supported by the (a) size of continents (b) the equal spacing of mountains (c) shapes of the continents (d) thickness of the mantle.

Group B

1. Briefly describe the theory of continental drift.

2. How is the theory of plate tectonics similar to the theory of continental drift?

appendix

A. How to Use This Book

B. Geology in the National Parks

C. Key for Identification of Minerals

D. International System of Measurement and

glossary

aa. Black lava.
absolute humidity. The weight of water vapor actually contained in a given quantity of air.

index

aa lava, 196
absolute humidity, **441**

aluminum ore, 507, 508
Andes Mountains, 223, 290
Andromeda, **6**
anemometer, **467**
aneroid barometer, 422–

13 Mount McKinley

24 Olympic   9 Glacier   26 Isle Royale

4 Mount Rainier   1 Yellowstone   15 Acadia

5 Crater Lake   6 Wind Cave

19 Grand Teton

12 Lassen Volcanic   Bryce Canyon

3 Yosemite   18   32 Canyonlands   10 Rocky Mountain   22 Shenandoah

25 Kings Canyon   16 Zion   7 Mesa Verde   23 Mammoth Cave

2 Sequoia   14 Grand Canyon   20 Great Smoky Mountains

31 Petrified Forest   8 Platt   17 Hot Springs

21 Carlsbad Caverns

27 Big Bend

28 Everglades

29 Virgin Islands

30 Haleakala

11 Hawaii Volcanoes

B. Geology in the National Parks

America's national parks are natural outdoor geological laboratories. There are no better places to see the results of geological processes. These areas set aside as national parks have been selected because they contain spectacular landscapes. An understanding of the natural forces that created this scenery can add much to the enjoyment of a visit to these special places.

Almost all national parks have special arrangements to help visitors understand and appreciate how the landscape was created. Marked trails and roads that will guide visitors to main features are found in every park. Many of the parks also provide conducted tours, evening campfire programs, nature hikes, and other activities designed to provide information about things to do and see. Each park usually has a visitor center where information can be obtained. The visitor centers frequently have exhibits and displays calling attention to the most interesting features of the park. A visit to one of the national parks also is an opportunity to enjoy outdoor recreational activities such as camping, picnicking, fishing, boating, swimming, and horseback riding.

A visit to any of the parks will be more enjoyable if it is well planned. Information leaflets, maps, and books are available for all of the parks. These can be very helpful in preparing for a trip to a park. The listing of the national parks that follows gives a brief description of the principle geologic features that can be seen. An address is also included that can be used for requests and questions about a particular park. The parks are listed in the order of their age beginning with Yellowstone which was declared a National Park in 1872. The location of each park is shown on the map on page 528. Below the description of each park is a section that gives references in the text to features found in the park. This section also includes text references to the park.

1. Yellowstone

Largest hydrothermal region in the world with about three thousand geysers, hot springs, mud pots, and volcanic vents; evidence of ancient volcanic activity in the form of many kinds of igneous rocks and petrified forests; a miniature version of the Grand Canyon with a waterfall twice the height of Niagara; mountain scenery and wilderness areas with abundant wildlife. (Open May 1–Oct. 31)

Address: Superintendent
Yellowstone National Park
Wyoming 83020

References in text:
geyser, 269, Fig. 12–22; 270, Fig. 12–23
hot springs, 269, Fig. 12–21; 270, Fig. 12–23
Yellowstone Park, 269, Fig. 21–21, Fig. 22–22; 193, Fig. 9–14
waterfalls, 259
volcanic vents, 197

2. **Sequoia** and **Kings Canyon**

(These two parks are side-by-side and can be visited together.)

Includes a part of the fault-block Sierra Nevada Mountains with rugged landscapes capped by Mt. Whitney; U-shaped valleys along with polished and grooved rock surfaces, erratic boulders, and moraines show the effects of glaciers during past ice ages; caverns dissolved out of marble deposits; groves of giant sequoias, largest trees in the world. (Open all year)

Address: Superintendent
        Sequoia and Kings Canyon
        National Parks
        Three Rivers, California
        93271

References in text:
block mountains, 220, Fig. 10–22
U-shaped valley formation, 278, Figs. 13–3A, 13–3B
Sierra Nevadas, 221, Fig. 10–23; 283, Fig. 13–11
glacial deposits, 284–286
erratic boulders, 284, Fig. 13–3B
moraines, 285, Fig. 13–15

3. **Yosemite**

Some of the world's most impressive and easily reached glaciated scenery; Yosemite Valley is probably the most spectacular ice-carved valley that exists with its sheer granite walls and waterfalls; many examples of rock domes formed by exfoliation; giant sequoia trees and mountain wilderness. (Open all year)

Address: Superintendent
        Box 577
        Yosemite National Park
        California 95389

References in text:
Yosemite Valley, 247, Fig. 11–22A
exfoliation domes, 237, Fig. 11–7
waterfalls, 259

4. **Mount Rainier**

Main feature is Mount Rainier which is a composite volcanic cone of great height in the Cascade Mountain Range formed by volcanic activity; active glaciers cover much of the mountain and have created many glacial features; dense forests surround the base of the mountain and contain much wildlife. (Open from May to October depending upon weather)

Address: Superintendent
        Mount Rainier
        National Park
        Longmire, Washington 98397

References in text:
composite volcanic cone (Mount Rainier), 198
glacial features, Chapter 13

5. **Crater Lake**

An unusual deep lake of great beauty lies in the caldera of an ancient volcano; walls of the crater around the lake show a variety of colors; volcanic features are common in the area; located in a region of forests and mountains. (Open all year)

Address: Superintendent
        Crater Lake National Park
        Crater Lake, Oregon 97604

References in text:
Crater Lake, 199, Fig. 9–26

6. **Wind Cave**

Limestone cavern with unusual honeycomb pattern of calcite deposits; name comes from strong air currents that alternately blow in and out of the cave mouth; prairie wildlife including bison. (Open all year)

Address: Superintendent
        Wind Cave National Park
        Hot Springs,
          South Dakota 57747

References in text:
limestone caverns, 271

7. Mesa Verde

Prehistoric Indian cliff dwellings; very large mesa rising above the surrounding countryside. (Open all year)
Address: Superintendent
Mesa Verde National Park
Colorado 81330
References in text:
mesa, 249, Fig. 11-26

8. Platt

Many cold springs, some with water containing unusual minerals; small herd of bison. (Open all year)
Address: Superintendent
Box 539
Platt National Park
Sulphur, Oklahoma 73086
References in text:
springs, 268, Fig. 12-20

9. Glacier

Large and unusually rugged section of the Rocky Mountains showing much of the history of this complex mountain range; many active glaciers, ice fields, and examples of glaciated landscapes; mountain plants and wildlife. (Open approximately June 15-September 15)
Address: Superintendent
Glacier National Park
West Glacier, Montana
59936
References in text:
glacial features, Chapter 13
ice field, 276, Fig. 13-1

10. Rocky Mountain

A section of the Rocky Mountains with lofty snow-capped peaks, glacier-carved valleys, lakes, and meadows; higher elevations with examples of tundra vegetation. (Open all year)
Address: Superintendent
Rocky Mountain
National Park
Estes Park, Colorado 80517
References in text:
glacier carved valley, 283, Fig. 13-11
glacial lakes, 287-288
tundra vegetation, 487, Fig. 23-9

11. Hawaii Volcanoes

Two active volcanoes that often produce flows of lava; many examples of land forms created by lava flows; wilderness areas with unusual plants and animals. (Open all year)
Address: Superintendent
Hawaii Volcanoes
National Park
Hawaii, Hawaii 96718
References in text:
Hawaii volcanoes, 197
lava flows, 196, Fig. 9-20

12. Lassen Volcanic

Location of the most recent volcanic eruption in the continental United States; Lassen Peak is the only active volcano in the continental United States; examples of volcanic features include cinder cones, thermal springs, and lava flows.
(Open approximately June 1-October 31)
Address: Superintendent
Lassen Volcanic
National Park
Mineral, California 96063
References in text:
volcanic block (Mt. Lassen), 197, Fig. 9-23
cinder cone, 198, Fig. 9-24
lava flows, 196, Fig. 9-20
hot spring (thermal), 269, Fig. 12-21

13. Mount McKinley

Highest mountain in North America; large area of the complex Alaska Mountain Range; active glaciers with many examples of glacial features; animals and plants of the subarctic

climate zone. (Open June 1 – September 10)
Address: Superintendent
Mount McKinley
National Park
Alaska 99755
References in text:
glacial features, Chapter 13
subarctic climates, 487 – 488

14. Grand Canyon

Spectacular gorge cut by the Colorado River exposing a giant slice through layers of sedimentary rocks; rock layers form a record of a large part of the earth's history. (South Rim open all year; North Rim open mid-May to mid-October)
Address: Superintendent
Grand Canyon National Park
Grand Canyon, Arizona
86023
References in text:
Grand Canyon, 221, Fig. 10 – 24

15. Acadia

A rugged glaciated region of shoreline whose shape is presently being altered by effects of wave erosion; many examples of submergent shoreline features such as bays, cliffs, sea caves, and a fiord; seashore plants and animals. (Open May through November)
Address: Superintendent
Box 338
Acadia National Park
Bar Harbor, Maine 04609
References in text:
submergent shoreline features, 301 – 302, Fig. 14 – 9A

16. Zion

A plateau that has been carved into steep-sided gorges and sheer cliffs by water and wind erosion; exposed rocks reveal a history going back more than 200 million years. (Open all year)

Address: Superintendent
Zion National Park
Springdale, Utah 84767
References in text:
Zion Canyon, 221, Fig. 10 – 24

17. Hot Springs

Numerous hot springs that have been used in the past to treat ailments; deposits of minerals made by chemical action of water from hot springs. (Open all year)
Address: Superintendent
Box 1219
Hot Springs National Park
Arkansas 71902
References in text:
hot springs, 268 – 269

18. Bryce Canyon

A horseshoe-shaped depression containing a large number of erosion features in the shape of spires, arches, natural bridges, and windows carved from brightly colored rock. (Open all year)
Address: Superintendent
Bryce Canyon National Park
Utah 84717
References in text:
Bryce Canyon, 221, Fig. 10 – 24
spires, arches, natural bridges (photo), 232

19. Grand Teton

A series of rugged peaks formed by erosion of a large fault-block mountain range; mountain landscape with glaciers, hot springs, and glacial features; wilderness areas include large herd of American elk. (Open all year)
Address: Superintendent
Box 67
Grand Teton National Park
Moose, Wyoming 83012
References in text:
fault-block mountains, 220, Fig. 10 – 22

glacial features, Chapter 13
hot springs, 268–269

20. Great Smoky Mountains

A section of the southern end of the very old folded Appalachian Mountain Range; name comes from haze that normally cover the valleys; covered by extensive hardwood forests. (Open all year).

Address: Superintendent
Great Smoky Mountains
National Park
Gatlinburg, Tennessee 37738

References in text:
folded mountains, 209, Figs. 10–3, 10–4
Appalachian Mountains, 224

21. Carlsbad Caverns

Largest and most beautiful limestone caverns yet discovered; many large rooms with examples of stalagmites, stalactites, and similar cave formations. (Open all year)

Address: Superintendent
Box 1598
Carlsbad Caverns
National Park
Carlsbad, New Mexico
88220

References in text:
Carlsbad Caverns, 271
limestone caverns, 271, Fig. 12–26

22. Shenandoah

A region along the crest of the very old Blue Ridge Mountains made of folded and fractured sedimentary rocks; extensive hardwood forest. (Open all year)

Address: Superintendent
Shenandoah National Park
Luray, Virginia 22835

23. Mammoth Cave

Large limestone cavern with many chambers and cave formations made of gypsum and travertine; unusual historical significance. (Open all year)

Address: Superintendent
Mammoth Cave
National Park
Mammoth Cave,
Kentucky 42259

References in text:
limestone caverns, 271, Fig. 12–26
Mammoth Cave, 271

24. Olympic

Located in the highest part of the youthful coastal mountain ranges that border the Pacific coast from California to Canada; contains the largest remaining section of Pacific Northwest rain forest; active glaciers. (Open all year)

Address: Superintendent
Olympic National Park
600 East Park Avenue
Port Angeles,
Washington 98362

References in text:
marine west coast climate, 488–489, Fig. 23–11
active glaciers, 278–284

25. Kings Canyon

See 2. *Sequoia and Kings Canyon National Parks.*

26. Isle Royale

Located on a large island in Lake Superior; no roads or vehicles in the park; must be explored by foot or boat; lava formations; northern wood forest wilderness. (Open late June to Labor Day)

Address: Superintendent
Isle Royale National Park
87 N. Ripley Street
Houghton, Michigan 49931

References in text: *lava formations,* 193–196

27. Big Bend

Combination of desert, mountains, and canyons in a remote area; volcanic mountains; evidence of ancient inland sea with many areas containing fossils; desert plants and animals. (Open all year)
Address: Superintendent
        Big Bend National Park
        Texas 79834

28. Everglades

Area of subtropical wilderness with extensive fresh water and salt water regions; mangrove forests and open prairies; many unusual kinds of wildlife.
(Open all year)

Address: Superintendent
        Box 279
        Everglades National Park
        Homestead, Florida 33030

29. Virgin Islands

Located on St. John Island, includes hills covered with tropical plants and sandy beaches with coral reefs. (Open all year).
Address: Superintendent
        Box 1707
        Virgin Islands National Park
        Charlotte Amalie, St. Thomas
        Virgin Islands 00802
References in text:
  coral reef, 305, Fig. 14–11A

30. Haleakala

One of the world's largest volcanic craters located at the summit of a huge volcanic mountain; rare plants and animals. (Open all year)
Address: Superintendent
        P.O. Box 456
        Haleakala National Park
        Kahului, Maui, Hawaii
        96732

31. Petrified Forest

Largest display of petrified wood known; also contains part of the brightly colored gullies and knobs that make up the Painted Desert badlands; desert plants and animals. (Open all year)
Address: Superintendent
        Petrified Forest
        National Park
        Holbrook, Arizona 86025

32. Canyonlands

Large and spectacular eroded region containing numerous canyons with natural arches, standing rocks, pinnacles, and other features carved from brightly colored rock. (Open all year)

Address: Superintendent
        Canyonlands National Park
        Post Office Building
        Moab, Utah 84532
References in text:
  natural bridges, arches, spires
    (photo), 232

C. Key for Identification of Minerals

The identification of particular mineral specimens is made simpler if a series of tables, or *key,* is used. The key in this book of 11 tables in which a number of the more important minerals are arranged systematically according to their characteristics.

Before attempting to use the key for identification of minerals, the student should thoroughly review the text material in Chapter 8 dealing with the characteristic properties of minerals (pages 160–169). In the key, the first step in arrangement of minerals is a division into two groups: those having metallic luster and those without metallic luster. The metallic luster group is further divided according to color; the non-metallic luster group is divided according to the color of the streak. The mineral species which fall into these groups are then arranged according to hardness.

The procedure in using the tables is a step-by-step determination of the properties of the specimen to be identified. First examine the mineral and determine its luster, color and streak, and hardness. Turn to the general classification list, page 537, and find the number of the table which applies. Then read carefully across the appropriate column of the table, and examine the mineral, making any required tests, to determine its properties as described there.

To illustrate the use of the tables, assume that we have a specimen of galenite. By examination, we see that it has metallic luster, which places it in the left column of the classification list. Next, we see that it is gray, which places it in the last group of the column. Now we determine the hardness and find that it is very soft. Turning to Table 2 as directed in the list, we find a detailed analysis of four minerals in the section labeled *gray, very soft.* In the righthand section of the *Key* column, distinctive features of each mineral are given as a help in distinguishing among the members of a small group. We determine that our sample does not have the low specific gravity of molybdenite and pyrolusite, nor is it sectile, like argentite. Galenite remains, and by elimination, we find that our sample is most likely to be this mineral. A check of the other characteristics listed, including testing for fusibility, helps to make the identification more certain.

Some additional information on the arrangement and terminology used in the tables will be helpful.

Luster. The presence of metallic luster is the first basis of classification in the key. Therefore it is of great importance to determine the luster of a mineral specimen. It is helpful to note that minerals with metallic luster are opaque; that is, they do not allow light to pass through them. However, some dark, non-metallic minerals appear opaque, and so some of these have been included in both the metallic and non-metallic tables in the key. Metallic luster is common to all true metals. In some minerals, the luster is indistinctly metallic, ranging between metallic and nonmetallic: this is called *submetallic.* The luster of minerals composed mostly of nonmetallic elements may be described as follows: *adamantine*, exceedingly brilliant; *vitreous*, glassy; *resinous*, with appearance of resin; *pearly*, with appearance of mother-of-pearl; *greasy or waxy*, with appearance of oiled or waxed surface; *silky*, similar to pearly, found in finely fibrous minerals; *dull*, without bright or shiny surface, earthy. A particularly bright luster may be called *splendent.*

Color, Streak, Hardness and Specific Gravity. The methods and scales used in determining these properties as given in Chapter 11 are referred to in the tables.

Cleavage and Fracture. Cleavage, the breaking of a mineral along the faces of its crystal planes, may be described in geometrical terms. A few of the more common terms used in the table are *cubic*, in the form of a cube; *rhombohedral*, in a form having six faces intersecting at angles other than 90°; *octahedral*, in a form having eight faces; *basal*, cleaving parallel to the base of the crystal; *clinopinacoidal*, in monoclinic crystals, cleaving parallel to the oblique and the vertical axes; *prismatic*, cleaving in forms with faces parallel to the vertical axis.

Most of the terms describing fracture are self-explanatory. Many minerals show *conchoidal* fracture: they break in smooth, curving surfaces resembling the inside of a shell; in *hackly*

fracture, the mineral breaks with thin points that catch the skin as one scrapes a finger across the surface.

Fusibility. The fusibility of a mineral is an important aid to identification. The temperature of fusion is not accurately known except for a few species. The approximate fusing points of different minerals, that is, their relative fusing points, can be determined by comparison. A small fragment of the mineral is heated with the oxidizing flame in the forceps or on a charcoal block. Seven species of minerals have been chosen as a scale of fusibility. Beginning with the most easily melted, they are stibnite—1; natrolite—2; garnet—3; actinolite—4; orthoclase—5; bronzite—6; quartz—7. If the specimen to be identified fuses as readily as stibnite, it has a fusibility of 1; if it cannot be rounded on the thinnest edges, like quartz, it is 7, and so on.

Other Characteristics. The table uses standard terms to indicate the ability of a mineral to transmit light: *opaque*, no light transmitted even through thin edges or layers; *translucent*, some light transmitted but objects appear indistinctly through the substance; *transparent*, objects distinctly seen through the substance. The **tenacity** of a mineral, that is, its behavior when an attempt is made to break, cut, hammer, or similarly alter it, is described as *brittle*, breaks under distortion; *sectile*, can be cut without crumbling; *malleable*, can be hammered into thin sheets; *flexible*, can be bent visibly without breaking; *elastic*, thin layers can be bent and spring back when released. Some brittle minerals are very rigid and strong and break with great difficulty. There are said to be *tough*.

The external forms and internal structures of minerals are described in the tables by a variety of terms. Some types of crystalline or amorphous masses are *acicular*, needlelike crystals; *botryoidal*, closely joined, small rounded masses, like grapes; *fibrous*, very thin crystals or filaments; *foliated*, plates or leaves which are easily separated; *granular*, grains, closely packed; *nodular*, round masses of irregular shape; *pisolitic*, small masses about the size of peas; *scaly*, thin scales or plates; *sheaflike*, crystals resembling a sheaf of wheat. Many similar terms are in general use in describing minerals. Further discussion of this subject may be found in any standard textbook of mineralogy.

The last column of each table in the key includes, as well as the name of the mineral species, its chemical formula and a roman numeral indicating the system in which the mineral crystallizes. The crystal systems are I, Isometric; II, Tetragonal; III, Hexagonal; IV, Orthorhombic; V, Monoclinic; and VI, Triclinic. See page 164. If no system number is given, the mineral is amorphous.

Classification of Minerals

Minerals with metallic luster

	table
Color red or brown	
Very soft	1
Hard	1
Color yellow	
Very soft	1
Soft	1
Hard	1
Color white or silver	
Very soft	1
Hard	1
Color gray	
Very soft	2
Hard	2
Color gray to black	
Very soft	2
Soft	2
Hard	2
Very hard	2

Minerals without metallic luster

	table
Streak reddish	
Very soft	3
Soft	3
Hard	3

Streak brown	
Very soft	3
Soft	3,4
Hard	4
Very Hard	4
Streak yellow	
Very soft	4
Soft	4
Hard	4,5
Very hard	5
Streak blue or green	
Very soft	5
Soft	5
Hard	5
Streak black	
Very soft	5
Soft	5
Hard	5
Streak gray or white	
Very soft	6,7
Soft	6,7
Hard	6
Very hard	6,7
Streak white	
Soft	8
Hard	9
Very Hard	9,10,11
Adamantine	10,11

MINERALS WITH METALLIC LUSTER

TABLE 1		KEY	LUSTER	COLOR	STREAK	HARDNESS, S.G., AND CLEAVAGE	CHARACTERISTICS	SPECIES
COLOR RED OR BROWN	VERY SOFT	STREAK WHITE	Pearly to metallic on cleavage	Light to dark brown	White	2.5 to 3 2.75 Basal, perfect	Infusible. Yields a little water in closed tube. Decomposes in strong sulfuric acid. Transparent to opaque. Brittle. Sectile, elastic. Foliated	MICA Phlogopite complex silicate V
COLOR RED OR BROWN	VERY SOFT	STREAK RED MALLEABLE	Metallic	Copper red, tarnishes black	Copper red	2.5 to 3 8.8 to 8.9 None Fracture hackly	Fuses at 780 C. Gives green solution in nitric acid, then blue in ammonia (DANGER). Malleable. Sometimes cubic; rounded branches	NATIVE COPPER Cu I
COLOR RED OR BROWN	HARD	STREAK DEEP RED	Submetallic to dull	Reddish black	Red	4 to 6 4.3 to 4.7 None Fracture uneven	Infusible; becomes magnetic on charcoal. Yields water in closed tube. Soluble in hydrochloric acid. Opaque. Brittle. Massive, botryoidal	TURGITE $2Fe_2O_3 + H_2O$ III
COLOR RED OR BROWN	HARD	COMPARE S.G.	Metallic	Reddish brown	Reddish brown	4 to 6 4.9 to 5.3 Basal Fracture uneven	Infusible. Yields no water in a closed tube. Sometimes magnetic. Opaque. Brittle. Massive, botryoidal	HEMATITE Red Ocher Fe_2O_3 III
COLOR RED OR BROWN	HARD	S.G. 6.4 TO 9.7	Submetallic to greasy	Gray, green, brown to black	Brownish black to olive green or grayish	5.5 to 6 6.4 to 9.7 None. Fracture uneven or conchoidal	Infusible. Gives coating of lead oxide with soda on charcoal. Soluble in acids. Nitric acid leaves fluorescent spot. Opaque. Brittle. Massive, botryoidal	URANINITE UO_2 I
COLOR YELLOW	VERY SOFT	DOES NOT TARNISH MALLEABLE	Metallic	Rich yellow to silvery yellow	Gold yellow	2.5 to 3 19.3 None	Fuses readily. Soluble only in aqua regia. Does not tarnish. Very malleable. Grains or nuggets	NATIVE GOLD Au I
COLOR YELLOW	SOFT	IRIDESCENT TARNISH	Metallic	Bronze yellow (brass)	Dark greenish black	3.5 to 4 4.1 to 4.3 One poor Fracture uneven	Gives copper on charcoal. Yields sulfur in test tube. Produces green solution with nitric acid. Often tarnished; iridescent	CHALCOPYRITE $CuFeS_2$ II
COLOR YELLOW	SOFT	BROWN TARNISH	Metallic	Bronze yellow	Dark grayish black	3.5 to 4.5 4.4 to 4.7 Basal Fracture uneven	Fuses to black, magnetic mass. Dissolves in hydrochloric acid to form H_2S. Often has dark brown tarnish. Opaque. Massive, granular	PYRRHOTITE $Fe_{1-x}S$ or Fe_5S_6 III
COLOR YELLOW	HARD	BRASS YELLOW COLOR	Metallic	Pale brass yellow	Brownish black to greenish black	6 to 6.5 4.8 to 5.2 Indistinct Fracture uneven	Fuses easily, leaving magnetic residue. Yields sulfur in a closed tube. Opaque. Massive or in cubes	PYRITE FeS_2 I
COLOR YELLOW	HARD	GRAYISH YELLOW COLOR	Metallic	Pale grayish yellow	Dark brownish black	6 to 6.5 4.9 to 5 Indistinct Fracture uneven	Fuses easily, leaving magnetic residue. Brittle. Yields sulfur in a closed tube. Brittle. Massive, fibrous, or crystalline	MARCASITE FeS_2 IV
WHITE OR SILVER	VERY SOFT TO HARD	COMPARE S.G.	Metallic (mirrorlike)	Silvery or steel gray	Red	1 to 6 4.9 to 5.3 None Fracture uneven	Infusible. No cracking when heated. Soluble in hydrochloric acid. Opaque. Brittle	HEMATITE Specular Fe_2O_3 III
WHITE OR SILVER	VERY SOFT TO HARD	COMPARE S.G.	Metallic	Silver white Bright white Black if tarnished	White	2.5 to 3 10 to 11 None Fracture hackly	Fuses readily. Dissolves in nitric acid. Opaque. Malleable. Cubic crystals	NATIVE SILVER Ag I
WHITE OR SILVER	HARD	S.G. 6 OR ABOVE	Metallic	Silver white to steel gray	Dark gray to black	5.5 to 6 6 to 6.4 Prismatic, distinct Fracture uneven	Fuses easily, leaving magnetic residue and arsenic coating. Tarnishes yellowish. Gives garlic smell after hammer blow. Opaque. Massive, wedgelike crystals	ARSENOPYRITE $FeAsS$ IV

MINERALS WITH METALLIC LUSTER

TABLE 2

SPECIES	CHARACTERISTICS	HARDNESS, S.G., AND CLEAVAGE	STREAK	COLOR	LUSTER	KEY
MOLYBDENITE MoS_2 III	Infusible. Yields sulfur fumes in oxidizing flame. Opaque. Sectile. Usually foliated. Hexagonal crystals; greasy, flexible plates	1 to 1.5 4.7 to 4.8 Basal, perfect	Blackish lead gray with bluish tinge	Lead gray	Metallic	SECTILE
PYROLUSITE MnO_2 II	Infusible. Yields chlorine with hydrochloric acid. Gives amethyst bead with borax. Opaque. Brittle. Radiating fibers or massive	2.5 * 4.8 Two cleavages. Prismatic Fracture uneven, splintery	Black	Iron black to steel gray	Metallic	BRITTLE
ARGENTITE Ag_2S I	Fuses to form silver button in oxidizing flame. Reacts to test for sulfur. Yields silver with soda. Tarnishes dull black. Sectile. Cubic crystals	2.5 7.3 Poor cubic and dodecahedral Fracture uneven or subconchoidal	Blackish lead gray	Blackish lead gray	Metallic	SECTILE
GALENITE PbS I	Fuses readily, producing lead and sulfur fumes. Opaque. Brittle. Cubic crystals	2.5 to 2.75 7.4 to 7.6 Cubic, perfect	Dark lead gray to black	Dark lead gray	Metallic (Dull if coated)	NOT SECTILE
MAGNETITE Fe_3O_4 I	Infusible. Soluble in hydrochloric acid. Magnetic. Striated. Granular	5.5 Octahedral parting Fracture conchoidal, uneven	Black	Dark gray to iron black	Metallic	MAGNETIC
GRAPHITE C III	Infusible. Insoluble. Greasy feel. Opaque. Sectile. Hexagonal crystals, fibrous	1 to 2 2 to 2.3 Basal, perfect Flexible scales	Black	Black	Metallic	GREASY
MANGANITE $MnO(OH)$ IV	Infusible. Gives amethyst bead with borax. Yields water in closed tube. Yields chlorine with hydrochloric acid. Translucent. Brittle. Striated prisms	4 4.2 to 4.4 Prismatic and basal poor; perfect side. Fracture uneven	Reddish brown and black	Iron black to steel gray	Submetallic	NOT MAGNETIC / STREAK BROWNISH RED AND BLACK WEAKLY MAGNETIC
ILMENITE Menaccanite $FeTiO_3$ III	Infusible. Pale yellow when held in hot oxidizing flame, turning to colorless or white when cooled. Weakly magnetic. Opaque. Brittle. Granular	5 to 6 4.1 to 5 None Fracture conchoidal	Brownish red and black	Iron black to brownish black	Submetallic to metallic	
CHROMITE $FeCr_2O_4$ I	Infusible. Becomes magnetic when heated. Gives green bead in borax after cooling. May be slightly magnetic. Opaque. Brittle. Massive or octahedral crystals	5.5 4.3 to 4.6 None Fracture uneven	Light grayish brown	Iron black	Submetallic	STREAK LIGHT GRAYISH BROWN SLIGHTLY MAGNETIC
URANINITE Pitchblende UO_2 I	Infusible. Gives coating of lead oxide with soda on charcoal. Soluble in acids. Opaque. Brittle. Massive, fibrous, botryoidal	5.5 9 to 9.5 † None. Fracture uneven or conchoidal	Gray to brown to black	Steel black	Metallic	S.G. 9 TO 9.5 (IF PURE)
RUTILE TiO_2 II	Infusible. Insoluble. Opaque. Brittle. Crystals or massive	6 to 6.5 4.2 Prismatic Fracture uneven	Gray to light yellow brown	Black, reddish brown or red	Metallic to adamantine	S.G. 4.2

* Crystal hardness 6 to 6.5   † If pure

Classification groupings:

- COLOR GRAY
  - VERY SOFT — S.G. BELOW 5
  - VERY SOFT — S.G. ABOVE 7
- COLOR GRAY TO BLACK
  - HARD / VERY SOFT — S.G. 4 TO 5
  - SOFT
  - HARD
  - VERY HARD

MINERALS WITHOUT METALLIC LUSTER

TABLE 3

Group	KEY	LUSTER	COLOR	STREAK	HARDNESS, S.G., AND CLEAVAGE	CHARACTERISTICS	SPECIES
STREAK REDDISH — VERY SOFT	S.G. LESS THAN 3	Dull	Red	Red	1 to 3; 2.0 to 2.6; None; Fracture earthy	Infusible. Insoluble. Colors blue in cobalt nitrate test. Opaque. Brittle. Usually massive like hard clay; pisolitic	BAUXITE $Al_2O_3 + 2H_2O$
STREAK REDDISH — VERY SOFT (S.G. 3 OR MORE)	BRITTLE	Dull	Red	Red	1 to 2.5; 3 to 4; None; Fracture earthy	Infusible. Becomes black and magnetic on charcoal. Yields water in closed tube. Soluble in hydrochloric acid. Opaque. Brittle. Massive or botryoidal	TURGITE Red Ocher $2Fe_2O_3 + H_2O$
STREAK REDDISH — VERY SOFT (S.G. 3 OR MORE)	SECTILE	Resinous	Red and orange	Red to orange	1.5 to 2; 3.5; Basal and clinopinacoidal; Fracture conchoidal	Volatile, combustible—burns with blue flame. Gives arsenic (garlic) odor. Transparent to translucent. Often has yellow spots. Sectile	REALGAR AsS V
STREAK REDDISH — SOFT	COMPARE S.G.	Adamantine to dull	Red to brown	Red to brown	3.5 to 4; 5.8 to 6.1; Octahedral; Fracture uneven	Fuses readily, yielding copper. Colors flame green; blue with hydrochloric acid. Translucent to opaque. Brittle. Often has green or blue spots	CUPRITE Cu_2O I
STREAK REDDISH — SOFT	COMPARE S.G.	Adamantine to dull	Red to brownish red	Scarlet	4; 8 to 8.2; Prismatic, perfect; Fracture uneven	Volatile. Yields sulfur fumes and mercury droplets. Transparent to opaque. Brittle to sectile. Hexagonal crystals, massive, granular, acicular	CINNABAR HgS III
STREAK REDDISH — HARD	STREAK ORANGE, S.G. ABOVE 5	Subadamantine	Red to orange	Orange	4 to 4.5; 5.4 to 5.7; Prismatic. Basal, perfect; Fracture conchoidal	Infusible—turns black. Soluble in acid. Translucent to opaque. Brittle. Pyramidal crystals rare, granular, foliated	ZINCITE ZnO III
STREAK REDDISH — HARD	YIELDS WATER IN CLOSED TUBE	Submetallic to dull	Reddish black	Red	4 to 6; 4.3 to 4.7; None; Fracture uneven	Infusible. Becomes black and magnetic on charcoal. Yields water in closed tube. Soluble in hydrochloric acid. Opaque. Brittle. Massive or botryoidal	TURGITE $2Fe_2O_3 + H_2O$
STREAK REDDISH — HARD	FLUORESCES GREEN	Vitreous-resinous	Yellowish, greenish, brownish	Reddish and brownish	5.5; 3.9 to 4.2; Imperfect; Fracture uneven	Fuses with difficulty. Yields zinc oxide with soda on charcoal. Gelatinizes with HCl. Fluoresces green. Translucent to opaque. Brittle. Massive	WILLEMITE Zn_2SiO_4 III
STREAK BROWN — VERY SOFT	S.G. 2.6	Dull	Yellow	Brown	1; 2.6; Basal; Fracture earthy	Infusible. Yields water in closed tube. Insoluble. Colors blue in cobalt nitrate test. Opaque. Brittle (clay)	KAOLINITE $Al_2Si_2O_5(OH)_4$ IV
STREAK BROWN — VERY SOFT	S.G. 3.6 TO 4	Dull	Brown	Brown	1 to 2.5; 3.6 to 4; None; Fracture earthy	Infusible. Turns black and magnetic on charcoal. Yields water in a closed tube. Opaque	LIMONITE Yellow Ocher $2Fe_2O_3 \cdot 3H_2O$
STREAK BROWN — VERY SOFT	S.G. 2 TO 2.5	Dull	Brown	Brown	1 to 3; 2.0 to 2.5; None; Fracture earthy	Infusible. Insoluble. Colors blue in cobalt nitrate test. Opaque. Brittle. Usually massive like hard clay; pisolitic	BAUXITE $Al_2O_3 + 2H_2O$
STREAK BROWN — SOFT	BECOMES MAGNETIC ON CHARCOAL	Dull to vitreous	Dark brown to gray	Brown	3.5 to 4; 3.8; Rhombohedral, perfect; Fracture conchoidal	Infusible. Turns black and magnetic on charcoal. Effervesces in hot acid. Translucent to near opaque. Brittle. Curved crystals	SIDERITE $FeCO_3$ III
STREAK BROWN — SOFT	REMAINS NONMAGNETIC AFTER HEATING	Resinous	Yellow to brown	Brown	3.5 to 4; 3.9 to 4.1; Perfect. Six cleavages; Fracture uneven	Infusible. Reacts to test for sulfur. Remains nonmagnetic after heating. Effervesces in hot acid, giving H_2S	SPHALERITE ZnS I

MINERALS WITHOUT METALLIC LUSTER

TABLE 4	KEY	LUSTER	COLOR	STREAK	HARDNESS, S.G., AND CLEAVAGE	CHARACTERISTICS	SPECIES
STREAK BROWN — SOFT	COLORS FLAME GREEN, FUSES READILY	Dull to adamantine	Red to brown	Brown	3.5 to 4 5.8 to 6.1 Octahedral Fracture uneven	Fuses readily, yielding copper. Colors flame green; blue with hydrochloric acid. Translucent to opaque. Brittle. Often has green or blue spots	CUPRITE Cu_2O I
STREAK BROWN — VERY HARD	BECOMES MAGNETIC WHEN HEATED	Submetallic	Iron black	Brown	5.5 4.3 to 4.6 None Fracture uneven	Infusible. Becomes magnetic when heated. Gives green bead in borax after cooling. May be slightly magnetic. Opaque. Brittle. Massive or octahedral crystals	CHROMITE $FeCr_2O_4$ I
	NOT MAGNETIC S.G. 4.2	Adamantine	Reddish brown	Light brown	6 to 6.5 4.2 Prismatic Fracture uneven	Infusible. Insoluble. Opaque. Brittle. Crystals or massive	RUTILE TiO_2 II
	NOT MAGNETIC S.G. 6.8	Adamantine to greasy	Brown to black	Light brown	6.5 6.8 Imperfect Fracture uneven	Infusible. Yields metallic tin with soda on charcoal. Not magnetic. Translucent to opaque. Brittle. Prismatic crystals; granular; fibrous	CASSITERITE SnO_2 II
STREAK YELLOW — VERY SOFT	COMPARE S.G.	Dull to vitreous	White to yellow or gray	Yellow	0.5 to 1.5 1.5 Fracture conchoidal	Infusible. Yields water in closed tube. Insoluble in acid. Transparent to opaque. Brittle. Massive, botryoidal. Earthy variety, diatomite	OPAL $SiO_2 + H_2O$ IV
		Dull	Yellow	Yellow	1 to 2.5 2.6 Basal Fracture earthy	Infusible. Yields water in closed tube. Insoluble. Colors blue in cobalt nitrate test. Opaque. Brittle (yellow clay)	KAOLINITE $Al_2Si_2O_5(OH)_4$ IV
		Dull	Yellow to brown	Yellow to brown	1 to 2.5 3.6 to 4 None Fracture earthy	Infusible. Turns black and magnetic on charcoal. Yields water in a closed tube. Opaque	LIMONITE Yellow Ocher $2Fe_2O_3 \cdot 3H_2O$ IV
		Dull	Bright yellow	Yellow	1 to 2 4.1 to 5.0 Basal, perfect Fracture earthy	Infusible. Cold borax bead is fluorescent green. Powder turns red brown in boiling nitric acid and dissolves to green solution. Opaque. Sectile. Scaly powder	CARNOTITE $K_2(UO_2)_2(VO_4)_2 \cdot 3H_2O$ IV
		Resinous to pearly	Yellow often with orange spots	Yellow	1.5 to 2 3.5 Perfect	Reacts to tests for arsenic and sulfur. Translucent. Slightly sectile; flexible plates. Massive, foliated, botryoidal	ORPIMENT As_2S_3 IV
		Resinous	Light yellow to brown	Yellow to white	1.5 to 2.5 2.0 to 2.1 Prismatic, basal, imperfect Fracture conchoidal	Burns with blue flame and odor of sulfur. Translucent. Brittle. Orthorhombic crystals; massive	NATIVE SULFUR S IV
STREAK YELLOW — SOFT	COMPARE S.G.	Dull to vitreous	Gray to dark brown	Gray to yellow and brown	3.5 to 4 3.8 Rhombohedral, perfect Fracture conchoidal	Infusible. Turns black and magnetic on charcoal. Effervesces in hot acid. Translucent to near opaque. Brittle. Curved crystals	SIDERITE $FeCO_3$ III
		Resinous	Yellow to brown	Light yellow to brown	3.5 to 4 3.9 to 4.1 Perfect. Six cleavages	Infusible. Reacts to test for sulfur. Remains nonmagnetic after heating. Effervesces in hot acid, giving H_2S	SPHALERITE ZnS I
STREAK YELLOW — HARD	COLOR BROWN TO BLACK	Dull to adamantine	Brown to black	Yellow	5.0 to 5.5 4 to 4.4 Prismatic, perfect Fracture conchoidal	Infusible. Turns black and magnetic on charcoal. Yields water in a closed tube. Opaque. Dark and light bands. Stalactitic; radiating plates	LIMONITE Goethite $2Fe_2O_3 \cdot 3H_2O$ IV

MINERALS WITHOUT METALLIC LUSTER

TABLE 5			KEY	LUSTER	COLOR	STREAK	HARDNESS, S.G., AND CLEAVAGE	CHARACTERISTICS	SPECIES
STREAK YELLOW	HARD		COLOR OCHER YELLOW	Submetallic	Ocher yellow	Ocher yellow	5 to 6 / None / Fracture conchoidal / 4.1 to 5	Infusible. Pale yellow when held in hot oxidizing flame, turning to colorless or white when cooled. Weakly magnetic. Opaque. Brittle. Granular	ILMENITE Menaccanite $FeTiO_3$ III
	VERY HARD		S.G. 2.6	Nearly dull	Brown to yellow	Light yellow	7 / 2.6 / None / Fracture conchoidal	Infusible. Insoluble. Translucent to opaque. Brittle to tough. Crystals six-sided with striations. Often with inclusions	QUARTZ Jasper $SiO_2 + Fe_2O_3$ III
STREAK BLUE OR GREEN	SOFT		COLOR BLUE	Dull to vitreous	Blue	Pale blue	3.5 to 4 / 3.8 / Perfect. Fracture conchoidal	Fuses readily. Colors flame green. Yields copper. Yields water in closed tube. Effervesces in acid. Translucent to opaque. Brittle, often velvety	AZURITE $Cu_3(OH)_2(CO_3)_2$ V
	VERY SOFT		S.G. 2.6 TO 3	Pearly to dull	White to dark green, black, brown, rose, yellow	Lighter green	1 to 2.5 / 2.6 to 3.0 / Basal, perfect. Fracture scaly, earthy	Fuses with difficulty. Yields water in closed tube. Translucent to opaque. Tough to brittle. Somewhat sectile. Usually finely foliated to massive	CHLORITE $MgFeAlSi_3 \cdot H_2O$ V
	SOFT		S.G. 3.9 TO 4	Adamantine, silky or dull to vitreous	Green	Pale green	3.5 to 4 / 3.9 to 4 / Basal, perfect. Fracture conchoidal	Fuses readily, coloring flame green. Leaves copper on charcoal. Yields water in closed tube. Effervesces in acid. Translucent to opaque; often banded. Brittle	MALACHITE $CuCO_3 + Cu(OH)_2$ V
	HARD		COMPARE S.G.	Vitreous	Greenish black	Grayish green	5 to 6 / 3.2 to 3.6 / Prismatic, perfect; two cleavages, often one parting	Fusible at 2.5–5. Insoluble. Translucent to opaque. Brittle. Massive, granular	AUGITE Pyroxene complex silicate V
STREAK BLACK	HARD			Submetallic to greasy	Gray, green, brown, black	Olive green	5.5 to 6 / 6.4 to 9.7 / None. Fracture uneven or conchoidal	Infusible. Gives coating of lead oxide with soda on charcoal. Soluble in acids. Opaque. Brittle. Massive, fibrous, botryoidal	URANINITE Pitchblende UO_2 I
	VERY SOFT		S.G. 2.6	Dull	Red or reddish brown to black or white	Black	1 to 2.5 / 2.6 / Micaceous. Fracture earthy	Infusible. Yields water in closed tube. Insoluble. Colors blue in cobalt nitrate test. Opaque. Brittle	KAOLINITE $Al_2Si_2O_5(OH)_4$ IV
	SOFT		S.G. 4.2 TO 4.4	Submetallic	Iron black	Black	4 / 4.2 to 4.4 / Prismatic and basal poor, side perfect. Fracture uneven	Infusible. Gives amethyst bead with borax. Yields water in closed tube. Yields chlorine with hydrochloric acid. Translucent. Brittle. Striated prisms	MANGANITE $MnO(OH)$ IV
	HARD			Submetallic to metallic	Iron black to brownish black	Black	5 to 6 / 4.1 to 5 / None. Fracture conchoidal	Infusible. Pale yellow when held in hot oxidizing flame, turning to colorless or white when cooled. Weakly magnetic. Opaque. Brittle. Granular	ILMENITE Menaccanite $FeTiO_3$ III
	HARD		COMPARE S.G.	Submetallic to greasy	Gray, green, brown to black	Brownish black to grayish	5.5 to 6 / 6.4 to 9.7 / None. Fracture uneven or conchoidal	Infusible. Gives coating of lead oxide with soda on charcoal. Soluble in acids. Opaque. Brittle. Massive, fibrous, botryoidal	URANINITE Pitchblende UO_2 I

MINERALS WITHOUT METALLIC LUSTER
STREAK GRAY OR WHITE

TABLE 6	KEY	LUSTER	COLOR	STREAK	HARDNESS, S.G., AND CLEAVAGE	CHARACTERISTICS	SPECIES
VERY SOFT	S.G. 2 TO 2.6 COLOR GRAY	Dull	Gray, red, white, brown	Gray	1 to 3 — 2.0 to 2.6; None; Fracture earthy	Infusible. Insoluble. Colors blue in cobalt nitrate test. Opaque. Brittle. Usually massive like hard clay; pisolitic	BAUXITE $Al_2O_3 + 2H_2O$
HARD	STREAK GRAYISH BROWN	Submetallic	Iron-black	Light grayish brown	5.5 — 4.3 to 4.6; None; Fracture uneven	Infusible. Becomes magnetic when heated. Gives green bead in borax after cooling. May be slightly magnetic. Opaque. Brittle. Octahedral crystals, massive	CHROMITE $FeCr_2O_4$ I
VERY SOFT	COLOR DARK GRAY	Dull	Gray or brown to black	Dark gray	1 — 2.6; Basal; Fracture earthy	Infusible. Yields water in closed tube. Insoluble. Colors blue in cobalt nitrate test. Opaque. Brittle (clay)	KAOLINITE $Al_2Si_2O_5(OH)_4$
HARD TO VERY HARD	COMPARE HARDNESS	Adamantine	Brown to black	Gray	6.5 — 6.8; Imperfect; Fracture uneven	Infusible. Yields metallic tin with soda on charcoal. Not magnetic. Translucent to opaque. Brittle. Prismatic crystals; granular; fibrous	CASSITERITE SnO_2 II
HARD TO VERY HARD	COMPARE HARDNESS	Submetallic to greasy	Gray, green, brown to black	Grayish	5.5 to 6 — 6.4 to 9.7; None; Fracture uneven or conchoidal	Infusible. Gives coating of lead oxide with soda on charcoal. Soluble in acids. Opaque. Brittle. Massive, fibrous, botryoidal	URANINITE UO_2 I
	COMPARE S.G.	Vitreous	Yellow green	Gray	6 to 7 — 3.25 to 3.5; One basal; Fracture uneven	Fuses at 3–3.5 to a magnetic mass. Yields water when strongly heated. Opaque. Brittle. Crystals darker with parallel striations	EPIDOTE $HCa_2(Al,Fe)_3Si_3O_{13}$ V
SOFT	COMPARE S.G.	Vitreous to dull	Gray to dark brown	Gray	3.5 to 4 — 3.8; Perfect, rhombohedral; Fracture conchoidal	Infusible. Turns black and magnetic on charcoal. Effervesces in hot acid. Translucent to near opaque. Brittle. Curved crystals	SIDERITE $FeCO_3$ III
SOFT	S.G. 1.5	Dull	Gray	Gray	.05 to 1.5 — 1.5; Fracture conchoidal	Infusible. Yields water in closed tube. Insoluble in acid. Transparent to opaque. Brittle. Massive, botryoidal. Earthy variety, diatomite	OPAL $SiO_2 + H_2O$
VERY SOFT	STRONG DOUBLE REFRACTION	Vitreous	White when pure	Gray to white	2 to 3 — 2.7; Rhombohedral, perfect; Fracture conchoidal	Infusible. Effervesces in dilute, cold hydrochloric acid. Fluoresces red, pink, yellow. Strong double refraction. Transparent. Brittle	CALCITE Iceland Spar $CaCO_3$ III
VERY SOFT	OFTEN TINTED OR DARKENED	Vitreous to dull or pearly	White when pure	Gray to white	2 to 3 — 2.7; Rhombohedral, perfect; Fracture conchoidal	Infusible. Effervesces in dilute, cold hydrochloric acid. Transparent to nearly opaque. Often tinted or darkened. Brittle	CALCITE $CaCO_3$ III
HARD	CONCHOIDAL FRACTURE	Vitreous	Black to dark gray, red	Gray to white	6 — 2.2 to 2.8; Fracture conchoidal	Fuses at 3.5–4 with intumescence. Insoluble in acids. Brittle, with sharp edges	OBSIDIAN complex silicate
HARD	PRISMATIC CRYSTALS YIELDS WATER IN CLOSED TUBE	Vitreous	Black	Gray to white	5 to 6 — 2 to 3.4; Cleavage angle 56° and 124°; Fracture subconchoidal to uneven	Fusible with difficulty. Yields water in a closed tube. Translucent to opaque. Brittle. Prismatic crystals, fibrous, granular, massive	AMPHIBOLE Hornblende complex silicate V
VERY HARD	S.G. 3.15 TO 4.3	Vitreous to resinous	Red, brown, yellow, green, black, white	Gray to white	6.5 to 7.5 — 3.15 to 4.3; None; Fracture uneven	Fuses at 3–3.5. Gelatinizes with HCl after fusing. Insoluble. Translucent to opaque. Brittle. Crystals 4, 6 or 8-sided, granular, massive	GARNET Pyrope complex silicate I

MINERALS WITHOUT METALLIC LUSTER
STREAK GRAY OR WHITE

TABLE 7

Species	Characteristics	Hardness, S.G., and Cleavage	Streak	Color	Luster	Key	Hardness group
QUARTZ Chert SiO_2 III	Infusible. Insoluble. Flake held in flame breaks up. Translucent. Brittle. Massive, botryoidal, nodular	7 — 2.6 to 2.7 — None — Conchoidal fracture	Gray to white	Gray, brown, black	Waxy to dull	USUALLY MORE BRITTLE WITH IMPURITIES	VERY HARD
QUARTZ Flint SiO_2 III	See above.	7 — 2.6 to 2.7 — None — Conchoidal fracture	Gray to white	Gray, brown, black	Waxy to dull	USUALLY DARKER IN COLOR THAN CHERT	VERY HARD
TOURMALINE complex silicate III	Mostly infusible. Gelatinizes with HCl after fusion. Insoluble. Transparent to opaque. Often zoned with bands. Very brittle. Prismatic crystals, often striated	7 to 7.5 — 2.9 to 3.2 — None — Fracture uneven or conchoidal	Gray to white	Black, red, green, pink	Vitreous	S.G. 2.9 TO 3.2	VERY HARD
TALC $Mg_3(OH)_2Si_4O_{10}$ V	Infusible. Insoluble. Swells, turns white, and gives violet color in cobalt nitrate test. Opaque to transparent. Sectile, plates flexible.	1 to 1.5 — 2.7 to 2.8 — Basal, perfect (micaceous) — Fracture uneven	White	Apple green to white, gray, etc.	Pearly to dull or greasy	COMPARE S.G.	VERY SOFT
CALCITE Chalk $CaCO_3$ III	Infusible. Effervesces in dilute, cold hydrochloric acid. Transparent to nearly opaque. Brittle	0.5 to 1.5 — 2.6 — Fracture earthy	White	White	Dull	COMPARE S.G.	VERY SOFT
GYPSUM $CaSO_4 + H_2O$ V	Fuses at 3. Yields water in closed tube. Phosphorescent. Fluorescent yellow. Transparent to opaque. Brittle, plates flexible. Varied crystal forms, fibrous	1.5 to 2 — 2.3 — Clinopinacoidal, perfect, two — Fracture conchoidal	White	White, gray, brown	Pearly, silky, dull, glassy	COLOR YELLOW	VERY SOFT
NATIVE SULFUR S IV	Burns with blue flame and odor of sulfur. Translucent. Brittle. Orthorhombic crystals, massive	1.5 to 2.5 — 2.0 to 2.1 — Prismatic, basal, imperfect — Conchoidal fracture	White	Yellow	Resinous	COLOR YELLOW	VERY SOFT
BAUXITE $Al_2O_3 + 2H_2O$	Infusible. Insoluble. Opaque. Brittle. Colors blue in cobalt nitrate test. Usually massive like hard clay; pisolitic	1 to 3 — 2.0 to 2.6 — None — Fracture earthy	White	White, gray, red, brown	Dull	COMPARE S.G.	VERY SOFT TO SOFT
MICA Muscovite complex silicate V	Infusible. Insoluble. Gives little water in closed tube. Transparent to translucent. Flexible, elastic plates or sheets	2 to 2.5 — 2.75 to 3 — Basal, perfect	White	White, light yellow, brown, pale green	Vitreous to pearly	COMPARE S.G.	VERY SOFT TO SOFT
CALCITE Iceland Spar $CaCO_3$ III	Infusible. Effervesces in dilute, cold hydrochloric acid. Fluoresces red, pink, yellow. Strong double refraction. Transparent. Brittle	2 to 3 — 2.7 — Rhombohedral, perfect — Fracture conchoidal	White	White when pure	Vitreous	TRANSPARENT SHOWS STRONG DOUBLE REFRACTION	VERY SOFT TO SOFT
CALCITE $CaCO_3$ III	Infusible. Effervesces in dilute, cold hydrochloric acid. Transparent to nearly opaque. Often tinted or darkened. Brittle	2 to 3 — 2.7 — Rhombohedral, perfect — Fracture conchoidal	White	White when pure	Vitreous to dull or pearly	OFTEN TINTED OR DARKENED	VERY SOFT TO SOFT
HALITE NaCl I	Fuses readily. Colors flame deep yellow. Soluble in water. Tastes salty. Transparent to translucent. Brittle	2.5 — 2.1 to 2.6 — Cubic, perfect — Fracture conchoidal	White	White to gray or brown	Vitreous, dull when moist	SOLUBLE IN WATER TASTES SALTY	VERY SOFT TO SOFT
CRYOLITE Na_3AlF_6 V	Fuses very readily, coloring flame yellow. Transparent to translucent. Almost invisible in water. Brittle	2.5 — 3 — Imperfect — Fracture uneven	White	White to brown	Vitreous or greasy	FUSES READILY COLORS FLAME YELLOW	VERY SOFT TO SOFT

MINERALS WITHOUT METALLIC LUSTER
STREAK WHITE   SOFT

TABLE 8	KEY	LUSTER	COLOR	STREAK	HARDNESS, S.G., AND CLEAVAGE	CHARACTERISTICS	SPECIES
FLEXIBLE, ELASTIC SHEETS	DARKER COLOR	Vitreous, pearly, splendent	Dark brown to black	White	2.5 to 3 2.7 to 3.4 — Basal, perfect	Fuses on thin edges. Decomposes in hot, strong sulfuric acid. Transparent to translucent. Flexible, elastic plates or sheets	MICA Biotite complex silicate V
	TRANSPARENT	Pearly to metallic on cleavage surface	Light to dark brown	White	2.5 to 3 2.75 — Basal, perfect	Fuses with difficulty. Decomposes in hot, strong sulfuric acid. Transparent to translucent. Flexible, elastic plates or sheets	MICA Phlogopite complex silicate V
	COLORS FLAME YELLOWISH GREEN INSOLUBLE IN ACIDS	Vitreous	White to bluish or brownish or reddish	White	2.5 to 3.5 4.3 to 4.6 — Basal, prismatic, perfect (diamond shaped cleavage). Fracture uneven	Fuses at 3. Colors flame yellow green. Insoluble in acid. Reacts to sulfur test with soda. May fluoresce orange after heating. Translucent. Brittle	BARITE BaSO₄ IV
	GREENISH COLOR FIBROUS	Silky or greasy	Green to yellow green	White	2.5 to 4 2.2 to 2.6 — None Fracture fibrous	Yields water in closed tube. Decomposed by hydrochloric acid. Yellow variety fluoresces cream yellow. Opaque to translucent. Flexible to sectile	ASBESTOS Chrysotile Mg₃Si₂O₅(OH)₄ V
	COLORS FLAME PURPLE RED	Pearly	Gray green, rose red and violet to white or pale yellow	White	2.5 to 4 2.7 to 3.3 — Basal, perfect (micaceous) Fracture scaly	Fuses and expands. Colors flame purple red. Yields little water in closed tube. Translucent to transparent. Tough	LEPIDOLITE complex silicate V
	REACTS LIKE BORAX UNDER BLOWPIPE HARDNESS 3 S.G. 1.9	Vitreous to dull	Colorless or white upon exposure to air	White	3 1.9 — Basal, perfect. Produces long splinters. Fracture conchoidal	Reacts like borax under blowpipe but with less swelling. Surface often chalky. Transparent to translucent. Brittle	KERNITE Na₂B₄O₇ + 4H₂O V
	COLORS FLAME RED, CRACKLES WHEN HEATED	Vitreous	White to bluish	White	3 to 3.5 — Basal, prismatic, perfect (diamond shaped cleavage). Fracture uneven	Fuses with difficulty. Colors flame red. Cracks when heated. After heating fluoresces bright green. Transparent to translucent. Brittle	CELESTITE SrSO₄ IV
	FUSES, YIELDING METALLIC LEAD, BUBBLES IN ACID	Vitreous to adamantine	White to gray	White	3 to 3.5 6.5 — Prismatic, imperfect Fracture conchoidal	Fuses readily yielding metallic lead with soda on charcoal. Bubbles in acid. Transparent to translucent. Brittle	CERUSSITE PbCO₃ IV
	FALLS TO PIECES WHEN HEATED WITH A BLOWPIPE ON CHARCOAL	Vitreous to dull	White to gray	White	3 to 4 2.9 to 3 — Prismatic, imperfect Fracture conchoidal	Falls to pieces when heated with a blowpipe on charcoal. Bubbles in acid. Transparent to translucent. Brittle. Fibrous	ARAGONITE CaCO₃ IV
	FUSES EASILY YIELDS WATER IN CLOSED TUBE	Vitreous or pearly on cleavage surface	White to yellow and red	White	3.5 to 4 2.1 to 2.2 — Perfect, one Fracture uneven	Fuses at 2.5. Yields water in closed tube. Swells, writhes in wormlike manner when heated. Transparent to translucent. Brittle. Shealike crystals	STILBITE Zeolite complex silicate V
	DOES NOT EFFERVESCE IN COLD, DILUTE ACID	Vitreous to dull or pearly	White to gray	White	3.5 to 4 *2.8 to 2.9 — Rhombohedral, perfect Fracture conchoidal	Infusible. Does not effervesce in cold dilute acid. Not fluorescent. Transparent to translucent. Curved surfaces when broken. Brittle	DOLOMITE MgCa(CO₃)₂ III
	DECREPITATES PHOSPHORESCENT WHEN GENTLY HEATED	Vitreous	White, green, violet, blue, brown, yellow	White	4 3 to 3.25 — Octahedral, perfect	Fuses at 3. Phosphorescent when gently heated. Decrepitates. Transparent to nearly opaque. Brittle	FLUORITE CaF₂ I

* When pure

TABLE 9

MINERALS WITHOUT METALLIC LUSTER
STREAK WHITE   VERY HARD TO HARD

KEY	LUSTER	COLOR	STREAK	HARDNESS, S.G., AND CLEAVAGE	CHARACTERISTICS	SPECIES
TURNS FLAME RED YELLOW (Ca)	Vitreous to almost resinous	Green, brown, yellow, white	White	5 3.2 — Basal, imperfect. Fracture conchoidal	Infusible. Soluble in acid. Turns flame reddish yellow. Fluoresces orange when heated. Opaque. Brittle. Prismatic crystals, granular, massive	APATITE $Ca_5(Cl,F)(PO_4)_3$ III
FUSES IN CANDLE FLAME	Vitreous	White	White	5 to 5.5 2.2 — Prismatic, perfect, two cleavages. Fracture uneven across the prism	Fuses in candle flame. Yields water in a closed tube. May fluoresce orange or blue to greenish white when heated. Transparent to translucent. Brittle	NATROLITE $Na_2Al_2Si_3O_{10}+2H_2O$ IV
FUSES WITH DIFFICULTY CLEAVAGE ANGLES 56° AND 124°	Vitreous	Black	White	5 to 6 2 to 3.4 — Prismatic, 56° and 124° Fracture subconchoidal to uneven	Fuses with difficulty. Yields water in closed tube. Translucent to opaque. Brittle. Prismatic crystals, fibrous, granular, massive	HORNBLENDE Amphibole complex silicate V
S.G. 2.71 TWIN STRIATIONS	Vitreous to pearly	Gray to greenish and reddish	White	5 to 6 2.71 — Basal, perfect. Fracture uneven	Fuses at 3.5. Insoluble. Translucent to near opaque. Often has bluish iridescence. Brittle	LABRADORITE Feldspar $(Na,Ca)Al_2Si_3O_{10}$ VI
YIELDS WATER IN CLOSED TUBE	Vitreous	White, light green to dark green	White	5 to 6 2.9 to 3.4 — Prismatic, perfect. Two cleavages	Fuses to black or white glass. Yields water in closed tube. Insoluble in acid. Transparent to opaque. Massive. One of two jade minerals	NEPHRITE Actinolite, Amphibole complex silicate V
POWDER GELATINIZES WITH HCl	Vitreous, resinous	Yellowish greenish, brownish	White	5.5 3.9 to 4.2 — Imperfect Fracture uneven	Fuses with difficulty. Yields zinc oxide with soda on charcoal. Gelatinizes with hydrochloric acid. Fluoresces green. Translucent to opaque. Brittle. Massive	WILLEMITE $ZnSiO_4$ III
FUSES AT 2.5 AND BLACKENS	Vitreous	Red and brown to gray	White	5.5 to 6.5 3.4 to 3.7 — Prismatic, perfect with two cleavages. Fracture uneven	Fuses at 2.5. Gives amethyst bead with borax. Insoluble. Blackens on exposure. Transparent to opaque. Tough, crystals brittle. Often with brown or black spots	RHODONITE $MnSiO_3$ VI
FUSES WITH INTUMESCENCE, GLASSY	Vitreous	Black to dark gray or red	White	6 2.2 to 2.8 — Fracture conchoidal	Fuses at 3.5 to 4 with intumescence. Insoluble in acids. Brittle, with sharp edges	OBSIDIAN complex silicate
LACKS TWIN STRIATIONS ON CRYSTAL FACES	Vitreous	White, red, green, flesh color	White	6 to 6.5 2.55 — Basal, perfect. Blocky Fracture conchoidal	Fuses at 5. Not affected by acids. Translucent to transparent. Brittle. Often very large crystals	MICROCLINE Feldspar $KAlSi_3O_8$ VII
WHEN PURE, S.G. 2.63, TWIN STRIATIONS	Vitreous to pearly	White to gray, red, green	White	6 to 6.5 * 2.63 — Basal, perfect Fracture conchoidal	Fuses with difficulty (4–5). Insoluble. Transparent to translucent. Brittle. Sometimes in thin, flat crystals	ALBITE Feldspar $Na_2Al_2Si_6O_{16}$ VI
BLOCKY CLEAVAGES TWO AT 90°, ONE GOOD	Vitreous to pearly	White to gray, red, green, etc.	White	6 to 6.5 2.5 to 2.65 — Basal, perfect. Blocky at 90° Fracture conchoidal	Fuses with difficulty. Insoluble. Not fluorescent after blowpiping. Translucent to opaque. Brittle. Short prismatic crystals	ORTHOCLASE Feldspar $KAlSi_3O_8$ V
S.G. 2.76 TWIN STRIATIONS	Vitreous	White, gray, red, brown, green	White	6 to 6.5 * 2.76 — Basal, perfect, two at right angles. Fracture conchoidal	Fuses at 5. Insoluble. Transparent to translucent. Brittle. Rare	ANORTHITE Feldspar $CaAl_2Si_2O_8$ VI
S.G. ABOVE 3 PURPLISH RED FLAME	Vitreous to pearly	White to gray and buff, green	White	6.5 to 7 3.1 to 3.2 — Perfect, prismatic 87° and 93° Splinters. Fracture uneven * When pure	Fuses at 3.5 with red purple flame. Insoluble. Fluoresces orange. Transparent to opaque. Brittle. Long, striated crystals	SPODUMENE Pyroxene $LiAlSi_2O_6$ V

COMPARE S.G.

FUSES WITH DIFFICULTY

MINERALS WITHOUT METALLIC LUSTER
STREAK WHITE   VERY HARD TO ADAMANTINE

TABLE 10 KEY	LUSTER	COLOR	STREAK	HARDNESS, S.G., AND CLEAVAGE	CHARACTERISTICS	SPECIES
FUSES, NOT AFFECTED BY ACIDS	Subvitreous to silky	Green to white; often spotty or patterned	White	6.5 to 7 3.3 to 3.5 Prismatic. Two cleavages Fracture splintery	Fuses readily. Not affected by acids. Colors sodium flame yellow. Tough. Massive, fibrous, granular	JADEITE Jade $NaAlSi_2O_6$
INFUSIBLE GREEN CRYSTALLINE GRAINS	Vitreous	Green	White	6.5 to 7 3.3 to 3.4 One fair, one poor Fracture conchoidal	Infusible. Gelatinizes with hydrochloric acid. May be magnetic. Transparent to translucent. Brittle	OLIVINE Chrysolite $(MgFe)_2SiO_4$ IV
FUSES GELATINIZES WITH HCl AFTER FUSION	Vitreous to resinous	Red, brown, yellow, green, black, white	White	6.5 to 7.5 3.15 to 4.3 None Fracture uneven	Fuses at 3–3.5. Gelatinizes with HCl after fusing. Insoluble. Translucent to opaque. Brittle. Crystals 4, 6, or 8-sided, granular, massive	GARNET Pyrope complex silicate I
S.G. ABOVE 4, SQUARE PRISMS AND PYRAMIDS RARELY IRREGULAR GRAINS	Adamantine	Colorless, brown, gray, blue, violet, reddish, green	Colorless	6.5 to 7.5 4.2 to 4.8 Two cleavages poor Fracture conchoidal to uneven	Infusible. Insoluble. Colors whiten when heated and glow briefly only once. Fluorescent yellow orange. Transparent to opaque. Brittle	ZIRCON SiO_2 II
MICROSCOPIC CRYSTALS ARRANGED IN SLENDER, BANDED, PARALLEL FIBERS	Waxy	White, red, green, blue, brown, gray, black	White	7 2.6 to 2.65 None Fracture conchoidal	Infusible. Flake held in flame breaks up. Dissolves with effervescence with soda on platinum wire. Translucent. Brittle	QUARTZ Chalcedony SiO_2 III
WITH WAVY COLOR BANDS	Waxy or vitreous	Red, brown, white, blue, yellow, gray, green	White	7 2.6 to 2.65 None Fracture conchoidal	See above	QUARTZ Agate SiO_2 III
WITH STRAIGHT COLOR BANDS	Waxy or vitreous	White, gray, brown, red, green	White	7 2.6 to 2.65 None Fracture conchoidal	See above	QUARTZ Onyx SiO_2 III
TRANSPARENT SIX-SIDED CRYSTALS	Vitreous	Colorless	White	7 2.65 to 2.7 None Fracture conchoidal	See above	QUARTZ Rock Crystal SiO_2 III
PURPLE OR VIOLET COLOR	Vitreous	Purple color due to traces of manganese or iron	White	7 2.65 to 2.7 None Fracture conchoidal	See above	QUARTZ Amethyst SiO_2 III

HARDNESS NOT BELOW 7   S.G. NOT ABOVE 2.7

MINERALS WITHOUT METALLIC LUSTER
STREAK WHITE

TABLE 11	KEY	LUSTER	COLOR	STREAK	HARDNESS, S.G., AND CLEAVAGE	CHARACTERISTICS	SPECIES
VERY HARD — S.G. NOT ABOVE 2.7	LIKE GLASS	Vitreous	Gray, black, brownish	White	7 2.65 to 2.7; None; Fracture conchoidal	Infusible. Insoluble. Flake held in flame breaks up. Dissolves with effervescence with soda on platinum wire. Translucent. Brittle	QUARTZ Smoky SiO_2 III
	MORE BRITTLE, OFTEN WITH IMPURITIES (WAXY TO DULL)	Waxy to dull	Gray, black, brown	White	7 2.6 to 2.7; None; Fracture conchoidal	See above	QUARTZ Chert SiO_2 III
	USUALLY DARKER IN COLOR THAN CHERT, COMPACT	Waxy to dull	Gray, black, brown	White	7 2.6 to 2.7; None; Fracture conchoidal	See above	QUARTZ Flint SiO_2 III
	STRIATED PRISMS, PYROELECTRIC, SHOWS DICHROISM, S.G. 2.9-3.2	Vitreous	Black, red, pink, green, brown	White	7 to 7.5 2.9 to 3.2; None; Fracture uneven or conchoidal	Mostly infusible. Insoluble. Gelatinizes with HCl after fusion. Transparent to opaque. Often zoned with bands. Very brittle. Prismatic crystals	TOURMALINE Complex silicate III
	S.G. BELOW 3	Vitreous	Greenish to bluish and yellow green	White	7.5 to 8 2.6 to 2.8; Basal, poor; Fracture conchoidal	Infusible. Does not decrepitate violently. Insoluble. Transparent to translucent. Brittle. Large, long, prismatic crystals, striated	BERYL $Be_3Al_2(SiO_3)_6$ III
HARDNESS ADAMANTINE	CRYSTALS PRISMS, CLEAVAGE BASAL, PERFECT	Vitreous	Yellow, white, blue, red, green	White	8 3.4 to 3.65; Basal, perfect, one cleavage; Fracture conchoidal, uneven	Infusible. Insoluble. Gives blue color in cobalt nitrate test. Transparent to opaque. Brittle	TOPAZ $Al_2SiO_4F_2$ IV
	HARDNESS 9, S.G. ABOUT 4	Vitreous to adamantine	Gray, brown, red, yellow, blue, black, pink	White	9 3.95 to 4.1; Basal, rhombic parting; Fracture conchoidal	Infusible. Insoluble. Powder gives blue color in cobalt nitrate test. Transparent to opaque. Brittle to tough	CORUNDUM Al_2O_3 III
	HARDNESS 10—SCRATCHES CORUNDUM	Adamantine	White, colorless, gray, blue, pink, black	White	10 3.5; Octahedral, perfect; Fracture conchoidal	Infusible. Insoluble. Often fluorescent. Translucent to opaque. Brittle. Crystals, grains, pebbles	DIAMOND C I

D. International System of Measurement and Temperature Units

METRIC UNITS

(UNIT SYMBOLS ARE IN PARENTHESES THE FIRST TIME ONLY.)

Metric System Prefixes

Greater than 1	Less than 1
tera (T) = 1,000,000,000,000	deci (d) = 1/10
giga (G) = 1,000,000,000	centi (c) = 1/100
mega (M) = 1,000,000	milli (m) = 1/1,000
kilo (k) = 1,000	micro (μ) = 1/1,000,000
hecto (h) = 100	nano (n) = 1/1,000,000,000
deka (da) = 10	pico (p) = 1/1,000,000,000,000

Commonly Used Metric Unit	Examples	English System Equivalents
Length: Meter (m)		39.37 inches (in)
	1 *kilo*meter (km) = 1000m	0.621 mile (mi)
	1 *centi*meter (cm) = 1/100m	0.394 in
	1 m = 100cm	
	1 *milli*meter (mm) = 1/1000m	0.0394 in
	1 m = 1000mm	
	1 micron (μ) = 1/1,000,000m (this is one exception to the use of the prefixes)	
Mass (Weight): gram (g)		0.0353 ounce (oz)
	1 *kilo*gram (kg) = 1000g	2.205 pounds (lb)
	1 *milli*gram (mg) = 1/1000g	.0000022 lb
	1 g = 1000mg	
Volume: Liter (L)		1.06 quarts (qt)
	1 *milli*liter (mL) = 1/1000 L	0.00106 qt
	1 L = 1000 mL	

Metric-English Equivalents

1 inch = 2.54 cm
1 foot (ft) = 30.48 cm
1 yard (yd) = 91.44 cm
1 mi = 1609.3 m
1 mi = 1.61 km
1 cm = 0.3937 in
1 cm = 0.0109 yd
1 m = 39.37 in
1 km = .621 mi
1 lb = 453.592 g
1 lb = 0.4536 kg
1 g = 0.002205 lb
1 g = 0.03527 oz
1 oz = 28.35 g
1 kg = 2.2046 lb

Temperature Scales
(IN DEGREES)

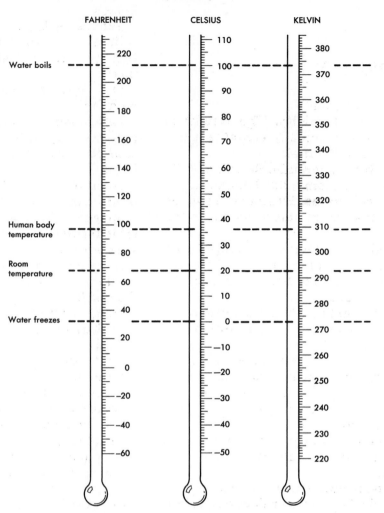

FAHRENHEIT	CELSIUS	KELVIN

Water boils

Human body temperature

Room temperature

Water freezes

Temperature Equivalents / Conversion Formulas

$1°F \cong 0.56°C$ $°F = 9/5°C + 32$

$1°C \cong 1.8°F$ $°C = 5/9(°F - 32)$

$1°K \cong 1°C$ $°K = °C + 273$

glossary

aa. Black lava.

absolute humidity. The weight of water vapor actually contained in a given quantity of air.

absolute magnitude. Brightness of a star viewed from a fixed distance.

abyssal plains. Extremely level regions that cover half of all the ocean floor.

adiabatic change. A temperature change that takes place without the addition or loss of heat to or from the surroundings.

advection fog. A fog produced by the horizontal movement of air over a cool surface.

agonic line. A line connecting locations of zero magnetic declination.

air mass. A large body of air that has about the same temperature and humidity throughout.

albedo. Average reflectivity of a body in space such as the moon or earth.

alluvial fan. A delta-shaped deposit formed when a stream deposits its load upon reaching a level surface.

alpha particle. An electrically charged particle produced by many radioactive substances and consisting of two protons and two neutrons.

alpine (or valley) **glacier.** A glacier formed from ice moving out from a mountain ice field.

altimeter. An instrument used to measure altitude.

amorphous. A non-crystalline solid having no orderly arrangement of molecules.

anemometer. An instrument to measure wind velocity.

aneroid barometer. A type of barometer that operates without the use of liquids.

annular eclipse. A type of partial eclipse of the sun in which a ring of light is seen around the edge of the moon.

anticline. The crest or upturn of a rock fold.

anticyclone. A large mass of high-pressure air, with winds moving out from its center in a clockwise whirl; a high.

aphelion. The point on the earth's orbit farthest from the sun.

apogee. The point in a satellite's orbit farthest from the object around which it revolves.

apparent magnitude. The observed brightness of a star as it appears from the earth.

apparent solar time. Time as determined by the apparent position of the sun.

applications satellites. A group of orbiting instruments constantly studying the earth from space.

aquaculture. The farming of living plants or animals in a body of water.

aquifer. A water-bearing layer of porous rock.

Arctic Circle. An imaginary line at $66\frac{1}{2}°$ north latitude, north of which the sun never rises on the winter solstice and never sets on the summer solstice.

arête. A sharp, narrow ridge separating two cirques or glacial valleys.

artesian well. A well located between two permeable rock layers, in which water rises under its own pressure.

ash. Smallest of the solid materials produced during a volcanic eruption.

asteroid. One of the many tiny planets between the orbits of Mars and Jupiter.

asthenosphere. A softer region of the lithosphere consisting of partially melted rock.

atmosphere. The layer of gases that surrounds an entire planet.

atoll. A roughly circular coral reef that encloses a lagoon.

atom. The smallest subdivision of an element that has all the properties of that element.

aurora. Streamers of glowing ionized gases in the earth's atmosphere.

axes of a crystal. Imaginary lines running from the center of a crystal face to the center of the opposite face.

axis of fold. An imaginary line running along the top of an anticline or along the bottom of a syncline.

bar. Underwater ridge just offshore.

barchan. A crescent-shaped dune.

barograph. A recording barometer.

barometer. An instrument to measure air pressure.

barrier island. Offshore bar that has been built up by wave action to a height above mean sea level.

barrier reef. A long narrow coral reef some distance from the shore.

basement complex. A term used to describe the earliest rock formation on the earth.

batholith. A large mass of intrusive igneous rock; its lower limit is at an unknown depth.

bead test. A color test in which a borax bead formed on a platinum wire loop is used to determine the chemical makeup of minerals.

bedrock. The unweathered solid rock of the earth's crust.

bench mark. Points of accurately known location and elevation which are usually identified by metal marker disks.

berm. Inner raised portion of a beach.

beta particle. A negative radioactive particle consisting of one electron.

blocks. Large chunks of rock thrown out in a volcanic eruption.

blowout. A depression whose origin can be traced to removal of material by wind. Also called *deflation hollow*.

body waves. Earthquake waves that travel through the body of the earth; primary and secondary waves.

breeder reactor. A kind of nuclear reactor that produces more nuclear fuel than it consumes.

butte. A flat-topped, steep-walled hill, usually a remnant of horizontal beds, smaller and narrower than a mesa.

caldera. A basin-like depression formed by the destruction of a volcanic cone.

calorie. Amount of heat required to raise the temperature of 1 gram of water to 1°C at near freezing temperatures.

capillary fringe. A zone in the rocks just above the water table in which water is drawn upward into the spaces by capillary tension.

capillary tension. Tendency for water to be drawn upward in very narrow spaces.

carbonation. The chemical combination of substances with carbon dioxide.

carbon dioxide cycle. A series of processes by which the balance of carbon dioxide and oxygen in the atmosphere is maintained.

carbonization. A process in which remains of living things are partly decomposed, leaving a residue of carbon.

casting. A trail of fossilized sand or mud left by worms.

cavern. An opening in rock formed when the dissolving action of ground water enlarges and connects cracks.

celestial north pole. That point in the sky directly over the earth's North Pole.

Cepheid variable. A variable star whose bright and dim phases are related to its true brightness.

chemical compound. A substance

formed by the combination of two or more different elements in a definite weight relationship.

chemical weathering. Natural chemical processes that alter the mineral composition of rock while reducing it to small fragments.

chromosphere. The sun's surface atmosphere, made up of glowing gases.

cinder cone. Cone formed by the explosive type of volcanic eruption; it has a narrow base and steep, symmetrical slopes of interlocking, angular cinders.

circumpolar whirl. High-altitude winds moving around the earth from west to east.

cirque. A bowl-shaped depression in a mountainside, formed at the head of a valley glacier.

cirrus clouds. A family of high-altitude clouds.

clay. Fine grains of silicate minerals also containing aluminum and water.

cleavage. The tendency of a mineral to break so that it yields definite flat surfaces.

climate. Average weather over a number of years at a particular location.

cloud seeding. Methods of artificially producing precipitation from clouds.

coalescence. Process by which small water droplets suspended in the air combine and grow into larger droplets.

cold front. A front at which a colder air mass thrusts up under a warmer air mass.

column. A structure formed when a stalactite and a stalagmite meet.

comet. A member of the solar system; composed of rocks and frozen gases, usually moving in a very eccentric orbit.

composite cone. Cone formed by intermediate type of volcanic eruption; consists of alternating layers of cinders and lava.

conchoidal. Shell-like surface produced by the fracture of certain minerals.

concretion. Rounded accumulation of mineral matter, often resembling petrified eggs.

condensation nucleus. The center around which a cloud droplet is formed.

conduction. Transfer of heat by direct contact.

conformity. Boundary between two rock layers deposited in succession.

constellation. An apparent grouping of stars in a recognizable pattern.

continental drift. The theory that the continents were not stationary but slowly drifted along the earth's surface.

continental glacier. A glacier that covers an entire land surface with a continuous sheet of ice. It is found only in the polar regions.

continental rise. A bulging accumulation of sediments at the base of the continental slope.

continental shelf. Relatively shallow ocean floor bordering a continental land mass.

continental slope. A zone of rapidly increasing depth in the sea between the edge of the continental shelf and the deeper ocean basin.

contour interval. The difference in elevation between two successive contour lines.

contour lines. Lines on a map joining points on the earth of equal elevation.

convection. Movement of a liquid or gas brought about by uneven heating.

coprolites. Fossil excrement.

core. The very dense innermost region of the earth.

Coriolis effect. The tendency of moving objects to veer to the right in the Northern Hemisphere and to the left in the Southern Hemisphere.

corona. A halo of faintly glowing

gases outside the chromosphere of the sun, usually visible only during a solar eclipse.

crater. The funnel-shaped pit at the top of a volcanic cone.

cratons. Precambrian rocks that form the centers of nearly all continents.

crest. The ridge of a wave, elevated above the surrounding water.

crevasse. A deep crack in a glacier.

cross bedding. Angled layers in sedimentary rocks.

crust. The outer layer of the solid earth.

crystal. A mineral form resulting from the arrangement of atoms, ions, or molecules in definite geometric patterns.

cumulus clouds. A family of clouds that resembles balls of cotton, usually having flat bases.

cyclone. A large mass of low-pressure air, with winds moving toward the center in a counterclockwise whirl; a low.

deflation. The effect of wind blowing away loose soil and rock particles.

delta. A triangular deposit at the mouth of a stream.

density. The weight per unit volume of any material.

desert pavement. A layer of close-fitting larger stones left on the desert floor after the lighter material has been blown away.

dew point. The temperature at which a given quantity of air reaches 100 percent relative humidity.

diatoms. Microscopic plants whose silica skeletons are often found in sea floor sediments.

dike. Solidified magma in vertical cracks or fissures.

dip. The angle of inclination of a rock bed or a fault from a horizontal surface, measured in degrees.

disconformity. Parallel horizontal rock layers separated by an erosional surface.

distillation. A method used to extract fresh water from sea water. The sea water is changed into vapor, then condensed into pure fresh water.

divide. A ridge or region of high ground that separates the drainage system of a stream.

doldrums. Belt of calm air under lower pressure near the equator.

dome mountains. Mountains that form when sedimentary beds are uplifted into broad, circular domes.

Doppler effect. The apparent effect on a train of waves if there is relative motion between the source and the receiver.

double stars. Two stars that move around one another, held together by mutual gravity.

drift. A general term for the material deposited by a glacier or by glacial meltwater.

drizzle. Precipitation caused when cloud or fog droplets fall to earth.

drumlin. A long, low mound of glacial till standing parallel to the direction of ice movement. They are generally steepest near the rounded end and gradually descend toward the narrow side.

dune. A hill or ridge of wind-blown sand.

earthshine. Illumination of the darker portion of the moon by reflection of light from the earth.

eclipse. The cutting off of light from one celestial body by another.

elastic rebound theory. The theory that earthquakes take place when the strain of many years of slow movement causes rocks to stretch and bend until they break and rebound (fault).

electromagnetic radiation. Energy that travels through space in the form of waves.

electrons. An atomic particle having a negative electrical charge.

element. A substance that cannot be decomposed by ordinary means to a simpler substance.

ellipse. Oval figure produced when a cone is cut in a diagonal direction.

emergent. Term used to describe shoreline features, resulting mainly from a drop in sea level or a rise of the land surface.

epicenter. The point or line on the earth's surface directly above the center or focus of an earthquake's origin.

epoch. Subdivision of a period in the geologic time scale.

equinox. A time when the sun is directly overhead at noon on the equator.

era. Largest unit of geologic time.

erosion. The removal of soil and rock fragments by natural agents.

erratic. A large boulder, deposited by glacial action, whose composition is different from that of the native bedrock.

escape velocity. The speed a rocket must reach in order to leave the earth's immediate gravitational field.

esker. A winding ridge of stratified drift deposited by streams running under or through glacial ice.

estuary. A shallow bay extending inland; a drowned river valley.

evaporites. Chemical sedimentary rocks formed by evaporation of water.

exfoliation. The splitting off of scales or flakes of rock as a result of weathering.

exosphere. The layer of the atmosphere, above the ionosphere, that merges with outer space.

extrusive rocks. Igneous rocks formed when magma flows out over the earth's surface.

fault. A fracture in a rock surface, along which there is displacement of the broken surfaces.

fault-block mountains. Mountains formed by faults that break part of the crust into large blocks which are then tilted.

faulting. The movement of rock layers along a break.

fetch. The extent of open water across which a wind blows.

fiord. A drowned glacial valley.

firn. The grainy ice in an accumulation of snow which later becomes glacial ice, also called *névé*.

fissure. An open fracture in the rock surface.

floe. A drifting section of arctic ice.

flood plain. The part of a stream valley which is covered with water during flood stage.

fluorescence. A luminescence given off by substances that are irradiated by ultraviolet or X-rays.

focus. The location below the earth's surface at which an earthquake originates; the source and center of the earthquake.

folded mountains. Mountains that result from the folding of rocks.

folding. The bending and breaking of horizontal rocks due to lateral pressures.

Foraminifera. Microscopic animals whose calcium carbonate skeletons often form sea-floor sediments.

fossil. Preserved remains or traces of prehistoric life on earth, buried in the rocks.

fracture. The way that a mineral breaks when it does not yield along a cleavage or parting surface.

fringing reef. A coral reef closely attached to the shore.

front. The boundary between two air masses.

frost action. Prying off of rock fragments by the expansion of freezing water in crevices.

fumaroles. Fissures or holes in volcanic regions, from which steam and other volcanic gases are emitted.

fusion. The union or blending of materials by chemical or nuclear reaction.

galaxy. An astronomical system composed of billions of stars.

gastroliths. Fossilized gizzard stones of reptiles.

geologic revolutions. Periods of marked crustal movement separating one geologic era from another.

geology. Scientific study of the solid part of the earth.

geomagnetic poles. The points on the earth's surface directly above the magnetic axis.

geosyncline. Very broad downfold in the earth's crust, extending for hundreds of miles.

geothermal energy. Energy supplied from heat in the earth's interior.

geyser. A hot spring that periodically erupts steam and hot water.

geyserite. A silicate mineral deposited around the opening of a geyser.

gibbous. The phase of the moon that is more than half full but less than full.

glaze ice. Formed when rain freezes upon striking the ground.

graben. A trough that develops when parallel faults allow the blocks between them to sink, forming broad valleys flanked on each side by steep cliffs. Also called *rift valley.*

gradient. The difference in elevation between the head and mouth of a stream.

gravity. A force that tends to pull every particle of matter toward every other particle.

great circle. Any circle drawn on the earth's surface that divides the sphere in half.

greenhouse effect. The process by which the atmosphere traps the sun's radiations.

Greenwich time. Time as measured on the prime meridian; used to determine longitude.

ground water. Water that penetrates into spaces within the rocks of the earth's crust.

gullying. A type of erosion in which deep channels form as water flows down a slope.

guyot. A flat-topped seamount.

gyre. Circular pattern of movement of surface water in the oceans.

hachures. A method of indicating a depression on a topographic map; short straight lines represent the direction that water would take in flowing down the slopes.

hail. Ice pellets formed by the successive freezing of layers of water.

half-life. The time interval during which half of a given amount of radioactive material disintegrates.

hanging valley. The valley of a tributary that enters the main valley from a considerable height above the main stream bed.

hardness (H). The resistance of a mineral to scratching.

headward erosion. Extension of the gullies that form the beginning or head of a stream.

highlands. Mountainous regions on the surface of the moon created by volcanic activity. They reflect more light than smoother parts of the moon's surface.

horn. Sharp peak formed where several arêtes join.

hot spots. Areas on the ocean floor where a continuous stream of magma erupts through the crust.

humidity. Water vapor in the air.

humus. Organic matter in the soil, produced by the decomposition of plant and animal materials.

hurricane. A destructive tropical storm over the Atlantic Ocean; such a storm over the Pacific Ocean is called a *typhoon.*

hydration. The chemical combination of substances with water.

hydrocarbons. Organic compounds containing carbon, hydrogen, and oxygen.

hydrologic cycle. The continuous process of water being evaporated from the sea, precipitated over the land, and eventually returned to the sea.

hydrosphere. The earth's envelope of water.

hygrometer. An instrument used to measure humidity.

icebergs. Large blocks of ice broken off from the leading front of a continental glacier ice shelf.

ice field. Formed where the amount of snow and ice that accumulates during a year is greater than the amount that melts during the warm season.

ice front. The advancing front wall of a glacier.

igneous rocks. Rocks formed by the solidification of magma.

index contours. Heavier and darker contour lines, usually every fifth line.

index fossils. Fossils that are found only in rocks of one particular period of geologic time; guide fossils.

international date line. An imaginary line at about 180° longitude; when it is crossed, standard time moves backward or forward by 24 hours.

intrusive rocks. Rocks formed by magma solidifying among other rocks below the earth's surface; also called *plutonic rocks.*

invertebrates. Animals without backbones.

ion. An atom or group of atoms that carries an electric charge.

ionosphere. The layer of the atmosphere above the stratosphere containing molecules of ionized gases.

island arc. A chain of islands lying close to the edge of a continent and adjacent to a deep submarine trench.

isobar. A line drawn on a weather map connecting locations of the same atmospheric pressure.

isogonic line. A line connecting locations of equal magnetic declination.

isostasy. Principle describing the state of balance of the earth's crust.

isotherm. A line on a map joining locations that record the same temperature.

isotopes. Atoms of the same element having different atomic weights.

jet streams. Swift, high-altitude winds at the edges of the circumpolar whirls.

joint. A fracture in a rock surface.

karst plain. A type of landscape characteristic of some limestone regions; drainage is mostly through underground streams in caverns.

kettle. A depression left after the melting of large blocks of ice buried in glacial drift.

knob. Large rock projection smoothed by glacial action.

laccolith. A domed mass of igneous rock formed by intrusion beneath other rocks.

lagoon. The body of water between an offshore bar and the mainland.

latent energy. Heat released when water vapor changes back to liquid water.

latitude. Location measured in degrees north and south of the equator.

lava. Liquid rock material that flows out on the surface of the earth from underground sources.

leaching. Removal of minerals from soil and rock in solution as water seeps down from the surface.

levee. The raised bank of a stream.

light year. The distance that light travels in one year.

lithosphere. The solid part of the earth.

load. The total amount of material carried by a stream.

lode. A large number of thick ore veins.

loess. An extensive deposit of wind-blown silt.

longitude. Location measured in degrees east or west of the prime meridian.

longshore current. Ocean current that moves close to and parallel with the shore.

luster (L). The appearance of the surface of a mineral in reflected light.

magma. Molten rock materials below the earth's surface.

magnetic declination. The difference, measured in degrees, between true north and magnetic north; also called *variation.*

magnetic storm. Electrical effects in the upper atmosphere usually caused by solar flares.

mantle. The thicker, more dense part of the earth beneath the crust.

map projection. A representation of the globe, or a portion of it, on a flat surface.

maria. Dark regions seen on the moon's surface.

massive. Rock masses without definite form (unstratified).

mass-wasting. Erosional processes chiefly caused by gravity.

meanders. Wide curves typical of well-developed streams.

mean sea level. The point midway between the highest and lowest tide.

mean solar time. Time as measured by the average position of the sun.

mechanical weathering. Natural processes which reduce rock to small fragments without changing its mineral composition.

meridian. An imaginary line extending from pole to pole that crosses the equator at a right angle.

mesa. A large, wide, flat-topped hill, usually a remnant of horizontal rock strata.

mesopause. Boundary between the mesosphere and thermosphere.

mesosphere. Layer of the atmosphere immediately above the stratosphere.

metamorphic rocks. Rocks that have been changed from their original form by great heat and pressure.

meteor. A rock-like particle in space; one of the smallest members of the solar system.

meteorite. A meteor that acually strikes the earth's surface.

meteorology. The science of the atmosphere.

mid-ocean ridges. A series of underwater mountain ranges in the ocean basins.

millibar. A unit for measuring air pressure; used in international weather observations.

mineral. Chemical compounds or uncombined elements found in rocks.

mirage. An optical illusion caused by the refraction of light as it passes through air layers of varying densities.

Mohorovicic discontinuity (Moho). The zone of contact between the crust and the mantle.

Mohs' scale. A numerical scale of hardness represented by ten minerals, ranging in hardness from 1 (talc) to 10 (diamond).

mold. The empty cavity of a fossil type left after the original organism has disappeared.

molecule. The smallest particle of a substance that can exist and still retain the properties of the substance.

monadnocks. Isolated hills or mountains of resistant rock rising above the general level of a peneplain.

monsoon. Seasonal wind that is

more pronounced over large continental areas near the equator.

moraine. A ridge or mound of boulders, gravel, sand, and clay carried by and deposited at the leading edge of a glacier.

mudflow. The movement down a valley of a large mass of mud and rock debris during a heavy rain.

native metals. Metals that exist in the pure form.

nautical mile. A distance equal to one minute of latitude; about 1.15 statute miles.

neap tides. Lower than normal tides, produced when the sun and moon are at right angles to the earth.

near-collision theory. A theory about the origin of the solar system; proposes that the planets were formed from matter drawn from the sun in a glancing collision with another star.

nebula. A cloud of gases or dust in space.

nebular theory. The theory that the solar system formed from the condensation of a large cloud of dust and gases.

neutrons. The electrically neutral parts of atoms.

neutron stars. A collapsed star whose atoms have been stripped down to the neutrons.

nitrogen cycle. The complex series of actions by which the nitrogen content of the atmosphere is maintained.

nodules. Black lumps of material on the sea floor, many of which contain valuable minerals.

nova. A star whose brightness suddenly increases many times the normal.

nuclear fission. A nuclear reaction in which heavy atoms split into lighter nuclei.

nuclear fusion. A nuclear reaction in which lighter-weight particles combine to form heavier nuclei.

nuclear reactor. Equipment used to control nuclear fission.

nucleus (of an atom). The central part of the atom, containing most of its total mass.

oblate spheroid. The shape produced if a sphere is slightly flattened; the true shape of the earth.

occluded front. A type of weather front created by an advancing cold front overtaking a warm front.

oceanography. The scientific study of the sea; also called oceanology.

offshore bar. A sand bar extending more or less parallel to the mainland.

oil pool. An underground pool of oil formed where there is an accumulation of oil and gas.

ooze. An extremely fine layer of sediment on the sea floor.

ore. A mixture of minerals containing at least one substance that can profitably be recovered.

outwash plains. Plains formed by the deposition of materials washed out from the leading edge of a glacier.

overturn. Sinking of chilled surface water in the sea or in a lake.

oxbow lake. A lake formed by the isolation of a meander from the main stream.

oxidation. The chemical combination of substances with oxygen.

ozone layer. A concentration of the gas ozone, lying largely in the region from 19 to 34 km (stratosphere) above the earth.

pack ice. A floating layer of ice completely covering the sea surface.

pahoehoe lava. Lava that has solidified, having a smooth, ropy, or billowy appearance.

paleontology. The science that deals with the study of fossils.

parallax. The apparent movement of

an object due to a change in the observer's position.

parallel fault. A fault in which there is evidence of both lateral and vertical motion.

parallel of latitude. An imaginary line drawn on the earth's surface parallel to the equator.

parasitic cones. Volcanic cones that develop at openings some distance below the main vent.

peat. Soft brown or black plant material formed in a decay process lacking oxygen.

peneplain. The flat surface left after landforms have been worn down by the agents of weathering and erosion.

penumbra. The lighter, outer part of a shadow.

perigee. The point of a satellite's orbit nearest the object around which it revolves.

perihelion. The point on the earth's orbit nearest the sun.

period. Subdivision of an era in the geologic time scale.

permafrost. A permanently frozen area in the deeper soil of the tundra.

permeability. The ability of water to penetrate and pass through rock.

Perseids. A meteor shower seen each year in mid-August.

petrifaction. A process in which the original substance of a fossil is replaced by mineral matter.

phosphorescence. The continued emission of light rays from certain minerals after exposure to ultraviolet light.

photosphere. A layer of brilliantly glowing gases beneath the chromosphere; the source of most of the radiant energy from the sun.

physiographic province. Large land areas separated from adjacent regions by fairly uniform physical features.

phytoplankton. Microscopic marine plants.

pitching fold. Folds whose axes slant downward at each end.

placer deposit. Mineral deposits in the gravel of a stream bed.

planetary circulation. The movement of the atmosphere over the entire earth.

plankton. Microscopic free-floating plants and animals in water.

plate tectonics. The theory that the earth's crust is made up of a number of separate rigid plates exposed to forces that cause them to move.

Polaris. The star very near the celestial north pole (north star).

polyp. Body form of a coral animal.

pothole. A rounded depression in the rock of a stream bed.

precession. A slight wobbling motion of the earth as it turns on its axis.

precipitation. Moisture that condenses and falls from the air. Also, separation of an insoluble substance from a solution.

pressure ridge. Formations of ice near the surface of ice due to uneven surface movements of an alpine glacier.

primary waves. The first waves of an earthquake, having a speed of 2.3 to 13.7 km per second; one of two body waves.

prime meridian. The 0° longitude line passing through Greenwich, England.

protons. An atomic particle having a positive electrical charge.

pseudofossils. Rock structures superficially resembling fossils but not actually made by prehistoric organisms.

psychrometer. A hygrometer that uses a wet-and-dry bulb thermometer.

pulsars. Stars that give off radio signals in regular pulses.

quadrangle. Detailed topographic maps commonly 7.5 minutes of latitude and 7.5 minutes of longitude. They are available from the Geological Survey, U.S. Department of Interior.

quasars (quasi-stellar radio sources). Stars that produce more energy than can be accounted for by any known process.

radar. A short-wave radio device that can be used to locate and track storms.

radiation. The process by which energy is given off in the form of rays.

radiation fog. A fog formed as a result of radiation heat lost by the earth. Also called *ground fog*.

radioactivity. The spontaneous breakdown of uranium and certain other radioactive elements into invisible radiations.

radiolarians. Protozoans that secrete siliceous shells of intricate design.

radiosonde. A set of electronic instruments that record and broadcast temperature, pressure, and humidity conditions at high altitudes.

radio telescope. An electronic device that focuses radio waves from outer space.

rays. Light streaks spreading out from some of the moon's craters.

red shift. A shift in the spectral lines of light from distant galaxies, caused by their receding motion.

reflector. A type of telescope in which a curved mirror produces an enlarged image.

refraction. The bending of a water wave when it approaches the shore at an angle, caused by the slowing down of the part that reaches shallow water first.

refractor. A type of telescope that uses only lenses to enlarge the image of the object viewed.

rejuvenation. Any action that tends to increase the gradient of a stream, thereby renewing youthful features of topography.

relative humidity. The ratio between the actual amount of water vapor in the air and the maximum amount it could hold at a given temperature.

relief. The irregularity in elevation of parts of the earth's surface; also the difference in elevation between the high and low points of a land surface.

residual boulders. Large rock fragments in their original position of the weathered bedrock.

resource. A natural material found in the earth that we can use to our advantage.

revolution. The movement of a celestial body in its orbit. See also *geologic revolution*.

Richter scale. A scale that measures earthquake magnitude as determined by seismograph records, ranging from 1 to 8.6.

rift. A narrow depression running along the center of the mid-ocean ridges.

rift valley. See *graben*.

rip currents. Swift currents moving away from the shore through a line of breakers.

rocks. A combination of different minerals found in the earth's crust.

rotation. The turning of a celestial body on its axis.

salinity. The total amount of dissolved solids in sea water.

satellite. Any celestial object that revolves around a larger object; a natural or artificial moon.

saturation value. The amount of water vapor a quantity of air can hold at any particular temperature.

scale. The mathematical comparison between actual distance and distance on a particular map.

SCUBA (self-contained underwater breathing apparatus). An apparatus used by an individual diver as an air supply.

sea-floor spreading. A process that creates new sea floor when plates separate and molten material fills in from below.

seamount. An underwater volcanic peak.

secondary waves. The second type of earthquake wave, having a speed of 3.2 to 4.0 km per second; one of two body waves.

section. A square region of land measuring one mile on each side.

sedimentary rocks. Rocks formed in layers from materials deposited by water, wind, ice, or other erosional agents.

sediments. Rock fragments produced by the weathering of solid rock.

seismograph. An instrument for detecting and measuring movements of the earth's crust.

shadow zone. A region on the earth's surface where a particular earthquake wave cannot easily be detected.

sheet erosion. A type of erosion in which water on a slope strips away exposed topsoil slowly and evenly.

shield. An exposure of Precambrian rocks over a wide area.

shield cone. A broadly based cone with a gentle slope, formed by the quiet type of volcanic eruption; composed of successive layers of lava. Also called *lava dome*.

sidereal year. The time required for the earth to complete one revolution in relation to the stars.

silicates. Minerals whose basic structure is a product of the joining of silicon and oxygen.

sill. Solidified magma in horizontal rock formations.

silt. Soil particles in an intermediate size range between clay particles and sand grains.

sink. A depression in the earth's surface formed by the collapse of the roof of an underground cavern.

sleet. Frozen raindrops.

slumping. A minor landslide involving only a small amount of loose material.

snow line. The elevation above sea level where snow remains all year.

soil. Mixture of loose rock fragments and materials produced by weathering.

soil creep. Slow movement of a mass of soil down a slope, caused by the soil's own weight.

soil horizon. One of the layers in a soil profile.

soil profile. Arrangement of layers in a mature soil.

solar flare. An unusually bright cloud of gases erupting from the sun, lasting for a few hours.

solar prominence. A streamer of glowing gas extending thousands of miles into the sun's atmosphere.

solar system. The sun and the celestial objects that revolve around it.

solstice. A time when the sun seems to reverse its apparent movement north or south of the equator.

spatter cones. Small cones that form in lava fields away from the main vent. Lava is spattered out through holes in a thin crust.

specific gravity (sp. gr). A number that relates the density of a substance to that of water.

spectrograph. A type of spectroscope that makes a photographic record of the spectrum of an object viewed through a telescope.

spectroscope. An instrument that separates a beam of light into its various component colors.

spectrum. The band of colors seen when light is separated by a prism.

spit. A sand bar extending outward from the shore.

spring. Ground water that comes

naturally to any exposed surface between the water table and the ground.

spring tides. Higher than usual tides produced when the moon and sun are lined up with each other.

squall line. A long line of heavy thunderstorms often advancing just ahead of a fast moving cold front.

stack. An isolated column of rock left standing after waves eroded a shoreline.

stalactite. A stony projection of minerals deposited from dripping water on the roof of a cavern.

stalagmite. A raised deposit of minerals deposited from dripping water on the floor of a cavern.

standard time. Time determined by division of the earth's surface into 24 time zones, the centers of which are approximately 15 degrees apart in longitude.

star cluster. A close grouping of stars, probably of a common origin.

stationary front. Either a cold or warm front that comes to a halt for a few days.

station model. A group of symbols on a weather map describing weather conditions at a particular location.

steppes. Regions bordering the tropical deserts, often extending into the middle latitudes.

stock. A small batholith.

strata. Rock layers or beds.

stratified rocks. Rocks that are found in parallel layers.

stratopause. The top boundary of the stratosphere.

stratosphere. The layer of the atmosphere above the troposphere, where the temperature remains fairly constant.

stratovolcanoes. Volcanoes with cones of alternate layers of lava as well as solid fragments.

stratus clouds. A family of clouds that form in layers.

streak. The color of a thin layer of any finely powdered mineral.

stream piracy. Capture of one stream by another as a result of erosion of the divide separating the stream valleys.

strike. The direction of a line along the edge of an inclined bed where it meets the horizontal plane. The strike is always at right angles to the dip.

strike-slip fault. A fault in which horizontal movement of one or both sides has occurred.

strip-mining. The method of mining in which layers of ore are exposed in strips along a mountain slope.

subduction zone. The region along plate boundaries where one plate is forced under another.

submarine canyon. An unusually deep-cut valley in the continental slope.

submergent. Term used to describe shoreline features resulting mainly from a rising sea level or from a drop in the land surface.

sub-tropical high. A belt of calm air under increased pressure; located about 30° north and south of the equator.

sunspots. Large dark areas of cooler gases on the sun's surface.

superposition. A principle that states that the overlying rock layer is younger than the layers below it.

supersaturation. A condition in which air reaches better than 100 percent relative humidity.

surface (long) waves. Earthquake waves that move over the surface of the crust.

swells. Groups of large ocean waves alike in size.

syncline. A trough or downturn of a rock fold.

taiga. A type of vegetation characteristic of subarctic climates.

talus. A mass of rock debris at the base of a steep mountain or cliff.

tarn. A lake formed in a glacial cirque.

temperature inversion. An unusual condition in which the layer of air at the surface is colder than the layer above it.

terraces. Step-like formations along the sides of a rejuvenated stream valley.

tetrahedron. Four-sided pyramid.

thermocline. A zone of rapid temperature change usually found near the surface of a body of water.

thermonuclear reaction. A nuclear *fusion* reaction taking place at high temperature.

thermosphere. Layer of the atmosphere above the mesopause.

tidal bore. A wave that passes upstream in a river from the sea as the tide rises.

tidal bulge. A bulge produced in the part of the sea facing the moon. This bulge is caused by the pull of the moon's gravity.

tidal oscillations. Very slow rocking motions in parts of the oceans, occurring in response to the tidal bulges.

till. Glacial deposits that have not been stratified or sorted by water action.

tombolo. Island connected to the shore by a ridge of sand.

topographic map. A map showing surface features of a portion of the earth.

topography. The physical features of a region.

tornado. A relatively small, destructive middle-latitude storm, usually originating over the land.

township. A square region of land measuring six miles on each side.

trade winds. Planetary winds in tropical areas blowing generally toward the equator.

transit. A crossing between the earth and sun by Mercury or Venus.

transpiration. The process by which plants release water vapor into the air.

travertine. A form of calcite deposited around the opening of a hot spring.

trenches. Deep fissure on the ocean floor, caused by faulting.

tributary. A stream that flows into a larger one.

trilobite. A common invertebrate animal that became extinct during the Permian period.

Tropic of Cancer. An imaginary line found parallel to the equator at $23\frac{1}{2}°$ north latitude; marks the most northern point where the sun is ever directly overhead.

Tropic of Capricorn. An imaginary line found parallel to the equator at $23\frac{1}{2}°$ south latitude; marks the most southern point where the sun is ever directly overhead.

tropical year. The time from one vernal equinox to the next.

tropopause. The upper boundary of the troposphere.

troposphere. The layer of the atmosphere closest to the earth's surface in which most weather activities occur.

trough. The depression on either side of a wave crest.

tsunami. A giant wave produced by disturbances on the sea floor; incorrectly called a "tidal wave."

tundra. A type of climate in the zone of transition between the subarctic regions and the icecaps.

turbidity currents. Strong currents produced by material sliding down the continental slopes.

umbra. The dark, inner part of a shadow.

unconformity. An eroded bedrock surface that separates younger strata from older rocks.

undertow. A current that runs constantly beneath a line of breakers as water from the breaking waves is pulled back to deeper water.

uniformitarianism. The principle that the same forces that changed rocks in the past are still operating today and causing the same kinds of changes.

upslope fog. A fog produced by the adiabatic cooling of air as it moves up a slope.

upwelling. Movement of deeper sea water to the surface.

variable star. Any star whose brightness appears to change from time to time.

varves. Annual, double layers of fine clay-like sediments.

vein. Ore deposits that have, as hot solutions, entered cracks in rock, then cooled.

ventifact. A stone smoothed by wind abrasion. Also called *dreikanter*.

vertebrates. Animals with backbones.

vertical faults. Several kinds of faults in which movement of the rocks is mainly vertical in direction.

volcanic bomb. Rounded rock fragments thrown out during a volcanic eruption.

volcanism. A general term including all types of activity due to movement of magma.

volcano. The vent from which molten rock materials move out onto the surface; includes the accumulation of volcanic materials deposited around the vent.

warm front. A front at which a warmer air mass overrides a colder air mass.

water budget. A comparison between the amount of water received from precipitation and the amount lost as water vapor passes back again into the atmosphere.

water gap. A valley with an existing stream that cuts across a mountain ridge.

watershed. A series of slopes that drain into a river system.

waterspouts. Tornadoes over the sea.

water table. The upper boundary of the ground water below which all spaces within the rock are completely filled with water.

wave-built terrace. A seaward extension of a wave-cut terrace, produced by the accumulation of debris from wave action.

wave-cut cliff. A cliff formed from wave erosion in which waves strike directly against the rock of the land.

wave-cut terrace. A level surface of rock under the water along the shore, formed as waves cut back the shoreline.

wavelength. The distance from one wave crest to the next, or from one wave trough to the next.

wave period. Time taken for two successive crests of a wave to pass a given point.

weathering. The natural disintegration and decomposition of rocks and minerals.

wind gap. An abandoned stream valley cutting across a ridge.

zone of aeration. A region beneath the ground surface where rocks contain both water and air.

zone of saturation. Region beneath the surface in which all spaces between rocks are filled with water.

zooplankton. Microscopic marine animals.

index

aa lava, 196

absolute humidity, **441**

absolute magnitude, *31,* 34, **35**

abyssal plains, 325–326

adiabatic changes, *443*

adiabatic cooling, 443–444

Adirondack Mountains, 221–222

advection fog, 448

aeration, zone of, 265, **266**

Agassiz, Lake, **288**

agonic lines, *189*

air, composition of, 420–421; gases in (Table), 421; pollution of, 426–**427;** pressure of, 421–424, 466; supersaturated, 444, 446; *see also* atmosphere

air masses, classification of, 458; cold, 458, **460;** North American, 458–**459;** origin of, 458; warm, 458, **460**

air pollutants, 426

Alabama Hills, **248**

Alaskan glaciers, **284,** 285, 290

albedo, 429

Aleutian Islands, 202–203

alluvial fan, **263**

alpha particle, 381

alpine glacier, 277, **278, 280, 283,** 290

altimeter, 423–424

altitude, temperature and, 482

altocumulus clouds, **445,** 446, 447

altostratus clouds, 446, **447**

aluminum, recycling of, **508**

aluminum ore, 507, 508

Andes Mountains, 223, 290

Andromeda, **6**

anemometer, **467**

aneroid barometer, 422–**423**

animals, microscopic, in sea water, 329–**330,** 343, 344, **345;** and nitrogen cycle, **424;** weathering and, 235

annular eclipse, **85**

Antarctic Current, 356

Antarctic Ocean, **317,** 318

Antarctica, **277**–278, 281, 487

antennas, radio telescope, 9–**10**

anthracite coal, 510

anticline, 208

anticyclones (highs), **465**

aphelion, *13*

apogee, *80,* 85

Appalachian Mountains, 209, 213, 224, 406, 408, 409

apparent magnitude, *31*

apparent solar time, 67

applications satellites, 102–105

aquaculture, 347–348

aquifier, **268**

Arctic Circle, **65**

Arctic Ocean, **317,** 318; ice layer in, 292, 337

Arcturus, spectra of, **34**

aretes, *281,* **282**

argon, 342, 421

artesian wells, **268**

asteroids, 56–**57**

asthenosphere, 184, **186,** 192

Atlantic Ocean, **317**–318, 326

atmosphere, 420; carbon dioxide in, **425,** 429; chemical balance of, 424–425; composition of, 420–421; continuous motion of, **353;** gases of, 421; layers of, 429–**431,** 432–433; moisture in, measurement of, 440, **441–442;** solar energy and, 427–**428,** 429; water vapor in, .439–442; *see also* air

atoll, *304;* formation of, **305**

atomic mass, 141–142; of hydrogen, 142; of oxygen, 142

atomic number(s), 141; Table of, 141

atomic theory, 138

atoms, electron arrangement in, 143–144; kinds of, 140–142; nucleus of, 28, 140, 513; structure of, 138–140

aurora australis (southern lights), 26

aurora borealis (northern lights), **26**–27

auroras, 26–27; frequency of, **27**

autumnal equinox, **65**

axis, of earth, 63, **64;** geographic, 188, 189; of rock fold, 210; of rotation, 189

bacteria, nitrogen-fixing, 424

banner clouds, **444**

barchans, **243,** *244*

barograph, **423**

(Note: Page numbers in **boldface** type indicate illustrations and those in *italic* type indicate definitions.)

barometer, aneroid, 422–**423;** mercurial, 422, **423;** Torricellian, **422**
barrier islands, 303
barrier reef, 304–305
basalt, 171, **173**
basaltic rocks, 171, **178**
basement complex, 381
basins, drainage, **257,** 258; lake, 287, 288; ocean, 323
batholiths, 193, 213, 221
bathythermograph, **319**
beaches, **299–300;** makeup of (Table), 299
bead tests for minerals, 167, **168,** 169
bedrock, 235–236
bends, 321
berm, **299**
beta particle, 382
Big Bang Theory of universe, **42**
Big Dipper, 4
bituminous coal, 510
Black Hills, 221–222
block lava, 196
blowout, *242*
body waves (earthquake), 215
bog, 288, **289**
borax, 289
breaker, wave, 362, **363**
breeder reactor, 515
breezes, land, 472; mountain, 472; sea, 472; valley, 472
bright-line spectrum, **32**
brown clay, 331
"brown" coal, 510
Buffon, Georges, 48
building materials, 507
Bunsen burner, 167
buttes, **249,** *250*

Caesar, Julius, 88

calcite, 271
calcium, 157
calcium carbonate, 329, 330
caldera, *199,* **199**
calendar, Gregorian, **88–**89; Julian, 88; Roman, 87–88; Universal, 89; World, **89**
calorie, *440*
Cambrian period, **398–399,** 402, **403**
Canadian Rockies, **276, 283,** 285
Canadian Shield, **401**
Cape Cod, 297
capillary fringe, **266**
capillary movement, zone of, **266**
capillary tension, 266
carbon dioxide, in air (Table), 421; in atmosphere, **425,** 429
carbon dioxide cycle, **425**
carbon-14 dating, 383
carbonation, *235,* 237, 270
carbonic acid, 234–235
Carboniferous period, 405–**406**
carbonization, *389*
Carlsbad Caverns (New Mexico), 270–271
Cascade Mountains, 221
Catskill Mountains, **248,** 250
caverns, 270–**271**
celestial north pole, **3,** 4
Celsius scale, 466
cement, 507
Cenozoic era, **398–400, 409**
cepheid variables, 31, 33
chain reaction, nuclear, 513
chalcedony, **155**
charcoal block tests, **168,** 169
chemical bonds, 144–145
chemical deposits in ocean, **330**
chemical formula, 144

chemical properties, of ocean, 319–320; of sea water, 335, 340–346
chemical rocks, 172
chemical weathering, 233–235, 297
chinook winds, **482**
chromosphere of sun, **23**
cinder cones, 197, 198
circumpolar whirl, **465–**466
cirques, **281–282**
cirrocumulus clouds, **445,** 446
cirrostratus clouds, **445,** 446
cirrus clouds, **445,** 446
clay, 238; brown, 331; in cement, 507; red, 330–331
cleavage of minerals, 164–**165**
cliffs, formation of, 177, **178**
climate(s), *479–496;* control of, 479–484; local, 494–496; maps of, 125; middle latitude, 484, **485,** 488, **489–491,** 492; North American, 492–496; polar, 484, **485, 487–488;** savanna, 486–**487;** and temperature, 479–484; tropical, 484, **485–487;** and weathering, 237–238; *see also* weather
clouds, banner, **444;** and cold front, **461;** formation of, **444, 445,** 446; high, 446; low, 446–448; middle, 446; seeding of, **450–451;** types of, 446, **447–**448; with vertical development, 448; and weather fronts, **461, 462**
coal, 172, 389, 507, 509; anthracite, 510; bituminous, 510; "brown," 510; formation of, 509–

510; seams of, 511; strip-mining of, **511;** supply of, 511, 515

coalescence of raindrops, **449**–450

coastal resources, 307–308

color of minerals, 163

Columbia ice field, **276**

Columbia Plateau, **196**

comets, 61–**62, 63**

communications satellites, **104**

composite cones, **198**

compound, *143*

concretion, 391

condensation nuclei, 446

condensation of water vapor, 442–443

conduction, heat transfer by, 433

conformity of rocks, 378

conservation, soil erosion and, **245**–**246,** 247; of water, 256

constellations, 3

contact metamorphism, *380,* **380**

continental collisions, 224

continental drift, theory of, *182*

continental glaciers, 277, 281, **283**

continental margins, **324**

continental rise, **325**

continental shelf, 324–**325**

continental slope, 325

continuous spectrum, **32**

contour interval, **128**–129

contour lines, **127,** *128,* **128, 129**

contour loop, 130

contour map, interpretation of, 129–130

contour plowing, **246**

convection, heat transfer by, 433, 434

Copernicus, Nicolaus, 12

copper ore, **506,** 507

coprolites, *390*

coquina, 172

coral reefs, 304–**305**

core of sun, 22–**23**

core samples, 320, **330**

Coriolis effect, 354–356, 358, **434,** 435, 465

corona of sun, **23**–24

Crab nebula, **41**

Crater Lake (Oregon), **199**

craters, on moon, **77, 79;** volcanic, 198

cratons, 401

Cretaceous period, **398–399, 408**

crevasses, glacial, **280, 281**

crop rotation, 247

cross bedding of rocks, **378**

crust of earth, **184**

crustal plates (*see* plate tectonics)

crystals, ice, 148, **450**–451; mineral, 163–**164;** sodium chloride, **146**

cumulonimbus clouds, **447,** 448

cumulus clouds, **444, 447,** 448

current(s), 320, 352, 353, **354**–359; Antarctic, 356; causes of, 353–354; deep, **359;** equatorial, **354,** 355; Japan, 356; Labrador, 356, 358; longshore, **300**–301; North Pacific, 356; rip, **363;** and temperature, 483; tidal, 366; turbidity, 354; undertow, 362–363

cyclone, wave, 461–462, **463–464,** 465

Cygnus, 3

dark-line spectrum, **32**

day, measurement of, 66–67

daylight, length of, **67**

Daylight Savings Time, 69–70

Dead Sea, 212

deflation, *242*

deflation hollow, 242

deltas, **263**

density, 16–17

depression contour, **130**

Descartes, René, 48

desert pavement, **242**

deserts, middle latitude, 489, **490;** tropical, 486

Devonian period, **398–399, 405**

dew, **443**

dew point, *442*–443

diatoms, **330,** 331

dikes, volcanic, 193, **194**

dinosaur tracks, **389**

dinosaurs, 408, 409

dip of rock bed, 209, **210**

direction, points of, 120–121

disconformity of rocks, **379**

divides, **257,** 258

doldrums, 435

domed mountains, 221–**222**

Doppler effect, *33, 33,* 41

Doppler shift, **34**

double stars, **38,** 39

drainage basins, **257,** 258

drizzle, 451

drumlins, 285, **286**

dunes, sand, **243**–244

dust cloud theory, 48

dust storms, **245,** 246

earth, 54; asthenosphere of, 184, **186,** 192; average distance from sun (Table), 63; average surface temperature (Table), 63; axis of, 63, **64;** crust of, **184;** crustal plates (*see* plate tectonics); diameter of (Table), 63; directions on, 120–122; elements in crust of, 141, 153, **154;** equatorial diameter of, 75; Eratosthenes' measurement of, 119–**120;** gravity of, 15–16; inner core of, 184; interior of, 183–185; layers of, 184; lithosphere of, 184, **185,** 420; locations on, 120–122; magnetism of, 188, **189**–190; mantle of, 184; maps of surface of, 122–125; mass of (Table), 63; models of, 116–117; motions of, 62–67; orbit of, **13,** 62–63; outer core of, 184; physical history of, 396–400; plate boundaries on, 185–187; revolution period (Table), 63; rotation of and winds, 434; rotation period (Table), 63; satellites of, 96–98; as seen from outer space, **2;** shape of, 117–118, 120; size of, 119–120

earthquakes, 213–220; control of, 220; damage caused by, 216–217; detection of, 213–216; elastic rebound theory of, **214;** epicenter of, 214, **216, 218;** faults and, 214; focus of, 214, 216; landslides and, 214, 241; locations of, 217–219; magnitudes of, 217; prediction of, 219; San Francisco (1906), 219; on sea floor, 354; volcanic eruptions and, 214; waves produced by, **183–184, 214**–216

earthshine, *82*

eclipse, annular, **85;** of moon, **83–84,** 85; of sun, **24, 83,** 84–85; total, **24,** 84–85

Einstein, Albert, 28

elastic rebound theory of earthquakes, **214**

electrical charge, 139, **140,** 145

electrical currents, magnetism and, 189–190

electromagnetic radiation, 7, 21

electromagnetic spectrum, 7, **8,** 9; and solar energy, 21–**22**

electromagnetic waves, 428

electronic weather instruments, 467–**468**

electrons, 140, 141; arrangement in atoms, 143–144; loss of, 144–**145;** repulsion of, **140;** sharing of, 144–145, 149; transfer of, 144–**145**

element(s), *141;* in earth's crust (Table), 141; percent by weight in earth's crust (Graph), **154**

ellipse, **13**

elliptical galaxy, **6**

elliptical orbit, 13, 80

energy, geothermal, 515–517; heat, 515; latent, 440; nuclear, 513–515; of sun (*see* solar energy); from tides, 516–517

Eocene epoch, **398–399, 410**

epicenter of earthquake, *214,* **216, 218**

equatorial currents, **354,** 355

equinox, *67;* autumnal, **65;** vernal, **65,** 66

Eratosthenes, 119–120; and earth measurement, **120**

erosion, *232;* and geologic time, 383–**384;** glacial, 281–283, 287–288; gravity and, 239, **240–241,** 242; headward, **257–258;** by running water, **256**–263; and sand, **242–243;** sheet, 245–246, 261; soil, and conservation, **245–246,** 247; water and, 253; wave, **296–300;** wind, **242–243**

erratics (boulders), **284**

escape velocity, *95,* 96

eskers, *286*

estuary, *302, 303*

evaporation, patterns of, 254–255; of sea water, 253, 338–339, 341, **342, 440**

evaporites, **172**

exfoliation, *236,* **237**

exosphere, 431

extrusive igneous rocks, 171, **173**

extrusive volcanism, 192

Fahrenheit scale, 466

fault-block mountain, **220**

faults, **210–212;** and earthquakes, 214; in mid-ocean ridges, **327;** San Andreas, 218–**219;** strike-slip, **210**

feldspars, 154, **156**

fertilizers, 507

fetch, *363*

Finger Lakes (New York), 287
fiords, 302, **303**
firn, 275–276
fish, 347–348
fissure flow, **196**
flame tests for minerals, 167, **168**
flood plain, **264**
flood stage, 263–264
floods, 263–**264,** 265; control of, 265
fluorescent minerals, 166, **167**
focus of earthquake, 214, 216
foehn, **482**
fog, advection, 448; radiation, 448; upslope, 449
folded mountains, 220
folding of rocks, **208**–209, **210**
food from sea water, 347–348
food chain in sea water, 344, **345**
Foraminifera, **330,** 331
fossil fuels, 509–513; coal (*see* coal); natural gas, 509, 511–513; petroleum, 507, 509, 511–513; supplies of, 511–513; underwater deposits of, 347
fossils, finding of, 390–391; formation of, 387, **388–390**; index, **386**–387; indirect evidence of, **389–390**; molds of, **390**; record of, 385–391
fractional scale, *127*
fracture of minerals, **165**
Franklin, Benjamin, 139, 357
fringing reef, 304
frost, 443
frost action, **233**
fuels (*see* fossil fuels)

fumaroles, *197*
Fundy, Bay of, **36⌐,** 367

galaxies, 4–7; Andromeda, **6;** barred spiral, 6; elliptical, **6;** irregular 6–7; Our Galaxy, 4–**5,** 6; spiral, **6**
Galileo, 117
gases, in air (Table), 421; dissolved, in sea water, 340, 342–343; molecular motion in, **147;** volcanic, 197, 340
gastroliths, *390*
Geiger counter, 167
gem minerals, 158, **162,** 506
geographic axis, 188, 189
geographic poles, 188, **189,** 190
geologic calendar, 396–397, **398–399,** 400; *see also* geologic eras
geologic epochs, **398–399,** 400, 409–412
geologic eras, Cenozoic, **398–400,** 409; Mesozoic, **398–400, 407**–408, 504, 509–510; Paleozoic, **398–400,** 402–407, 509; Precambrian, 397, **398–399,** 400, **401**–402
geologic maps, 125
geologic periods, **398–399,** 400
geologic revolution, 397, 408
geologic time, and deposition of sediments, **384**–385; erosion and, 383–**384;** measurement of, 381–385; radioactivity and, 381–383; salt in sea water and, 385; *see also* geologic eras

geology, *207*
geomagnetic poles, *188,* **188**
geosyncline, 213, 223
geothermal energy, 515–517
geyser basins, **270**
geyserite, 269
geysers, **269–270**
glacial drift, 283–284
glacial erratic boulders, **284**
glacial meltwater, 286
glacial periods, 289–290, **291**
glacial till, **282, 284**
glaciated rocks, **282**–283
glaciers, *213,* 275–292; Alaskan, **284,** 285, 290; alpine, 277, **278, 280, 283,** 290; continental, 277, 281, **283;** crevasses in, **280, 281;** cycle of formation of, **282;** deposits left by, 283–285; erosion by 281–283, 287–288; formation of, 275–278; of Greenland, 278, 281; ice front of, 278; lakes created by, 287–288; movement of, **278–280,** 281; pressure ridges in, **280, 338;** relative speeds of flowage in, **278;** valley, 277
glaze ice, **451**–452
Glomar Challenger, **316**
gneiss, **175**
gold, 504, 506; in sea water, 347
Gold Rush (1849), 504
graben, *210,* **212**
gradient, stream, 258
Grand Canyon, **221,** 401
granite, **171,** 238
grantitic rock, 171
graphic scale, *127*
graphite, 176, 401
gravitational force, 14–15,

93–95

gravity, 14–17; barrier of, 93–95; center of, 79–**80;** of earth, 15–16; and erosion, 239, **240–241,** 242; of moon, 78; mutual, 39; pit of, 94; and space travel, 93–95; specific, of minerals, 165–166; and stress in rocks, 213; of sun, 98; and weight, 16–17

great circle, **122**

Great Lakes, formation of, **287**–288

Great Red Spot of Jupiter, 59

greenhouse effect, **429,** 430, 439

Greenland glaciers, 278, 281

Gregorian calendar, **88**–89

Gregory, Pope, 88

ground moraine, 285, 287

ground water, 254, **266,** 267

Gulf Stream, 355–356, **357**–359, 483; movement of water in, **358**

gullies, **257**

gullying, **245**

guyots, *326*

gypsum, 172

gyres, *352*

hachure lines, *130,* **130**

hailstones, **451,** 452

hair hydrometer, 442

half-life of uranium, 382, **383**

halite, 172, 507–508

Halley's Comet, 62, **63**

hanging valleys, **282,** 283

hardness of minerals, 165

Hawaiian Islands, 326–**327**

headward erosion, **257–258**

heat energy, 515

heating, air movements and, **433**–434

helium, 29, 37, 39; in air (Table), 421

highlands of moon, 77, **79**

highs (anticyclones), **465**

Himalaya Mountains, 224

H. M. S. Challenger, 316

horizons of soil, **239**

horn peaks, 281, **282**

hornblende, **157**

Horse Head nebula, **37**

hot spot, *326*

hot springs, 268–**269, 270**

humid continental climates, 490–**492, 493**

humid subtropical climates, 490

humidity, *440;* absolute, **441;** relative, 441, 442; saturation value of, 441

humus, 238–239

hurricanes, 472–**473, 474**

Hutton, James, 207

hydration, *235*

hydrocarbons, *509; see also* fossil fuels

hydrogen, 29, 39; in air (Table), 421; atomic mass of, 142; isotopes of, **142**

hydrographic maps, 125

hydrologic cycle, 253–**255,** 256

hydrosphere, 420

Ice Age, 306, 409

ice ages, 289–290, 324; causes of, 290–292

ice crystals, 148, **450**–451

ice field, **276**

ice floes, 337–338

ice front of glacier, 278

icebergs, 281, 328

icecap climates, 487

igneous rocks, 169–**171, 173;** extrusive, 171, **173;** intrusive, 170–**171;** massive, **176,** 177

index contours, 129, **130**

index fossils, **386**–387

Indian Ocean, **317,** 318

inferior mirage, **432**–433

intermediate volcanoes, 195

International Date Line, 70

intrusive igneous rocks, 170–**171**

intrusive volcanism, 192, **193,** 200

invertebrates, **403**

ion(s), *145,* 149; sodium chloride, 145–**146**

ionic bond, 147

ionosphere, 431

iron, 77, 189, 507; ores of, 507, 508

irregular galaxies, 6–**7**

island arcs, **202,** 223

isobars, **470**

isogonic lines, *189*

isostasy, *212*–213

isotherms, *480,* **481**

isotopes, 142–143; of hydrogen, **142**

Japan Current, 356

jet streams, 465–**466**

joints of rocks, 177, **210**

Julian calendar, 88

Jupiter, 57–**58,** 59; characteristics of (Table), 63; escape velocity from, 95; gravity of, **100;** Great Red Spot of, 59; travel to, 100

Jurassic period, **398–399,** 408

karst plain, **271**

Kepler, Johannes, 12, 117; his laws of planetary motion, 12–14

kettles, *286*

knob, glaciated rock, 282–**283**

Krakatoa, 194, **195**

Kruger 60 (double star), **38**

Labrador Current, 356, 358

La Brea tar pits, **388**

laccoliths, 193, **200,** 221

lagoon, 303

lakes, basins of, 287, 288; formation of, **267;** life history of, 288–289; origin of, 287–288; ox-bow; **262;** salt, 288–289

land breezes, 472

landforms, guide to study of, 127–128; locating of, 130–131; minor, origin of, **247**–249; structure of, 248

Landsat (applications satellite), **103**

landslides, 214, 241, 288; underwater, 328

latent energy, 440

lateral moraines, **284**

latitude, *121;* parallels of, *121,* **121;** and temperature (Table), 480

launch windows, 100

lava, 170, 171, 192; aa, 196; block, 196; pahoehoe, **196;** plateaus of, **196**

leaching, *235,* 239

lead, ores of, 506, 507

leap year, 88, 89

Lembert Dome (Yosemite National Park), **283**

levees, 264

light rays, absorption of by sea water, **336**

light waves, 7, 33

light year, *4–5,* 6

lightning, 474–**475**

lignite, 510

limestone, 172, **174,** 176, 401; in cement, 507

limestone column, **271**

limestone rock, 270, **271**

limestone sink, **270,** 271

liquids, molecular arrangement in, 147; molecular motion in, **147;** *see also* water

lithosphere, 184, **185,** 420

local climates, 494–496

lode, *506*

lodestone, 166

loess, **244,** 245

longitude, 121; meridians of, 121–**122**

longshore currents, **300**–301

looming mirage, 433

luster of minerals, 163

magma, *170*–171; origin of, 191–192

magnetic declination, *188–***189**

magnetic disturbances of sun, 24–25

magnetic field, **190**

magnetic poles, **190**

magnetic storm, 25–26

magnetism, of earth, 188, **189–190;** of minerals, 166

magnetite, 166

magnitude, absolute, *31,* 34, **35;** apparent, *31;* of earthquakes, 217; scale of, 31

main sequence stars, **35,** 37

Mammoth Cave (Kentucky), 270–271

mantle, of earth, 184

map projections, 123–125; of plane surface, **125**

maps, climate, 125; contour, interpretation of, 129–130; of earth's surface, 122–125; geologic, 125; hydrographic, 125; political, 125; relief, 125; star, **3;** topographic, 127–131, **132–133;** use of, 125, 127; weather, 125, 422, **470**–471

marble, **175,** 176

maria of moon, 78, **79**

marine west coast climate, 488–**489**

Mars, **54–55,** 56; characteristics of (Table), 63; knowledge gained from exploration of, 101–102; rocks of, 138; travel to, **98**

mass-wasting, *241*

massive rocks, 176, 177

mature mountains, **248,** 249

mature soil, **239**

mean sea level, *128*

mean solar time, 68

meanders, 261–**262**

mechanical weathering, 233, 237, 238, 296–297

median moraines, **284**–285

Mediterranean climates, 489–490, **491**

mercurial barometer, 422, **423**

Mercury, **51–52;** characteristics of (Table), 63

meridians, 121–**122**

mesa, **249,** 250

mesopause, 430, **431**

mesosphere, 430, **431**

Mesozoic era, **398–400, 407**–408, 504, 509–510

metal ore minerals, 158, **160–161,** 505–507

metals, 505–507; recycling of, **508**–509

metamorphic rock, 170, 172, **175,** 176, **177**

meteorites, 61, 76, 77, 79, 331

meteorology, *420*

meteors, **61**

mica, **156,** 157

middle latitude climates, 484, **485,** 488, **489–491,** 492

mid-ocean ridges, 186–**187, 190,** 200, 201, 212, **324;** faults in, **327**

Milky Way, 4

minerals, *153*–169; cleavage of, 164–**165;** color of, 163; crystals of, 163–**164;** fluorescent, 166, **167;** fracture of, **165;** gem, 158, **162,** 506; hardness of, 165; identification of, 158, 163–169; luster of, 163; magnetism of, 166; metal ore, 158, **160–161,** 505–507; nonmetallic, 507–508; optical properties of, **167;** phosphorescent, 166; radioactive, 167; rock-forming, 158, **159;** in sea water, 346–347; specific gravity of, 165–166; streak of, 163; structure of, **154;** tests for identification of, 167, **168,** 169; types of, 153–158

minimum energy orbit, **98**–100

minutes, 121

Miocene epoch, **398–399,** 411

mirages, **432**–433

Mississippian period, **398–399,** 405–406

mixing in sea water, 345

Mohorovicic discontinuity (Moho), 184

Moh's scale, 165, 166

molds, fossil, **390**

molecule(s), *143*–149; chemical bonds in, 144–145; forces holding, 145–149; relative motion of, **147;** water, **144;** 147–**148**

monadnocks, **248,** 249

monsoons, **483**

moon, 75–89; albedo of, 429; and calendar, 87–89; collisions on, **77**–78; craters of, **77, 79;** diameter of, 75, 78; distance from earth, 80; eclipse of, **83–84,** 85; escape velocity from, 95; gravity of, 78; highlands of, 77, **79;** history of, 75–78; maria of, 78, **79;** orbit of, 76, 79–81; phases of, **81–82,** 83; rise of, 81; rocks of, 76, 138; soil on, **79;** stages in development of, 75–78; surface of, **77, 78,** 79; temperature of, 78–79; and tidal bulges 85–**86;** visible, 78–79

moraines, **284–285,** 287

mountain breezes, 472

mountains, domed, 221–**222;** fault-block, **220;** folded 220; formation of, 220–224; life cycle of, **248,** 249; mature, 248, 249; peneplain of, **248,** 249; plate tectonics and building of, 222–224; synclinal, **208**–209; volcanic, 221, 326; youthful, **248,** 249

mudflow, 241

muds in ocean sediment, 331

mutual gravity, 39

native metal, *505*

natural gas, 509, 511–513; underwater deposits of, 347

nautical mile, *121*

navigation charts, 125

neap tides, **87, 88**

near collision theory, 48

nebula, *38,* 48; Crab, **41;** Horse Head, **37**

nebular theory of solar system, 48, 49

Neptune, 60; characteristics of (Table), 63

neutron stars, 40

neutrons, 140–142, 513

névé, 275–276

Newton, Isaac, 14, 120; his law of gravitation, 15

Niagara Falls, **259,** 383–**384**

nickel, 189

Nile River delta, **263**

nimbostratus clouds, **445,** 448

NIMBUS (weather satellite), 104, 468

nitrogen, 148; in air (Table), 421

nitrogen cycle, **424**–425

nitrogen-fixing bacteria, 424

North American air masses, 458–**459**

North American climates, 492–496

North American continent, building of, 412–**413**

North Atlantic Drift, 356, 358, 483

North Atlantic Ocean, 317–318; floor of, **324**

North Equatorial Current, 355

North Pacific Current, 356

North Pacific Ocean, **317**

North Pole, 63, **65**

North Star (Polaris), **3**, 4, **118**

northern lights, **26**–27

Nova Hercules, **40**–41

nova stars, 40–41

nuclear energy, 513–515

nuclear fission, 513–514

nuclear fusion, **28–29**, 515

nuclear reactor, **514**; breeder, 515

nuclear rockets, 96

nucleus, atomic, 28, 140, 513

oblate spheroid, *120*

obsidian, 171, **172, 173**

ocean basin floor, **324**–327; *see also* sea floor

ocean farming, 347–348

ocean waves, 359–365; breakers, 362, **363**; crest of, 360, **361**; giant, 363–365; height of, 360, **361**; period of, 360–**361**; refraction of, 361–**362**, 363; speed of, 360–361; swells, *360,* **360**; tidal, 354, 363; trough of, 360, **361**; tsunamis, 363–**365**; wavelength of, 360–**361**; wind and, 359–360

oceanography, 316

oceans, 316–331; basins of, 323; chemical deposits in, **330**; chemical properties of, 319–320; currents in (*see* currents); depths of, 318–**319**; floor of (*see* sea floor); humans beneath, **321**–322, **323**; and light rays, 336; pollution of, 342; surface temperature of, 337–339; temperatures of, 319; underwater exploration of, 316, **321**–322, **323**; water movements of, 320; of the world, **317**–318; *see also* sea water

oil pool, 511

oil shale, 512

Oligocene epoch, **398–399**, 410–411

olivine, 157

Olympus Mons, **54**

ooze on sea floor, 331

orbit, of earth, **13**, 62–63; elliptical, 13, 80; minimum energy, **98**–100; of moon, 76, 79–81; planetary, **51**; of rocket, 96–**97**; of satellite, 96–**97**

Orbiter (Space Shuttle), **105–106**

Ordovician period, **398–399**, **403–404**

ores, 158, 505–507

organic rocks, 172

organic sediments, 329–**330**

Our Galaxy, 4–**5**, 6

outer core of earth, 184

outwash plain, 286

overturn in sea water, 345

oxbow lake, **262**

oxidation, *235*

oxygen, 148; in air (Table), 421; atomic mass of (Table), 141; atomic number of (Table), 141

ozone, 421

ozone layer, 430

Pacific Ocean, **317,** 318, 326

Pacific Standard Time, 69

pack ice, 337, **338**

pahoehoe lava, **196**

Paleocene epoch, **398–399**, 409

paleontology, *385*

Paleozoic era, **398–400**, 402–407, 509

Palomar Observatory, **9**, 31

Pangaea, 182, 402, 405, 407, 509

parallax, 29–**30**

parallel faults, 210, **212**

parallels of latitude, *121,* **121**

parasitic cones, 198

peat, 510

Pegasus, 3

Pelée, Mount, eruption of, **195**

peneplain, **248**, 249

Pennsylvanian period, **398–399**, 405–**406**

penumbra, **84**

perigee, *80,* 87

perihelion, *13*

permafrost layer, 487

permeability of rock, **266**

Permian period, **398–399**, **406–407**

Perseids, 61

Perseus, 3

petrification, 388, **389**

petroleum, 507, 509, 511–513; underwater deposits of, 347

phosphorescent minerals, 166

photosphere of sun, **23**

photosynthesis, 424

phyllite, **175,** 176

physical properties of sea water, 335–339

phytoplankton, 344

pitching fold, **210**

placer deposits, **506**

plains, abyssal, 325–326; life cycle of, 249–250; outwash, 286

planets, 11, 12; characteristics of (Table), 63; circulation system of, 433; Earth (*see* earth); inner, **51**–55; Kepler's laws of motion of, 12–14, knowledge gained from exploration of, 100–102; life on other, 101; miniature, 56–**57**; outer, 57–60; sizes of, **64**; travel to, **98–100**; *see also specific planets*

plankton, 344, **345**

plants, and photosynthesis, 424; in sea water, 329, **330**, 343–344; weathering and, **235**

plate tectonics (theory), 182–185, 212; and earth's magnetism, 188–190; and mountain building, 222–224; volcanism and, 200, **201**–203

plateaus, lava, **196**; life cycle of, **249**–250

platinum, 506

Pleistocene epoch, **398**–**399**, 411–**412**

Pliocene epoch, **398**–**399**, 411

Pluto, 60; characteristics of (Table), 63

plutonic rocks, **171**

polar Atlantic air masses, **459**

polar Canadian air masses, **459**

polar climates, 484, **485**, **487–488**

polar easterlies, 435

polar front, 465–466

polar Pacific air masses, **459**

Polaris (North Star), **3**, 4, **118**

political maps, 125

pollution, air, 426–**427**; of sea water, 342

polyconic projection, *125*

polyps, coral, 304

pothole, **261**

Precambrian era, 397, **398**–**399**, 400, **401**–402

precession, **66**

precipitation, 254–255; causes of, 449–451; measurement of, **452**; rainfall patterns, **484**, **494**; and temperature, 483–**484**; types of, **451**–452; *see also* rain

pressure, air, 421–424, 466; of sea water, 321–322

pressure ridges, **280**, **338**

primary earthquake wave, 215

prime meridian, *122*, **122**

prominences of sun, 25

protons, 140–142, 513

pseudofossils, 391

psychrometer, 441–**442**

Ptolemy, 12

pulsar stars, 40

Pulsating Theory of universe, **42**

pumice, 171

pyroxene, **157**

pyrrhotite, 166

quadrangles, 127

Quarternary period, 409, 411

quartz, 237, 238; structure of, **154**; types of, **155**

quartzite, **175**, 176

quasars, 43

radar, 468

radiation, electromagnetic, 7, 21; heating by, 433

radiation fog, 448

radio telescope antennas, 9–**10**

radio telescopes, 9–**10**, 43

radio waves, 9

radioactive decay, rate of, 381–**382**

radioactive minerals, 167

radioactivity, geologic time and, 381–383

radiolaria, 329–**330**, 331

radiosonde, 467–**468**, 470

rain, 457; causes of, 449–451; coalescence of drops of, **449**–450; and ice crystals, **450**–451; *see also* precipitation

rain forest climate, **485**–486

rain gauges, **452**

rainbow, **431**, 432

rainfall, pattern of, **484**; over North America, **494**

rapids, 259, 260

recycling of metals, **508**–509

red clay, 330–331

red giants (stars), 35, 37

red shift, 41

reefs, barrier, 304–305; coral, 304–**305**; fringing, 304

reflector telescope, **9**

refraction of waves, 361–**362**, 363

refractor telescope, **9**

relative humidity, 441, 442

relief, *129*

relief maps, 125

residual boulders, **236**

resources, *504;* renewable and nonrenewable, 504

revolutions, rock layers and, 377

Rhine River, 212

Richter scale, 217

rift valley, 210, **212**

rip currents, **363**

ripple marks in rocks, **378**

rivers, formation of, 256–**257**, 258–260

rocket engine, 94; escape velocity of, *95,* 96; thrust of, **95**

rockets, 94–97; fuel-oxidizer type, 95–96; nuclear, 96; orbit of, 96–**97;** *Saturn V,* **96**

rocks, 153, 169–178; basaltic, 171, **178;** carbonation and, 270; chemical, 172; conformity of, 378; cross bedding of, **378;** deforming of, **208;** dip of, 209, **210;** disconformity of, **379;** and early history, 379–381; extrusive igneous, 171, **173;** faulting in (*see* faults); folding of, **208**–209, **210;** glaciated, **282**–283; granitic, 171; igneous, 169–**171, 173, 176;** intrusive igneous, 170–**171;** joints of, 177, **210;** layers of, 376–**377,** 378–379; limestone, 270, **271;** Martian, 138; massive, **176,** 177; melting temperature of, 192; metamorphic, 170, 172, **175,** 176, **177;** moon, 76, 138; organic, 172; permeability of, **266;** plutonic, **171;** residual boulders, **236;** and revolutions, 377; ripple marks on, **378;** sedimentary (*see* sedimentary rocks); strata of, 176–**177;** stress in, 212–213; strike of, 209–**210;** types of, 169–172; unconformity of, 378, **379;** weathering of, 235, **236–237**

Rocky Mountains, 290, 408,

512; Canadian, **276, 283,** 285

Roman calendar, 87–88

Sahara desert, **486**

salinity, *341;* graph of, 341

salt(s), dissolved in sea water, 337–**342;** for Table (*see* sodium chloride)

salt lakes, 288–**289**

San Andreas fault, 218–**219**

San Francisco earthquake (1906), 219

sand, as building material, 507; grains of, 299–300; wind erosion and, **242–243**

sand bars, **300,** 302–303

sand dunes, **243**–244

sand spit, 300–**301**

sandstones, 171, **174,** 385–386

sandy soils, 238

Sargasso Sea, 356

sargassum, 356

satellites, applications, 102–105; communications, **104;** earth, 96–98; orbit of, 96–**97;** planetary (Table), 63; weather, **104**–105, **468**

saturation, zone of, **266**

saturation value, *441*

Saturn, 59–**60;** characteristics of (Table), 63

Saturn V rocket, **96**

savanna climates, 486–**487**

scale of map, 127

schist, **175;** 176

scientific laws, *12–13*

SCUBA, **321**

sea arches, 298

sea breezes, 472

sea caves, 298

sea cliffs, **301**

sea floor, earthquakes on, 354; features of, **323**–324; nodules on, 347; sediments on, 319, 320, 325–**330,** 331; spreading of, 186–**187,** 188, 190, 212, 327; volcanic eruptions on, 221, 326, 328, 354

sea level, changing of, 305–**306,** 307; mean, *128*

Sea of Serenity, **78**

sea water, 335–348; chemical properties of, 335, 340–346; density of, **339;** desalting of, 346; dissolved gases in, 340, 342–343; dissolved salts in, 337, 339–**342;** distillation of, 346; evaporation of, 253, 338–339, 341, **342, 440;** food from, 347–348; food chain in, 344, **345;** life in, 343–346; light rays absorbed by, **336;** microscopic organisms in, 329–**330,** 343, 344, **345;** minerals in, 346–347; mixing in, 345; movements of, 320, 352–354; overturn in, 345; physical properties of, 335–339; plants in, 329, **330,** 343, 344; pollution of, 342; pressure of, 321–322; salinity of, *341;* solar energy and, 253, 335–336, 352–353, **440;** substances dissolved in (Table), 341; temperature and depth of, **339;** thermocline of, **339;** upwelling in, 345, **346;** *see also* oceans

seamounts, 326
seasons, passage of, 10–**11,** 63
secondary earthquake wave, 215
seconds, 121
sedimentary rocks, 169, **170**–172; conglomerate, 171, **174;** fragmental, 171; stratified, **176;** types of, 171–172
sediments, 169, 213, 219, 221; deposition of, **384**–385; microscopic organisms and, 329–**330;** organic, 329–**330;** on sea floor, 319, 320, 325–**330,** 331; stream, 288, **289**
Seismic Sea Wave Warning System, 365
seismograph, 214–**215**
shadow zone, **184**
shale, 172, **174,** 176
Shasta, Mount, **198**
Sheep Mountain (Wyoming), **210**
sheet erosion, 245–246, 261
shield cones, 197, **198**
Shiprock volcano, **200**
shoreline, emergent, **301;** features of, **301–302,** 303; submergent, 301–**302**
sidereal year, 66
Sierra Nevada Mountains, 220–**221, 283,** 408, 409, 504
silica, 329–330
silicates, *153*
sills, volcanic, **193**
silt, 238, 331
Silurian period, **398–399, 404–405**
sky, night, 2–4, **5**

slate, **175,** 176
sleet, 451
slumping, **241**
smog, **427**
snow, 451, 457
snow line, *276*
sodium chloride, crystals of, **146;** formation of, 144–**145;** ions of, 145–**146**
soil(s), *232;* formation of, 238–239; horizons of, **239;** lunar, **79;** mature, **239;** profile of, **239;** sandy, 238
soil creep, *240*
soil erosion and conservation, **245–246,** 247
solar energy, 27–28, 49; and atmosphere, 427–**428,** 429; changes in amount of, 290–291; electromagnetic spectrum and, 21–**22;** reflection of (Table), 428; and sea water, 253, 335–336, 352–353, **440;** trapping of, 429
solar flares, 25–**26**
solar system, 47–48; dust cloud theory of, 48; formation of, **48**–49; near collision theory of, 48; nebular theory of, 48, 49; origin of, 47–50; path of members in, **13;** plan of, 12–14; *see also* planets, sun, universe
solar time, apparent, 67; mean, 68
solids, molecular arrangement in, 146–147; molecular motion in, **147**
sonic depth recorder, **319**
sound waves, 33, 319
South Atlantic Ocean, **317**
South Equatorial Current,

355
South Pacific Ocean, **317**
southern lights, 26
Space Shuttle, **105–106**
space travel, gravity and, 93–95; planetary, **98–100,** 101–102; *see also* rocket engine, rockets
spatter cones, 198
specific gravity of minerals, 165–166
specific heat of water, 482
spectrograph, 33
spectrum, bright-line, **32;** continuous, **32;** dark-line, **32;** electromagnetic, 7, **8,** 9
spiral galaxy, **6**
spring tides, **87, 88**
springs, **268;** hot, 268–**269, 270**
squall line, 460
stacks, **298**
stalactite, **271**
stalagmite, **271**
standard time, 68–70; zones of, **69**
star cluster, **38**–39
star streaks, **4**
starlight, analysis of, 33
stars, absolute magnitude of, *31,* 34, **35;** birth of, 37–39; brightness of, 3, 30–34; cepheid variables, 31, 33; death of, 39–41; distance from earth, measurement of, 29–**30;** double, **38,** 39; life cycle of, **36;** locations of, 3; magnitude of, 31–35; main sequence, **35,** 37; map of, **3;** neutron, 40; nova, 40–41; pulsars, 40; red giants, 35, 37; super giant, **35,** 37; tempera-

ture of, 34, **35;** variable, 39; white dwarfs, **35, 36,** 37

station model, symbols used in, **469,** 470

Steady State Theory of universe, **42**

stellar parallax, measurement of, 29–**30**

steppes, 486, 489

strata of rocks, 176–**177**

stratified drift, 284

stratocumulus clouds, **447,** 448

stratopause, 430, **431**

stratosphere, 430, **431**

stratovolcanoes, 198

stratus clouds, **445,** 448

streak of minerals, 163

stream piracy, **258**

streams, load of, 262–263; rejuvenated, 260; sediment in, 288, **289;** terraces of, **260;** youthful, 259

stress in rocks, 212–213

strike of rock layer, 209–**210**

strike-slip fault, **210**

strip cropping, **246**

strip-mining, **511**

Stromboli volcano, 195

subarctic climates, 487–**488**

subduction zone, *187*

submarine canyon, **325**

submarine vehicles, 322–**323**

subpolar lows, 435

subtropical highs, **435**

sulfur, 507

summer solstice, 63, **65**

sun, average surface temperature of (Table), 63; chromosphere of, **23;** core of, 22–**23;** corona of, **23**–24; diameter of (Table), 63; eclipse of, **24, 83,** 84–85; energy of (*see* solar energy); gravity of, 98; magnetic disturbances of, 24–25; mass of (Table), 63; nuclear reactions in, **28–29;** path of, 10–**11;** photosphere of, **23;** prominences of, 25; revolution of (Table), 63; rising and setting of, 66–67; rotation period (Table), 63; structure of, 21–27; temperature of, 22, **23;** and tides, **87;** vertical ray of, **65**

sundial, 67–68

sunspots, **24–25**

super giant stars, **35,** 37

superior mirage, **432,** 433

supernovas, 41

superposition, principle of, 377, 378

supersaturated air, 444, 446

surfacelong earthquake wave, 215

Surtsey, 200–**201**

swamps, formation of, **267**

swells, *360,* **360**

synclinal mountains, **208**–209

syncline, **208**

taiga forest (Canada), **488**

talus slope, **240**

tarn, **281,** 287

telescopes, optical, 9, 43; radio, 9–**10,** 43; reflector, **9;** refractor, **9**

temperature, and air pollution, 426–**427;** and altitude, 482; Celsius scale, 466; and climate, 479–484; Fahrenheit scale, 466; latitude and (Table), 480; of moon, 78–79; ocean currents and, 483; of oceans, 319; precipitation and, 483–**484;** of stars, 34, **35;** of sun, 22, **23;** wind and, 483

temperature inversion, *427,* **427**

terminal moraines, **285,** 287

terracing, **246**–247

Tertiary period, **398–399,** 409

tetrahedron, **154**

thermocline of sea water, **339**

thermometers, 466, **467**

thermonuclear reaction, *28*

thermosphere, 430–**431**

thrust of rocket engine, **95**

thunderstorms, 457, 460, **474–475**

tidal bore, 366, **367**

tidal bulges, moon and, 85–86

tidal currents, 366

tidal oscillations, *366*

tidal power, 516–517

tidal waves, 354, 363

tides, 352, 365–**366,** 367; energy from, 516–517; moon and, 85–87; neap, **87, 88;** power of, 366–367; spring, **87, 88;** sun and, **87**

time systems, 67–68

TIROS, **468**

topographic conditions, ice ages and, 291–292; and weathering, 238

topographic maps, 127–131, **132–133**

topographic sheets, 127

topography, *127*

tornadoes, **475**–476
Torricelli, Evangelista, 422
trade winds, 435
transpiration, **254,** 440
traps, petroleum, **511**
travertine, 269
trenches, ocean, **202,** 325
Triassic period, **398–399, 408**
tributaries, **257**
trilobites, **403**
Tropic of Cancer, 63, **65**
Tropic of Capricorn, **65,** 66
tropical Atlantic air masses, **459**
tropical climates, 484, **485– 487**
tropical continental air masses, **459**
tropical Gulf air masses, **459**
tropical Pacific air masses, **459**
tropical year, *66*
tropopause, 430, **431**
troposphere, 430, **431**
tsunamis, 363–**365**
tundra climates, **487**
turbidity currents, 354
Tycho, **79**

umbra, *84,* 85
unconformity of rock, 378, **379**
undertow, 362–363
uniformitarianism, *207*
United States, major climatic regions of, **495**
Universal Calendar, 89
universe, Egyptian concept of, **117;** Greek model of, **12;** Hindu concept of, **117;** middle-ages concept of, **117;** models of, 11–12, 41–**42,** 43;

movement in, 10–12; origins of, 41–43; *see also* solar system
upslope fog, 449
upswelling in sea water, 345, **346**
uranium, 381; atomic mass of (Table), 141; atomic number of (Table), 141; conversion into lead, 382; half-life of, 382, **383;** reserves of, 514– 515
uranium-235, nuclear fission of, 513
Uranus, 60; characteristics of (Table), 63
Ursa Major, 3, **6**

valley breezes, 472
valley glacier, 277
valleys, hanging, **282,** 283; rift, 210, **212;** U-shaped, **283;** V-shaped, **258,** 259
variable stars, 39
varves, **384**–385
veins, mineral, 506
ventifacts, **243**
Venus, 52, **53**–54; characteristics of (Table), 63; flight to, **98–100**
verbal scale, *127*
vernal equinox, **65,** 66
vertebrates, 404
vertical faults, **210**
Vesuvius, 195
volcanic activity (*see* volcanic eruptions)
volcanic ash, 197
volcanic blocks, **197**
volcanic bombs, **197**
volcanic cinders, 197
volcanic cones, 202
volcanic dikes, 193, **194**

volcanic dust, 197
volcanic eruptions, 194– 196; beneath earth's surface, results of, 193–194; and earthquakes, 214; explosive, 104, **195;** fissure flow, **196;** products of, 196– 200; quiet, **194;** on sea floor, 221, 326, 328, 354
volcanic gases, 197, 340
volcanic islands, 326–**327**
volcanic mountains, 221, 326
volcanic peaks, 326–**327**
volcanic sills, **193**
volcanism, extrusive, 192; intrusive, 192, **193,** 200; and plate tectonics, 200, **201**–203
volcanoes, 193, **194–198,** 199–200; active, 194, 199; eruptions of (*see* volcanic eruptions); extinct, 199–200; intermediate, 195; stratovolcanoes, 198

water budget, 254; conservation of, 256; cycle of, 253–**255,** 256; and erosion, 253; ground, 254, **266,** 267; makeup of, 144; molecules of, **144,** 147–148; running, erosion by, **256**–263; sea (*see* oceans, sea water); specific heat of, 482; use of in industry (Table), 256
water farming, 347–348
water gap, **260**
water table, 265–**266,** 267– 268

water vapor, 421; in atmosphere, 439–442; condensation of, 442–443; transpiration of, 254, 440

waterfalls, 259, 260

watershed, 257, **258**

waterspouts, **476**

wave-built terrace, **297,** 298, **300**

wave-cut cliff, **297**–298

wave-cut terrace, **297,** 298, **300, 308**

wave cyclone, 461–462, **463–464,** 465

wave erosion, **296–300**

wavelength, *7, 7,* 8, 9; of ocean waves, 360–**361**

waves, earthquake, **183**–**184, 214**–216; electromagnetic, 428; light, 7, 33; ocean (*see* ocean waves); radio, 9; sound, **33,** 319

weather, air mass, 458; causes of changes in, 457; forecasting of, 470–471; instruments used to observe, 466, **467–468;** local, 471–476; symbols in station models, **469,** 470; *see also* climate

weather fronts, clouds and, 461, **462;** cold, **460, 461;** formation of, 459, **460**–461; polar, 465–466; stationary, 461; storm centers formed by, 461, **462–463,** 464–465; warm, **460,** 461

weather maps, 125, 422, **470**–471

weather satellites, **104**–105, **468**

weathering, *232;* and animals, 235; chemical, 233–235; 297; climate and, 237–238; mechanical, 233, 237, 238, 296–297; and plants, **235;** results of, 235–239; of rock, 235–**236–237;** topographic conditions and, 238; types of, 232, **233–234,** 235

Wegener, Alfred, 182

weight, gravity and, 16–17

westerlies, 435

white dwarfs (stars), **35,** 36, 37

whitecaps, 363

wind(s), chinook, **482;** deposits made by, 243–**244,** 245; direction of, recording of, **467;** earth's rotation and, 434; measurement of speed of, **467;** monsoons, **483;** movement of in cyclone, **465;** and ocean currents, 353–355; and ocean waves, 359–360; patterns of, **355,** 435–436; and temperature, 483; trade, 435

wind erosion, **242–243**

wind gap, **260**

Winnipeg, Lake, 288

winter solstice, **65,** 66

World Calendar, **89**

year, measurement of, 63

Yellowstone National Park, **269,** 270

Yellowstone River, **258**

Yosemite National Park, **283**

Yosemite Valley, **247**

youthful mountains, **248,** 249

zero declination, line of, 189

zinc, ores of, 506, 507

zone of aeration, 265, **266**

zone of capillary movement, **266**

zone of saturation, **266**

zooplankton, 344